SCIENCE TRANSLATED

MEDIAEVALIA LOVANIENSIA

SERIES I / STUDIA XL

KATHOLIEKE UNIVERSITEIT LEUVEN
INSTITUUT VOOR MIDDELEEUWSE STUDIES
LEUVEN (BELGIUM)

SCIENCE TRANSLATED

LATIN AND VERNACULAR TRANSLATIONS OF SCIENTIFIC TREATISES IN MEDIEVAL EUROPE

Edited by

Michèle GOYENS
Pieter DE LEEMANS
An SMETS

LEUVEN UNIVERSITY PRESS
2008

ISBN 978 90 5867 671 9
D/2008/1869/7
NUR: 613-684

CONTENTS

PREFACE

The English language seems to become the *lingua franca* of science nowadays, at least in most domains. Vulgarization of science is the privilege of the mother tongue of each language community, as we see in the translations and summaries of scientific articles in thematically general journals and magazines. Scientific languages have been developed for specific purposes; they had to meet criteria such as monoreference, semantically homogeneous morphological series, and absence of connotations.

The Middle Ages witnessed a similar situation. However, the medieval *linguae francae* of science and religion were Greek and Latin, which by then had become languages without native speakers. In the 14th century already, Nicole Oresme captures adequately the relation between those languages and the vernaculars:

> Grec estoit en resgart de latin quant as Romains si comme est maintenant latin en regart de françois quant a nous. Et estoient pour le temps les estudians introduiz en grec et a Rome et ailleurs, et les sciences communelment bailliees en grec et en ce pays le langage commun et maternel, c'estoit latin.
>
> 'For Romans, Greek had the same position with respect to Latin as has Latin nowadays with regard to French. In those days, students were introduced to Greek in Rome and elsewhere, and science was commonly rendered in Greek in those regions where the mother tongue was Latin.'
>
> (Nicole Oresme, prologue of the translation of Aristotle's book on Ethics and Politics[1])

Thus, vernacular languages were confronted with specific problems when they had to render scientific concepts expressed in Greek or Latin and they often had to elaborate particular lexical fields. Translators played a very important role in this process.

The present volume contains a selection of papers that were presented at the international conference *Science Translated. Latin and Vernacular Translations of Scientific Treatises in Medieval Europe*, organized by the *Institute for Medieval Studies* of the University of Leuven (K.U.Leuven),

1. Nicole Oresme, *Le livre de Ethiques d'Aristote*, ed. A.D. Menut (New York, 1940), p. 99.

May 27-29, 2004. First of all, the study of medieval translations has a long history at the K.U.Leuven. Since 1973, the University has been hosting the *Aristoteles Latinus* Centre, the editing project (supported by the International Union of Academies) of all medieval Greek-Latin translations of the *Corpus Aristotelicum*. Moreover, translations of Latin texts into French have been the central research topic at the unit of French diachronic linguistics. This tradition was initiated by Willy Van Hoecke, who developed a specific methodology for the diachronic analysis of linguistic topics by use of translations. In 2000, both units decided to unite forces and to collaborate on a research project on the development of the biological vocabulary in medieval Greek-Latin and Latin-French translations[2]. This project was followed, in 2002 and 2005 respectively, by two other projects on the medieval translations of Pseudo-Aristotle's *Problemata Physica*; the first project studied the expression of movement and change, whereas the second focuses on the development of the medical vocabulary[3]. In organizing a conference on Latin and vernacular translations of scientific treatises, it was our purpose to put our own research in a larger perspective and to open the discussion with colleagues who are working in related fields.

A specific input for this conference was provided by the *Centre for Translation, Communication and Cultures* (CETRA) at the K.U.Leuven. From its own background, viz. research into the cultural functions of translation, mainly with respect to contemporary issues, this Centre expressed its interest in the study of the medieval translations as a means to mirror and to confront present-day insights. This particular approach has truly enriched our own more linguistically and historically oriented research. We are therefore grateful to José Lambert to highlight, in a short introduction, the possibilities of mutual interaction.

2. Research project funded by the Research Council of the K.U.Leuven, nr OT/01/06, *Translating biological treatises in the Middle Ages: a comparative study of translators ant their strategies in the learned language and the vernacular* (2000-2004; promotors: W. Van Hoecke, † J. Brams, M. Goyens, B. Van den Abeele).

3. Research projects G.0110.03, funded by the Research Foundation – Flanders, *From source- to goal-language in the Middle Ages: expressing motion and change in the Medieval translations of Aristotle's «Problemata»* (2002-2006; promotors: M. Goyens, † J. Brams, W. Van Hoecke, P. De Leemans); OT/05/15 *Science in text and context. The medieval translations of Aristotle's «Problemata» and the development and positioning of the medical terminology* (2005-2009; promotors: M. Goyens, P. De Leemans, B. Van den Abeele, J. Lambert).
See also De Leemans P. – Goyens M. (eds.), *Aristotle's Problemata in Different Times and Tongues* Mediaevalia Lovaniensia, Series I/ Studia XXXIX (Leuven, Leuven University Press, 2006).

The essays presented in this volume are divided into two sections. The first section deals with translations from Greek, Arabic and Hebrew into Latin, the second with translations into the vernacular (French, Italian, Dutch). Whereas most essays are concerned with the Middle Ages, some authors have also considered the Renaissance. The texts treated in the essays cover various topics (such as medicine, zoology, meteorology, and mathematics) and are studied from different points of view. More specifically, some focus on terminology, whereas others deal with the reception of the translations or situate them in their institutional and historical context. These different points of view illustrate that translating was and is not just a mechanical transposition of a text into another language by an author. Rather, source-text and translation are both the result of specific contexts and thus have to meet specific criteria. The translator does not live in an ivory tower but interacts (or at least tries to), not only with his source-text but also with his public. The essays in this volume shed a light on the complex web that is formed by medieval translations with respect to this broader context.

This volume would not have come about without the support of several institutions and persons. We express our gratitude to the Research Council of the K.U.Leuven and the Research Foundation – Flanders for their financial support of our research projects as well as of the conference, which also enabled the publication of this volume. We gratefully acknowledge the other sponsors of the workshop for their kind grants: the Faculty of Arts, the Institute of Philosophy, the Department of Linguistics, the *Aristoteles Latinus*, the Institute of Medieval Studies, the Department of History, and CETRA.

Last but not least, we thank all persons, among whom the collaborators of our research projects, who have assisted in the organisation of the conference and/or in the preparation of this volume. Special thanks go to Werner Verbeke, secretary of the Institute for Medieval Studies, on whose experience, once more, we were allowed to fall back.

This volume is dedicated to emeritus Prof. Willy Van Hoecke, who conceived and developed the diachronic linguistic research from Latin to French based on translations. We would like to acknowledge our colleague sincerely, whose valuable support and remarks continue to back up our own research.

Michèle GOYENS, Pieter DE LEEMANS & An SMETS

José LAMBERT

MEDIEVAL TRANSLATIONS AND TRANSLATION STUDIES: SOME PRELIMINARY CONSIDERATIONS

> *Translation is always a shift not between two languages but between two cultures* (Umberto Eco)
>
> *You may be wondering why a historian of early modern Europe should have chosen to lecture about translation. The subject seems to be attracting more and more interest these days, especially under the banner of 'Translation Studies', which is on the way to becoming a new discipline. I should like to argue that history deserves a large place in Translation Studies and that studies of translation deserve a large place in history* (Peter Burke)

It is not at all difficult to collect impressive quotes and slogans about translation from the work of established scholars or from the ideas of great intellectuals (according to I.A. Richards, translation is even supposed to be one of the most sophisticated linguistic activities of the human brain). The trouble is that these generally impressive statements tend not to be compatible with each other (*anything goes?*), and for rather simple reasons. They are usually the result of interesting intuitions, but hardly ever stem from systematic investigation nor familiarity with research on the field. This is of course a paradox when they are proposed by established researchers.

Whoever is interested in translation or in intercultural relations will feel excited when great names such as Umberto Eco, or Peter Burke, a distinguished authority in historiography from Cambridge University[1], are fascinated by the role played by translation(s) in culture. The trouble is that in the case of translation (Eco) or in the case of Translation Studies (Burke), such authorities often show good intentions rather than

1. On April 26, 2005, Peter Burke offered his first KB Lecture at the Koninklijke Bibliotheek (KB) – National Library of the Netherlands in The Hague, on the occasion of his research stay at NIAS, the *Netherlandic Institute for Advanced Studies in the Humanities and Social Studies*; cf. Burke P., *Lost (and Found) in Translation: A Cultural History of Translators and Translating in Early Modern Europe,* (Wassenaar, 2005-2006), NIAS (KB Lecture 1). The Dutch translation *Ik vertaal, dus ik ben* was published by Uitgeverij Bert Bakker in 2005.

expertise. Notwithstanding the development of a successful new market for scholarly books, translation as well as Translation Studies remain strange *no man's lands* in academic scholarship.

There can be no doubt any more about the links between culture and translation. However, the current volume, which deals with *Science Translated* in the European Middle Ages, demonstrates, in almost any page, that medieval translation is hardly ever a story between *two* languages or *two* cultures, notwithstanding Eco's quote. In fact, most translators and translations are struggling between *several* languages (Latin, often also Greek, sometimes Arabic, and one or more vernaculars) and between *several* cultural frameworks. According to recent insights[2], both theoretical and historical, a high percentage of translations systematically tend to reduce the visibility of the languages and cultures involved. When referring to two languages and cultures, this is often one of the techniques used in order to hide links with further languages and cultures. In our contemporary world, it is not at all sure that our bookshops display any international bestseller that reflects – only – bilateral relationships between languages (or cultures). Neither medieval nor modern (nor ancient) worlds are that bilateral, and quite a few translation events involve more than two languages and cultural traditions.

The happy observation about any statement on translation, be it famous or not, is its promotional value in terms of research: it is time for many disciplines in the Humanities, first of all for Translation Studies itself, to recognize translation(s) as a challenge for research rather than for statements.

The openness to Translation Studies as an upcoming discipline and to the interaction between history and translation (which can be found e.g. in Burke's *Lost (and Found) in Translation,* then in *Cultural Translation in Early Modern Europe*; or in Rener's *Interpretatio*)[3] is in itself an event from the point of view of Medieval Studies. However, unless some of the

2. Lambert J., 'La Traduction dans les sociétés monolingues', in: H. Kittel, A.P. Frank, N. Greiner, T. Hermans, W. Koller, J. Lambert & F. Paul (eds), *Übersetzung – Translation – Traduction. Ein Internationales Handbuch zur Übersetzungsforschung. An International Encyclopedia of Translation Studies. Encyclopédie internationale de la recherche sur la traduction,* Handbücher zur Sprach- und Kommunikationswissenschaft (Berlin & New York, 2004 -...), Vol. 1 & 2; here vol. 1, p. 69-85.

3. Burke, *Lost (and Found)...*; Burke P. & Ronnie Po-chia Hsia, *Cultural Translation in Early Modern Europe* (Cambridge, 2007); Rener F. M., *Interpretatio: Language and Translation from Cicero to Tytle* (Amsterdam, 1989).

basic tools and underlying conceptual frameworks[4] become known, translation will remain a *no man's land.*

There are no reasons for rejecting the recommendations issued by established scholars as long as they insist on the complex embedding of translation, discourse and language within the dynamics of (inter)cultures. From the moment we agree on the necessity to recognize translation as a cultural phenomenon – or as more than just a mechanical operation between languages that can be given to technicians or simply to machines, it belongs and will for ever belong to the history of cultures, languages, societies, literary and other genres, political discourse, or legal discourse (which the European Union and all interpreters exemplify). Much more than one book or one 'history' will be needed in order to clarify the various roles that translators, translations and translated discourse (or genres, or religious and legal texts) have played in the making and the remaking of societies, during wars, on islands or in military headquarters, in the international press or in the distribution of the Bible, in business or in legislation, – or even in literature, and above all in the various masterworks for children. It is even relevant to criticize scholarship for having (too) heavily invested in the literary traditions of translation while neglecting crucial societal areas such as legislation, religion, political discourse, etc. A lot needs to be done in areas such as law (from the Romans to the Code Napoléon, and up to modern international organizations such as the European Union), book markets and media, oral and written and business communication, or simply: in the area of any research, about Science, before and since Universities have spread out. This is precisely why *Science Translated* is also an invitation into interdisciplinarity.

Along the lines of wise historiography, the title places particular emphasis on the enormous task ahead for researchers: it promises continuity in research while indicating a few channels. It happens to focus on one of the strategic areas of European culture, i.e. scholarship. Scholarship in the Middle Ages? What's in a name? We have no strong reasons for avoiding the concept of 'scholarship', except because it may imply the

4. Besides rich bibliographical tools (such as John Benjamin's electronic *Translation Studies Bibliography,* Saint Jerome's *Translation Studies Abstracts*), scholars have now access to: (a) journals with a programmatic status such as *Target* (1989-) or *The Translator* (1995-); (b) an impressive series of Handbooks (such as Kittel et al., *Übersetzung – Translation – Traduction…*); (c) theoretical-methodological work where history and historical (including medieval) material are recognized as an unavoidable component of conceptualization. Specialized websites such as Anthony Pym's (http://www.tinet.org/~apym/welcome.html), Gideon Toury's (http://www.tau.ac.il/~toury/) and others indicate how the new (inter)discipline is developing within the international university.

idea of a profession, which does not make sense in medieval environments. But in medieval traditions 'science' refers to one of the key areas in society – the centre of knowledge – that has helped shaping the entire cultural tradition in Western Europe and – probably – far beyond. The cultural and intellectual heritage of our continent can and will never be a simple issue, the question will be however what it has to do with translation.

It is not at all by hazard that the title *Science Translated*, being short and almost syntax independent, leads to various layers of questions. What exactly is meant by the association of two – not so simple – words? Most titles of scholarly publications are selected in view of a series of assumptions, or they suggest a few (hypo)theses.

For most scholars from any discipline, the very idea that 'science' has been translated may look a bit surprising: in modern worlds translations often tend to be a matter of routine and habit that make us unaware of them, as in the case of legal texts, business texts, or the European Union. For medieval intellectuals, the translation of 'scientific' texts may simply have been a matter of necessity: how would the Western medieval world have got access to Aristotle, or to the Bible, without translations? *Necessity*, that is what many contemporary scholars tend to assume about translation in any culture ('translation fulfils a need; without any need, why would we translate at all?'). In modern society, translations are often rendered invisible[5], for marketing reasons and as a component of diplomatic or commercial behaviour; just like the *lingua franca*, modern translations have opportunities/the chance to be shocking. In our contemporary society, in the age of globalization, the worldwide use of English as a *lingua franca* (as in business life) is more shocking than the (often invisible) use of translations (e.g. in the European Union): the use of English is supposed to hinder the dissemination of the various national languages. Since one of the reasons why *lingua franca* and translations tend to generate negative attitudes and resistance seems to be their success, maybe their *ubiquity*[6], it is

5. The idea of the translator's 'invisibility' is familiar now among translation scholars (but not at all to the man in the street) since Larry Venuti's 1995 book (*The Translator's Invisibility. A History of Translation,* London, 1995). In fact, the shifts from more visible into more or less invisible translations as well as the dynamics underlying such strategies had been explored much earlier in Gideon Toury's work, as one of the key strategies in any translation. Furthermore, it is not unknown that the German romanticists, e.g. Schleiermacher, had already linked the fluctuations in translation traditions belonging to large cultural frameworks.

6. Kittel et al. (eds), *Übersetzung – Translation – Traduction…* It is mainly due to Armin Paul Frank that the theme of the ubiquity of translations in modern societies has become so prominent in the important handbook *Übersetzung* in the famous collection published by de Gruyter.

important to have deeper insights into the dissemination of translation: how often and how systematically was science – indeed – translated. Was it indebted to a kind of internationalization long before there was any 'nationalism'[7]? From which language(s) have translations been organized and undertaken, by whom, where, why? And how? According to modern theories, any attempt to analyze translation phenomena in depth requires a minimum of descriptive (statistical?) evidence: the so-called hows and why's are supposed to reflect very different options in situations where they abound from the situations where they are exceptional. Whether in modern or in ancient times and contexts, translating is or was a real job (whether paid or unpaid), and there are or were not many chances that, without any justification, anyone will/would translate (or have translated) any of those many sophisticated, technical texts, as in the case of scholarly writing. Translating may offer some excitement, some fun, but fun is no sufficient explanation for the long lists of translated texts and books that have circulated between countries and centres of knowledge. Whether in oral or in written traditions, whether in print or in electronic format, translations are supposed to offer qualities and advantages that the actual *(local)* repertoire of texts and documents do not really offer: conflicts are even supposed to be unavoidable between translations and 'original writing'[8], translations are supposed 'to fulfil a need'[9]. Given the fact that we deal, *hic et nunc*, with 'scientific' texts, the issue can hardly be accepted as an innocent topic for reflection or observation: throughout the history of mankind, in any tradition and any place, translations cannot be explained as mere accidents; and whatever motivations may have been, translations may also have generated particular and probably spectacular consequences. The more since our own modern concepts and ideas and rules may have been borrowed from the repertoires of medieval texts. After all, our contemporary Constitutions – in many a country, the legal tradition is going back to the *Code Napoléon*, whatever this may mean – and it is also well-known that Napoléon's teams have studied and reworked the Roman legal traditions in view of their – then – pioneering Code.

On the basis of the two key words in the title *Science Translated*, a whole network of questions is starting to emerge. They may relate to

7. Anderson B., *Imagined Communities. Reflections on the Origin and Spread of National Communities* (London, 1983).

8. Lambert, 'La Traduction…'.

9. Toury G., *In Search of A Theory of Translation* (Tel-Aviv, 1980); Toury G., *Descriptive Translation Studies and Beyond,* Benjamins Translation Library 4 (Amsterdam & Philadelphia, 1995).

other areas of medieval intellectual knowledge – in synchronic terms – or to modern times, and also to previous traditions – in diachronic terms. In the age of globalization, it is worthwhile to wonder whether our own knowledge would necessarily depend on translation(s): how could there be any international/global language of science without the systematic interaction between languages and, hence, the mediating role of translations? Since any reader, from more ancient and more recent times, is more or less accustomed to make use of translations (in the supermarket, on television, in religion, in children's books), there are good reasons for investigating what appears to be one of the whereabouts of medieval scholarly knowledge.

Science Translated is not at all the final answer to Peter Burke's questions nor to the (much more sophisticated) programmes worked out since the 1980's within Translation Studies, but it provides at least an important and very new collection of essays on the *translatio studii* from the late Roman period up to the Renaissance. The entire volume as well as its individual contributions provide a platform for answers to profound underlying questions that are well-known nowadays: about 'the art' of translation, about the paradigmatic components of any translational event; about the *users* of translation, about its different *markets*, as Bourdieu would put it, about the vernacularization and the standardization of written and oral discourse, about the migration or mobility of genres. Whether translations are 'good' or 'faithful' is not at all the first issue of the debates in *Science Translated.* As most scholarly disciplines, Translation Studies has an almost unlimited repertoire of questions that goes far beyond the individual translator, his language or discourse, his text. What may have been largely neglected so far, however, is the exact function fulfilled by translation in the establishment of cultures and societies, – or the role fulfilled by translation(s) and other intercultural verbal operations in the 'discursive construction' of societies[10]. The main reason why *Science Translated* deserves to be recognized in the new programme(s) on 'Cultural Translation' is exactly here: rather than simply enriching and completing our insights into the microscopic world of languages and translations and translators – which it does – it also confirms how

10. Lambert J., 'The Status and Position of Legal Translation: a Chapter in the Discursive Construction of Societies', in: F. Olsen, A. Lorz & D. Stein (eds.), *Language and Law: Key Perspectives* (New York, forthcoming); Shavit Z., 'Fabriquer une culture nationale. Le rôle des traductions dans la constitution de la littérature hébraïque', *Actes de la Recherche en Sciences Sociales* 44 (2002), p. 21-32; Shavit Z., *Past without Shadow: Constructing the Past in German Books for Children* (New York, 2005).

(West-)European Science has been shaped with(in) translation, or rather, how translation has occupied a central, almost monopolized position through the centuries in the formulation of knowledge, and how 'scholarship' – in its modern bureaucratic terminology – has been transformed and vernacularized – as Benedict Anderson would have put it – from intercultural ('international'?) traditions into the various modern so-called national cultures[11]. One of the merits of Anderson's *Imagined Communities* has been to make us aware of the (inter?)continental internationalization processes that have preceded, centuries ago, the establishment of the Nation-State models: Nation-States, according to Anderson's analysis, are the result of the vernacularization of intercultural discourses, – they are not the opposite, i.e. the simple basis on which contemporary internationalization and globalization were going to spread out. The circulation of ideas through the Middle East and around the Mediterranean areas long before and after Jesus Christ, the almost mythological respect for Aristotle and other Greek or Middle East theoreticians and philosophers have given shape to the first canonized centre of scholarly knowledge in Europe, i.e. universities.

Science Translated is in fact devoted to one of the heavy paradoxes of our cultural tradition, i.e. the conflict (or combination?) between continuity and diversity in the history of Western knowledge. Continuity seems to be narrowly connected with the question of *lingua franca* (Latin, Greek and ...), and also with translation. One does not go without the other, – and it is true that research on translation – during the Roman period, during Renaissance as well as in the 20th century – can never ignore the competition between translation phenomena and the *lingua franca*. It is supposed to be obvious that universities have opted for the *lingua franca* model rather than for translation (or for diversity): they have never really discussed how their universalistic claims could have been smuggled into the various ('vernacular') texts. However, the hidden component behind any *lingua franca* tradition is – unavoidably, necessarily – translation: from one *lingua franca* into the other (as illustrated by Bible Translation as well as the Aristotle tradition), between *lingua franca* and vernacular, between (and against) various vernacular traditions. It is one of the merits of *Science Translated* to show, within every sentence or paragraph as well as on the macroscopic level, how the battle between vernacular(s) and *lingua franca* is going on, e.g. from the very moment any conceptual problem requires clarification or diplomatic options (taboo

11. Anderson, *Imagined communities...*

options seem to be linked with the use of vernacular, probably also with the use of the *lingua franca*, as in the case of modern medicine; and medieval translators seem to refer to such conflicting options in their 'glosses'). Within these tensions and conflicts, learned people have good reasons for preferring the language(s) of the scholar, and trying to avoid (as much as possible) mixing it up (too much) with everyday discourse. The combination of high and low culture that is now forever linked with the work of Bakhtine, seems to pop up in many kinds of writing, and also in the translator's discourse. The multilayeredness of the translator's words and worlds deserves to be explored as a test of the interaction between more international (West-)European worlds and the more local ones: it is supposed to be a key to the dynamics of the Middle Ages, in particular when favouring the distribution of encyclopedia: where exactly does the world of knowledge come from, and how is it being packaged and formulated?

To take one example, the circulation of e.g. the *Aristoteles Latinus* (and of all Aristotle's texts) is a remarkable cultural phenomenon which indicates the relevance of concepts such as 'markets' and 'symbolic values' (in Pierre Bourdieu's terms, or in Itamar Even-Zohar's). Within the more or less borderless world – 'borderless' mainly for Learned people? – a repertoire of Great Books (Texts) and Authors and Concepts has been circulating: when necessary, the transfer from Arabic or from Greek into Latin was no real problem; the transfer into vernacular was much more problematic, it was waiting for (its) Martin Luther ('Man soll dem Volke ins Maul sprechen'). But even among the Learned people, popular wisdom was widespread, and the establishment of an almost permanent and dynamic World of Learning was starting, a real popular encyclopedia of knowledge in medieval style.

The observation that an 'international' encyclopedia develops – in the *lingua franca*, and/or in translation: one does not go without the other – is itself a remarkable phenomenon of the cultural functions fulfilled by translations (and/or by the *lingua franca*). The circulation of ideas between societies, cultures and universities was made (rather) simple with the aid of the *lingua franca*, and students, professors and intellectuals were able to move from one centre to the other, making use also of translations[12]. One of the prominent theoreticians of translation, Anthony Pym,

12. Pym A., 'Negotiation theory as an approach to translation history. An inductive lesson from fifteenth-century Castile', in: Y. Gambier & J. Tommola (eds.), *Translation and Knowledge* (Turku, 1993), p. 27-39.

makes use of medieval translations in order to explain and clarify our contemporary Localization Industries phenomenon, in which texts circulate from country to country and from language to language as (anonymous) matrixes that are adapted to the local requirements and legislation: the 'Mobile Text'[13] belongs to the age of mobility in space and time, but the prehistory of books and book cultures indicates that electronic communication provides a renaissance to medieval models. In order to better understand our modern culture, translation experts will need to help renew the study of mobile texts in the Middle Ages, and it is predictable that this will benefit as well the understanding of modern culture[14].

Complex approaches to the medieval great texts were heavily indebted to 'the absence of the original', which is one of the other cruxes in modern Translation Studies. The authority of the writer(s) as well as its opposite, i.e. anonymity or instability in the text tradition, seem to be specific, neither for the Middle Ages nor for Modern Times. The free circulation of more or less identifiable texts has been one of the backbones of our Western culture long before the Internet, and translation is one of its instruments, an instrument that has not really been investigated, neither for the Middle Ages nor for modern times[15].

Several contributions in this volume tackle the question of scientific terminology. It seems that we are witnessing here the birth of scientific concepts and terminology, either in Latin or in various vernaculars. In the Dutch tradition, it is well-known that Simon Stevin (17th century) struggled in favour of the 'local language' (Dutch) and against the empire of the (Latin and French) *lingua franca*. Without re-examining the history of the making of European national languages, it can be assumed that scientific terminology in everyday discourse would not have developed all over Europe without translations as bridges between the international lexicon and the 'local' idiom. Various contributions about medical discourse, about ichthyology, etc. in this volume may shed light on the development of the scholarly language that was due to conquer European intellectuals' minds in the future.

Rather than the mere production of translations, or the world of translators and their audiences, or their influence on the development of

13. Pym A., *The Moving Text. Localization, Translation, and Distribution* (Amsterdam – Philadelphia, 2004).

14. See e.g. Lambert J., 'De la mobilité des littératures considérées dans leurs rapports avec les sociétés', *Francophonies d'Amérique* X (2000), p. 23-41; Ong, W., *Orality and Literacy. The Technologizing of the Word*, New Accents (New York, 1982).

15. Rener, *Interpretatio…*

intellectual life, *Science Translated* offers a contribution to the analysis of the making of scholarly discourse in early Europe, on the basis of common knowledge concentrated in the great texts circulating between societies, between the *lingua franca* culture and translations, in their permanent struggle/cohabitation with the new languages-to-come, those of the Lutheran Age. Whatever the success of Martin Luther's was going to be in the religious world, and whatever the future of the new national languages was going to become, the various translations under observation in this volume all are milestones in the continuity and in the conflicts of our intellectual discourse during a few centuries. The remnants of such battles are surviving in our 21st century talks with doctors, pharmacologists, lawyers, priests.

For all these reasons, this volume about translation in the dissemination of scholarly discourse is a contribution to the study of the origins of modern scholarly traditions.

K.U. Leuven

Charles BURNETT

SCIENTIFIC TRANSLATIONS FROM ARABIC: THE QUESTION OF REVISION

In memory of Jef Brams

In 1997 Jef Brams published 'L'édition critique de l'*Aristote latin*: Le problème des révisions'[1]. In this he pointed out that the 'Medieval Aristotle' consisted largely of revisions by William of Moerbeke of previous translations, whether those of Boethius (e.g. *Analytica Secunda*, *De sophisticis elenchis*) or those of his medieval predecessors (e.g. *Physics*, *De generatione et corruptione*, *De anima*, *Parva naturalia*, *Metaphysics*, and *Nicomachean Ethics*). There is no question here of William *pretending* to make translations of works which he merely revised. His contemporary, Roger Bacon, reports that William himself promised 'to revise ('immutare') all the existing translations', and, when they were lacking, 'beat out ('cudere') new ones'[2]. The 'problem' that Jef addressed in his article, was how to edit revisions. Ideally, one should know the particular form of the text that was revised, and the specific changes made by the reviser. He showed that different editors adopted different solutions, largely on the basis of different situations regarding the texts. Thus, the editors of the revisions of the *Analytica Secunda* and *De sophisticis elenchis* (Lorenzo Minio-Paluello and Bernard Dod)[3], opted to reconstruct the hypothetical manuscript of William himself: the manuscript of the previous translation that he had in his hands and into which he had inserted corrections, additions and so on. Since these revisions had a very

1. Brams J., 'L'édition critique de l'*Aristote Latin*: Le problème des révisions', in: S.G. Lofts & P.W. Rosemann (eds.), *Éditer, traduire, interpréter: essais de méthodologie philosophique* (Louvain-la-Neuve, 1997), p. 39-53.

2. Roger Bacon, *Opera quaedam hactenus inedita*, III, Compendium philosophiae, ed. J.S. Brewer (London, 1859), p. 472: 'omnes translationes factas promisit immutare et novas cudere varias'. Bacon's choice of verbs is a little unexpected. 'Immutare' literally means 'to change', and does not in itself imply correction or checking against the original text; 'cudere' is used mainly for 'beating out coins'/'coining'–'excudere' is more commonly used for 'composing'.

3. In all cases in the next paragraph the reference is to the relevant edition in the *Aristoteles Latinus* series.

limited diffusion, there was no 'disseminated version' whose readings had to be shown. Williams revision of Grosseteste's translation of the *Nicomachean Ethics*, however, presents a different situation: the manuscript tradition divides into two branches, one being that of the manuscript brought to Toledo by Peres Gonzales Gudiel (Toledo, Bibl. cap., 47.10), the other being the exemplar of the manuscripts copied at the university of Paris ('recensio Parisiaca'). Rather than reconstruct Williams text that was the origin of these two branches, the editor, René-Antoine Gauthier, opted to provide the exemplar of the Paris manuscript, both because the large number of manuscripts representing this branch facilitated this, and because this was the text used by the readers and commentators of the Middle Ages. The case of the *Metaphysics* is different again. Here, some books of Williams are revisions of an earlier translation (the 'translatio anonyma'), while others are translated afresh. For consistency, therefore, Gudrun Vuillemin-Diem had to try to establish what William himself wrote. In fact, it became obvious that William made two versions. Since his first version exists in only one manuscript, its readings are given in the *apparatus criticus*, while the second version, which was the one that was diffused, was the basis of the edition. For the *Physics*, to which, naturally, Jef devoted most attention, again two versions exist, but in this case it is the *first* version that was used in the university of Paris and elsewhere; the second has survived only in one manuscript, from Madrid, and in some marginal readings in a manuscript from Ravenna (subsequently incorporated into two manuscripts in Florence). Jef decided to present the first version, because of its wider diffusion, and the nature of the revisions in the Madrid manuscript enabled him to reconstruct the first version as it was written by the reviser.

One could quote further examples, and describe other revisions made by William to his own translations, as in the case of various books of the *De animalibus* and the *De anima*[4]. But I would now like to turn to the situation in translations from Arabic that show signs of revision. I should say at this point, that by 'revision of a translation' we assume, in this case, correction of a translation in respect to the original text, or by the translator himself (this is implied in Jef's use of the word). Revisions

4. For *De gen. an.* see: *De generatione animalium. Translatio Guillelmi de Moerbeka*, ed. H.J. Drossaart Lulofs, Aristoteles Latinus XVII 2.V (Bruges-Paris, 1966), esp. p. XIX-XXXIII; for *De mot. an.* and *De inc. an.*, see: *De motu animalium. De progressu animalium. Translatio Guillelmi*, ed. P. De Leemans, Aristoteles Latinus XVII 2.II-III (Turnhout, forthcoming); for *De an.*, see Wielockx R., 'Guillaume de Moerbeke réviseur de sa révision du «De anima»', *Recherches de Théologie ancienne et médiévale*, 54 (1987), p. 113-185.

of the text for stylistic reasons, abbreviations or expansions of the text, without recourse to the original, will not be considered. I will leave aside, for example, the edition of 'Ali ibn abi-rijal's *Book on the Judgements of the Stars*, published in 1551, in which Antonius Stupa avowedly undertakes to excise all the medieval barbarisms, while using only the earlier Renaissance edition of the Latin text to do this[5].

The most fully documented of revisions of Arabic texts are those of the works of Avicenna. All the sections of the *Shifa'* that were translated in the twelfth century show evidence of a revision[6]. In the case of the *De anima*, Simone Van Riet divided the manuscripts into those that gave 'A-readings' and those that gave 'B-readings'. She hesitated over whether the 'A-readings' or the 'B-readings' were prior, but opted to base the text on the 'B-readings' because their manuscripts were earlier. Dag Hasse, however, has shown that it is the 'A-readings' that are found in the earliest users of the text, including its translator, Dominicus Gundissalinus[7]. In the *Physics* of the *Shifa'*, again, Van Riet detected 'A-readings' and 'B-readings', but conjectured that the 'B-readings' are revisions of the 'A-readings', and were based on a re-reading of the Arabic text. These revisions are frequent in the first book, but rare for the rest of the text: the reviser only got so far in his project. Van Riet based her text on the manuscripts that give the supposedly earlier 'A-readings'. The *Metaphysics* of the *Shifa'*, similarly, appears in two forms, which Van Riet called, tentatively, the 'texte ancien' and the 'texte revu'. Once again, the revised version occasionally shows knowledge of the Arabic, but Van Riet chose the 'texte ancien' as the basis for her edition. More extensive are the differences between the two versions of the *Liber celi et mundi*, also associated with Dominicus Gundissalinus[8]. Here the differences between versions *alpha* and *beta* extend as far as chapter 11 (of sixteen chapters). *Beta* is much fuller than *alpha*, but since the original Arabic text has not been identified, it is difficult to know which reflects best the Arabic original, or whether they are based on different Arabic manuscripts. The manuscripts of *alpha* are older, but the first quotations (including an extensive quotation by Daniel of Morley in ca. 1175) are

5. *Albohazen Haly filii Abenragel Libri de iudiciis astrorum, summa cura et diligenti studio de extrema barbarie vindicati, ac Latinitati donati, per Antonium Stupam Rhoetum Praegalliensem*, Basel, 1551, sig. a2v-a3r.

6. All the references in the following paragraph are to the relevant volume of the series *Avicenna Latinus*.

7. Hasse D.N., *Avicenna's* De anima *in the Latin West: The Formation of a Peripatetic Philosophy of the Soul* (London, 2000).

8. Pseudo-Avicenna, *Liber celi et mundi*, ed. O. Gutman (Leiden, 2003), p. xxxi-xxxv.

of the *beta* version. The editor, Oliver Gutman, gives both versions in parallel columns, where they differ from each other.

It would be tempting to see the twelfth-century translations from the *Shifa'* and of the *Liber celi et mundi* which replaced the corresponding section on cosmology in the *Shifa'*, as part of the same enterprise – a programme of translation taking place in Toledo under the aegis of Dominicus Gundissalinus. One would then like to see the revisions as due in turn to a single enterprise. But Simone Van Riet and Marie-Thérèse d'Alverny, who had studied the translations of Avicenna most extensively, were reluctant to attribute all these translations to the same translator (or group of translators) and did not, as far as I am aware, commit themselves to the conclusion that the revisions were the work of the translators themselves, or of other scholars, and whether the same scholars revised the different parts of the *Shifa'*. In this respect the editors of William of Moerbeke's revisions seem to be more confident in their conclusions than those of Avicenna. Perhaps we should wait until all the parts of the twelfth-century Latin *Shifa'* have been edited, before making a detailed analysis of the vocabulary and style of the text and the revisions (the remaining section of the *Physics* and the *Logic* should be published soon)[9].

At the same time as Gundissalinus was supervising the translation of Avicenna in Toledo, Gerard of Cremona was translating a large number of works from Arabic into Latin. Danielle Jacquart has shown that, at least in one case – Rhazes, *Liber ad Almansorem* – Gerard was revising an older translation, of which the unrevised version survives in some manuscripts[10]. The Arabic text which, we are told, attracted Gerard to Toledo in the first place, was Ptolemy's *Almagest*, and Paul Kunitzsch has shown that, at least in respect to the star table, the Latin translation was revised[11]. The second version is certainly more accurate mathematically than the first, and is the one that Kunitzsch uses as his base text.

I would now like to turn to three Arabic texts I have been directly involved with; namely, the *Great Introduction to Astrology* and *Great Conjunctions* of Abu Ma'shar, and the *Introduction to Astrology* of al-Qabisi. Here I must pay tribute to another scholar – Richard Lemay –

9. The second book of the *Physics* has now been published: *Avicenna Latinus, Liber primus naturalium, tractatus secundus, De motu et consimilibus*, eds. S. Van Riet, J. Janssens & A. Allard (Brussels, 2006).

10. See Jacquart D., 'Note sur la traduction latine du *Kitab al-Mansuri* de Rhazes', *Revue d'Histoire des Textes* (Paris, 1994), p. 359-74, reprinted in Jacquart, *La science médicale occidentale entre deux renaissances (XIIe s.-XVe s.)* (Aldershot, 1997), article VI.

11. See Kunitzsch P., *Der Sternkatalog des Almagest*, 3 vols (Wiesbaden, 1986-91), II, p. 1-10.

who died on 11 May, 2004. His life work was the edition of the Arabic and the two Latin translations of Abu Ma'shar's *Great Introduction*[12]. He established that the more literal of these translations, that made by John of Seville in 1131, had been transmitted in two versions: an original translation, and a revised one. One manuscript even showed, in his opinion, the revisions in the form of annotations that were later incorporated into the text[13]. His editorial policy was to provide the original version in the text, and to devote a separate apparatus to the changes introduced by the revision. Two chapters were not so much revised as retranslated: these he edited separately[14]. Lemay conjectured that the first version was made by John of Seville, while Gerard of Cremona was responsible for the revised version.

When I came to edit the Latin text of the *Great Conjunctions*[15], which often accompanies the *Great Introduction* in the manuscripts, and, though anonymous, appears to be by the same translator, I discovered that this translation, too, had survived in two translations: one is represented in full only by a manuscript now in Vienna, but copied in Toledo in 1248 A.D. (Vienna, Österreichische Nationalbibliothek, 5478), while a Munich manuscript (Bayerische Staatsbibliothek, Clm. 125) follows this version in its first book, and marginal notes in a Vatican manuscript (B.A.V., Vat. lat. 5713) are taken from this version[16]. The other version is found in the rest of the Munich manuscript and in all the other extant copies (i.e. 18 manuscripts and the Renaissance printed text). It seemed more likely that the 'vulgate' version (which I called the 'C' version) was a revision of the Vienna version (the 'V' version). Moreover, the reviser had not only used a different Arabic manuscript from that used by the original translator but had also added a set of glosses explaining both linguistic and doctrinal matters. I was, therefore, faced with a dilemma. Should I edit the first version or the revised version? In favour of the latter procedure were the facts that the revised version was the one used by scholars from the twelfth century onwards, and that the glosses, which

12. Abū Ma'shar, *Liber introductorii maioris ad scientiam judiciorum astrorum*, ed. R. Lemay, 9 vols (Naples, 1995-6).

13. Abū Ma'shar, *Liber introductorii…*, ed. Lemay, IV, p. 291; the manuscript in question is London, British Library, Harley, 3631.

14. Abū Ma'shar, *Liber introductorii…*, ed. Lemay, IV, p. 252-264.

15. Abū Ma'shar, *On Historical Astrology: The Book of Religions and Dynasties (On the Great Conjunctions)*, ed. and tr. K. Yamamoto & C. Burnett, 2 vols (Leiden, 2000).

16. The arguments of the next three paragraphs are set out in full in Burnett C., 'The Strategy of Revision in the Arabic-Latin Translations from Toledo: The Case of Abu Ma'shar's *On the Great Conjunctions*', in: J. Hamesse (ed.), *Translators at Work: Their Methods and Manuscripts* (Louvain-la-Neuve, 2002), p. 51-113, 529-540.

I wanted to include in an apparatus criticus, depended directly on this text. In favour of the former were the precedents of the editions of Avicenna and Abu Ma'shar's *Great Introduction*, in which the supposed earlier versions were the basis of the edition. I also took into consideration the fact that the revised version was already available in two Renaissance printed editions (Augsburg 1489 and Venice 1515). After much hesitation, and a false start, I opted to provide the first version in the text, in spite of the that fact the glosses occasionally did not follow the wording of this text. I followed Lemay's example in indicating in a separate apparatus the changes introduced by the reviser.

Although the evidence is not conclusive, it seems likely that the same scholars were involved in the original and revised versions of both the *Great Introduction* and the *Great Conjunctions*. Both the changes and the glosses are more extensive in the *Great Conjunctions*. What needs further investigation is whether, indeed, two scholars are involved, rather than one scholar (presumably John of Seville) who revised a text that he had earlier translated. The grounds for attributing the revision of the *Great Introduction* to Gerard of Cremona are purely circumstantial, and have not yet been corroborated by comparison with works securely attributed to Gerard.

Jef Brams pointed out that, in the case of the extant versions of William of Moerbeke's translations, what we appear to have is simply transcripts of successive working copies of the scholar, rather than versions which William would have considered as definitive[17]. The situation is probably analogous with the two works of Abu Ma'shar: we have as it were 'snapshots' of two stages in the elaboration of a translation; these stages have become 'fossilised' by the sheer chance selection of the manuscript tradition. In the case of the *Great Conjunctions* there are hints of at least one further stage in the variants in the first book as transmitted by the Munich manuscript. A much more complicated situation, however, can be seen in the case of al-Qabisi's *Introduction to Astrology*, of which over 200 Latin manuscripts survive, which show, between them, a great variety of versions of a single translation[18].

At first sight the impression one gets from looking at the manuscripts and the thirteen Renaissance printed editions is that we are dealing with another translation by John of Seville. But when one looks closely at the earliest manuscripts, in most instances one does not find his name, and

17. Brams, 'L'édition critique…'.

18. Al-Qabisi (Alcabitius), *The Introduction to Astrology*, ed. and tr. C. Burnett, K. Yamamoto & M. Yano (London-Turin, 2004).

they differ significantly from each other. I shall briefly describe those versions which appear to have originated in the hands of scholars who knew Arabic and had al-Qabisi's Arabic text available[19].

1) An early version of the text survives uniquely in a Barberini manuscript from the Vatican (Città del Vaticano, B.A.V., Barb. 236). This version exhibits some Graecizing features: e.g. it uses the Greek word for 'introduction' in the title (*Liber ysagogarum Alcabicii ad iudicia*), and it has the full complement of the letters of the Greek alphabet (transliterated into Latin letters) assigned to each of the signs of the zodiac. Its transliteration of Arabic terms differs from that of the other manuscripts, but is still correct. It is characterised by the predominance of the use of 'e' for both short and long 'a' in Arabic and the use of 'a' for the *'ayn*. The transliterations, are, on the whole, accurate, but inconsistent, as if the translator has not yet become familiar with Arabic astrological terms: e.g. by writing 'eldarbigen' (IV [18]) he shows that he reads the middle letter of *al-darijan* as 'b' (as if it had a single dot underneath it, instead of two), while later on in the same paragraph he read the same letter correctly as 'i' in transliterating the plural of the same word as 'darigeneth'. What became the standard transliteration of the same term reads the letter in yet another way: as an 'n' (the same letter with one dot above it): 'dorungen'. This version incorporates into the text versification of some of the astrological doctrine. It also gives some unique translations of the Arabic. Moreover, it is the only manuscript to include a date: 8 January 1181, though, unfortunately, it is difficult to interpret the significance of this. It does not, however, give the name of a translator. The only other text I know that includes the Greek letters (this time in Hebrew transliteration) for each of the signs of the zodiac is an astrological work written by Abraham Ibn Ezra in the 1140s: the *Principium sapientiae*[20].

2) Another early version of the text shows an Arabicizing tendency. This is MS Oxford, Bodleian Library, Ashmole 369 of the late twelfth century. In this manuscript planets and signs of the zodiac keep their Arabic names: Saturn is usually called 'zohal', Mars and Venus are occasionally

19. The following paragraphs are an abbreviation of the discussion of the relationship between the manuscripts of the *Introduction* in Al-Qabisi (Alcabitius), *The Introduction...*, ed. Burnett e.a., p. 210-220.

20. See Hübner W., 'Tierkreiszeichen, Himmelsrichtungen, Säfte, Buchstaben und die sieben Zonen bei Abraham Ibn Ezra', in: J. Althoff, B. Herzhoff & G. Wöhrle (eds.), *Antike Naturwissenschaft und ihre Rezeption* 12 (Trier, 2002), p. 111-128; see also Dornseiff F., *Das Alphabet in Mystik und Magie*, second ed. (Leipzig-Berlin, 1925), p. 84-85 for the appearance of the letter-zodiac combinations in Greek astrological works.

called 'almarech' and 'adzohare' respectively, and Aries and Taurus, on their first mention, are called 'alhamel' and 'astaur'. Arabic names for the planets and signs of the zodiac are found in the quotations from al-Qabisi's work in the *Liber Archachelis* in another twelfth-century manuscript: Paris, BNF, lat. 16208, fol. 26v-32v, which suggests that this may be an early feature. But they may also be an affectation, retained to give the text an exotic flavour, rather than a sign that the Ashmole manuscript gives a primitive version of the Arabic. For the manuscript gives a poor text, and the transliterations were already known in the Latin tradition, since they appear in the earliest astrological texts of Arabic origin, the late tenth-century *Alchandreana*[21]. The Ashmole manuscript, too, does not mention the name of the translator.

3) On palaeographical evidence the oldest manuscript would seem to be Pommersfelden, Schlossbibliothek, 60. Unfortunately, because of the loss of some quires, more than half of the text of the *Introduction* is missing. However, a very close relation has survived in a fifteenth-century manuscript (Wrocław [Breslau], Biblioteka Uniwersytecka, Ac. IV F 23), which gives a complete text, which is not, however, free from contamination from another tradition. This version exhibits neither the Graecisms of the Barberini manuscript nor the Arabisms of the Ashmole. Once again, however, neither of these manuscripts mention the name of the translator.

4) Another discrete group of early manuscripts is formed from two Oxford manuscripts, Oxford, Merton College, 259 and Oxford, Bodleian Library, Bodley 430. Both manuscripts were apparently written in the last quarter of the twelfth century, and, again, neither manuscript mentions the name of the translator. Moreover, the two manuscripts show different affiliations. The Merton manuscript is an important collection of the works of John of Seville, including the version of the *Toledan Tables* which is sometimes attributed to him[22]. The Bodley manuscript, on the other hand, is one of only three manuscripts of the *Liber trium iudicum*, a collection of works on astrological judgements made in Northeast Spain by Hermann of Carinthia and Hugo of Santalla and dedicated, in this manuscript, to Hugo's patron, Michael, the bishop of Tarazona. This manuscript is English, and, besides including a copy of Roger of Hereford's astrology, exhibits the use of the Insular form of 'g', especially in transliterations from Arabic.

21. Juste D., *Les* Alchandreana *primitifs* (Leuven-Paris, 2006).

22. See Pedersen F.S., *The Toledan Tables*, 4 vols (continuous pagination) (Copenhagen, 2002), p. 194-196.

5) The earliest manuscript to include the name of a translator – in this case, 'Iohannes yspalensis' – is Kraków, Biblioteka Jagiellońska, 578, which has the distinction of being the only early manuscript of the *Introduction* written in the Iberian peninsula, and may date from the last quarter of the twelfth century. Is this John of Seville's version? There are several manuscripts that belong to the same family as this manuscript (β). These exhibit different readings from what I shall call the 'Vulgate version'. In particular, they avoid using Arabic words in transliteration, most conspicuously at the beginning of the text, where 'nithac' is replaced with 'zodiacus' (though the Jagellonian manuscript, perhaps incorporating a gloss, gives both). The text in the β family begins with a Christian equivalent of the Muslim invocation of God (the *basmala*): 'In nomine domini (occasionally with the addition 'nostri Ihesu Christi')'. In this it agrees with the non-revised versions of Abu Ma'shar's *Great Introduction* and *Great Conjunctions* of which the former is explicitly attributed to John of Seville.

6) The majority of manuscripts, as well as the printed editions, show what I call the 'Vulgate' version. This is the version in which the text begins with the Arabic word 'nitaq' (= girdle)[23]. The manuscripts and printed editions of the Vulgate version consistently ascribe the translation to John of Seville. The clearest differences between families are the translations exhibited in this version as opposed to manuscripts in the β family. Some manuscripts give the text of the Vulgate with variants from the β family in the margins; others provide the reverse.

So the editor is faced with a rather complicated situation. Ideally, one would wish to edit all six of these versions, and print them in parallel columns. However, if one is wishing to provide a single text, which version should one choose as the base text? That of the earliest manuscript? The earliest looking version? The 'Vulgate version', attributed to John of Seville? The text which corresponds most closely in terminology and style with other works by John of Seville? In fact, I chose none of these options, but rather, a manuscript in which an early thirteenth-century scholar had already done much of the editorial work himself.

This manuscript is Città del Vaticano, B.A.V., Reg. lat. 1285, which contains a collection of astronomical and astrological texts that have been carefully copied by a single hand from an exemplar which already had an apparatus of glosses. The copy of the *Introduction* provides a good text

23. Only one manuscript, Berlin, Staatsbibliothek-Preussischer Kulturbesitz, lat. fol. 307 (ca. 1300), retains in the text the correct interpretation of this word: 'id est cingulus'. In the rest of the tradition 'cingulus' has been replaced by (or banalized into) 'circulus'.

of the Vulgate version, but it has, in turn, been carefully corrected by a scholar who evidently was checking the text against the Arabic original. He detected and marked the Latin words which were not in the Arabic (e.g. the glosses within the text, most of which are present in all the versions I have mentioned so far). He also added phrases from the Arabic which are missing in the other manuscripts, or retranslated certain Arabic phrases, in the margin. The retranslations are usually prefixed by 'in al.' (on one occasion this abbreviation is spelt out as 'in alio libro'), and many of these readings imply a more careful reading of the Arabic manuscript, or the replacement of a translation of one Arabic word with a translation of another that looks very similar to the first[24]. It is also possible that the editor/reviser had access to a different Arabic manuscript from that of the original translator, and that this is what 'in al.' refers to. In general the revision tends to make the Latin translation more Arabic-looking and more literal. Arabic words are retained in transliteration, or the literal meaning in the Arabic is retained in the Latin: e.g. in the case of *qiyam* (literally 'rising, getting up' etc., but used also for acceding to the throne) the word 'electio' is replaced by 'elevatio'.

In my edition of the *Introduction* I have, therefore, used the text of the Vatican manuscript (together with another early representative of the Vulgate text, Berlin, Staatsbibliothek-Preussischer Kulturbesitz, lat. fol. 307) for the edition, and indicated in one apparatus all the corrections, alternate translations and glosses added in that manuscript. In another apparatus I have indicated the readings of manuscripts representing all the other versions listed above. Finally, a third apparatus signals the differences between the Latin text and the Arabic text published in the same volume. There are plans to publish a Medieval English translation of the *Introduction* together with a Latin text unencumbered by these extensive *apparatus critici*, but in this case a different choice will probably be made: namely, the Latin text which approximates most closely to that implied by the English translation.

Every text, therefore, makes its own demands, and no rules can be universally applied when faced with the choice of editing an 'original' translation, or one of its revisions. It is incumbent on the editor to explain clearly what he or she has chosen to do, and why such decisions have been made.

Warburg Institute London

24. It is less likely that 'in al.' refers to a Latin manuscript, because, in most cases, no Latin manuscript has been identified as having these readings.

Specimina

Following the example of Aristoteles Latinus, *Codices*, p. 113-233 and 791, and Suppl., p. 25-42, and Avicenna Latinus, *Liber de anima I-II-III*, p. 137*-156* and *Philosophia Prima*, p. 142*-152* I provide here parallel texts of the different versions of Abu Ma'shar's *Great Introduction* (1) and *Great Conjunctions* (2), and al-Qabisi's *Introduction to Astrology* (3). Italics indicate the occurrence of a different term or phrase between the versions; an ellipse in square brackets indicates the absence of a term or phrase that occurs in other versions.

1. Abu Ma'shar, *Great Introduction*

1.1. *Incipit*

'John of Seville's Version'[25]	**Revision**[26]
In nomine domini nostri Iesu Christi. Incipit liber *in quo est* maior introductorius Abumasar *astrologi ad scientiam iudiciorum* astrorum, *et tractatus eius super eadem iudicia cum disputatione rationali et auctentica et figure signorum atque nature.*	[...] Incipit liber introductorii maioris *in magisterio scientie* astrorum *editione Abumasar et interpretatione Johannis Hispalensis ex arabico in latinum.*
In nomine Dei pii et misericordis. Laus Deo qui creavit celum et terram cum omnibus que in eis sunt *ex* mirabilibus suis, et posuit stellas ornamentum et lucernas, posuitque eas in significatione ut preberent ductum rationalibus. Et posuit terram receptaculum et in ea sumptus suos mensuravit. Non est igitur alius Deus preter eum solum et nullum habet participem.	In nomine Dei pii et misericordis. Laus Deo qui creavit celum et terram cum omnibus que in eis sunt [...] mirabilibus suis, et posuit stellas ornamentum et lucernas, posuitque eas in significatione ut preberent ductum rationalibus. Et posuit terram receptaculum et in ea sumptus suos mensuravit. Non est igitur alius Deus preter eum solum et nullum habet participem.
Hic est liber quem edidit Gafar qui dictus est Abumasar astrologus Albalaghi in magisterio introitus ad scientiam iudiciorum astrorum, et sunt.viii. tractatus.	Hic est liber quem edidit Gafar qui dictus est Abumasar astrologus Albalaghi in magisterio introitus ad scientiam iudiciorum astrorum, et sunt.viii. tractatus.

25. Abū Ma'shar, *Liber introductorii...*, ed. Lemay, V, p. 2.
26. Readings from Abū Ma'shar, *Liber introductorii...*, ed. Lemay, VI, p. 388-391.

Et in unoquoque tractatu dicemus numerum eius differentiarum, si Deus voluerit. Nunc autem narremus differentias primi tractatus et sunt.vi.

Prima differentia, in inicio libri et capitibus.vii.

Secunda, in inventione scientie iudiciorum astrorum. Tercia, in qualitate operis planetarum in hoc mundo. Quarta, in immaginibus et naturis, et compositione *naturalium*.

...

Dixit Gafar qui vocatur Abumasar: fuit causa que me provocavit ut ederem librum introductorium ad scientiam iudiciorum astrorum quia vidi editores sapientie *ad hoc* tendere ad scientiam rerum *ut possint edere ac proferre sapientiam*. Et cum hoc effectum fuerit eis, tunc perficitur eorum inquisitio *et intentio*; quia *perfectio* intentionis sapientum est perfectio scientie eorum ad que tendunt. Cum igitur invenissem plures libros quos ediderunt quidam antiqui ex civibus magisterii iudiciorum, et nullum librum inter eos *invenissem* in quo esset confirmatio huius artis per *rationem sufficientem*, nec in quo esset inicium eorum que necessaria sunt inspicienti in hac scientia; dumque vidissem eos diversos ac discordes in *essentia iudiciorum*, ita ut quidam dicerent quod nullum opus esset virtuti motus planetarum in hoc mundo et alii dicerent quod esset eis opus in mutatione temporis et eius corruptione tantum...visum est mihi edere hunc librum introductorium ad scientiam iudiciorum eo ordine quo soliti erant sapientes ordinare libros suos in expositione scilicet eorum quorum scientia necessaria est in libris eorum – preponere scilicet que debent preponi, et postponere que debent postponi.

Et in unoquoque tractatu dicemus numerum eius differentiarum, si Deus voluerit. Nunc autem narremus differentias primi tractatus et sunt.vi.

Prima differentia, in inicio libri et capitibus.vii.

Secunda, in inventione scientie iudiciorum astrorum. Tercia, in qualitate operis planetarum in hoc mundo. Quarta, in immaginibus et naturis, et compositione *elementatorum*.

...

Dixit Gafar qui vocatur Abumasar: causa que me provocavit ut ederem librum introductorium ad scientiam iudiciorum astrorum fuit quia vidi editores sapientie [...] tendere ad scientiam rerum *et ad inventionem sapientiarum et ordinationem earum*. Et cum hoc effectum fuerit eis, tunc perficitur eorum inquisitio [...]; quia *completio* intentionis sapientum est perfectio scientie eorum ad que tendunt. Cum igitur invenissem plures libros quos ediderunt quidam antiqui ex civibus magisterii iudiciorum, et nullum librum inter eos *vidissem* in quo esset confirmatio huius artis per *rationes sufficientes*, nec in quo esset inicium eorum que necessaria sunt inspicienti in hac scientia; dumque vidissem eos diversos ac discordes in *iudiciis*, ita ut quidam dicerent quod nullum opus esset virtuti motus planetarum in hoc mundo et alii dicerent quod esset eis opus in mutatione temporis et eius corruptione tantum...visum est mihi edere hunc librum introductorium ad scientiam iudiciorum eo ordine quo soliti erant sapientes ordinare libros suos in expositione scilicet eorum quorum scientia necessaria est in libris eorum – preponere scilicet que debent preponi, et postponere que debent postponi.

1.2. Book 3, chapter 4

'John of Seville's Version'[27]	**Revision**[28]
Iam prediximus in precedentibus ex significationibus planetarum et Solis super res que accidunt in hoc mundo, et quod non sit in hoc mundo compositio alicuius nature ex naturis nisi per causam motus Solis et participationem planetarum ei. Et nunc volumus narrare proprietatem significationis Lune super [...] aquam et terram.	Iam prediximus in precedentibus ex significationibus planetarum et Solis super res que accidunt in hoc mundo, et quod non sit in hoc mundo compositio alicuius nature ex naturis nisi per causam motus Solis et participationem planetarum ei. Et nunc volumus narrare proprietatem significationis Lune super *accessionem et recessionem et super alias res. Et dixit Philosophus quod magis universalis significatio Solis est super ignem et aerem, et magis universalis significatio Lune est super* aquam et terram.
Facta est autem significatio Solis et Lune in hoc mundo fortior significationibus ceterorum planetarum, et plus apparens, propter duas causas. Una, quod Sol *clarior est planetis. Quia quedam eorum, etsi sunt maioris quantitatis*, tamen sunt longiores a nobis; quedam vero, etsi sunt prope nos, sunt tamen minoris quantitatis. Et Luna est prope nos plus ceteris. Secunda vero causa est quod planete sint lumina *lucentia* et non sunt eis radii. Ideoque quod apparet in hoc mundo ex opere eorum apparet per fortitudinem motus eorum atque claritatis. Luminaribus vero sunt radii fortis operationis hoc mundo ipsaque operantur in nobis per motus ac radios eorum, et ipsa applicant naturas planetarum *ad* hunc mundum in quatuor elementis.	Facta est autem significatio Solis et Lune in hoc mundo fortior significationibus ceterorum planetarum, et plus apparens, propter duas causas. Una, quod Sol *maior est stellis quantitate et est medie longitudinis a nobis, et Luna est propinquior stellis nobis. Sed cetere stelle, etsi ex eis quedam sunt quibus est magnitudo*, tamen sunt longiores a nobis; quedam vero, etsi sunt prope nos, sunt tamen minoris quantitatis. Et Luna est prope nos plus ceteris. Secunda vero causa est quod planete sint lumina *splendida* et non sunt eis radii. Ideoque quod apparet in hoc mundo ex opere eorum apparet per fortitudinem motus eorum atque claritatis. Luminaribus vero sunt radii fortis operationis hoc mundo ipsaque operantur in nobis per motus ac radios eorum, et ipsa applicant naturas planetarum *in* hunc mundum in quatuor elementis.

27. Abū Ma'shar, *Liber introductorii...*, ed. Lemay, V, p. 102.
28. Readings from Abū Ma'shar, *Liber introductorii...*, ed. Lemay, VI, p. 468-469.

1.3. *Explicit*

'John of Seville's Version'[29]

Rursum, si gradus Lune esset ignotus, acciperemus *a parte in ascendentem* essetque quatuor signa et tres gradus. Post hoc adderemus desuper quod *abiret* Sol in signo suo; deinde proiceremus hoc de inicio signi Solis recto ordine finireturque numerus in.xx. gradu Leonis sciremusque quod Luna esset ibidem.

Perfectus est liber introductorii maioris in magisterio scientie astrorum, *translatus a Iohanne Hispalensi. Explicit.*

Revision[30]

Rursum, si gradus Lune esset ignotus, acciperemus *ab ascendente in partem* essetque quatuor signa et tres gradus. Post hoc adderemus desuper quod *abieret* Sol in signo suo; deinde proiceremus hoc de inicio signi Solis recto ordine finireturque numerus in.xx. gradu Leonis sciremusque quod Luna esset ibidem.

Perfectus est liber introductorii maioris in magisterio scientie astrorum, *edicione Abimaser et interpretatione Iohannis Yspalensis ex arabico in latinum sub laude Dei et eius auxilio* [+.71. Scriptus est liber iste anno domini nostri Ihesu Christi.1171. mense Aprili MS London, Harley 3631, fol. 57v].

2. Abu Ma'shar, *Great Conjunctions*

2.1. Book 1, chapter 1[30]

V version (Vienna, Österreichische Nationalbibl., 5478 and Munich, Bayerische Staatsbibl., Clm. 125)

In nomine domini Ihesus Christi.

Hic est liber in *quo continentur plura collectiva* de significationibus individuorum superiorum super accidentia que efficiuntur in mundo *effectus* et corruptionis de presentia *scilicet* eorum respectu ascendentium inceptionum coniunctionalium

C version (Vulgate)

[...]

Hic est liber in *summa* de significationibus individuorum superiorum super accidentia que efficiuntur in mundo *generationis* et corruptionis de presentia [...] eorum respectu ascendentium inceptionum coniunctionalium et aliorum. Et

28. Abū Ma'shar, *Liber introductorii...*, ed. Lemay, V, p. 386.
29. Readings from Abū Ma'shar, *Liber introductorii...*, ed. Lemay, VI, p. 677-678.
30. Abu Ma'shar, *On Historical Astrology...*, ed. Yamamoto & Burnett, II, p. 3.

et aliorum. Et sunt octo tractatus, [...].63. differentie. Editus a Iaphar astrologo quid dictus est Albumasar.

...

Differentia prima in premissione inceptionum universalium multarum utilitatum.

Scientia significationum individuorum circularium super effectus inferiores accipitur a motibus naturalibus, eo quod sint sensui *propiores* significationibus individuorum superiorum. Cum enim motus naturales non excedant.3. divisiones, consideratus est ordo nexus circulorum planetarum *superiorum*, et proprietates motuum eorum volubilium. *Que etiam* dividuntur in.3. divisiones, quarum prima est planetarum altiorum ordinatorum super luminare maius, et secunda divisio est luminaris maioris, tertia vero divisio est planetarum inferiorum *cuspidum* qui sunt *scilicet* infra luminare maius. Relata est itaque unaqueque divisio individuorum altiorum ad unamquamque divisionem motuum naturalium propter fortitudinem affinitatis eorum ad illas, et propter successionem effectus *vestigiorum* eorum in mundo *effectus* et corruptionis.

sunt octo tractatus *et*.63. differentie. Editus a Iaphar astrologo quid dictus est Albumasar.

...

Differentia prima in premissione inceptionum universalium multarum utilitatum.

Scientia significationum individuorum circularium super effectus inferiores accipitur a motibus naturalibus, eo quod sint sensui *propriores* significationibus individuorum superiorum. Cum enim motus naturales non excedant.3. divisiones, consideratus est ordo nexus circulorum planetarum [...], et proprietates motuum eorum volubilium. *Ergo ille* dividuntur in.3. divisiones, quarum prima est planetarum altiorum ordinatorum supra luminare maius, et secunda divisio est luminaris maioris, tertia vero divisio est planetarum inferiorum [...] qui sunt *positi* infra luminare maius. Relata est itaque unaqueque divisio individuorum altiorum ad unamquamque divisionem motuum naturalium propter fortitudinem affinitatis eorum ad illas, et propter successionem effectus *impressionum* eorum in mundo *generationis* et corruptionis.

2.2. Book 8, chapter 1[31]

V version

Et si fuerit *afirdaria* Mercurii, significat *salutem regibus* Babilonie communiter et filiis eorum et <additamentum eorum> in honore et *exaltatione* et *regno*. Et *preponent reges filios eorum* et *facient bonum erga servos et servientes suos*, et *forte incarcerabuntur quosdam suorum servorum* et *afficient eos*

C version

Et si fuerit *alfirdaria* Mercurii, significat *sanitatem regis* Babilonie communiter et filiis eorum et additamentum eorum in honore et *nobilitate* et *dominio*. Et *servient se ex filiis suis* et *benefacient suis et servientibus suis et vassallis suis*, et *forsitan capient aliquos eorum et castigabunt eos et facient eos*

31. Abu Ma'shar, *On Historical Astrology...*, ed. Yamamoto & Burnett, II, p. 313-314.

penis et mulctabunt eos. Deinde evadent ab illo et multiplicabuntur itinera regum civium Babilonie, et superabunt inimicos suos quocumque iverint et capient civitates et multiplicabuntur sapientes et religiosi et vaticinatores et astronomici et scriptores et consequentur homines nocumentum propter *causam inquisitionis reddituum ab eis*, et duritiam senum super eos, et multiplicabuntur rumores terribiles et falsi super portas regum et consequentur cives climatum infortunium preter cives Persie quibus augmentabitur bonum, et multiplicabuntur angustia et tristitia et cogitatio et *luxuria* et fraus et proditio, et consequentur homines *angustiam* et necessitatem et effundetur [in]iustitia in homines et benefacere, et precipient reges fodere flumina et fabricare civitates et *multiplicabitur mercatura* in terra Babilonie, et multiplicabuntur res aquatice sicut margarite et pisces et aves aque, et fortasse *tristabuntur* reges *propter res occultas*.

pectare. Deinde evadent ab illo et multiplicabuntur itinera regum civium Babilonie, et superabunt inimicos suos quocumque iverint et capient civitates et multiplicabuntur sapientes et religiosi et vaticinatores et astronomici et scriptores et consequentur homines nocumentum propter *quesitum eorum in portatico*, et duritiam senum super eos, et multiplicabuntur rumores terribiles et falsi super portas regum et et consequentur cives climatum infortunium preter cives Persie quibus augmentabitur bonum, et multiplicabitur angustia et tristitia et cogitatio et *fornicatio* et fraus et proditio, et consequentur homines *angustias* et necessitatem et effundetur iustitia in homines et benefacere, et precipient reges fodere flumina et fabricare civitates et *fiet merces cara* in terra Babilonie, et multiplicabuntur res aquatice sicut margarite et pisces et aves aque, et fortasse *suspicantur* reges *de rebus occultis*.

2.3. *Explicit*[32]

V version

Completus est liber coniunctionum, *ordinatio* Albumasar Iafar filii Machometi *Albalichi*, *quod est gentile vel patrium*, cum laude Dei et auxilio eius, et maledictio Domini super Machometum et super socios eius.

C version

Completus est liber coniunctionum, *ex dictis* Albumasar Iahfar filii Machometi *Albalhci*, [...] cum laude Dei et eius auxilio, et maledictio Domini super Machometum et super socios eius.

32. Abu Ma'shar, *On Historical Astrology*..., ed. Yamamoto & Burnett, II, p. 317.

3. Al-Qabisi, *Introduction to Astrology*

3.1. *Incipit*[33]

MS *V* (Vaticano, B.A.V., Barb. 236)	**MS *M* (Oxford, Merton College, 259)**	**MS *J* (Kraków, Bibl. Jagiellońska, 578)**	**MS *U* (Vaticano, B.A.V., Reg. lat. 1285)**[34]
Incipit liber *ysagogarum Alcabicii ad iudicia.*	[...]	[...]	*Incipit introductorius Alcabici ad iudicia astrorum, interpretatus a Iohanne Yspalensi.*
Incipit prologus.	Incipit prologus.		Incipit prologus *Alcabici.*
Postulata a domino prolixitate vite, *Sceyph Eldeule*, id est *gladii* regni, et durabilitate sui honoris, custodia quoque eius *operum* sive bonorum atque extensione sui imperii, exordiamur id quod volumus narrare.	Postulata a domino prolixitate vite, *Ceif Handdaula*, id est *gladius* regni, [...] durabilitate sui honoris, custodia quoque eius *opum* sive bonorum atque extensione sui imperii, exordiamur id quod volumus narrare.	*De* prolixitate vite *Ceufadaule*, id est *gladius* regni, et durabilitate sui honoris, custodia quoque eius *opum* sive honorem (*sic*) atque extensionem sui imperii, exordiamur id quod volumus narrare[35].	Postulata a domino prolixitate vite Ceyphaddala, ^id est *gladius* regni^, et durabilitate sui honoris, custodia quoque eius *opum* sive bonorum atque extensione sui imperii, exordiamur id quod volumus narrare.
Cum vidissem conventum quorumdam antiquorum ex auctoribus magisterii iudiciorum astrorum edidisse libros quos	Cum vidissem conventum quorumdam antiquorum ex auctoribus magisterii iudiciorum astrorum edidisse libros quos	*Incipit prologus.* Cum vidissem conventum quorumdam antiquorum ex auctoribus magisterii <iudiciorum astrorum> edidisse	Cum vidissem conventum quorumdam antiquorum ex auctoribus magisterii iudiciorum astrorum edidisse libros quos

33. Al-Qabisi, *The Introduction...*, ed. Burnett e.a., p. 225.
34. The passages absent in the Arabic that *U* indicates by bracketing them with the words 'glosa' or 'vacat', are here bracketed with ^...^. The extensive glosses (giving alternative translations, or explanations) are omitted here, but are fully recorded in the edition.
35. *J* rubricates this sentence as if it is a title.

vocaverunt *introductiones* huius magisterii, id est iudiciorum astrorum, set quosdam ex eis non fuisse scrutatos diligenter universa que necessaria sunt in eodem magisterio […] quosdam vero ea que necessaria sunt protulisse prolixe, et quia quod necessarium est in eo periisse cernerem, quosdam quoque in ordinatione eorum que protulerunt non incessisse tam (*sic*) discipline conspexissem, cepi edere hunc librum, et posui eum introductorium atque collegi in eo ex dictis antiquorum quicquid necessarium est huic magisterio secundum […] introductorium. Et non introduxi ratiocinationes disputationi sive defensioni eorum que protulimus necessarias, cum sint in libro Ptolomei […] *alarbaa macilet*, id est quatuor tractatuum, et in libro meo quem edidi in confirmatione magisterii iudiciorum astrorum et in destructione epistole *Binhali*, in annullatione eius ex ratiocinatione que	vocaverunt *introductorios* huius magisterii, id est iudiciorum astrorum, set quosdam ex eis non fuisse scrutatos diligenter universa que necessaria sunt in eodem magisterio *de* his que conveniunt introductorio, quosdam vero ea que sunt necessaria protulisse prolixe, […] quia quod necessarium […] in eo periisse cernerem, quoddam (*sic*) […] in ordinatione eorum que protulerunt non incessisse itinere discipline conspexissem, cepi edere hunc librum. […] Posui eum introductorium atque collegi in eo ex dictis antiquorum quicquid necessarium est huic magisterio secundum […] introductorium. Et non introduxi ratiocinationes disputationi sive defensioni eorum que protulimus necessariis (*sic*) cum sint in libro Tholomei *qui appellatur alarba macalet*, id est quatuor tractatuum, et in libro meo quem edidi in confirmatione magisterii iudiciorum astrorum et in destructione epistole *Aisse*	libros quos vocaverunt *introductorios* huius magisterii, id est iudiciorum astrorum, set quosdam ex eis non fuisse scrutatos diligenter universa que necessaria sunt in eodem magisterio *in* his que conveniunt introductorio, quosdam vero ea que necessaria sunt protulisse prolixe, et quia quod necessarium est in eo periisse cernerem, quosdam quoque in *ordine* eorum que protulerunt non incessisse *in* itinere discipline conspexissem, cepi hunc librum edere, et posui eum introductorium atque collegi in eo ex dictis antiquorum quicquid necessarium est huic magisterio secundum *modum* introductorium. Et non introduxi ratiocinationes disputationi sive defensioni eorum que protulimus necessarias, cum sit in libro Ptolomei *qui appellatur alarbamachalet*, id est quatuor tractatuum, et in libro meo quem edidi in confirmatione magisterii iudiciorum astrorum et in	vocaverunt *introductorios* huius magisterii, id est iudiciorum astrorum, set quosdam ex eis non fuisse scrutatos diligenter universa que necessaria sunt in eodem magisterio de his que conveniunt introductorio, quosdam vero ea que necessaria sunt protulisse prolixe, et quia quod necessarium est in eo periisse cernerem, quosdam quoque in ordinatione eorum que protulerunt non incessisse itinere discipline conspexissem, cepi edere hunc librum, et posui eum introductorium atque collegi in eo ex dictis antiquorum quicquid necessarium est huic magisterio secundum *modum* introductorium. Et non introduxi ratiocinationem (*sic*) disputationi sive defensioni eorum que protulimus necessarias, cum sint in libro Ptholomei *qui appellatur alarbamacalet*, id est quatuor tractatuum, et in libro meo quem edidi in confirmatione magisterii iudiciorum astrorum et in destructione epistole

ad hoc possint sufficere. Et divisi eum in quinque differentias.	*Benhali* in annullatione eius ex ratiocinatione que ad hoc possint sufficere. Et divisi eum in quinque differentias.	destructione epistole *Aiza Aveneli* in annullatione eius ex ratiocinatione que ad hoc possint sufficere. Et divisi eum in.5. differentias,	*Aissebenhali*[36] in annullatione eius ex ratiocinatione que ad hoc possint sufficere. Et divisi eum in quinque differentias.
	Incipit liber introductorius Alcabizi qui Abdilazi dicitur, i.e. servus dei, ad totam astronomiam.		
Prima differentia *est* in esse circul<i sign>orum essentiali *et* accidentali. Secunda differentia *est* in naturis planetarum.vii. et quid sit illis proprium et quid significent. Tertia differentia *est* in his que accidunt planetis.vii. in semetipsis et quid accidat eis ab invicem. Quarta differentia *est* in expositionem nominum astro<logo>rum. Quinta *est* in universitate partium, et expositione esse eorum in gradibus.	Prima differentia, in esse circuli signorum essentiali *vel* accidentali. Secunda differentia, in naturis planetarum.vii. et quid sit illis proprium et quid significent. Tertia [...], in his que accidunt planetis.vii. in semetipsis et quid accidat eis ab invicem. Quarta [...], in expositione nominum astrologorum. Quinta [...], in universitate partium, et *in* expositione esse eorum in gradibus.	*prima quorum (sic) est* in esse circuli signorum essentiali *et* accidentali. Secunda [...], in naturis planetarum.7. et quid sit illis proprium et quid significent. Tertia [...],in his que accidunt planetis.7. in semetipsis et quid accidat eis ab invicem. Quarta [...], in explanatione *vel expositione* nominum astrologorum. Quinta [...], in universalitate partium, et *in* expositione esse eorum in gradibus.	Prima differentia, in esse circuli signorum essentiali *et* accidentali. Secunda differentia, in naturis planetarum.vii. et quid sit illis proprium et quid significent. Tertia differentia, in his que accidunt planetis.vii. in semetipsis et quid accidat eis ab invicem. Quarta differentia, in expositione nominum astrologorum. Quinta differentia, in universitate partium, et expositione esse eorum in gradibus.
Prima differentia, in esse circuli signorum essentiali vel accidentali.	[...]	*Explicit prologus incipit liber primus* *Capitulum primum de.12. signorum (sic)*	*Prima differentia, in esse circuli signorum essentiali vel accidentali.*

36. The ‘h’ has been expunged.

In nomine domini. Nitac, id est zodiacus circulus, signorum dividitur in.12. partes equales […] et hee.12. partes *equales vocantur* signa et referuntur ad ymagines que sunt sub eodem circulo zodiaco, que sunt Aries, Taurus, Gemini, Cancer, Leo, Virgo, Libra, *Scorpius*, Sagittarius, Capricornus, Aquarius, *Piscis*. Et unumquodque istorum signorum dividitur in.30. partes equales que gradus vocantur. Et *unusquisque* gradus dividitur in.60. minuta, et minutum in.60. *secunda* […]; similiter que *secuntur*, *tertia scilicet et quarta*, *ascendendo* usque ad infinita.	*Sytac (sic) circuli* signorum dividitur in partes.xii. equales secundum divisionem circuli signorum. Et he.xii. partes *vocantur* signa et referuntur ad imagines que sunt sub eodem zodiaco circulo, que sunt Aries, Taurus, Gemini, Cancer, Leo, Virgo, Libra, *Scorpius*, Sagittarius, Capricornus, Aquarius, *Pisces*. Et unumquodque istorum […] dividitur in.xxx. partes equales que vocantur gradus et gradus dividitur in.lx. minuta, et minutum in.lx. *secunda*, et *secundum* in.lx. *tertia*; similiter que *sequuntur, quarta scilicet et quinta, crescendo* usque ad infinita.	*In nomine domini zodiacus circulus id est nitac id est circulus* signorum dividitur in.12. partes equales secundum divisionem circuli signorum et hee.12. partes *vocantur* signa et referuntur ad imagines que sunt sub eodem zodiaco circulo, que sunt Aries, Taurus, Gemini, Cancer, Leo, Virgo, Libra, *Scorpius*, Sagittarius, Capricornius, Aquarius, *Piscis*. Et unumquodque istorum signorum dividitur in.30. partes equales que gradus *dicuntur* et gradus dividitur in.60. minuta, et minutum in.60. *secundas*, et *secunda* in.60. *tertias*; similiter *que succedunt, quarta scilicet et quinta minuenda* usque ad infinita.	*Nithac, id est circulus* signorum, dividitur in.xii. partes equales secundum divisionem circuli signorum et hee.xii. partes *dicuntur* signa et referuntur ad ymagines que sunt sub eodem zodiaco circulo, que sunt Aries, Taurus, Gemini, Cancer, Leo, Virgo, Libra, *Scorpio*, Sagittarius, Capricornus, Aquarius, *Piscis*. Et unumquodque istorum signorum dividitur in.xxx. partes equales que gradus vocantur. Et gradus dividitur in.lx. minuta, et minutum in.lx. *secunda*, et *secundum* in.lx. *tertia*; similiter que *sequuntur*, *quarta scilicet et quinta*, *ascendendo* usque ad infinita.

3.2. Chapter 1[37]

MS *V*	MS *M*	MS *J*	MS *U*
Et unumquodque signorum habet propriam significationem	Et unumquodque *eorum* signorum habet propriam significationem	Et unumquodque signorum habet propriam significationem in his	Et unumquodque signorum habet propriam significationem in his

37. Al-Qabisi, *The Introduction…*, ed. Burnett e.a., p. 242.

in his que significant ex *creatione sive moribus* hominum, sementum et regionum et cetera.	in his que significantur ex *creatione sive* moribus hominum, sementum et regionum et *ceterorum.*	qui significantur ex *creatione sive ex* moribus hominum, sementum et regionum et cetera.	que significant ex *membris et* moribus hominum, sementum et regionum et cetera.
Aries habet ex corpore hominis caput et faciem, et ex regionibus Bebil et Feriz, id est Babyloniam et *Persidam*, et Adrabigen et *Palestinam*; *et ex litteris a. et n.*	Aries habet ex corpore hominis caput et faciem, et ex regionibus Bebil et Feriz, i.e. Babiloniam et *Persidam* et Arabigen et *Phalastin.*	Aries habet ex corpore hominis caput *scilicet* et faciem, et ex regionibus Bebil et Fe<ri>z, id est Babiloniam et *Persiam* et Adrobigen, Philistin.	Aries habet ex corpore hominis caput et faciem, et ex regionibus Bebil et Feriz, id est Babiloniam et *Persidam*, et Adrabigen et Falastin.
Taurus habet arbores que plantantur, et ex corpore hominis nodum gutturis et collum et ex regionibus [...]	Taurus habet arbores que plantantur, et ex corpore hominis collum et gutturis nodum, et ex regionibus Egeuuet et Almehin et Handen et *Haracerad.*	Taurus habet arbores que plantantur, et ex corpore hominis collum [...] gutturis nodum, et ex regionibus Ezemech et Almeyut et Haulen et Harachera.	Taurus habet arbores que plantantur, et ex corpore hominis collum et gutturis nodum, et ex regionibus Ezeuuet et Almehin et Handen et *Aracerad.*
[...] Iurgen et Armeniam, *et Adrabigen et Gilen* id est Egyptum, et *Meucan; et ex litteris.go.*	Signum Geminorum est largum, boni animi; habet ex corpore hominis humeros [...] brachia et manus, et ex regionibus Iurgen et Armeniam, *Wahtrabigen* et *Helen*, Egiptum et *Barcha.*	Signum Geminorum est largum, boni animi; habet ex corpore hominis humeros et brachia et manus, et ex regionibus Ungent et Armeniam, *et Adrabigen* et *Helen*, Egiptum et *Barcha.*	Signum Geminorum est largum, boni animi; habet ex corpore hominis humeros et brachia et manus, et ex regionibus Iugen et Armeniam, *Wahtrabigen* et *Helen, id est* Egyptum et *Barcha.*
Cancer habet ex arboribus eas que *fuerint equales longitudine*, et ex corpore hominis pectus, [...] *et* stomacum, costas, splen	Cancer habet ex arboribus eas que *sunt equalis longitudinis,*[38] et ex corpore hominis pectus, *cor et* stomacum, costas, splen	Cancer habet ex arboribus eas que *fuerint equales longitudine*, et ex corpore hominis pectus, costas, [...] stomacum, splen et	Cancer habet ex arboribus eas que *fuerint equales longitudine*, et ex corpore hominis pectus, [...] stomachum, costas, splen

38. Other manuscripts have ‘mediocris longitudinis’.

et pulmonem, et ex regionibus Armeniam Minorem et orientalem plagam *Churasen* et *Sin*, et habet participationem in *Balech* et *Adrabigen*; *et ex litteris d. et p.*	et pulmonem, et ex regionibus Armeniam Minorem et orientalem plagam *Hurachcen* et *Acin*, et habet participationem in *Barach* et *Ahtrabigen*.	pulmonem, et ex regionibus Armeniam Minorem et orientalem plagam Butachen et Athin, et habet participationem *Baurach* et *Arahge*.	et pulmonem, et ex regionibus Armeniam Minorem et orientalem plagam *Hurachcen* et *Acin* et habet participationem in *Barach* et *Atrabigen*.

3.3. Explicit[39]

MS *V*	MS *M*	MS *J*	MS *U*
Pars *cotonis* a Mercurio in Venerem.	[...]	Pars *alchoto id est bambabasii* a Mercurio in Venerem.	Pars *bambasii* a Mercurio in Venerem.
Pars ciceris a Venere in Solem.			
Pars *elusce*, quod est quoddam genus seminis albi in similitudinem [...] lini et eo utuntur medici [...], a *Mercurio* [...] in Venerem.	Pars *alcincen*, quod est quoddam genus seminis albi in similitudine *seminis* lini et utuntur eo medici, fitque ex eo unguentum in medicinis utile, a *Mercurio in Iovem vel* in Venerem.	Pars *alchonicen*, quod est quoddam genus seminis, alba in similitudinem *seminis* lini et utuntur ei (*sic*) medici, fitque ex ea unguentum [...] medicinis utile, a *Saturno in Iovem vel* in Venerem.	Pars *alcicem*, ^quod est quoddam genus seminis albi in similitudine *seminis* lini et utuntur eo medici, fitque ex eo unguentum in medicinis utile^, a *Saturno in Iovem vel* in Venerem.
Pars *bathigh*, qui sunt butefles magni atque maturi et crocei *coloris*, a Mercurio in Saturnum.	Pars *albathigh*, qui sunt butefles magni atque maturi et crocei, a Mercurio in Saturnum.	Pars *abathegh*, qui sunt butefles *vel memolones* magni atque maturi et crocei, a Mercurio in Saturnum.	Pars *abbathigh*, ^qui sunt butefles magni atque maturi et crocei^, a Mercurio in Saturnum.

39. Al-Qabisi, *The Introduction*..., ed. Burnett e.a., p. 362-364.

Pars lilii a Saturno in Iovem vel Venerem.	[…]	[…]	[…]
Pars ciborum acrium a Saturno in Martem.	Pars ciborum acrium a Saturno in Martem.	[…]	Pars ciborum acrium a Saturno in Martem.
Pars ciborum dulcium a Sole in Venerem.	Pars ciborum dulcium a Sole in Venerem.	[…]	Pars ciborum dulcium a Sole in Venerem.
Pars ciborum saporis *apii et* huiusmodi herbarum […] habentium a Marte in Saturnum.	Pars ciborum saporis *apii vel* huiusmodi herbarum saporem habentium a Marte in Saturnum.	Pars ciborum saporis *apii vel* huiusmodi herbarum saporem habentium, a Marte in Saturnum.	Pars ciborum saporis *acuti ^vel* huiusmodi herbarum saporem habentium^, a Marte in Saturnum.
Pars ciborum amarorum a Mercurio in Saturnum.	Pars ciborum amarorum a Mercurio in Saturnum.	Pars ciborum amarorum a Mercurio in Saturnum.	Pars ciborum amarorum a Mercurio in Saturnum.
Pars medicaminum *egerentium dulcium* a Sole in Lunam.	Pars medicaminum *calidorum* a Sole in Lunam.	Pars medicaminum *vel ciborum* […] *calidorum* a Sole in Lunam.	Pars medicaminum *purgantium dulcium* a Sole in Lunam.
Pars medicaminum *egerentium acrorum* a Saturno in Iovem.	Pars medicaminum *acrium a Sole in Lunam vel* a Saturno in Iovem.	Pars medicaminum *et ciborum acrium* a Saturno in Iovem.	Pars medicaminum *acrium* a Saturno in Iovem.
Pars medicaminum salsorum a Marte in Lunam.	Pars medicaminum salsorum a Marte in Lunam.	Pars medicaminum *vel ciborum* salsorum a Marte in Lunam.	Pars medicaminum salsorum a Marte in Lunam.
Pars *venatorum* a *geusahar*, id est a Capite Draconis, in Saturnum.	Pars *venenatorum* a *geuzahar*, id est a Capite Draconis, in Saturnum.	Pars *venenatorum* a *capite Drachonis que est geuzehar* in Saturnum.	Pars *venatorum* a *geuzahar*, id est a Capite Draconis, in Saturnum.
Et proiciuntur universa hec ab ascendente revolutionis.	Et proiciuntur universa hec ab ascendente […]	Et proiciuntur universa hec ab ascendente […]	Et proiciuntur universa hec ab ascendente *revolutionis.*
Hec sunt *universa* que *ceciderunt* nobis ex partibus. Introduximus	Hec sunt *universa* que *ceciderunt* nobis ex partibus. Introduximus	Hec sunt *omnia* que *accidunt* nobis ex partibus. Introduximus	Hec sunt *universa* que *ceciderunt* nobis ex partibus. Introduximus

quoque has partes novissimas, etsi est in eis narratio debilis, ne *dimittamus* aliquid quod possit esse *introductio* ad magisterium iudiciorum astrorum. *Quia proferamus illud.*	quoque has partes novissimas, etsi in eis sit narratio debilis, ne *dimitteremus* aliquid quod posset esse introductorium ad magisterium iudiciorum astrorum *quin proferremus illud.*	quoque has partes novissimas, etsi est in eis narratio debilis, ne *dimittemus* aliquid quod possit esse introductorium ad magisterium iudiciorum astrorum *quam (sic) illud.*	quoque has partes novissimas, etsi est in eis narratio debilis, ne *dimitteremus* aliquid quod possit esse introductorium ad magisterium iudiciorum astrorum *quin proferamus eum.*
Perfectus [...] introductorius liber *Alcabisii* ad magisterium iudiciorum astrorum *octavo die mensis Ianuarii tertie indicionis annis domini perfectis.1181. Explicit Deo gratias.*	Perfectus est introductorius *Abdilazit, id est servi gloriosus* (*sic*) [...] *Dei, qui dicitur Alcabiczi* ad magisterium iudiciorum astrorum, *cum laude Dei et eius adiutorio.*	Perfectus est *liber* introductorius *Abdileziz, id est servi gloriosi scilicet Dei, qui dicitur Alcabitius* ad magisterium iudiciorum astrorum, *cum laude Dei et eius adiutorio, interpretatus a magistro Iohanne Yspalensi feliciter.*	Perfectus est introductorius *Abdilaziz, id est servi gloriosi, scilicet Dei, qui dicitur Alcobizi*, ad magisterium iudiciorum astrorum, *cum laude Dei et eius adiutorio, interpretatus a Iohanne Hyspalensi.*

Carla DI MARTINO

LE BONHEUR PERDU: NOTE SUR LA TRADUCTION LATINE MÉDIÉVALE DU *TALKHÎS KITÂB AL-ḤISS WA-L-MAḤSÛS* (*EPITOMÉ* DU *LIVRE DU SENS ET DU SENSIBLE*) D'AVERROÈS

Introduction

Les traducteurs médiévaux étaient conscients de l'importance de leur travail, et, ici comme ailleurs, ils suivaient des modèles classiques. Le modèle classique de traduction était, évidemment, celui des traductions du grec en latin, théorisé d'une part par Boèce[1], d'autre part par Saint Jérôme[2], modèle qui privilégiait la méthode de traduction *verbum de verbo* et qui a été suivi, par exemple, par un grand traducteur comme Burgundio de Pise. Mais les premiers à traduire de l'arabe en latin furent aussitôt confrontés au problème qui avait bouleversé, trois siècles auparavant, leurs collègues arabes: la langue arabe a une syntaxe fort différente des langues occidentales, et traduire mot à mot de l'arabe en latin n'est pas comme traduire du grec!

Dans la préface de sa traduction du *Liber Imbrium* de Jafar Indus, vers la moitié du XIIe siècle, Hugo de Santalla se plaint ouvertement de l'*implacabilis diversitas linguarum*, et son collègue, Hermannus de Carinthia, dans la préface de sa traduction du *Liber Introductorius* d'Abu Mashar dénonce 'quam grave sit ex tam fluxo loquendi genere quod apud Arabes est Latine orationi congruum aliquod commutari atque in hiis maxime que tam artam rerum imitationem postulant'.

En revanche, Richard l'Anglais, en 1220 environ (ce sont les années des premières traductions des œuvres d'Averroès) trouve que ce *fluxum*, la prolixité de l'arabe, est un avantage, et justement à propos du *De Anima,* il oppose la *difficultas et obscuritas* des traductions du grec, *cuius modus est brevis*, à l'*explanatio et claritas* des traductions de l'arabe[3].

1. Cf. Boèce, *In Isagogen Porphyrii Commentorum Editio secunda*, chap. 1, éds G. Schepps – S. Brandt (Vienna-Leipzig, 1906), p. 135.
2. Hieronymus, *Epistula LVII*.
3. Cf. Burnett Ch., 'Translating from Arabic into Latin in The Middle Age: Theory, Practice, and criticism', in: *Éditer, traduire, interpréter: essais de méthodologie philosophique*. Études réunies sous la direction de S.G. Lofts et P.W. Rosemann (Louvain-la-Neuve, Louvain-Paris, 1997), p. 55-78.

L'*Epitomé* du *De Sensu* d'Averroès est une œuvre idéale pour tester la valeur et l'intérêt des traductions médiévales de l'arabe en latin, car nous avons la chance d'en posséder l'original arabe, une traduction juive et deux traductions latines. Une comparaison entre le texte arabe et le texte latin de ce traité nous montre immédiatement à quel point, dans les traductions de l'arabe en latin, la frontière entre traduction et exégèse est fragile.

Traduction et exégèse dans l'étude de la tradition grecque, arabe et latine des œuvres d'Aristote

Le *corpus* philosophique grec fut traduit quasi entièrement du grec en arabe, souvent par l'intermédiaire du syriaque, au IXe siècle[4]. En conséquence de ce mouvement de traduction, le monde arabe produit un *corpus* très important d'œuvres 'aristotéliciennes': d'une part des commentaires des traités d'Aristote connus (tous, sauf la *Politique*), comme les épitomés, paraphrases et commentaires littéraux d'Averroès, d'autre part des œuvres indépendantes directement mais librement inspirées d'Aristote, comme ce fut le cas d'Alfarabi, d'Ibn Bâjjah et des livres du *Shifâ'* d'Avicenne. Une partie importante de ces textes passa à l'Occident latin en environ un siècle, notamment grâce au mouvement des traductions des XIIe et XIIIe siècles, accompagna et parfois précéda la réception dans le monde latin de l'œuvre d'Aristote, traduite, elle aussi, du grec en latin pour la première fois (à l'exception du *corpus* logique déjà traduit par Boèce) dans la même période.

L'étude de la tradition textuelle du *corpus* aristotélicien, et en particulier des œuvres de psychologie d'Aristote qui seront étudiées ici, à savoir le *De Anima* et les *Parva Naturalia*, nous confronte donc à trois classes de textes:

1. Les traductions. Plusieurs cas sont possibles:

– du grec en arabe, que l'original soit conservé, comme en partie pour le *De Anima* d'Aristote et la paraphrase de Thémistius, ou non, comme

4. Pour une étude du mouvement de traductions du grec en arabe aux IXe-Xe siècles, voir Gutas D., *Greek Thought, Arabic Culture. The Graeco-Arabic Translation Movement in Baghdad and Early 'Abbâsid Society (2nd-4th / 8th-10th centuries)*, (London – New York, 1998). Pour la tradition arabe du *De Anima* et des *Parva Naturalia* voir: Elamrani-Jamal A., 'Le *De Anima* dans la tradition arabe', *Dictionnaire des Philosophes Antiques*, I (Paris, 2003), p. 346-358; Di Martino C., 'Notice sur la tradition arabe des *Parva Naturalia*', *Dictionnaire des Philosophes Antiques*, I (Paris, 2003), p. 260-263.

pour la paraphrase du *De Anima* éditée et étudiée par Arnzen[5]. Tous ces textes sont du plus grand intérêt d'un double point de vue lexical et doctrinal, car il s'agit de la première formulation en langue arabe d'un vocabulaire philosophique technique.

- du grec en latin; pour les textes qui nous intéressent, l'original grec est toujours conservé, comme dans les deux traductions latines, la *Vetus* et la *Nova*, du *De Anima* et des *Parva Naturalia* d'Aristote, la paraphrase du *De Anima* de Themistius et la troisième partie des commentaires du *De Anima* d'Aristote par Simplicius et Philopon.
- de l'arabe en latin, que l'original arabe soit conservé, comme pour le *De Anima* d'Avicenne ou l'*Epitomé* des *Parva Naturalia* d'Averroès, ou non, comme pour le *Grand Commentaire* du *De Anima* par Averroès. À la différence de ce qui s'était passé au moment des traductions du grec en arabe, le lexique arabo-latin a dû confronter et s'adapter au lexique gréco-latin et à une langue philosophique latine qui avait déjà une longue tradition.
- du grec en arabe et de l'arabe en latin, comme c'est le cas de la version du *De Anima* contenue dans le *Grand Commentaire* du *De Anima* par Averroès. Cette version présente notamment des différences très importantes par rapport à la tradition grecque. Un seul exemple: le texte commenté par Averroès parle déjà, au chapitre IV du livre III, de trois intellects et non, comme le fait le texte grec, de deux.

2. Les commentaires en langue grecque, arabe et latine des textes d'Aristote. Ces textes s'efforcent d'expliquer la *littera* d'Aristote; en conséquence, il est évident que, lorsqu'ils commentent un texte en traduction, souvent à l'aide d'autres textes qui sont, eux aussi, des traductions, ils sont fortement influencés par les choix du traducteur.

3. Les textes de psychologie 'libres', à savoir des textes relevant de tous les genres littéraires dans lesquels la philosophie classique arabe, aux Xe-XIe siècles, et ensuite surtout l'Occident latin aux XIIe-XIVe siècles s'exprima en matière de doctrine de l'âme: épitomés, paraphrases, 'discussions' (je pense à l'œuvre homonyme d'Avicenne, sorte de recueil de *quaestiones*[6]), traités, *quaestiones*, *expositiones*, *sententiae*.

5. Cf. *Aristoteles' Kitâb al-nafs: eine verlorene spätantike Paraphrase in arabischer und persischer Überlieferung. Arabischer Text nebst Kommentar, quellengeschichtlichen Studien und Glossaren*, éd. R. Arnzen, Aristoteles Semitico-Latinus 9 (Leiden – New York – Köln, 1998).

6. Cf. *Kitâb al- Mubâḥathât*, 'A. Badawî, *Arisṭû 'inda 'l-'Arab. Dirâsat wa-nuṣûṣ ġayr manšura* (Cairo, 1947), p. 122-238.

La tradition arabe des *Parva Naturalia* et *l'Epitomé* du *De Sensu* d'Averroès.

En particulier, les traités aristotéliciens connus dans la tradition arabe sous le titre du premier d'entre eux, le *Kitâb al-ḥiss wa-l-maḥsûs* ou *Kitâb al-ḥâss wa-l-maḥsûs*, 'Livre du sens et du sensible', et en Occident sous le nom de *Parva Naturalia*, ont un rôle capital dans la formation des notions devenues classiques dans la science psychologique arabe et des auteurs principaux de cette tradition, dont de nombreux textes passèrent à l'Occident latin, notamment grâce aux traductions des XIIe et XIIIe siècles. À côté d'une traduction du grec des *Parva Naturalia* d'Aristote et du *De Sensu* d'Alexandre, elles sont à l'origine d'une tradition d'exégèse et de réflexion dont le *De Sensu* d'Albert le Grand et la *Sententia libri de sensu* de Thomas d'Aquin sont les témoins directs les plus importants.

Aucun texte complet de la traduction arabe des *Parva Naturalia* ne nous est parvenu, mais en 1974, S. Pinès[7], en étudiant les citations du *De divinatione per somnium* dans *al-Risâla al-Manâmiyya* d'Ibn Sînâ, grâce à une comparaison avec l'*Epitomé* du *De Sensu* d'Ibn Rushd, a établi que le texte des *Parva Naturalia* connu et utilisé par les auteurs arabes est le témoin d'une rédaction différente de celui transmis par la tradition grecque.

Cette conclusion a été en partie confirmée, en partie remise en discussion par la récente découverte, par H. Daiber, d'une traduction acéphale de notre texte. En 1986, H. Daiber[8] avait déjà attiré l'attention des chercheurs sur l'existence, dans un ms. de la Bibliothèque Raza a Rampur, en Inde[9], d'un texte qui pourrait éclairer considérablement l'histoire de notre traité. L'édition de ce précieux texte est maintenant le sujet de la thèse de Mme Hansberger à Oxford, et H. Daiber en a lui-même publié, en 1997, une première description[10].

7. Pinès S., 'The arabic recension of *Parva Naturalia* and the philosophical doctrine concerning veridical dreams according to al-risâlâ al manâmiyya and others sources', *Israel Oriental Studies* (1974), p. 104-153, réimprimé dans *The collected works of Shlomo Pines*, vol. II (Leiden, 1986), p. 104-153.

8. Daiber H., 'New Manuscript Findings from Indian Libraries', *Manuscripts of Middle East* 1 (1986), p. 26-48; Daiber H., 'Semitische Sprachen als Kulturvermittler zwischen Antike und Mittelalter. Stand und Aufgaben der Forschung', *Zeitschrift der Deutschen Morgenländischen Gesellschaft* 36 (1986), p. 292-313.

9. À savoir le ms. n° 1752, 7r-54v.

10. Daiber H., 'Salient Trends of the Arabic Aristotle', in: G. Endress & R. Kruk (éds) *The Ancient Tradition in Christian and Islamic Hellenism. Studies on the Transmission of Greek Philosophy and Sciences dedicated to H.J. Drossaart Lulofs* (Leiden, 1997), p. 29-41, voir p. 36-41.

Le texte, intitulé *Kitâb al-ḥiss wa-l-maḥsûs*, est composé de trois chapitres: un premier chapitre, acéphale, porte sur le sens et le sensible; un deuxième chapitre se divise en trois sections, l'une sur la mémoire et la réminiscence, la deuxième sur le sommeil et la veille, la troisième sur les songes, tout comme dans l'*Epitomé* d'Ibn Rushd; un troisième chapitre porte sur la longueur et la brièveté de la vie, comme le troisième chapitre de l'*Epitomé* d'Ibn Rushd et le dernier livre dans la tradition grecque.

Ce texte fut bien connu et largement utilisé par les auteurs arabes. Cependant, l'*Epitomé* des *Parva Naturalia* d'Averroès est le seul texte arabe consacré à l'exégèse des *Parva Naturalia* qui nous soit parvenu et le seul qui soit traduit en latin. Il porte sur six des neuf traités de la tradition grecque: *De Sensu et sensibilibus*, abrégé dans le premier livre de l'*Epitomé* d'Averroès; *De memoria et reminiscentia, De Somno et vigilia, De somniis* et *De divinatione per somnium* dans le deuxième; *De longitudine et brevitate vitae* dans le troisième, et il est organisé comme un véritable traité sur les sens internes et un essai de 'psychologie appliquée'.

Dans l'œuvre d'Averroès, ce texte est le témoin d'une étape particulière de sa pensée psychologique, qui privilégie les opérations sensibles – ainsi, la prophétie et l'obtention du bonheur dépendent de la vie sensible et non de la pensée intellectuelle. Cette position est moins célèbre que la doctrine du *Grand Commentaire* du *De Anima,* mais elle eut néanmoins une influence, importante et pas encore assez étudiée, sur les auteurs latins. Certaines doctrines soutenues par Averroès dans l'*Epitomé*, et disparues ensuite de son *Grand Commentaire* du *De Anima*, ont eu une influence importante sur Albert le Grand: par exemple, la notion de 'spirituel' dans la *Summa de Creaturis* et dans le *De Anima* et l'idée d'une prééminence de la mémoire parmi les facultés sensibles dans les *Parva Naturalia* d'Albert le Grand[11]. En particulier, l'œuvre qu'Albert le Grand consacra aux *Parva Naturalia,* texte peu étudié mais du plus grand intérêt, se comprend beaucoup mieux si on le lit à la lumière de l'*Epitomé* d'Averroès, dont il a repris la démarche et la méthode, et si on le considère, à l'instar du texte d'Averroès, comme un *Epitomé* (et non comme une paraphrase, comme c'est le cas du *De Anima* d'Albert qui reprend plutôt Avicenne).

Ce texte d'Averroès a été traduit deux fois en latin: la *Translatio Vulgata*, un travail des années 1225-30 attribué à Michel Scot et conservé dans

11. Cf. Di Martino C., 'La perception spirituelle. Perspectives de recherche pour l'histoire des Parva Naturalia dans la tradition arabo-latine', *Veritas* 52 (2007), p. 21-35.

plus de soixante manuscrits, et la *Translatio Parisina*, une révision tardive de la *Vulgata*[12]. Il donna naissance à une tradition d'exégèse et de réflexion en langue latine sur des sujets de psychologie appliquée (la perception et la sensation, l'imagination reproductrice et l'imagination créatrice, les souvenirs, la signification et l'interprétation des rêves, la prophétie, mais aussi les conditions physiologiques de la vie et les raisons de la mort).

La *Translatio Vulgata* se présente comme une traduction *verbum de verbo* de notre traité. On y retrouve des termes qui ont fait couler beaucoup d'encre, comme *intentio*, traduction de l'arabe *ma'nâ*, ainsi que des termes qui sont silencieusement et définitivement entrés dans le lexique psychologique latin, comme *investigatio per rememorationem*, traduction de l'arabe *tadhakkur*. Les deux cas méritent notre attention.

La notion de *ma'nâ* répond à une exégèse précise des passages d'Aristote portant sur la problématique de la sensibilité par accident[13]. Il prend un sens technique, qu'on pourrait ainsi définir: *ma'nâ* est la signification de cette chose sensible individuelle dans le contexte sensible où elle se trouve à un moment donné et pour le sujet précis individuel qui la perçoit en ce moment même. Elle exprime une qualité relative, et non une qualité essentielle.

La traduction médiévale de *ma'nâ* par *intentio* est une traduction étymologique: *ma'nâ,* du verbe *'anâ*, 'signifier', comme *intentio* du verbe *intendere*. On pourrait traduire le terme en français par 'signifié', 'signification'.

Les traductions modernes proposées ne sont pas toutes convaincantes:
- entité: cette traduction ne me semble pas convenir, car le *ma'nâ* est une qualité,
- idée: non plus, car le *ma'nâ* est une qualité sensible.
- notion: cette traduction pourrait convenir, mais avec le risque de perdre l'aspect technique du mot arabe *ma'nâ*.

12. Le texte arabe est édité dans: *Averroès, Talkhîṣ kitâb al-ḥiss wa-al-maḥsûs* (*Epitome* du *De Sensu*), éd. H.A. Blumberg, Mediaeval Academy of America (Cambridge, 1972); les deux traductions latines sont éditées ensemble (la *Parisina* en bas de page) dans: *Averrois Compendia librorum Aristotelis qui Parva Naturalia vocantur*, éds A.L. Shields & H.A. Blumberg, Mediaeval Academy of America (Cambridge, 1949). Traduction anglaise dans: *Epitome of Parva Naturalia*, Translated from the original Arabic and the Hebrew and Latin versions, trad. Blumberg H., Mediaeval Academy of America (Cambridge, 1961).

13. Cf. Di Martino C., 'Ma'ani/*intentiones* et sensibilité par accident', in: M.C. Pacheco & J.F. Meirinhos (eds). *Intellect and Imagination in Medieval Philosophy*, Rencontres de Philosophie Médiévale 11 (Turnhout 2006), p. 507-521.

Michel Scot, dans ses traductions d'Averroès, traduit le même mot *ma'nâ* parfois par *forma*, parfois par *intentio*, parfois par *intellectus*. *Ma'nâ* peut en effet indiquer la forme par opposition à la matière (la perception est la saisie de la forme sans la matière); techniquement, *ma'nâ* est l'objet propre de la mémoire (la faculté cogitative distingue le *ma'nâ* d'une chose de l'image et la mémoire le perçoit et le conserve); le *ma'nâ* d'un mot est, enfin, son sens, sa 'notion'.

Un autre cas intéressant est le cas de la traduction du mot grec ἀνάμνησις (*anamnesis*). Normalement, ce terme est traduit du grec en latin par *reminiscentia*, et du grec en arabe par *al-quwa al-dhâkira* ou *al-dhikr*, de la racine **dhkr*, traduite à son tour en latin par *virtus rememorativa, rememoratio* et d'autres mots de la même famille. Dans la version arabe des *Parva Naturalia*, et par conséquent dans l'*Epitomé* d'Averroès, correspond au mot grec ἀνάμνησις le terme arabe *tadhakkur*, *masdar* (infinitif) de cinquième forme, dérivé de la même racine **dhkr*, ce que les grammairiens appellent la forme 'inchoative' ou 'causative'. Si le grec *anamnesis* indiquait une *dunamis,* une capacité de l'âme, que la tradition de l'Antiquité tardive a étudiée avec la faculté de la mémoire, et plus spécifiquement comme la mémoire humaine, la traduction arabe indique non plus une faculté mais un procès, le procès du souvenir qui consiste justement dans une sorte de recherche pour faire revenir à la mémoire une donnée perdue, ce qui en effet traduit bien la *ana-mnesis* (*ana* indiquant un mouvement de retour). Ainsi la traduction arabe a explicité la signification du mot grec… Le traducteur latin a bien choisi de traduire par une paraphrase, *investigatio per rememorationem*, que l'on retrouve dorénavant dans tous les textes latins portant sur la réminiscence (voir en particulier Albert le Grand et Thomas d'Aquin).

Ici, la paraphrase d'un mot, *investigatio per rememorationem* pour l'arabe *tadhakkur,* répond à un souci précis d'exégèse et de clarté. Mais, à d'autres endroits, le traducteur se contente de la paraphrase lorsque, on peut le supposer, le texte arabe est particulièrement dur à rendre en langue latine. Cependant, toutes les différences entre le texte arabe et le texte latin ne peuvent pas s'expliquer par ces seuls éléments linguistiques…

Le bonheur perdu: sur trois passages du *De Somniis* d'Averroès

Le troisième chapitre du livre II de *l'Epitomé* expose en continuité les sujets traités par Aristote dans ses deux traités *De Somniis* et *De divinatione per somnum*, tout comme dans la version arabe de Rampur.

Or, dans cette section, Averroès parle trois fois du bonheur, à chaque fois à propos de la prophétie, alors qu'Aristote n'en parle pas, ni dans son texte grec ni dans sa version arabe[14]. On laissera ici de côté la question d'établir pourquoi Ibn Rushd intègre ce sujet à l'étude de la prophétie: cela n'étonnera pas par ailleurs les connaisseurs d'Alfarabi par exemple, qui dans son traité sur *Les idées des habitants de la cité vertueuse* parle continuellement de l'obtention du bonheur.

Ce qui retiendra notre attention se passe plus tard, au moment de la traduction de l'*Epitomé* de l'arabe en latin. En effet, dans les deux versions latines, la *Vulgata* et la *Parisina*, les trois passages sur le bonheur sont omis. Est-ce que cela est dû à des difficultés linguistiques? Cette explication paraît fort improbable.

Par contre, la traduction juive de l'*Epitomé,* achevée par Mosé ben Tibbon en 1254, contient les trois passages[15].

1. Au tout début de la section sur les rêves et sur la divination, Averroès se propose d'étudier 'la nature des rêves et, parmi les divines perceptions qui n'ont rien à voir avec l'entendement humain, de celles qui appartiennent à la même classe que les rêves'. 'Parmi ces perceptions – continue Averroès – certaines sont nommées rêve, certaines divination, certaines révélation'. Entre ces perceptions, il n'y a pas de différence de nature, mais de degré, *bi-l-aqalli wa-l-akthari*, c'est-à-dire, explique Averroès, que leurs causes sont différentes. C'est par ailleurs justement pour cela qu'elles ont des noms différents, car pour les gens ces phénomènes dérivent de causes différentes:

En effet on croit que les rêves (*ru'yâ*) viennent des anges, la divination (*kahâna*) des djinn, la révélation (*waḥy*) de Dieu, qu'il soit loué, soit sans intermédiaire soit par un intermédiaire déterminé. ...encore, on considère que ... les connaissances (contenues) dans la	*Sermo de quiditate sompnii sufficiet, quia cause eorum non differunt, nisi secundum magis et minus, sed tamen differunt secundum nomina propter hoc, quod vulgus dicit. Dicunt enim quod sompnia sunt ab angelis et divinationes a demonibus et prophetie a Deo, aut cum medio aut sine medio.*

14. Les indications sur le texte arabe des *Parva Naturalia* contenu dans le ms. de Rampur m'ont été gentiment communiquées par R. Hansberger, qui en prépare l'édition critique et à qui s'adressent mes vifs remerciements.

15. *Averrois Cordubensis, Compendia librorum Aristotelis qui* Parva Naturalia *vocantur*. Textum hebraicum recensuit et adnotationibus illustravit, éd. H. Blumberg (Cambridge, Massachussets, 1954).

révélation portent sur des notions universelles, comme par exemple la connaissance de la quiddité (*mâhiyyah*) du bonheur, et la connaissance des choses grâce auxquelles on obtient le bonheur. Quant aux rêves et aux divinations, en revanche, les connaissances qui y sont acquises portent sur des informations contingentes. Et Aristote ne parle de tout cela qu'à propos des rêves[16].

Et Aristoteles non fuit locutus nisi tantum de sompniis[17].

Averroès n'adopte pas cette classification. Après avoir exposé l'opinion courante, il entame l'étude de ces trois phénomènes, rêve, divination et révélation, selon une seule et unique distinction entre rêves véridiques et rêves non véridiques. Et étudier les rêves signifie en premier lieu établir quelle est leur cause et quelle faculté de l'âme en est responsable.

Tous les rêves, véridiques et non véridiques, explique Averroès, sont des opérations de la faculté imaginative. Les rêves véridiques sont des rêves grâce auxquels nous appréhendons quelque chose dont l'existence nous était inconnue auparavant, et cela sans aucune connaissance précédente sur le sujet et sans y réfléchir: Averroès compare ce genre de connaissance à l'acquisition des 'propositions premières'. La cause de ces rêves est l'intellect en acte (*bi-l-fa'li*), dont l'émanation coule sur l'imagination humaine.

La divination et la révélation se réalisent de la même manière. Une fois établie la cause commune de ces trois phénomènes, Averroès en étudie le contenu, en se posant la question suivante: que peut-on apprendre dans les rêves? Le contenu de la révélation doit être du même genre.

Averroès commence par établir ce qu'on ne peut pas apprendre dans les rêves: la connaissance théorétique, car dans ce cas l'intellect en acte devrait entrer en communication directe avec l'intellect humain, sans l'intermédiaire de l'imagination. 'Quant aux rêves', continue Averroès, il est clair que 'pour leur nature', puisqu'il s'agit d'opérations de l'imagination, ils ne peuvent rien avoir à faire avec des connaissances scientifiques, mais 'ils portent uniquement sur des choses futures'. C'est exactement là la nature et la finalité de la prophétie (*nubuwwa*).

Une fois cela établi, Averroès s'attache au plus épineux des problèmes. Si la prophétie est du même genre que les rêves, c'est-à-dire une opération

16. Éd. citée p. 67. Toutes les traductions de l'arabe en français sont de l'auteur.
17. Éd. citée p. 95.

psychique qui, quel que soit le rôle de l'intellect, passe forcément par la faculté imaginative, qui est une faculté matérielle et donc capable uniquement de connaissances particulières, alors comment est-il possible qu'elle dévoile des connaissances, tels que les rêves véridiques et les visions prophétiques, dont la cause serait un 'intellect privé de matière', capable pour sa nature de concevoir et donc de communiquer uniquement des notions universelles? Comment un intellect immatériel saurait-il concevoir et communiquer des formes individuelles concernant le temps, le lieu, une certaine classe de personnes ou un individu de cette classe? Averroès commente:

La discussion de ces sujets, même si cela est difficile à mener au niveau de la compréhension humaine, doit atteindre les limites les plus extrêmes que la nature de la connaissance humaine puisse permettre, car ceci et ceci seul est la vraie essence du bonheur[18]

Sermo enim de istis, quamvis sit valde difficilis, tamen rectum est pervenire ad hoc secundum quod possumus in natura nostra[19].

Alors que le traducteur juif nous donne une traduction littérale de l'intégralité du passage, le traducteur latin, qui paraphrase ce passage plutôt que de le traduire, a omis la référence finale au bonheur.

Averroès ne répond pas vraiment à cette double interrogation, et le débat qui s'ensuit dépasse en tous cas nos intérêts actuels. Lorsqu'il revient à son sujet, les rêves véridiques, comme par exemple les visions prophétiques, c'est pour expliquer que, parmi ces visions prophétiques, il y en a certaines qui contiennent des éléments suffisants permettant leurs interprétations, certaines qui ne les contiennent pas, et que parfois ces visions sont directement compréhensibles pour la personne même qui les reçoit, et parfois pas. Dans ce dernier cas, il faut s'adresser à un interprète, dont le profil idéal est ainsi tracé par Averroès:

... L'interprète est celui qui a la capacité de connaître l'imaginaire qui est commun à l'ensemble des peuples, et celui qui est propre à chaque population et par classes de personnes, car

Interpretator autem est homo qui habet animam paratam naturaliter ad distinguendum similitudines rerum et sompniorum. Et est ille cui largitur intellectus intentiones corporales, cui

18. Éd. citée p. 74-5.
19. Éd. citée p. 104.

les peuples se distinguent en ce domaine par deux aspects: 1) à cause de la nature, c'est-à-dire selon les facultés de leur âme, et 2) à cause de l'imaginaire et des opinions dans lesquelles ils ont grandi, et qu'ils ont l'habitude de partager dès leur naissance: le principe premier, les anges, et l'essence (*jawhar*) du bonheur humain[20].

assimilantur in sompno intentiones spirituales; et de necessitate debet scire similitudines universales omnibus gentibus et similitudines proprias cuilibet genti et cuilibet modo hominum. Gentes enim diversantur in hoc duobus modis, quorum unus est naturaliter, scilicet secundum virtutes eorum et secundum entia eis propria in suis regionibus; secundus autem est secundum similitudinem et opiniones secundum quas nati sunt secundum leges suas, et receperunt fidem a nativitate, scilicet de primo principio et angelis et resurrectione[21].

Conclusion

Une comparaison systématique entre le texte arabe original de *l'Epitomé* du *De Sensu* d'Averroès et sa traduction latine révèle:

- un lexique psychologique précis et attentif, dans l'effort constant du traducteur de rendre fidèlement les nuances lexicales de l'arabe par l'introduction de nouvelles locutions.
- des omissions et des ajouts, qui peuvent parfois s'expliquer par la difficulté du texte arabe, parfois par un choix doctrinal du traducteur.

Ainsi, dans le dernier passage analysé, le traducteur latin a non seulement encore omis la référence finale au bonheur, qui sentait peut-être à ses yeux le paganisme, mais il l'a remplacé en outre par la notion toute chrétienne de résurrection.

20. Éd. citée p. 85.
21. Éd. citée p. 117.

Michelle Reichert

HERMANN OF DALMATIA AND ROBERT OF KETTON: TWO TWELFTH-CENTURY TRANSLATORS IN THE EBRO VALLEY

When I initially broached this topic, it was my intention to demonstrate what I believed to be the importance of the institutional affiliations of the two twelfth-century scholars, Hermann of Dalmatia and Robert of Ketton, in determining the kinds of philosophical and scientific works they chose to translate. I had taken as my working hypothesis that their association with the Benedictine order, and more specifically their punctual collaboration with Cluniac prelates, such as Peter the Venerable, somehow facilitated the production and transmission of the kinds of scientific translations and redactions made by these two Spanish-based scholars. However, in the course of my research, what became apparent is rather the relative *inconsequence* of these institutional affiliations, at least those properly considered ecclesiastical. Thus, it has been necessary to query how two scholars could have translated, unimpeded, the kinds of mystical Islamic texts, especially alchemical, we find punctuating their translation of scientific texts at the height of the Crusades, and amid concerns of the Church over the ground being gained by such heretical movements as the anti-ecclesial Petrobrussian movement. Paradoxically, it appears to be Robert of Ketton, archdeacon of Pamplona and later Canon of Tudela – thus the more institutionally entrenched church figure – and not Hermann of Dalmatia – the more itinerant and apparently secular of the two – who veers most towards the production of occult works, such as the *Liber latitudinis clavis stellarum* (attributed to Hermes and first translated into Arabic in 743), a translation of *Morienus,* one of the earliest treatises on alchemy translated from Arabic into Latin, entitled the *Liber de compositione alchemie*, and possibly the *Turba Gallica*, inspired by the tenth-century *Turba philosophorum*, and which Paulette Duval convincingly suggests was written by Robert of Ketton. She suggests that the *Turba Gallica* inspired, in its turn, the composition of the *Conte du Graal* written by the twelfth-century Champenois writer, Chrétien de

Troyes[1]. Indeed, some of the iconography of the *Conte du Graal*, such as the female figure bearing a luminous *grail*, finds its expression in the eleventh and twelfth-century mural paintings of northern Spain, and specifically Navarre where Robert of Ketton is believed to have been associated with the ruling family in his official duties. In the course of my research, I have noted what appear to be references to Robert of Ketton in Chrétien de Troyes' *Erec et Enide*, and Erec's heritage in Rutlandshire, Robert's presumed geographical place of origin. Indeed, *Erec et Enide* abounds with alchemical references and symbolism, and in Chrétien's subsequent works we find strategically placed references to Spain and an emphasis on the predictive and astrological aspects of what Chrétien calls 'le meilleur des arts': the art of astronomy. It is not surprising that Chrétien de Troyes would have alluded to Robert of Ketton in a text rich in alchemical symbolism given that Robert is allegedly the first to have introduced this kind of mystical alchemy, with its particularly Islamic flavor, to Latin Europe. Further, it is reasonably probable that Chrétien's patron, Phillip of Flanders, could have had access to the *Turba Gallica*, given that he was related through his mother-in-law with some of the ruling members of northern Spain and Tudela, where Robert eventually became canon, or through his pilgrimage in the late 1170s to Santiago de Compostela[2]. Indeed, Chrétien de Troyes refers strategically throughout his works to cities in northern Spain.

However, like Robert of Ketton, Hermann of Dalmatia, along with Hugo of Santalla who translated into Latin a text later entitled *Hermetis Trimegesti Liber de secretis naturae et occultis rerum*[3], is among the first of the Latin authors to have demonstrated a familiarity with hermetic texts of the Arabic tradition patently unknown to most of their contemporaries. In Hermann's *De Essentiis*, written in 1143, we find a summary of pseudo-Apollonius' account of the discovery of the *Tabula Smaragdina* or 'Emerald Tablet' – one of the most important medieval alchemical texts – which, according to Charles Burnett, editor of the *De Essentiis*, appears to have been known at this time by only one other Latin scholar, Hugo of Santalla, who translated the *De secretis naturae* which contains the *Emerald Tablet*, spuriously attributed to Apollonius

1. Duval P., *Recherches sur les structures de la pensée alchimique et leurs correspondances dans le conte du Graal de Chrétien de Troyes* (Lille – Paris, 1975), p. 239-240.

2. Duval, *Recherches*..., p. 534-535 and p. 532.

3. Haskins C.H., *Studies in the History of Mediaeval Science* (New York, 1960), p. 79-80.

of Tyana[4]. Burnett leans in favor of Hermann's direct acquaintance with the Arabic text as opposed to his use of Hugo of Santalla's Latin translation. Perhaps this preoccupation with the Arabic tradition and transmission of hermetic texts is the 'secret knowledge' to which Hermann refers in the introduction to the *De Essentiis* as being shared by him and Robert. In any event, it is telling, as Burnett notes, that Dominicus Gundissalinus, possibly archdeacon of Segovia, quotes extensively the *De Essentiis* in his cosmological treatise, the *De Processione Mundi.* Gundissalinus is admittedly one of the first Latin authors in whose works appear the *Epistles,* a mystical Islamic work by the *Ikhwan al-Safa* or 'Brethren of Purity', and which reflects the influence of the alchemical corpus of the tenth-century scholar Jabir Ibn Hayyan, and the ninth-century astrological writings of Abu Ma'shar, of which the *Introductorium* or 'Greater Introduction to Astrology' is Hermann's most important translation.

What sets apart the translators and scholars, indeed the 'cosmologists', of twelfth-century Spain from their northern counterparts other than the ready accessibility of the Arabic works themselves and their linguistic ability to work directly from the original Arabic? Further, what in their institutional affiliations allowed them the freedom to adopt a more secular approach to their intellectual pursuits than those associated with the same religious orders in the north?

The reception of the works of a certain Master Odo or Eudes of Champagne by his northern ecclesiastical contemporaries as quasi-heretical appears to be representative of the fate of a scholar with pursuits similar to those of the Spanish-based Hermann and Robert. Odo of Champagne, a champenois astrologer, most likely associated with the counts of Champagne (perhaps Count Henry II), dedicated to William of White Hands, brother of count Henry the Liberal of Champagne and uncle of King Phillip Augustus of France, a treatise in defense of his astrological art of prediction, whose principle source of cosmology is Abu Ma'shar's *Introductorium maius.* William of White Hands was bishop of Chartres from 1164-1168, but Bishop of Rheims at the time Odo of Champagne wrote his treatise (1176-1202). Yet all that remains of this text are the substantial portions of it cited by Hélinand de Froidmont in his *Chronicon*, which includes a fifty-chapter vituperative refutation of Odo's *Libellus.* In Hélinand's opinion the ideas put forth in

4. *Hermann of Carinthia's De Essentiis*, ed. & tr. C. Burnett (Leiden, 1982), p. 38-39.

this work were tantamount to heresy[5]. Perhaps Odo's association with a strong secular lord, such as the Count of Champagne, explains his ability to formulate, unimpeded, the view of secular cosmology put forth in his *Libellus*, and considered damnable in the eyes of more traditional theologians such as Hélinand. Hermann and Robert may also have benefited from some similar kind of secular patronage, or protection, which would have allowed them to pursue freely their intellectual activities. Yet it is also reasonably inferable that such scholars as Hermann of Dalmatia and Robert of Ketton may have benefited from the position of the Spanish church in twelfth-century Spain. Its resistance to the reforms promoted by the powerful prelates of northern France, and the reluctance of its local prelates to relinquish their claim to the rich cultural plurality of a society in which the members of the three monotheistic communities continued to collaborate, despite the waning of Hispano-Islamic society, provided a unique atmosphere of relative intellectual freedom and intercultural florescence.

Cluny, despite its normative role in Spain, is partially responsible for the *translatio studii* of Latin translations of Arabic texts beyond the Pyrenees. Cluny took a particularly active interest from the time of the abbacy of Odilon in the re-conquest of Spain; and it is known that one of the purposes of Peter the Venerable's trip to Spain, undertaken in 1141, was to reform those Benedictine abbeys and priories that had previously adopted the Cluniac rule, and to call to order those which had relapsed. During Peter's inspection rounds in Spain, Cluny received from Alfonso VII the donation of the monastery of San Pedro of Cardeña and San Salvador of Bodinio, as well as substantial privileges over the Abbey of Carrión and the monastery of Saint Vincent of Salamanca. Yet according to James Kritzeck, it was only a matter of time before such religious houses relapsed and left the Cluniac fold, as did San Pedro of Cardeña near Burgos, which had originally been ceded to Cluny by Alfonso VII in substitution for arrears on an annual census donation of two thousand *metcales* owed to the order by his family. According to Marie-Thérèse d'Alverny, the monks of Cardeña went as far as appealing to Rome for an annulment of the donation made to Cluny. It was also during this visit to Spain that Peter enlisted Hermann of Dalmatia and Robert of Ketton to work on his translation of the Koran and other Islamic texts (collectively know as the *Collectio Toledana*). However, formidable monetary incentives were required in order to allay the two translators' reluctance to leave aside,

5. d'Alverny M.-T., 'Astrologues et théologiens au XIIe siècle', *Mélanges offerts à M.-D. Chenu*, Bibliothèque thomiste 37 (Paris, 1967), p. 40.

even momentarily, their Latin translations of Arabic astronomical and mathematical texts. It has been suggested that Robert's archdeaconry was given him in direct reward for his collaboration on the *Collectio Toledana*. In Robert and Hermann's epistolary exchange recorded by Charles Burnett in his edition of the *De Essentiis*, Robert appears to lament the responsibilities devolved to him when he complains to Hermann that he, Robert, must expose himself to the 'public wrestling grounds' and 'endure the hostile attacks of fellow competitors', while Hermann keeps himself 'free for the leisures of private studies'. Although Hermann insists that he and Robert still share the same 'mind, life and soul'[6], the beginning of Robert's archdeaconry appears to be, not necessarily the end of their friendship, but certainly the point of departure for independent intellectual endeavors. By 1143 Hermann had left Spain for the other side of the Pyrenees, and we find him at Toulouse completing his translation, from an Arabic text, of Ptolemy's *Planisphere*, dedicated to Thierry of Chartres, and a little later in the same year, in Béziers where he purportedly wrote his *De Essentiis*. As Marie-Thérèse d'Alverny notes, these toponymical identifications have been called into question by P.M. Alonso who contends Hermann was not in Toulouse, but in Tolosa de Guipuzcoa, a port city in Spain whose Basque name has been mistakenly confused by historians with that of Béziers[7].

Yet what precisely were Hermann of Dalmatia and Robert of Ketton's institutional affiliations, and how do these appear to have influenced their work? What we know of the sketchy itineraries of the two translators tells us that Hermann of Dalmatia (variably known as Hermannus Sclavus, Hermannus de Carinthia or Hermannus Secundus in contrast with the Reichenau monk Hermann Contractus) was born in the Duchy of Carinthia and was most likely educated among the Benedictines. In his introduction to the *Planisphere*, he refers to Thierry of Chartres as his 'most loving teacher' (*diligentissime preceptor*), and it has generally been assumed that his higher education took place in the northern French schools. Whether this meant Chartres proper, Paris or elsewhere, is subject to dispute, for as Peter Dronke demonstrates in his revisionist view of the School of Chartres, many of its chancellors were not continually present in Chartres[8]. In any case, Burnett somewhat nuances the strong

6. *Hermann...*, ed. & tr. Burnett, p. 6.

7. d'Alverny M.-T., 'Deux traductions latines du Coran au Moyen Âge', *Archives d'Histoire doctrinale et littéraire du Moyen Age* (Paris, 1947-1948), p. 81, note 1.

8. Dronke P., 'New Approaches to the School of Chartres', *Annuario de Estudios Medievales* 6 (1969), p. 117-40.

Chartrian influences seen by some scholars to be the basis for Hermann's independent work, the *De Essentiis*. Burnett says that he has found no 'clear case of a citation of a contemporary author in the *De Essentiis*'[9]. Although some of the sources dear to Chartrian scholars such as Plato's *Timaeus,* the *Asclepius*, and Macrobius and Calcidius' *Commentaries* are sources of inspiration to Hermann, he sets himself apart from the northern cosmologists by his attempt to construct a philosophical system based on a deep understanding of the Arabic tradition.

The only other confirmed ecclesiastical connection we have with regard to Hermann is his association with Peter the Venerable, leader of the Cluniac movement, who in 1141 enlisted him to work on the Toledan Collection alongside Robert of Ketton, Peter of Toledo and a certain Mohammed about whom we know nothing other than the fact that he was responsible for assuring that the translations from the Arabic were faithful and correct renderings of their originals. James Kritzeck has claimed that none of these translators had any known prior connection to Cluny[10], but he also suggests that Nájera, located in the Castilian Rioja, below the Ebro River, is the most likely identification for the site where Robert and Hermann were working at the time Peter the Venerable enlisted them for the translation project[11]. Nájera was precisely the site of the first great Cluniac possession in Spain, the monastery of Santa María de Nájera[12]. Further, Charles Burnett leans in favor, subsequent to J.-M. Lacarra's article on the French colonization of Navarre and Aragon[13], of associating Hermann's move south with the French, and particularly the Cluniac interest in the newly conquered areas of Spain.

Robert of Ketton's institutional affiliations are somewhat more clearly known. An Englishman presumed to be from Ketton in the county of Rutlandshire, he is known to have studied in Barcelona under Plato of Tivoli (circa 1136), who translated, among other works, Ptolemy's astrological treatise the *Quadripartitum* (1138), with the help of the Jewish scholar Savasorda or Abraham bar Hiyya, Theodosius' *Spherica,* and the astronomy of al-Battani[14]. Robert of Ketton, like his teacher, also translated the astronomical works of al-Battani. In 1143, after completing the joint

9. *Hermann*..., ed. & tr. Burnett, p. 21.

10. Kritzeck J., *Peter the Venerable and Islam* (Princeton, New Jersey, 1964), p. 59, note 32.

11. Kritzeck, *Peter*..., p. 34.

12. Kritzeck, *Peter*..., p. 12.

13. Lacarra J.M, 'A propos de la colonisation *franca* en Navarre et en Aragon', *Annales du Midi* LXV (1953), p. 331-342.

14. Haskins, *Studies*..., p. 11.

translation project commissioned by Peter the Venerable, Robert was appointed archdeacon of Pamplona (or specifically the Valdonsella)[15], and later Canon of Tudela. Based on the archival discoveries made by two Spanish historians, d'Alverny indicates that Robert was attached in some capacity to the King of Navarre, Garcia Ramirez. Indeed, Robert appears in numerous charters and witness lists in Spain from 1145 to 1152, often serving in an official capacity for King Ramirez, such as he did in July 1149 when he drew up the peace treaty between King Ramirez and Ramon Berenguer IV[16]. Duval has also advanced the idea that Robert of Ketton was in the service of King Ramirez, and continued to serve the Navarrese court, citing a donation made by King Ramirez's son Sancho in 1157 to the Collegiate Church of Santa Maria of Tudela for 'the salvation of his parents soul, and for love of my beloved friend Master Robert, canon of Tudela'[17]. She also advances the idea that Robert died between 1185 and 1186, citing an obituary in Pamplona dated April 10th for Magister 'Robertus, racionerus canonicus'. She further explains Robert's burial in Pamplona and not Tudela, where he finished his days as Canon, by the fact that the cathedral of Tudela was partially under reconstruction from 1185 and, as is reasonably inferable, unable to accommodate new burials.

Yet there is an inexplicable discrepancy between the image we have of Robert, as transmitter to Latindom of the rich treasures of Arabic scientific and philosophical learning, and that of the translator of the Koran for Peter the Venerable. Indeed, in his dedicatory letter to Peter the Venerable transmitting his translation of the *Fabulae Saracenorum* and the Koran, he refers to Islam as a 'perverse sectarian doctrine, covering up a scorpion', and 'a foul and poisonous thing' which 'destroys by trickery, ministers of the law of the Christian faith'[18]. As Norman Daniel has demonstrated, Robert does not even seem to be able to translate the word 'Moslem' in his Latin version of the Koran, preferring circumlocutions derived from the Latin verb 'credere'[19]. Both James Kritzeck and Richard Fletcher see the attitudes reflected in Robert's dedicatory letters as indicative of the spirit in which Peter the Venerable dictated that the Toledan

15. d'Alverny M.-T., 'Translations and Translators', in: R.L. Benson & G. Constable (eds.), *Renaissance and Renewal in the Twelfth Century* (Cambridge, 1982), p. 449.

16. Martin Duque A.J., 'El inglés Roberto, traductor del Coran', *Hispania* 22 (1962), p. 496; Goñi Gaztambide J.M., 'Los obispos de Pamplona del sigle XII', *Antología annua* 13 (1965), p. 135-358-64. Cited in d'Alverny, 'Translations...', p. 449.

17. Duval, *Recherches...*, p. 470.

18. Kritzeck, *Peter...*, p. 63-64.

19. See Iogna-Prat D., *Ordonner et Exclure* (Paris, 2000), p. 342.

Collection was to be executed. Indeed, such a vituperative and polemic tone is found in Peter's own summary of the collection: *Summa haeresis Sarracenorum*[20]. Yet Robert's letter remains rather ambiguous. He misquotes the date of Mohammed's first revelation and refers to Moslems as 'God's beautiful portion of the human race' ('generis humani pulcra portio'), who are so hated by Christian priests that the latter do not even deem them worthy of conversion[21]. It is plausible that this ambiguity reflects the tenuous attitude Hermann and Robert may have taken in their delicate balance as translators of Arabic astronomical and mathematical treatises or commentaries, a welcome *translatio studii* in most twelfth-century Christian circles, and then as translators of works of Arabic astrology and mystical alchemy which were not as readily assimilative to the purposes of Christian theology. Charles Haskins tells us that their initial studies in the 'inner treasures of Arabic learning were at first carried out in secret'. Indeed, as Hermann says in his prefatory letter to Robert in the *De Essentiis*, translated by Charles Burnett:

> You remember, I think, that while we went forth from our inner sanctuaries into the public festival of Minerva, the multitude of people milling around were gaping at us with open mouths, not valuing us so much as individuals as admiring the trappings and decorations which long vigils, and our most earnest labour had acquired for us from the depths of the treasuries of the Arabs. At that time I began to have a very deep sense of pity concerning those men who were so impressed by these outward appearances: How much they would value the undergarments, if it were lawful for them to look at them![22]

It seems justified to ask whether these 'undergarments', or this 'secret knowledge' referred to by Hermann are not, in fact, what motivates their interest in Arabic scientific and mathematical texts. Hermann's only known independent redaction, the *De Essentiis,* does, in fact, integrate hermetic texts, Koranic citations and Muhammadan legend[23], along with purely scientific and philosophical texts in a way which forms an original kind of cosmology. His cosmology was a conflation of basically three disciplines – astronomy, astrology and forms of mystical theology – and was centered on the belief that observation and understanding of the natural world, whether sublunary or celestial, leads to an understanding of the Divine. Although Hermann and Robert were writing at a time when

20. Fletcher R., *The Cross and the Crescent* (London, 2003), p. 129.
21. Kritzeck, *Peter*..., p. 64.
22. *Hermann*..., ed. & tr. Burnett, p. 71.
23. Burnett C., 'Arabic into Latin in Twelfth-Century Spain: the Works of Hermann of Carinthia', *Mittellateinisches Jahrbuch* 13 (1978), p. 129.

the boundary between astronomy and astrology was tenuous, no one among their Northern French counterparts appears to be constructing the same kind of synthesis as Hermann and Robert between hermetic cosmology and alchemy on the one hand and predictive or judicial astrology and the natural and mathematical sciences on the other. It is true that the Latin *Asclepius,* a translation of a Greek Hermetic text, was known to and used by Thierry of Chartres, Bernard Silvestris, John of Salisbury, Alan of Lille and others throughout the twelfth-century. Yet the manner in which they assimilated the hermetic cosmology of the *Asclepius* into Christian doctrine can hardly be deemed unorthodox or even very innovative. As Burnett has suggested, Hermann, in particular, was attracted to astrology as a 'key to the explanation of all physical phenomena within the universe', and in a secular way his Northern counterparts were not.

Thus, the key intellectual impetus for Hermann of Dalmatia and Robert of Ketton's intellectual endeavors appears to lie well outside the purview of the standard twelfth-century canon of contemporary translations which preoccupied their ecclesiastical analogues. Indeed, their composite form of cosmology can be elucidated more by their Arabic sources than by their Christian affiliations, and does not necessarily reflect an interest in gaining access to the original Greek works, but rather in exploring the philosophical and spiritual use made of these works by subsequent Arabic thinkers.

The close professional association and amicable complicity known to have existed between Hermann of Dalmatia and Robert of Ketton allows certain assumptions to be made with respect to the missing pieces of their respective biographies and itineraries, especially with respect to their shared access to specific works and scholars. For instance, we have no proof that Robert of Ketton had any first-hand contact with the School of Chartres or Thierry of Chartres, as did Hermann. However, the latter highly commends some of the translations made by him and Robert as part of his recommendations to the chancellor of Chartres as to which text books should figure in a program of scientific education. It is even possible that Thierry's *Heptateuchon* included translations made by Robert and Hermann.

In a recent article, Burnett has suggested that Hermann may have had access to the library of the Banū Hūd, which was transferred from Zaragosa to Rueda Jalón after the fall of the former city to the Christians in 1118. This he postulates on the basis of the fact that Hermann knew the works of Theodosius and Archimedes, and made versions of Euclid's *Elements* and Al-Khwarizmi's astronomical tables, all of which are believed to have belonged to the Banū Hūd library transferred to Zaragoza by Saif al-Dawla, the last of the dynasty originally established in Zaragoza. Bishop Michael of Tarazona apparently also had access to this library, for it was from its impressive collection of Arabic-language works that he chose those he commended to Hugo of Santalla to be translated into Latin[24]. In a prior article, Burnett had already postulated a connection between Hugo of Santalla and Hermann, based on the striking similarity of their technical vocabulary in Hugo's translations of Jafar Indus' work on rain and Hermann's *Liber Imbrium*. Thus it seems reasonably inferable that Hermann could also have had a connection with Bishop Michael of Tarazona (1119-1151), most likely a French Clunisian, who gained access for Hugo of Santalla to works from the Banū Hūd library.

The most important of recent evidence with regard to Hermann's connection to various intellectual peers was presented by Burnett who suggests that Thierry of Chartres (also 'Thierry the Breton'), the great Chartrian teacher and chancellor of the Cathedral of Chartres, was the point of convergence and source of dissemination in the northern part of France of the treatises and Latin translations of Arabic works by Hermann, Adelard of Bath, and Dominicus Gundissalinus, author of the *De processione mundi*[25].

It is also known that Hermann's disciple Rudolph of Bruges dedicated his *Cum celestium sperarum* to John David, whom he refers to as his 'most dear master'. Plato of Tivoli also devoted one of his translations to this same John David, and given that Robert of Ketton studied in Barcelona around 1136 with Plato of Tivoli[26], and that Hermann's pupil Rudolph of Bruges and Plato shared the same patron, we may reasonably assume that there existed some relationship between Hermann and Plato as well[27].

Finally, what seems to characterize what has been termed the 'Renaissance of the Twelfth Century' is the emergence of more loosely configured circles of scholars, whether regular or secular clerics, whose institutional affiliations were attenuated by their association with, or protection by, a strong local lord. The courts of Henry Plantagenet and count Henry the Liberal of Champagne are good examples of such instances in which

24. Burnett C., 'Translations, Scientific, Philosophical, and Literary (Arabic)', in: E.M. Gerli (ed.), *Medieval Iberia, An Encyclopedia* (New York – London, 2003), p. 802.

25. Burnett C., 'Metaphysics in the Twelfth Century on the Relationship among Philosophy, Science and Theology', *Textes et Études du Moyen Âge* 19 (2004), p. 41-65.

26. Kritzeck, *Peter*..., p. 62.

27. Burnett, 'Arabic into Latin...', p. 14.

courtly patronage fostered the development of a more culturally composite and secularized literary production and general scholarship. Yet the fractured political entities of northern Spain, vying for dynastic control over a culturally, linguistically and religiously diverse population, attenuated even further – at least during this short interlude in history – the ecclesial constraints on scholars experienced by their northern counterparts. First and foremost, such reform movements as that led by Cluny in an effort to replace the Mozarabic rite with the Roman one was never a widespread or popular movement since it aimed at eradicating the scriptural and liturgical traditions which constituted the very fabric of the Mozarabic church's historical and spiritual identity. If the Cluniac reform movement succeeded, it was largely due to the intercession of specific monarchs and the upper layers of Luso-Hispanic society. Unlike the Cistercian movement which recruited a large number of peasant *conversi* and had a far broader popular appeal than the Cluniac movement, Hispano-Cluny, it could be said, was an aristocratic phenomenon bolstered by dynastic ties between the ruling families of Burgundy and other parts of France with those of northern Spain. Yet paradoxically, I believe it is a closer study of the particular relationship between Cluny, its prelates, churches, and monastic houses with the ruling families of northern Spain that will eventually reveal the reasons behind the sudden effervescence of trans-linguistic scholarship during this period of conjuncture, and the unique phenomenon that constituted the cultural and intellectual florescence of twelfth-century Spain in which our translators participated.

José Manuel FRADEJAS RUEDA

SHADHĀNIQĀT AL-BALANSIYYA OR *SHADHĀNIQĀT AL-BAḤRIYYA*: ON THE ARABIC TEXT AND THE LATIN TRANSLATIONS OF THE *CALENDAR OF CORDOVA*

0. Introduction

In 1961 Charles Pellat published a new trilingual edition of the Arabic text known as the *Calendar of Cordova* and Gerard of Cremona's Latin translation[1]. Pellat based his edition on R. Dozy's[2] and, with the aid of the Latin text, was able to add a French translation of the Arabic text, thus offering a very useful piece of research. Four years later, the French Arabist François Viré published a very interesting article in which he explained the ornithological and falconry data present in the monthly agricultural epilogue. These explanations allowed Viré to propose some corrections to the text, as some of the words were *très mal rendues* and *défigurées* due to

> l'incompétence d'un tardif traducteur en latin, [...], étranger au pays, peu versé, semble-t-il, dans l'art du vol et, certainement, très embarrassé par la langue andalouse, un arabe teinté, inévitablement, de roman[3].

Furthermore, he proposed an alternative translation for some passages of the Arabic text of the *Calendar of Cordova*, because 'la fauconnerie [est] un domaine peu familier à l'orientaliste'[4].

1. Romance Loanwords in Arabic

Certainly, in Arabic several bird-names can be traced back to Romance languages, or even to Latin, some of which can be found in the

1. Pellat Ch., *Le Calendrier de Cordoue* (Leiden, 1961).
2. Dozy R., *Le Calendrier de Cordoue de l'année 961. Texte arabe et ancienne traduction latine* (Leiden, 1873).
3. Viré F., 'La Volerie dans l'Espagne du Xe siècle à travers le calendrier de Cordoue', *Arabica* 12 (1965), p. 306-314, here p. 306.
4. Viré, 'La Volerie...', p. 306.

Calendar itself. This is the case of *gābya* (pl. *gawābī*), which is a loanword directly taken from the Latin *gavia* 'gull'. Gerard of Cremona was unable to identify it and rendered it as *algaguab*[5], a literal adaptation to Arabic. Another case is that of *farāṭil* (sing. *farṭāl*), derived from the Ibero-romance *pardal* 'sparrow'. In this case the translator did a fine job when he translated it as *passeres*[6]. Moreover, Viré himself reminds us that many terms used by Arab falconers, mostly from the Maghreb, are of Romance origin. This can be illustrated by *tarshūn*, which is the Arabic rendering – through the distorsion of *tarshūl* / *ṭarshūl* – of *torzuelo* ~ *terçuelo* 'male hunting bird'[7], or even *nblī*, if we accept Viré's hypothesis that this word is a cognate of the Latin *nebula*[8]. Even an old eighth-century Arabic treatise on falconry, al-Ġiṭrīf's *Kitāb ḍawāri aṭ-ṭayr*, makes use of a Romance loanword to designate the *Falco peregrinus pelegrinoides*: *balanjrīniyya*, which is the Arabic for the Latin *pelegrinus*[9].

Nevertheless, one of the Arabic words is mistakenly explained by Viré as a Romance loanword. In this contribution I will try to demonstrate that it is, in fact, an Arabic word that was misspelled in the Arabic text on which Gerard of Cremona based his Latin translation of the *Calendar*.

In the farming, hunting, and weather information for the months of January, March and April there are mentions of the *shaḏḫāniqāt al-balansiyya*[10]. These were rendered into Latin as 'falcones Valentie'[11] and 'falconum valentiorum'[12], that is, Valencian falcons. What Gerard of Cremona did was to interpret *balansiyya* as the place-name *Balansya* (Valencia). This place-name is also cited on January 22nd, when the Catholic Church celebrates the nameday of Saint Vincent, who was martyred in the city of Valencia in 304[13]. Therefore, translating *balansiyya* as *Valentia* was not a grave error, not even the work of an incompetent translator because, as Viré himself admits, 'le rapprochement avec la ville de Valence était séduisant'[14]. In any case, Viré proposed to amend *balansiyya* and to translate it in a very different way.

5. Pellat, *Le Calendrier…*, p. 145.
6. Pellat, *Le Calendrier…*, p. 91.
7. Viré, 'La Volerie…', p. 309 n. 1.
8. Viré, 'La Volerie…', p. 312 n. 1.
9. al-Ġiṭrīf, *Traité des oiseaux de vol (Kitāb ḍawāri aṭ-ṭayr). Le plus ancien traité de fauconnerie arabe* traduit, introduit et annoté par F. Viré et D. Möller (Nogent-le-Roi, 2002), p. 84 and note 127.
10. Pellat, *Le Calendrier…*, p. 37, 61 and 77.
11. Pellat, *Le Calendrier…*, p. 37 and 61.
12. Pellat, *Le Calendrier…*, p. 77.
13. Pellat, *Le Calendrier…*, p. 33, line 23.
14. Viré, 'La Volerie…', p. 307 n. 1.

His main argument is not based on linguistics, but on a profound knowledge of falconry and ornithology. So he begins by explaining which birds of prey were available to a tenth-century Andalusian falconer:

> Mis à part les faucons inconnus de l'avifaune hispanique, comme le Sacre, le Lanier et le Gerfaut que, seuls, les hauts personnages fortunés pouvaient acquérir des oiseleurs-marchands [...], les campagnards d'al-Andalus ne disposaient, sur place, pour fournir leurs équipages de vol, que de deux races géographiques du faucon Pèlerin (*Falco peregrinus*; en espagnol: halcon comun [sic]), de l'Autour et de l'Épervier; ce sont, précisément, ces quatre rapaces que nous trouvons mentionnés au fil des mois du *Calendrier*[15].

At this point Viré reaches the conclusion that

> avec le *shadhāniq al-balansiyya*, nous reconnaissons le Pèlerin de la sb./sp. *Brookei* Sharpe, plus faible de taille que les autres races, sédentaire commun et nicheur dans toute l'Espagne et le Portugal[16],

and goes on to explain the strategy employed by falconers hunting with this kind of falcon, which

> suivant l'épagneul dans sa quête et, dès l'arrêt de ce dernier, *balançait sur la remise, se balançait, tenait la balance*, c'est-à-dire tenait l'air en 'saint-esprit' au-dessus du groupe chien-gibier ... Cette tactique de *tenir la balançoire* est propre au Pèlerin [...] On retrouve les expressions de 'balancer sur la remise', 'se balancer sur l'espaigneul', 'tenir la balançoire', en synonyme de 'se soutenir dessus le chien' dans la plupart des théreuticographes médiévaux, et tous les lexicographes les ont relevées.

The conclusion Viré reaches, is that *balansiyya* is, in fact, the plural of *balansī*, a Romance loanword – I should add a Gallo-Romance loanword because all of Viré's data are of French origin – taken from *balancier*, which itself is derived from the Latin *bilancia* 'balance'. According to Viré, *balansī* is the adjective used to specify the species of the Peregrine falcon that lives in the Iberian Peninsula and which should be identified with the *Falco peregrinus brookei*.

Although I agree that the falcon mentioned in the months of January, March and April is the *Falco peregrinus brookei*, I totally disagree with his attempt to derive *balansiyya* from *balancier*.

15. Viré, 'La Volerie...', p. 307 n. 1.

16. The editors printed what they took from the Madrid manuscript enclosed between angled brackets.

2. The *Liber Regius* version

For nearly a century and a half, until the early 1980s, the only known Latin version of the *Calendar* was that of Gerard of Cremona published in 1838 by Guillaume Libri[17], reedited by R. Dozy in 1873 and again by Ch. Pellat in 1961. However, in 1981 J. L. Martínez Gázquez and Julio Samsó edited another Latin version based on a manuscript housed in the Museum Episcopal of Vic (ms. 167, fol. 1r-8r), in northern Spain, which has the title of *Liber Regius siue descriptio temporum anni*. It is a thirteenth-century version, which can easily be dated between 1228 and 1235. The *terminus post quem* of 1228 is determined by the fact that patron saint of October 4th is 'Sancti Francisci confesoris [sic] primi de ordine minorum, ciuitate Assissi'. It is known that Saint Francis of Assisi died in 1226 and was canonized two years later. Thus, this translation, or at least this copy, cannot have been made before 1228 unless it can be demonstrated that such information is an interpolation. The *terminus ad quem* of 1235 is based on a marginal note on fol. 18v:

> Ne mireris, lector, precedentem numerum annorum maiorem subsequencium esse, quia tunc quando fuit scriptum hoc tabulatum currebat annus Domini M CC XXXVus.

Besides this thirteenth-century copy, Martínez Gázquez and Samsó found another in the Biblioteca Nacional at Madrid (ms. 6036). This is a fourteenth-century manuscript which lacks the initial portion of the prologue, but which is of great importance as it helps to fill in the places where the Vic copy is unreadable.

Martínez Gázquez and Samsó believe that the *Liber Regius* version was produced in Catalonia and they base their judgement on two arguments. First, the two versions make use of different calendars of saints' days. In Gerard of Cremona's translation the calendar of saints' days is a pan-Hispanic one that conforms to the Mozarabic rite[18] whereas in the *Liber Regius* it has been replaced by a Catalonian calendar that follows the Roman rite[19].

17. Libri G., *Histoire des sciences mathématiques en Italie depuis la renaissance des lettres jusqu'à la fin du 17e siècle* (Hildesheim, [1838–41] 1967).

18. A Spanish version was published by Simonet F., 'Santoral hispano-mozárabe escrito en 961 por Rabi ben Zaid, obispo de Iliberis', *Ciudad de Dios* 5 (1871), p. 105-116 and 192-212.

19. Martínez Gázquez J., '¿Una nueva traducción latina del Calendario de Córdoba del 961 en un manuscrito del siglo XIII?', in: *Unidad y pluralidad en el mundo antiguo: Actas del VI Congreso Español de Estudios Clásicos* (Madrid, 1983), p. 373–379, here p. 375.

The second argument is the existence of two Catalanisms, which I believe to be rather weak evidence. The two words are *butheca* and *bleta*. *Butheca* is an Arabism meaning 'melon, watermelon', an adaptation of *al-baṭṭīka* (in Gerard of Cremona's text it is rendered as *albateka*). Martínez Gázquez and Samsó[20] accept Corominas and Pascual's opinion that it is a Catalan Arabism[21], because the Castilian form *albudeca*, attested for the first time in 1537, derives from the Catalan *budeca*, documented for the first time in 1252, while the correct Castilian form is *badea*[22]. However, in the *Libro de Alexandre*, a Castilian poem written before 1228, we find the word *budefa* 'low quality watermelon', which shows a different way of rendering the Arabic khā (خ) as <f>, not unusual in the Hispano-Romance dialects[23].

The second proposed Catalanism is to be found in the translation of the Arabic word *silq* as *bleta* (Gerard of Cremona, *sicla*[24]). Martínez Gázquez and Samsó believe that it looks like Catalan – 'es forma sospechosamente catalana'[25]. They base their reasoning on what they state that Corominas and Pascual say. Yet what these two lexicographers say is that 'las formas modernas son tomadas del catalán y de la lengua de Oc'[26]. I believe the anonymous translator has simply used a Latin word, common to most of the Latin and Romance-speaking world, to refer to a very common vegetable: the *Beta vulgaris* 'Swiss chard'[27].

3. Gerard of Cremona's version versus the *Liber Regius*

Both Gerard of Cremona's version and the *Liber Regius* share the same contents, but there are some differences, mainly in wording, which

20. Martínez Gázquez J. & Samsó J., 'Una nueva traducción latina del calendario de Córdoba (siglo XIII)', in: J. Vernet (ed.), Textos y estudios sobre astronomía española en el siglo XIII (Barcelona, 1981) p. 9-78, here p. 52 n. 210. This work was also published as a book *Una nueva traducción latina del calendario de Córdoba (siglo XIII)* (Barcelona, 1982).

21. Corominas J. & Pascual J.A., *Diccionario crítico etimológico castellano e hispánico* (Madrid, 1981-1991), I, p. 123a.

22. According to Corominas and Pascual's *Diccionario*... (I, p. 451b), *badea* is first recorded in 1423.

23. Steiger A., *Contribución a la fonética del hispano-árabe y de los arabismos en el ibero-románico y el siciliano* (Madrid, 1932), p. 218.

24. Pellat, *Le Calendrier*..., p. 173.

25. Martínez Gázquez & Samsó, *Una nueva traducción*..., p. 74 n. 355.

26. Corominas & Pascual, *Diccionario*..., I, p. 600a.

27. In the Italian translation of Moamin we find 'frondi de sicla, cioè de bleta' (Glessgen M.-D., *Die Falkenheilkunde des 'Moamin' im Spiegel ihrer volgarizzamenti: Studien zur Romania Arabica* (Tübingen, 1996), I, p. 157 – Moamin III, 12, 11).

lead us to believe that the two translations descend from different versions of the Arabic text. Some fragments lead us to believe that the *Liber Regius* paraphrases Cremona's translation, while others point to an abridgement. This can be noticed, for example, in the list of saints' days for the month of March:

Gerard of Cremona	*Liber Regius*
in ipso est Christianis festum Emeterii et Celidonii. <*Et sepulcra eorum sunt in civitate Calagurra*>[28]	Emeterii et Celidonii, Calahorra[30]
<*in ipso est Christianis festum Gregorii domini Rome*>[29]	Gregorii pape[31]

or in the falconry information for the month of May:

Gerard of Cremona	*Liber Regius*
Et ponuntur in muta accipitres et falcones, et remanent in muta usque ad principium Augusti aut ad finem eius, secundum quantitatem virtutis eorum et sanitates ipsorum, et egrediuntur pulli asipheti (*id est cristarelle*) et accipitrum ex ovis suis, et vestiuntur pennis usque ad triginta dies[32].	et mutantur falcones et ancipitres pennas suas et exeunt pulli sparuerini de ouis[33],

or in the November farming information:

Gerard of Cremona	*Liber Regius*
et multiplicantur holera hyemis, sicut caules et sicla et napi et baucie et porri et rafanus. Et in ipso colliguntur arundines, et coagulantur fabe autumnales, et plantatur in agusto. *Et in ipso cooperiuntur viridia, et citrus et musa et sambucus, ut non noceat eis pruina.* Et in ipso colliguntur flores croci[34].	et colliguntur harundines et fructificant <fabe> autumnales qua augusto seminantur et multiplicantur olera hyemalia, uidelicet caules Romanorum et bleta et porra et rafani et flos croci colligitur[35].

28. Pellat, *Le Calendrier...*, p. 53. Pellat uses italics to indicate that the text enclosed between angled brackets does not appear in the Arabic text he edits.
29. Pellat, *Le Calendrier...*, p. 57.
30. Martínez Gázquez & Samsó, *Una nueva traducción...*, p. 35.
31. Martínez Gázquez & Samsó, *Una nueva traducción...*, p. 35.
32. Pellat, *Le Calendrier...*, p. 91.
33. Martínez Gázquez & Samsó, *Una nueva traducción...*, p. 48.
34. Pellat, *Le Calendrier...*, p. 173.
35. Martínez Gázquez & Samsó, *Una nueva traducción...*, p. 74.

Certainly, the *Liber Regius* may resemble a paraphrase of Gerard of Cremona's translation, especially when it changes the order in which the information is offered, as can be seen in the farming information for the month of February. Gerard of Cremona's version ends with the return of storks and swifts to the towns – 'et in ipso veniunt ciconie et yrrundines ad habitationes'[36]. In the *Liber Regius* this is found in the middle, together with the information about the cranes' return to the river islands – 'grues ad insulas conuertuntur et ueniunt ciconie et hyrundines'[37]. Nevertheless, the *Liber Regius* did not achieve its aim to put all the ornithological information together, as the farming epilogue begins 'aues amicantur'[38].

Yet, in some other places both Latin translations coincide whereas they differ from the Arabic text: the two of them add information not found in the known Arabic text. At the end of the month of April the *Calendar* says 'peacocks, storks and many other birds lay eggs'; in the Latin versions we are told what the birds do after they have produced eggs – they incubate them:

Gerard of Cremona	*Liber Regius*
et pariunt ova pavones et ciconie et multe aves, <*et incipiunt incubare ovis*>[39]	et pariunt pauones et ciconie et incipiunt multe ex auibus fouare oua[40].

In the month of September Gerard of Cremona – 'et incipiunt sparagi primitivi in montibus'[41] – and the *Liber Regius* – 'et inueniuntur sparagi primitiui montani'[42] – inform us that the first asparagus is ready to be harvested. This is not to be found in the Arabic calendar.

Both Latin versions tell us that Dacianus martyred Saint Vincent, Saint Sabina and Saint Cristeta in the Spanish city of Abule[43] / Abela[44], which is the city of Ávila. The Arabic one, on the other hand, claims the martyrdom took place in the city of Guadalajara[45].

36. Pellat, *Le Calendrier*..., p. 49.
37. Martínez Gázquez & Samsó, *Una nueva traducción*..., p. 33.
38. This would seem to be a better translation than the 'frangunt ova aves' offered by Gerard of Cremona (Pellat, *Le Calendrier*..., p. 49), which Viré tried to amend ('La Volerie...', p. 308 n. 1).
39. Pellat, *Le Calendrier*..., p. 53.
40. Martínez Gázquez & Samsó, *Una nueva traducción*..., p. 35.
41. Pellat, *Le Calendrier*..., p. 145.
42. Martínez Gázquez & Samsó, *Una nueva traducción*..., p. 65.
43. Pellat, *Le Calendrier*..., p. 155.
44. Martínez Gázquez & Samsó, *Una nueva traducción*..., p. 68.
45. Pellat, *Le Calendrier*..., p. 155.

In any case, I believe that Gerard of Cremona and the *Liber Regius* share a common Arabic ancestor. I do not believe that both versions were made from the same original, but from two different, yet related witnesses slightly dissimilar from the *unicum* known to us.

4. *Balansiyya* or *baḥriyya*?

Returning to the starting point of this article, the *Liber Regius* may help us to decide if we should read *sha<u>dh</u>āniqāt al-balansiyya* and interpret it as Viré has done or if we can propose another reading. Let us have a look at the places of the *Calendar* where the *sha<u>dh</u>āniqāt al-balansiyya* are mentioned, and study how they were translated by Gerard of Cremona on the one hand, and the *Liber Regius* on the other:

Calendar	Cremona	*Liber Regius*
wa-taltazimu l-sha<u>dh</u>āniqāt al-balansiyya a'shāsha-hā wa-ta'<u>kh</u>u<u>dh</u>u fī l-sifād[46].	et faciunt falcones Valentie nidos suos, et incipiunt coire[49].	Falcones marini nidificant et calcant[52].
wa-fīhi tabīḍu l-sha<u>dh</u>āniqāt al-balansiyya bi-l-jazā'ir wa-taḥḍunu <u>th</u>alā<u>th</u>īna yawm[an] [ilā awwali abrīl][47].	et in ipso ponunt ova falcones Valentie in insula, et incubant super ova triginta diebus usque ad principium Aprilis[50].	et pariunt falcones marini, subponunt oua triginta diebus[53].
wa-fīhi ta<u>kh</u>ruju firā<u>kh</u> l-sha<u>dh</u>āniqāt al-balansiyya min bayḍi-hā <u>th</u>umma taktasī bi-l-rīsh ilā <u>th</u>alā<u>th</u>īna[48].	et in ipso egridiuntur pulli falconum valentiorum ex ovis suis; deinde vestiuntur pennis usque ad trigenta dies[51].	et exeunt pulli falconum de ouis et cooperiuntur pennis post dies triginta[54].

It should be noted that on two of the three occasions the Arabic text speaks of *sha<u>dh</u>āniqāt al-balansiyya*, Cremona's translation says *falcones Valentie*, and the *Liber Regius* reads *falcones marini*. The third time, the *Liber Regius* felt no need to specify the species of falcon; the author

46. Pellat, *Le Calendrier...*, p. 37.
47. Pellat, *Le Calendrier...*, p. 61.
48. Pellat, *Le Calendrier...*, p. 77.
49. Pellat, *Le Calendrier...*, p. 37.
50. Pellat, *Le Calendrier...*, p. 61.
51. Pellat, *Le Calendrier...*, p. 77.
52. Martínez Gázquez & Samsó, *Una nueva traducción...*, p. 27.
53. Martínez Gázquez & Samsó, *Una nueva traducción...*, p. 38.
54. Martínez Gázquez & Samsó, *Una nueva traducción...*, p. 43.

has been discussing the breeding of falcons and the other species mentioned in the *Calendar* is a non-breeding one: *shadhāniqāt al-libliyya* – *falconem allebiati*[55], *falcones deblensi*[56]. According to the way the falcons are named in the *Liber Regius*, I am confident that the Arab text used by the anonymous translator read *shadhāniqāt al-baḥriyya*. Let us examine if such a reading is possible.

The *baḥriyya* falcons are mentioned in several Arabic treatises on falconry. When al-Ġiṭrīf describes the different species of falcons according to their colour, in his *Kitāb ḍawāri aṭ-ṭayr*, he says that the black peregrines are the *baḥriyya*[57]. In the anonymous tenth-century *Kitāb al-bayzara* the chapter devoted to the peregrine falcon – *shāhīn* – mentions a species under the name of *baḥri*[58]. In the fourteenth-century *Uns al-mala' biwaḥsh al-falā* by Muḥammad ibn Mankalī an-Nāṣirī, the *baḥriyya* falcons are mentioned in the chapter devoted to the peregrine falcon (*shāhīn*)[59], and in the same century Khudāyār Khān 'Abbāsī, in his *Qawānin al-ṣayyād*, says 'un altro tipo de *shāhīn* nero è il *baḥrī*' and 'il *baḥrī* è una specie de *shāhīn*'[60].

In al-Ġiṭrīf and the *Kitāb al-bayzara* the word *baḥri* was translated into French as *marin*, and in ibn Mankalī as *ultramarin*. Al-Ġiṭrīf goes even further when he says that *baḥriyya* falcons are those 'qui naissent et croissent dans les îles bordant les littoraux'[61], and the *Kitāb al-bayzara* tells us that they 'sont ceux qui martellent aux abords de la mer'[62]. This coincides with the *Qawānin al-ṣayyād*: 'che nidifica presso il mare' and 'ma vive e nidifica preferibilmente sulle coste e nelle isole'[63].

The Spanish medieval version of Muḥammad ibn ᶜAbdallāh ibn ᶜUmar al-Bāzyār's *Kitāb al-jawāriḥ*, a translation finished on April 9th, 1250, names the two species of falcons, and says that:

> La una manera dízenles mariniegos porque se engendran a partes de la mar e dízenles en arávigo *bahariet*. La .ii[a]. dízenles en lengua persiana cozteniet, e éstos son los que se engendran en las pennas e en los montes; e

55. Pellat, *Le Calendrier...*, p. 145.
56. Martínez Gázquez & Samsó, *Una nueva traducción...*, p. 65.
57. al-Ġiṭrīf, *Traité...*, p. 85.
58. Viré F. (tr.), *Le Traité de l'art de volerie (Kitāb al-bayzara) rédigé vers 385/995 par le Grand-Fauconnier du calife fāṭimide al-ᶜAzīz bi-llāh* (Leiden, 1967), p. 100.
59. Ibn Manglî, *De la chasse: commerce des grands de ce monde avec les bêtes sauvages des déserts sans onde*, translated by François Viré (Paris, 1984), p. 129 and 131.
60. Guizzo D., 'Il lessico ornitologico nel *Bāz-nāma*', in: Khʷushḥal Khān Khatak, *Il libro del falcone* (Venezia, 2001), p. 112-124, here p. 115.
61. al-Ġiṭrīf, *Traité...*, p. 85.
62. Viré, *Le Traité...*, p. 103.
63. Guizzo, 'Il lessico...', p. 115.

la manera de los *baharís* son mejores que la manera de los coztenís, e depártense los *baharís* de los coztenís en que son mayores e an las cabeças más blancas, o más escontra amariellas e la color más clara, e los coztenís son en contra d'estos en que son menores e an las cabeças vermejas e la color más turbia[64].

Therefore, the epithet *baḥri*, sometimes nominalized, was used by Arab falconers to denote a species of the Peregrine falcon. This substantiates the meaning offered by Daniele Guizzo 'varietà del falco pellegrino che vive e nidifica presso le coste'[65]. Here Viré seems to contradict himself: in the endnotes to his translation of ibn Mankalī's treatise he says that the *baḥri* falcons 'sont les migrateurs venant hiverner en climat méditerranéen; en fauconnerie, on les dit *passagers* ou *pris de passage*; ils arrivent des pays nordiques et de Sibérie où ils nichent'[66], while in the footnotes to the *Calendrier de Cordoue* he attributed this migrational behaviour to the *neblí*: 'à côté du faucon Pèlerin sédentaire et nicheur [...], l'Espagne méridionale connaît des Pèlerins étrangers, hivernants réguliers, venant de Scandinavie et de la Baltique [...] et de Sibérie'[67].

Another problem, which I will not deal with, is which falcons are to be identified with the epithet *baḥri*, as Viré and Möller, in their translation of al-Ġiṭrīf, say:

> Tout ce passage consacré aux 'Pèlerins noirs' décrit, en fait, les deux espèces méditerranéennes de faucons [...]. La première est le faucon d'Éléonore (*Falco eleonorae*) propre aux îles de la Méditerranée [...]. La seconde espèce est le faucon Concolore (*Falco concolor*) propre à l'Égypte[68].

But Viré himself, in a previous work, seems to identify the breeding peregrine falcon of Egypt not with the *Falco concolor* but with the Barbary falcon when he says 'l'Égypte connaît [...] comme nicheuse sédentaire, la race *Falco peregrinus pelegrinoides*'[69].

In Spanish the Arabism *baharí* (*bafarí* in Portuguese) has been used at least since 1250, the date of the first documentation of the word in Spanish used to name the *Falco peregrinus brookei*. Therefore, when the *Liber Regius* mentions the *falcones marini*, I believe that its anonymous

64. Moamín, *Libro de los animales que cazan*, ed. J.M. Fradejas Rueda (Madrid, 1987), p. 14 (italics mine).
65. Guizzo, 'Il lessico...', p. 115.
66. Ibn Manglî, *De la chasse...*, p. 272 n. 219.
67. Viré, 'La Volerie...', p. 312 n. 1.
68. al-Ġiṭrīf, *Traité...*, p. 85 n. 137.
69. Viré, *Le Traité...*, p. 100 n. 1.

translator did not read *shadhāniqāt al-balansiyya* as Gerard of Cremona did but *shadhāniqāt al-baḥriyya*, a very well attested name in Arabic for a variety of the Peregrine falcon.

Now, the difficult part is to explain how *al-baḥriyya* (البحريّة) was converted or corrupted into *al-balansiyya* (البلنسيّة), a form not found anywhere else.

5. On Arabic script

It is a well-known fact that the Arabic script is rather confusing. Many Arabic letters are only distinguished by means of the diacritical dots they carry over or under them. Seventeen out of twenty-eight letters, which can be grouped in eight different sets, are identical and can be distinguished only by means of the diacritical dots: ج ح خ د ذ ر ز س ش ص ض ط ظ ع غ ف ق. This becomes more complicated by the fact that twenty-five letters have different figures depending on whether they are isolated, initial, medial or final. Let us have a look at the letters jīm, .hā' and khā':

	initial	medial	final	isolated
jīm	جـ	ـجـ	ـج	ج
ḥā'	حـ	ـحـ	ـح	ح
khā'	خـ	ـخـ	ـخ	خ

It is clear that everything depends on where the dots are placed. Yet in early and medieval times the use of the dots was not compulsory; they were used sporadically, so there was ample scope for scribal errors. The diacritical dots could be omitted, misplaced, or even be generated by spots of ink accidentally dropped from a pen[70]. To make things even worse, Arab scribes from al-Andalus used the dots in a different way from the scribes of Baghdad or Damascus. For example, an Andalusian scribe would mark the letter fāh (ڢ) with a dot below the letter and the letter qāf (ق) with one above, while the Damascus scribes would do just the opposite: a dot above for the fāh (ف) and below for the qāf (ڢ).

Although many scribes were very careful when writing, others were careless and anyone not well versed in the idiosyncrasies of a given scribe could easily mistake one letter for another. There are numerous

70. Latham J.D., 'Arabic into Medieval Latin', *Journal of Semitic Studies* 17 (1972), p. 30-67, here p. 33.

examples in Arabic manuscripts. Let us examine a few cases taken from the *Calendar* itself.

On March 11th we are informed of the change of season. The Arabic text refers to it by means of the word *nāb* (ناب) 'canine'. Gerard of Cremona translated it as *porta*[71] and the *Liber Regius* as *ianuam*[72]. Both of them misread the first letter, and read *bāb* (باب) 'door'. This mistake could be easily made independently by both of them; it all depends on where the diacritical dot is placed: for the letter nūn (ـنـ) above the letter, while for bā' (ـبـ) below.

The feast of Saints Emeterius and Celedonius, who were martyred in the town of Calahorra, is celebrated in the month of March. For Emeterius the Arab text reads *endirash* (اندراش) but in light of the Latin translations the correct form should be *emtirash* (امتراش)[73]. Both words agree on the first – 'alif (ا) – and the last three letters – rā', 'alif, shīn (ـراش) —; the third one, a final dhāl or a medial tā' could be easily mistaken (د / ـتـ); the only difference between both forms being the initial nūn (ـنـ) and the initial mīn (ـمـ).

Viré himself located some cases were Gerard of Cremona had misread the Arabic text. One of them is to be found in the month of July, where a quarry for the falcon is mentioned using the word *shaqāshiq*. This bird was unknown to Gerard of Cremona, who rendered it as *assacassik*. Pellat translated it as *merle d'eau* (*Cinclus cinclus*)[74] which, according to Viré, is an impossible quarry for any falcon. But if we read *saqāsiq* (sing. *saqsāq*) instead of *shaqāshiq* (sing. *shuqshūq* / *shaqshāq*), we will be confronted with a more likely quarry, whether it is a lapwing (*Vanellus sp.*) or a plover (*Pluvialis sp.*)[75]. All we have to do is to delete three small dots above the Arabic letter to convert a shīn (ش) into a sīn (س).[76]

The month of April is the right season to sow privet, basil, cauliflowers, green-beans and rice. For rice Gerard of Cremona uses *rizus*[77], but in the *Liber Regius* we are confronted with *ador*, an unknown Arabism[78]. Here the anonymous translator either misread his original or

71. Pellat, *Le Calendrier...*, p. 55.
72. Martínez Gázquez & Samsó, *Una nueva traducción...*, p. 35.
73. This correction was proposed by Pellat in his *Le Calendrier...*, p. 53 note *h*.
74. Pellat, *Le Calendrier...*, p. 118.
75. Viré, 'La Volerie...', p. 311 n. 3.
76. In this case the *Liber Regius* omitted any specification of the water birds, contents itself by saying that *multiplicantur aues aquarum* (Martínez Gázquez & Samsó, *Una nueva traducción...*, p. 57).
77. Pellat, *Le Calendrier...*, p. 76-77.
78. Martínez Gázquez & Samsó, *Una nueva traducción...*, p. 43.

his Arabic text misspelt *al-ruzz*. A dāl (د) is replaced by a rā' (ر), which in turn was interpreted as a zāy (ز). The substitution of dāl for rā' can be explained because they are very similarly shaped letters; the difference between rā (ر) and zāy (ز) is the simple presence or absence of a dot above the graph.

So there is ample scope to believe not in a misreading on the part of Gerard of Cremona, but in a miswriting in the Arabic text on which Gerard of Cremona based his translation. *Al-balansiyya* (البلنسيّة) and *al-baḥriyya* (البحريّة) may look much alike, as they have the same three letters at the beginning of the word – 'alif, lām, bā' (البـ) – as well as the two last ones – medial yā' and tā' marbūta' (ـيـة). They differ only in the central part of the word: *al-balansiyya* has a sequence of medial lām, nūn and sīn (ـلنسـ); in *al-baḥriyya* we find ḥā' and rā' (ـحر). Given the fact that in cursive written by a careless hand the letters may be confused, miswritten or later misread, as I have already shown, the letters nūn (ن) and rā' (ر) can be easily misinterpreted. However, it is almost impossible to give a paleographic explanation for the replacement of ḥā' (ح) by lām (ل) and sīn (س). I am unable to explain this tiny change, just as I cannot explain how Gerard of Cremona read *ramādī* 'ash-gray'[79] three times instead of *ramāḍi* 'igneus', which is the reading of the *Liber Regius*[80]. I cannot explain how a ḍād (ض) was confused with a dāl (ذ), unless it can be demonstrated that somebody was reading aloud the Arabic text and Gerard of Cremona misinterpreted the reader. There could be many more palaeographic explanations, even psychological ones, but these are neither very sound nor easy to discover and formulate.

6. Conclusion

I must conclude that Gerard of Cremona's Latin translation was based on a defective Arabic text that can be corrected using the version contained in the *Liber Regius*, even though the *Liber Regius* was made from an Arabic text belonging to a different branch than the one used by Gerard of Cremona. This is possible because both share a common ancestor.

Finally, Gerard of Cremona's translation of Valencian falcons was based on a corrupted reading. Therefore, in the Arabic text published by

79. Pellat, *Le Calendrier…*, p. 109, 129 and 139.

80. Martínez Gázquez & Samsó, *Una nueva traducción…*, p. 54, 60 and 63. *Ramāḍi* is the name that the Arabs give to the summer rain.

Dozy and Pellat the expression *sha<u>dh</u>āniqāt al-balansiyya* (شدانقات البلنسيّة) should be corrected to read *sha<u>dh</u>āniqāt al-baḥriyya* (شدانقات البحريّة).

Universidad de Valladolid

Ilya DINES

THE TEXTUAL AND PICTORIAL METAMORPHOSES OF THE ANIMAL CALLED *CHYROGRILLIUS*

0. Introduction

The small animal called *shafan sela* (Fig. 1), or alternatively *Hyrax Syriacus* or *Procavia Capensis Syriacus*[1], hides itself among the rocks in the Middle East, and is not particularly noteworthy in itself except for two rather curious facts. First, from a zoological point of view this

Figure 1. SHAPHAN SELA - *Hyrax Syriacus, Procavia Capensis Syriacus*

animal is the smallest elephant in the world; and second, according to a wide-spread but incorrect belief, the country of Spain took its name from the name of this animal. These facts are well known to two categories of scientists, zoologists and linguists. I am not in a position to add something to the zoological research. Rather, this contribution presents the result of linguistic research, more precisely an account of the history

1. France P., *An Encyclopedia of Bible Animals* (Tel Aviv, 1986), p. 42-43.

of the 'translations' of the name of this animal into other languages. In addition, it also examines how these 'translations' were depicted by medieval painters.

1. The Bible and the beginning of mistranslation

Shafan sela is mentioned four times in the Bible, as can be seen in the *Table* in the appendix[2]. It first appears in *Leviticus* and *Deuteronomy*, where the *shafan* together with the hare is mentioned in the long list of unclean animals which are not allowed to be eaten. In the *Psalms* the *shafan* is said to live among the rocks; and in *Proverbs* the *shafan* is listed among the '...four very little things of the earth, and they are wiser than the wise'. This is the sum total of what the Bible says about the animal. The *shafan*'s metamorphosis begins in the third-second century BC, when the Hebrew Bible was translated into Greek in Alexandria, a redaction known as the *Septuagint*. For some unknown reason, the word *shafan* was translated as χοιρογρύλλιος, literally a 'grunting pig'. This word is very rare and as far as I know, it does not appear in earlier Greek literature. Most likely, it referred to a hedgehog (*echinos*) or a porcupine (*akantachoiros*); at least, this is how the term was explained in the fifth century AD by Hesychius of Alexandria, the composer of the largest dictionary of rare words[3]. We find an analogous interpretation in the dictionary of the Byzantine lexicographer Suidas, who lived in the tenth century[4]. In fact, the word *chirogrillus* was incomprehensible at the time of the *Septuagint* translation; this is demonstrated by the fact that in the famous Codex Alexandrinus (codex *a*)[5] the occurrence of *shafan* in the *Psalms* is translated as λαγός, a hare[6]. This translation may be due to the proximity of these two animals, the *shafan* and the hare, in *Leviticus* and *Deuteronomy*. Perhaps the author of the Codex Alexandrinus considered them to be similar animals. The Syro-Hexapla, an early seventh century translation of Origen's third century work, translates the occurrence of *shafan* in the *Psalms* by representing it as two animals, the *chirogrillus* and the hare,

2. For the reader's convenience I have replaced the text of the Hebrew Bible in the Table by that of the Douai-Reims version. The word 'shafan' is retained instead of the 'cherogrillus' which is used in this version. The same Douai-Reims version was used for all biblical quotations in the text of this article.

3. From Liddell H.G. & Scott R., *A Greek-English Lexicon* (Oxford, 1897), p. 1732.

4. Suidas, *Lexicon*, ed. A. Adler (Leipzig, 1935), v. 4, p. 835.

5. London, British Library, MS Royal 1 D V-VII.

6. *Septuaginta: id est Vetus Testamentum graece iuxta LXX interpretes*, ed. A. Rahlfs (Stuttgart, 1935), v. 2, p. 313.

probably in an attempt to reconcile these two readings[7]. We also find an analogous variant in the quotations of the later Western Church Fathers[8].

In his translation of the Bible into Latin at the end of the fourth century, Jerome retained the reading *chirogrillus* in *Leviticus* and *Deuteronomy*, while in the *Psalms*, following a redaction of the *Vetus Latina*, he translated the word *chirogrillus* as *ericii* (hedgehogs)[9], and in *Proverbs* he translated it as *lepusculus*, literally a 'small hare'. Jerome, who knew Hebrew and lived in Palestine, was certainly familiar with that animal, as can be read in his own works[10]:

> ... petra refugium erinaciis. Pro quo in Hebraeo positum est '*sphannim*' et omnes τοις χοιρογρυλλιοις voce simili transtulerunt exceptis Septuaginta, qui '*lepores*' interpretati sunt. Sciendum autem animal esse *non maius ericio, habens similitudinem muris et ursi*, unde et in Palestina *αρκομυς* dicitur. Et magna est in istis regionibus huius generis abundantia semperque in cavernis petrarum et terrae foveis habitare consuerunt.
>
> '... *rocks are a refuge for hedgehogs*. For which in Hebrew the term '*sphannim*' is used and all translated it in a similar way as '*chirogrillus*', except the Septuagint, which has it as 'hares'. It should be noted that this animal is not bigger than the hedgehog and is similar to the mouse and bear, and in Palestine it is called '*arkomus*'. There is an abundance of this animal in these regions, and they always live in rocky caverns and in holes in the ground.'

And elsewhere[11]:

> ... statuit supra petram pedes meos et: petra refugium leporibus, pro quo multi 'erinacios' legunt, animal paruum et fugax et [peccatorum] sentibus praegravatum. Sed ideo Iesus spinis coronatus est et nostra delicta portavit et pro nobis doluit...
>
> '*And he put my feet upon a rock and*: *rocks are a refuge for the hares*, which many call 'hedgehogs' - a small, timorous animal which is burdened by feelings of sin. But, the same way Jesus was crowned with spines and bore our sins and suffered for us...'

In his *Commentariorum in Matheum* he observes[12]:

> Petra refugium leporibus, siue erenaciis. Timidum enim animal in petrae cavernas se recipit: et cutis aspera, et tota armata iaculis, tali se

7. *Origenis Hexaplorum*, ed. F. Field (Hildesheim, 1964), v. 2 p. 260.
8. E.g. Augustinus, *Enarrationes in Psalmos* 70,3 (PL 36,879).
9. Sabatier P., *Bibliorum Sacrorum Latinae versiones antiquae: seu vetus Italica* (Remis, 1743), v. 2, p. 203.
10. Jerome, *Epistulae*, Epist.106,65,6 (CSEL 55,281-2).
11. Jerome, *Epistulae*, Epist.130,8,2 (CSEL 56,187.6-9).
12. Jerome, *Commentariorum in Matheum* 7,25 (CCSL 77,47).

> protectione tutatur. Vnde et Moysi dicitur eo tempore quo de Aegypto fugerat et Lepusculus Domini erat: Sta in foramine petrae, et posteriora mea videbis.
>
> '*Rocks are a refuge for hares or hedgehogs*. This timorous animal lives in rocky caverns and protects itself by its rough skin and by all its armed spines. Thus it was said to Moses at the time when he escaped from Egypt and was the Lepusculus ('young hare') of God: *Thou shalt stand upon the rock and thou shalt see my back part.*'

Rather than clarifying the nature of this unknown animal, we see that Jerome's interpretations confuse the matter even further, since the 'hare-hedgehog' acquires the characteristics of the mouse and bear. In another work, Jerome confuses the word *shafan* with the Hebrew word *shafam* 'moustache'[13]: 'Safan: chirogryllius lepus uel ericius sive *labium* eorum' ('Safan: chirogryllius, hare, or hedgehog, or their lips').

The reason why Jerome translated a Greek text so freely can be found in the fact that he was much more interested in the 'spirit' than the 'letter' of his translation. Indeed, the main characteristic of the *chirogrillus* in *Leviticus* and *Deuteronomy*, i.e. it is an unclean animal, is quite incompatible with the important role this animal plays in the *Psalms* and *Proverbs*. As I have shown elsewhere[14], in these books the *chirogrillus* is an animal with an emphatically positive reputation, which represents Christ and Moses in the eyes of Jerome.

Following Jerome, other Church Fathers commenting on the *Psalms* and *Proverbs*, including Augustine, Venerable Bede, and Hraban Maur, essentially repeat Jerome's explanations, and thanks to these exegetical efforts, the *shafan-chirogrillus* was finally 'divided into three': the *chirogrillus* itself, the hare, and the hedgehog[15]. Meanwhile, the word *chirogrillus* continued to sound strange to the Latin ear, and the zoological nature of the animal remained obscure. Consequently, towards the end of the seventh century, the famous Codex Amiatinus (codex **A**)[16] translates the word as *corcodillus*, i.e. a crocodile, in the relevant chapter of *Deuteronomy*, which clearly shows the scribe's bafflement over the nature of the *chirogrillus*[17].

13. Jerome, *Liber Interpretationis Hebraicorum Nominum* (CCSL 72,85;128).

14. See note 15.

15. For the detailed study of the subject, see Bauer J., 'Lepusculus Domini. Zum altchristlichen Hasensymbol', *Zeitschrift fur Katholische Theologie* 79 (1957), p. 457-466; Bauer J., 'Hase', *Reallexikon für Antike und Christentum* (Stuttgart, 1986), B. 13, p. 662-678. See also my 'The Hare and its Alter Ego in the Middle Ages', *Reinardus* 17 (2004), p. 73-84.

16. Florence, Biblioteca Mediceo-Laurenziana, cod. Amiatinus I.

17. *Biblia Sacra: iuxta Vulgatam versionem*, ed. B. Fischer e.a. (Stuttgart, 1983), p. 255.

In the eighth century, archbishop Theodulph of Orleans changed the reading *ericius* in *Psalm* 103 (codex θ) to *cuniculus*, a rabbit[18]. The reason for this change is that Theodulph believed that this text involves the same animal as the verse in *Proverbs*. Born in Spain, which was the only European country where rabbits were found in the eighth century[19], Theodulph chose this variant for his translation, suggesting this animal is similar to the hare. The *chirogrillus* continued in this indeterminate trinity condition for most of the Middle Ages, at least until the beginning of the thirteenth century.

2. Representation in art

2.1. Hare

Having discussed the history of the mistranslations of 'shafan', we turn to the history of depictions of the *chirogrillus* in art. Theoretically being a combination of a hedgehog, hare, mouse, and bear, this animal does not lend itself particularly well to artistic depiction, but medieval painters needed to depict it in order to illustrate the Bible. The easiest way to avoid depicting a strange composite animal was to opt for a single animal, and most often illuminators opted for the hare. The earliest depiction of the *chirogrillus* as a hare in Latin manuscripts that I am aware of is found in the illustration of *Psalm* 103 of the famous ninth-century Utrecht Psalter[20] and its eleventh-century copy, the Harley Psalter[21]. This painter provided very detailed illustrations of scenes based on the *Psalms*. In the right part of the *Psalm* 103 scene, there are 'lions roaring after their prey', a deer standing on the high mountain, and a cave from which a small hare is peeping, illustrating verses 18; 21: 'montes excelsi cervis petra refugium *erinaciis,* catuli leonum rugientes ut rapiant et quaerant a Deo escam sibi' ('The high hills are a refuge for the harts, the rock for the urchins. The young lions roaring after their prey, and seeking their meat from God').

We can be quite certain that the painter responsible for this scene did not use the text of the Utrecht Psalter itself, since that work has the reading *ericii.* Rather, he must have copied the animal from an earlier illuminated edition of the *Psalms*, presumably a Greek one. All extant Greek

18. *Biblia Sacra...*, p. 901.
19. See Van Dam P., 'Status Loss Due to Ecological Success. Landscape Change and the Spread of the Rabbit', *Innovation* 14, 2 (2001), p. 158-173, especially p. 159.
20. Utrecht, Universiteitsbibliotheek, MS 32, f. 59v.
21. London, British Library, MS Harley 603, f. 51v.

illustrations of *Psalm* 103 are later than the Utrecht Psalter, but those which illustrate verse 18 depict the hare[22], as is the case in this scene from a Psalter from the beginning of the twelfth century[23].

An analogous example appears in a Psalter produced at Bury St. Edmunds at the beginning of the eleventh century[24]. This image illustrates verses 25-26 of *Psalm* 103:

> hoc mare magnum et spatiosum manibus; illic reptilia quorum non est numerus animalia pusilla cum magnis illic naves pertransibunt draco iste quem formasti ad inludendum ei…
>
> 'So is this great sea, which stretcheth wide its arms: there are creeping things without number: Creatures little and great. There the ships shall go. This sea dragon which thou hast formed to play therein.'

One of the animals that appear in this picture is a hare. We might assume that the hare is just a representation of the *animalia pusilla*, much like the elephant exemplifies big animals. However, considering the trinity nature of the *chirogrillus*, we can be sure that the hare was chosen by the painter because of the proximity of verses 25-26 to verse 18. The fact that the hare began to represent the *chirogrillus* in both the works of the Church Fathers and in the illustrations of *Psalm* 103 raised the animal's status considerably. *Psalm* 103 is a creation psalm, a mini-*Genesis sui generis*, in which David thanks God and enumerates everything created by him, including, as we saw, the animals. As *Genesis* does not list the animals that were created by God or named by Adam, *Psalm* 103 became a sort of a model book for the painters. For instance, there is a creation scene from a French bible of the twelfth century[25], where the only animals represented are birds, lions, a hare under the rock, and deer – in other words, this scene is completely based on *Psalm* 103 verses 16-24. Thus the hare entered into creation scenes as the representation of the *chirogrillus*.

In an earlier paper devoted to the problem of Creation and Naming of the Beasts scenes in various types of medieval manuscripts[26], I have shown that with few exceptions, painters draw only animals with a positive

22. Dufrenne S., *Tableaux synoptiques de 15 psautiers médiévaux à illustrations intégrales issues du texte* (Paris, 1978), no pagination, see the table of Psalm 103.

23. De Wald E.T., *The Illustrations in the Manuscripts of the Septuagint* (Princeton, 1941), no pagination, see for the illustration of Psalm 103.

24. Vatican, Biblioteca Apostolica Vaticana, MS Regin. Lat. 12, f. 108r. For illustration see Klingender F.D., *Animals in Art and Thought to the End of the Middle Ages* (London, 1971), p. 198.

25. Paris, BNF, MS lat. 10, f. 3v; for illustration see Cahn W., *Romanesque Manuscripts: The Twelfth Century* (London, 1996), v. 1 fig. 36.

26. Dines, 'The Hare and its Alter Ego…', p. 77-80.

connotation in Creation scenes, and that the animals closest to God creating the animals or to Adam naming them have a high reputation in Christian doctrine. Examples are the lion, bull, sheep, and deer. The hare, that is, the real zoological hare, is listed as an unclean animal in *Leviticus* and *Deuteronomy*, and has generally been associated with lechery and homosexuality since antiquity. This miserable reputation only became worse during the Middle Ages[27], and thus it seems very unlikely that the hare would appear in Creation and Naming scenes, or even occupy a prominent place in them. Yet the hare does appear prominently in these scenes. A Naming scene from an English bestiary of the end of the thirteenth century may serve as an example (Fig. 2). Here we see Adam giving names

Figure 2. Oxford, Bodl. Lib. MS e Musaeo 136 f. 3v
Reproduced by permission of the library

27. See Abraham C.K., 'Myth and Symbol: The Rabbit in Medieval France', *Studies in Philology* 60 (1963), p. 589-597, especially p. 592-595; Boswell J., *Christianity, Social tolerance, and Homosexuality: gay people in Western Europe from the beginning of the Christian era to the fourteenth century* (Chicago, 1980), p. 253 and 317; Lauzi E., 'Lepre, donnola e iena: contributi alla storia di una metafora', *Studi Medievali* 29.2 (1988), p. 539-559; Schweitzer E.C., 'Chaucer's Pardoner and the Hare', *English Language Notes* (1967), p. 247-50.

to the animals. A hare leaping into Adam's lap occupies a place as Adam's favourite animal, which is likely to be named first. We can confidently conclude that the hare owes all of its interpretation *in bono* to the positive interpretation of the *chirogrillus*, which was first transformed into pseudo-*lepus* and *lepusculus*; or, what amounts to the same thing, that illustrations of the hare as an animal with positive connotations are, in fact, illustrations of the *chirogrillus* depicted as a hare.

2.2. Hedgehog and Porcupine

The hedgehog did not come to represent the *chirogrillus* in illustrations of the *Psalms* or in Creation and Naming scenes. This was partly because commentaries of the Church Fathers preferred the interpretation of the *chirogrillus* as a hare, and partly because the reputation of the hedgehog was even worse than that of the hare. It is well known that in the *Physiologus* and bestiaries, the hedgehog climbing on the vine and carrying away a grape on its spines symbolized the Devil stealing Christian souls[28]. Nonetheless, the features of the *chirogrillus* also show through in the hedgehog image. In the famous Montecassino manuscript of Hraban Maur's *De rerum naturis*[29], written in the first quarter of the eleventh century, there is a peculiar illustration to Chapter VII.8[30] – *De pecoribus et iumentis*[31]. It visually represents all animals mentioned by name in that chapter: the sheep, bull, horse, camel, etc. Among these, in the second row from the top, a hare is listed. The hare chapter in Hraban's text says:

> Lepus levipes quia velociter currit. Vnde et Graece pro cursu lagos dicitur, velox enim animal et satis timidum. Significat autem lepus homines timentes Deum, qui non in semetipsis, sed in Creatore suo fiduciam habent. Vnde legitur in Salomone: Lepusculus pleps invalida, quae collocat in petra cubile suum. Vnde dicitur in Psalmo: Petra refugium leporibus et erinaciis. Petra autem Christus est. Hinc de Moyse scribitur, quod Lepusculus Domini Moyses steterit in foramine petrae, quia in passione redemptoris, spem salvus suae habuit.
>
> 'The hare is called light-footed because it runs so swiftly. It is a swift creature, and fairly timid. The hare represents men who fear God, and

28. See for example *Physiologus Latinus: éditions préliminaires, versio B.*, ed. F.J. Carmody (Paris, 1939), p. 26-27; McCulloch F., *Mediaeval Latin and French Bestiaries* (Chapel Hill, 1962), p. 124-125.
29. Montecassino, Biblioteca dell'Abbazia, cod.132, f. 171r.
30. PL 111, 205.
31. The picture is illustrated in Reuter M., *Text und Bild im Codex 132 der Bibliothek von Montecassino 'Liber Rabani de originibus rerum': Untersuchungen zur mittelalterlichen Illustrationspraxis* (Munich, 1984), pl. 45.

who put their trust not in themselves but in the Creator. So we read Solomon's words: *Lepusculus are but a feeble folk, yet make their houses in the rocks.* Whence the Psalmist says: *The rocks are a refuge for hares and hedgehogs.* For the rock is Christ. It is written of Moyses that he, the *Lepusculus* of God, shall stand in the cleft of the rock, because he hoped for salvation through the passion of our Redeemer.'

The text of this chapter is very interesting: the first two sentences are taken from the chapter on the hare in Isidore's *Etymologiae*, XII.I.23; but the following sentences are a continuation of the discussion about the nature of the *chirogrillus* which we saw in the works of the Church Fathers[32]. This hare chapter, surprisingly, continues as follows, without any rubric or gap:

> Erinatius vero est quem vocamus ericium animal omnino timidum, natura providente semper armatum, cuius cutem invicem setarum sudes accutissimae densis simeque communiunt. Cui tamen non sufficit nativa munitio, sed ne aliqua fraude possit intercipi refugium habet, semper in saxis. Huic competenter aptatus, qui peccatis suis hispidus futura iudicia metuens petram Christum firmissimum noscitur habere refugium.
>
> 'The hedgehog is a very timid animal, but always provided by nature with armor; its skin is covered with very thick, sharp bristles. But it does not rely only on its natural armor; lest any harm should come to it, it always hides among the rocks. It may fairly be likened to the man who, fearing to be judged for his sins, knows that he has a firm refuge in the rock of Christ.[33]'

This continuation about the hedgehog is not a chapter devoted to the latter animal, since the hedgehog's own chapter appears in Book VIII. This text, which was adopted from Cassiodorus[34], provides the same explanation of the nature of the *chirogrillus* as the one we saw in the section on hares earlier in the chapter. The illustration shows an animal which is discussed in the second part of this chapter; this is situated in the first row from the bottom, the small creature that looks like a hare-hedgehog with small spines. This cannot be an unsuccessful effort to draw a hedgehog, since a perfectly normal illustration of the hedgehog appears in the chapter about the hedgehog in Book VIII[35]. Most probably

32. Hraban Maur used a paraphrase of Venerable Bede, *Super Parabolas Salomonis* 3,30 (PL 91,1026) and Jerome, *Commentariorum in Matheum* 7,25 (CCSL 77,47).

33. The translation is taken from Barber R., *Bestiary: Being an English Version of Bodleian Library Oxford. MS Bodley 764 with All the Original Miniatures Reproduced in Facsimile* (Woodbridge, 1993), p. 66-67.

34. Cassiodorus, *Expositio Psalmorum*, 18 (CCSL 98, 933-4).

35. Fol. 199r; the picture is illustrated in Reuter, *Text und Bild…*, pl. 50.

the illuminator was trying to depict a porcupine. Evidence for this may be found in an analogous illustration from a fifteenth century manuscript of *De rerum naturis*, where the place of this hare-hedgehog is occupied by an animal which looks like a porcupine[36]. As I stated before, it was difficult to draw a combination of a hare-hedgehog in one animal, but at least one attempt has been made by medieval painters.

2.3. Squirrel

During the eleventh and twelfth centuries, the interpretation of the *chirogrillus* did not change: Church Fathers and encyclopaedists followed the long-established tradition. Thus, in his *Elementarium* from the middle of the eleventh century, the Italian lexicographer Papias (following Jerome) describes this animal as a big hedgehog[37]; in the twelfth century the famous philosopher Alain of Lille, a representative of the Chartres school, describes this creature as a hedgehog and a rabbit[38]:

> Ericius vel erinacius dicitur *cuniculus* secundum quosdam, qui alio nomine dicitur *cyrogrillus*.
>
> '*Ericius* or *erinacius*, which is known by the other name as *cyrogrillus,* is also called rabbit according to some authorities.'

At the end of the twelfth century the chaplain of Richard the Lionheart, Raoul (Radulfus) Ardens, interprets this animal as being similar to the hedgehog[39]. At the beginning of the thirteenth century Alexander Nequam interprets the *chirogrillus* as a rabbit, *cunig*[40], in his literal commentary on the Bible, *Corrogationes Promethei*. In his commentary on the *Proverbs* entitled *Hortus Deliciarum Salomonis*, written in 1225-56, Hermann of Werden (Hermannus Werdensis) interprets this animal much as Jerome did, i.e. as a combination of a mouse, hare, and hedgehog[41].

36. Vatican, Bibl. Apostolica Vaticana, Cod. Pal. lat. 291, f. 77v. The picture is illustrated in Reuter, *Text und Bild*..., pl. 46.

37. I would like to thank Prof. W. Edwards and B. Merrilees for their help in checking Paris, BNF, MS lat. 8844 and Papias, *Elementarium doctrinae erudimentum* (Venice, 1496; reprint Torino, 1966), which were not available to me.

38. Alain of Lille, *Distinctionibus Dictionum Theologicalium*, E (PL 210,780).

39. The manuscripts of this author were not available to me; I have this quotation from Vincent of Beauvais, *Speculum quadruplex sive Speculum maius* (Douai, 1624; reprint Graz, 1964-1965), *Speculum Naturale*, XIX.35.

40. From Hunt T., *Teaching and Learning Latin in Thirteenth-Century England* (Cambridge, 1991), v. 1, p. 241.

41. Hermann of Werden, *Hortus Deliciarum Salomonis*, v. 9506-9510; from Pitra J.B. (ed.), *Spicilegium Solesmense* (Graz, 1963), v. 3 p. 74-75.

Bartholomew the Englishman, who finished his *De proprietatibus rerum* in about the year 1245, again interprets the *chirogrillus* as a porcupine[42]:

> Herinacius idem est quod cirogryllus, animal parvum et spinosum, habitans in cavernis, simile hericio sed dicitur esse majus ...
>
> '*Herinacius* is the same as *cirogryllus* – a small, spiny animal that lives in pits, similar to a hedgehog but it is said to be bigger ...'

Vincent of Beauvais briefly describes the whole history of the misinterpretation, presents many sources, tends towards the hedgehog version, and claims that the animal is not a rabbit[43]; Konrad of Megenberg defines the *chirogrillus* as a big hedgehog[44].

At the beginning of the thirteenth century a strange new identification of the *chirogrillus* appears in England, which was not based on the physical similarity between *shafan* and the animal in question: the *chirogrillus* came to be identified as a squirrel. This interpretation of the *chirogrillus* most likely came from France, but the earliest dictionary giving such an explanation for this animal dates from the middle of the fourteenth century[45]: 'Hic *cirogrilus*.li- gallice *eskuireus*'.

The well-known *Dictionarius Firmin le Ver* (middle of the fifteenth century) gives[46]:

> Cirogrillus-erinacius spinosus, maior ericio. Herisson ou porc espi, communement on le dit pour escureuis.
>
> '*Cirogrillus* - a spiny *erinacius*, bigger than a hedgehog. Hedgehog or porcupine, commonly used to refer to the squirrel.'

The fact that the word *chirogrillus* began to be understood to refer to the squirrel can easily be explained by linguistic factors: the word is very similar to the Latin word for squirrel, *sciurellus* or *sciurus*, and to the medieval word for the animal, *scurulus*. As far as I have been able to determine, this word was first interpreted as 'squirrel' in the *Dictionarius* of John Garland (Johannes de Garlandia) (1180–1252), an English-born grammarian and poet who spent most of his life in France. His *Dictionarius*

42. Bartholomaeus Anglicus, *De rerum proprietatibus* (Frankfurt, 1964), 18.61.

43. Vincent of Beauvais, *Speculum Naturale*, XIX.35.

44. Konrad von Megenberg, *Das Buch der Natur*, ed. F. Pfeiffer (Stuttgart, 1861), III.17.

45. Montpellier, Bibliothèque de la faculté de médecine, MS H 236, from *Duo glossaria: anonymi Montepessulanensis dictionarius: le glossaire latin-français du Ms. Montpellier H236*, ed. A. Grondeux, and *Glossarium Gallico-Latinum: le glossaire français-latin du Ms. Paris Lat. 7684*, ed. B. Merrilees & J. Monfrin (Turnhout, 1998), p. 45.

46. *Firmini Verris dictionarius*, ed. B. Merrilees & W. Edwards (Turnhout, 1994), p. 67.

was the first dictionary of its kind in England. It was written in Paris around the year 1220, a period in which the dictionary genre was flourishing, but the work became particularly popular in England. The word *chirogrillus* appears in this dictionary in the chapter on *pelliparii*, as the squirrel was important because of its beautiful fur[47]:

> Pelliparii vendunt pelles deliciosas cuniculorum et *cyrogrillorum* et *esperiolorum*, qui minores sunt cyrogrillis secundum Ysidorum, et lutric[i]orum et mustellarum.
>
> 'Skinners sell delicious furs of rabbits and of *cyrogrilli* and of squirrels, which are smaller than *cyrogrilli* according to Isidore, and of otters and of weasels.'

This text does not explain what 'cyrogrillus' actually means, and is even slightly muddled because *esperiolus* also means 'squirrel' and it is not clear why they are described as 'minores sunt cyrogrillis', but it is confirmed by the vernacular glosses which appear in different manuscripts of the *Dictionarius* that 'cyrogrillus' refers to a squirrel and not to any other animal. Another manuscript gives the gloss 'cyrogrillus – *esquiews*'; earlier manuscripts give vernacular glosses which explain 'cyrogrillorum' as 'de grans equireus', 'esqurels,' or 'ecquer(er)us'[48]. A little after Garland's *Dictionarius*, the Third Family of bestiaries appears in England, the most problematic one of all the bestiaries, partly based on *Nominales* – the Latin-French-English dictionaries of that epoch. These bestiaries include a chapter on the *chirogrillus*, which again is understood to be a squirrel and also drawn as such. The first part of the chapter provides something like a linguistic interpretation of the word[49]:

> *Cirogrillus*, qui et *pirulus* alio nomine dicitur a *pir*, quod est ignis. Est enim pusillum animal et indignabundum adeo, ut quandoque prae ira deficiat. Cirogrillus dictus eo quod captus ad manum subjicitur. Cyros enim manus dicitur.
>
> '*Cirogrillus*, which is called also by another name *pirulus*, from *pir*, that is fire. It is a small animal and so angry that sometimes it dies of rage. It is called *Cirogrillus* since it can be caught by hand and tamed, for the hand is called *cyros*.'

47. From Hunt, *Teaching and Learning*..., v. 1, p. 197-198.

48. Cambridge, Gonville and Cajus College, MS 136 (s. xiii); MS 385 (s. xiii); Worcester, Cathedral Chapter Library, MS Q.50 (s. xiii); from Hunt, *Teaching and Learning*..., v. 2, p. 130, 152.

49. Cambridge, Fitzwilliam Museum, MS 254, f. 23r; Cambridge, University Library, MS Kk. IV. 25, f. 75r; Oxford, Bodleian Library, MS Douce 88, f. 95v; London, Westminster Abbey, MS 22, f. 28v; Oxford, Bodleian Library, MS e Musaeo 136, f. 27r.

The second part describes characteristics and pseudo-characteristics of the squirrel. The visualization of the mistranslation of *chirogrillus* as a squirrel can be seen in the single Fourth Family bestiary, written in England in the fifteenth century and partly based on the above-mentioned work of Bartholomew the Englishman[50] (Fig. 3, the lower part). In the chapter on the *grillus*, i.e. the cricket, Bartholomew says it is a small animal like a *chirogrillus*. Following this explanation, the illuminator illustrates the chapter on this animal with pictures of two animals – a unique case, as far as I know, in medieval bestiaries[51]: a picture of a hedgehog (most probably, since as I mentioned earlier Bartholomaeus explains *chirogrillus* as a porcupine) and a picture of a squirrel (possibly, because of his own knowledge that the *chirogrillus* is a squirrel). The same logic makes the illuminator draw two animals in the preceding chapter devoted to the *glis*, the dormouse. Possibly because the plural of *glis*, i.e. *glires*, sounds very similar to *chirogrillus*, the painter illustrated this chapter with pictures of a dormouse and a porcupine (Fig. 3, upper part).

2.4. Dog-like animal

We find an illustration of the *chirogrillus* that is very different from those described above in Thomas of Cantimpré's *Liber de natura rerum* (c. 1240), in which the *chirogrillus* looks like a dog. The reason for that is not readily apparent, but it might be explained by the fact that Thomas does not define or describe the *chirogrillus* with any precision, and thus left the painter free to depict the animal as he saw fit[52].

> *Cirogrillus* animal est lege prohibitum. Est autem animal parvum et naturaliter debile et in hoc plerumque mirabile, quia licet debile sit, tamen infestum et rapax est et mortiferum ceteris animantibus terrae.
>
> '*Cirogrillus* is an animal prohibited by law. It is a small animal and naturally weak and extraordinary mostly in that, although it is weak, it is dangerous, rapacious and pernicious to other terrestrial animals.'

An analogous depiction of the *chirogrillus* is found in manuscripts of Jacob van Maerlant's *Der naturen bloeme*, who followed Thomas[53]:

50. Cambridge, University Library, MS Gg. VI. 5, f. 10r.

51. The only exceptions are several so-called 'double chapters', e.g. *sirena et onocentaurus*; *onager et simia*; and *mustela et aspis* in the *Physiologus* and several early bestiaries. For discussion of double chapters see Baxter R., *Bestiaries and Their Users in the Middle Ages* (London, 1998), p. 35-37.

52. Thomas Cantimpratensis, *Liber de Natura Rerum*, ed. H. Boese (Berlin-New York, 1973), IV. 24.

53. Jacob van Maerlant, *Der Naturen Bloeme*, ed. E. Verwijs (Leiden, 1878), p. 65-66, II, v. 1199-1204.

Figure 3. Cambridge, University Library MS Gg. vi. 5 f. 10r
Reproduced by permission of the library

Cyrogillus es een dier bor groet,
Dat Moyses in die wet verboet.
Ende al eist cranc ende cleene,
Nochtan eist noesel al ghemene
Ende vreselijc tallen dieren;
So fel es et van manieren.

'Cyrogrillus is an animal not very large,
that Moses prohibited in the Laws.
And even if it is little and small,
nevertheless it is the most harmful
and dangerous of all animals,
it is so cruel by nature.'[54]

We can see the depiction of this description in Figure 4, namely in the form of a poorly-defined animal that looks somewhat like a dog[55].

3. Conclusion

In this article I have discussed in some detail the consequences of the mistranslation of the biblical word *shafan sela* from the moment of its transformation into the *chirogrillus* up to the interpretative 'splitting' of the animal into several different animals: hare, hedgehog, porcupine, squirrel, etc[56]. I have also shown what visual interpretations correspond to these various translations. Of course, the names of other animals have suffered various mistranslations and misinterpretations by

54. I would like to thank Prof. Luuk Houwen for this translation.

55. The Hague, Koninklijke Bibliotheek, MS 76 E4, f. 15r; an analogous illustration appears in another Maerlant manuscript (KA XVI f. 52v).

56. The story of misinterpretation of chirogrillus did not end in medieval times. Thus, Du Cange C.F., *Glossarium mediae et infimae latinitatis* (Niort, 1883-1887), v. 2, p. 310, quotes John Garland's *Opus Synonymorum* an. 1495 (s.l., s.n.) with commentaries of Galfridus Anglicus, a Dominican Friar of the middle of the 15th century, who explains the *chirogrillus* as a wild cat, *cattus silvestris*: 'Cyrogrillus est animal habitans sub terra: et dicitur proprie *cattus silvestris*, dispositus ad modum canis, et dicitur a cyros, quod est manus, et grillus quod est cattus, quasi habens pedes et manus ad modum catti' ('The cyrogrillus is an animal that lives beneath the ground and strictly speaking it is called a *forest cat*. It resembles a dog; its name comes from *cyros*, which is hand, and *grillus,* which is cat, as it has feet and hands in the manner of a cat'). Obviously, this text is not written by John Garland himself, since it does not appear in the manuscripts and printed editions of this author, see for example PL 150, 1577A-1590C; Diefenbach L., *Glossarium Latino-Germanicum mediae et infimae aetatis* (Frankfurt am Main, 1857), p. 122, supplies other obscure 15th century German interpretations of the word, as e.g. sea cat, ermine etc. I was not able to check the mentioned references myself and I am not aware of any example of visualization of those interpretations.

Figure 4. The Hague, Koninklijke Bibliotheek, 76 E4 f. 15r
Reproduced by permission of the library

medieval authors and artists as well, but I doubt very much whether any other animal can boast the same number of metamorphoses as the *chirogrillus*.

Hebrew University of Jerusalem

Appendix: Table

Source	PLACE	TEXTS
HEBREW BIBLE	**LEV. 11.5-6**	The **shafan** which cheweth the cud, but divideth not the hoof, is unclean. The **hare** also: for that too cheweth the cud, but divideth not the hoof.
	DEUT. 14.7	But of them that chew the cud, but divide not the hoof, you shall not eat, such as the camel, the **hare**, and the **shafan**: because they chew the cud, but divide not the hoof, they shall be unclean to you.
	PS. 103.18	The high hills are a refuge for the harts, the rock for the **shafans**.
	PR. 30.26	The **shafans**, a weak people, which maketh its bed in the rock.
SEPTUA	**LEV. 11.5-6**	Choirogrullios
	DEUT. 14.7	Choirogrullios
	PS. 103.18	Choirogrullios **α [Lagos]**
	PR. 30.26	Choirogrullios
HEX	**PS. 103.18**	Choirogrullios **[Lagos]** **[Choirogrullius and Lagos]**
VULGATA	**LEV. 11.5-6**	Chyrogryllius
	DEUT. 14.7	Choerogyllius **A [corcodillum]**
	PS. 103.18	Montes excelsi cervis Petra refugium **ericiis θ [cuniculi]**
	PR. 30.26	**Lepusculus** plebs invalida Quae conlocat in petra cubile suum

Outi MERISALO and Päivi PAHTA

TRACING THE TRAIL OF TRANSMISSION: THE PSEUDO-GALENIC *DE SPERMATE* IN LATIN

0. Introduction

Our contribution deals with the pseudo-Galenic treatise on human generation and embryology known as *De spermate*. The text has its roots in the intellectual traditions of learned medical and natural-philosophical writings of the antiquity, but the actual origin of the text and the earliest phases of its transmission are not known. In modern scholarship the text is usually referred to as a Latin translation. It has been suggested that it may have been composed in late antiquity; according to Kudlien[1] it may be a deliberate forgery under Galen's name produced in the Neoplatonist circles connected with the school of Alexandria. The connections of the text with Arabic natural-philosophical traditions of writing have also been discussed by Burnett[2].

De spermate is at present known to survive in 42 Latin manuscript copies, some of them containing only fragments of the text, and in several sixteenth-century prints. Apparently there is only one medieval translation of the text into a European vernacular: a Middle English translation survives in a unique copy in the fifteenth-century Cambridge, Trinity College R.14.52, edited by Pahta[3]. There is no modern edition of the Latin *De spermate*, nor has its textual tradition been hitherto extensively studied.

In this article we shall focus on the Latin version of the text, paying special attention to the text history of the work. Our cautious remarks on the Latin tradition are based on earlier research and our examination

1. Kudlien F., 'The Seven Cells of the Uterus: The Doctrine and its Roots', *Bulletin of the History of Medicine* 39 (1965), p. 415-423.

2. Burnett C.F., 'Physics before the *Physics*: Early Translations from Arabic of the Texts concerning Nature in MSS British Library, Additional 22719 and Cotton Galba E IV', *Medioevo* 27 (2002), p. 53-110, and 'The Chapter on the Spirits in the *Pantegni* of Constantine the African', in: C.F. Burnett & D. Jacquart (eds.), *Constantine the African and 'Ali ibn a-'Abbas al Magusi: The 'Pantegni' and Related Texts* (Leiden, 1994), p. 99-120.

of twenty of the extant manuscript copies. We are planning to publish the Latin text; this contribution is thus a report of work in progress.

1. Origin

The origin of *De spermate* is not known. It may have been composed in late antiquity and transmitted to the Latin West through Arabic; according to Burnett, an Arabic origin is also possible[4]. At some stage the text was falsely ascribed to Galen. This confusion is perhaps understandable, as Galen had in fact written on this topic too. His authentic text, *Peri spermatos*, was translated into Syriac by the well-known translator Hunain ibn Ishaq in 840-1, into Arabic maybe by one of his pupils, and finally into Latin in the fourteenth century by Niccolò da Reggio[5]. In Latin manuscripts Galen's text also occurs under the title *De spermate*. In modern scholarship Galen's text is commonly referred to as *De semine*, but sometimes even today the titles *De spermate* and *De semine* are used for both texts, which continues to create confusion[6].

It has been suggested that *De spermate* is a deliberate forgery produced in the hope of spreading ideas under Galen's influential name; Kudlien[7] argues for a possible origin in the Neoplatonist circles connected with the school of Alexandria, where some forgeries of Galen were apparently

3. Pahta P., *Medieval Embryology in the Vernacular: The Case of* De spermate, Mémoires de la Société Néophilologique de Helsinki LIII (Helsinki, 1998).

4. For a discussion of the origin of the text and its Arabic and Greek connections, see Burnett, 'Physics...' and 'The Chapter...'. See also Pahta, *Medieval Embryology...*, p. 95-98. A Greek copy of the text in Oxford, Bodleian Library, Miscell. 20 (f. 375) is mentioned in Passalacqua V.T., 'Microtegni seu de Spermate', *Corpus scriptorum medicorum infimae latinitatis et prioris medii aevii* 1 (1958), p. 305. The text bears the title *Peri spermatos*. Passalacqua does not, however, comment on the relationship of the Greek and Latin versions. Diels H., *Die Handschriften der antiken Ärzte I: Hippokrates und Galenos*, Abhandlungen der königlichen preussischen Akademie der Wissenschaften (Berlin, 1905), p. 70, lists this text as one of three extant Greek copies of Galen's *Peri spermatos*. The manuscript was not examined for this study.

5. The Greek text is edited in Galen, *On Semen*, ed. P. De Lacy, Corpus medicorum Graecorum 5.3.1 (Berlin, 1992). De Lacy bases his edition on two manuscripts, Moscow, gr. 466, fifteenth century, only Book 1; BNF, gr. 2279, sixteenth century, Books 1-2.

6. For example, in the Wellcome catalogue of manuscripts the title *De spermate* is used for referring to both texts (Moorat S.A.J., *Catalogue of Western Manuscripts on Medicine and Science in the Wellcome Historical Medical Library I: Manuscripts Written before 1650 A.D.* (London, 1962), p. 185), though the title *De semine* is also used for the authentic treatise by Galen (Moorat, *Catalogue...*, p.182).

7. Kudlien, 'The Seven Cells...'.

produced around the third century[8]. The text seems to be a compilation based on several earlier sources, citing a variety of ancient medical and philosophical authors as sources of knowledge, including false references to works by Hippocrates and Aristotle. There is nothing in the contents to suggest that the compiler would have drawn his information directly from Galen's authentic text[9].

De spermate may have been translated into Latin in the early Salernitan period when interest in theoretical medical and natural-scientific knowledge was growing in the West, and translation and copying of texts on these topics began to increase. There is no conclusive evidence suggesting that this text was one of the works translated from Arabic by Constantine the African in the second half of the eleventh century[10]. Some scholars have suggested that the text was translated at the school of Toledo from a translation by Hunain[11].

2. The Latin *De spermate* manuscripts

Manuscript evidence shows that *De spermate* was widely known from the twelfth century onwards, enjoying considerable popularity especially in the thirteenth and fourteenth centuries. The Latin text, or extracts from it, is now known to survive in 42 manuscripts[12] and in several sixteenth-century prints, displaying considerable variation.

The majority of the extant Latin copies are from the thirteenth and fourteenth centuries, whereas there are only few copies apparently dating

8. The influence of Neoplatonism is evident in *De spermate*, especially in the lengthy *de anima* passage, which gives particular emphasis to Porphyry (c. 234 – c. 305 BC), one of the outstanding Neoplatonist philosophers, and to his *Isagoge*. See Kudlien, 'The Seven Cells…'.

9. Although on many points concerning the origin and constitution of semen the text takes a similar view as Galen in *De semine*, the range of topics addressed in *De spermate*, the arrangement of the material, the level of detail, and some of the views expressed on other topics, notably the anatomy of the womb, differ drastically from Galen's work.

10. For a discussion, see Pahta, *Medieval Embryology…*, p. 96, and Burnett, 'Physics…' and 'The Chapter…'.

11. See Vandi L., 'Dalla biblioteca di Giovanni di Marco da Rimini, il manoscritto galenico D XXV 2 della Malatestiana di Cesena', *Scriptorium* 51.2 (1997), p. 325.

12. In addition to the 38 manuscripts listed in the study of the vernacular text in Pahta, *Medieval Embryology…*, the following should be considered: Oxford, Bodleian Library, Ashmole 1471; Munich, Bayerische Staatsbibl., Clm 490; Gdańsk, Mar F. 41 C 4c; Berlin, SBB-PK, lat. fol. 638. We are grateful to Monica H. Green for the identification of the Oxford ms. and to Dominique Stutzmann for the discovery of the Berlin manuscript. The Gdańsk and Munich (Clm 490) manuscripts were found thanks to the remarkable *Scriptorium* web site (www.scriptorium.be). Charles Burnett very kindly provided us with microfilm copies of the Munich ms.

from the twelfth or fifteenth centuries. In manuscripts the translation is variously ascribed to Constantine the African, Copho of Salerno, or Arnald of Villanova. The attributions to Copho and Arnald solely rest on manuscript ascriptions. Some additional evidence has been interpreted as pointing to the possible role of Constantine as the translator. The most important source concerning Constantine's life and works, *De viribus illustribus* by Peter the Deacon, lists items called *Tegni* (i.e. *Tekhne*) and *Microtegni* (*Mikrotekhne*) among texts translated by Constantine. The first of these probably refers to Galen's *Ars parva*. It is possible that the second item refers to *De spermate*, which in several extant manuscripts is called *Microtegni*[13].

Despite its popularity, *De spermate* apparently never became part of compulsory core reading in medieval university faculties of medicine. It is mentioned, for example, in Giles of Rome's *De formatione corporis humani in utero*, and in Pietro d'Abano's *Conciliator*, where it is referred to as a forgery[14]. Several of the Latin manuscripts contain other works written by or attributed to Galen. The fact that *De spermate* was often printed together with Galen's authentic works must have continued to

13. Peter the Deacon's work survives in Codex Cassinensis 361 (autograph) and an early printed edition of 1655. For further details, see Bloch H., *Monte Cassino in the Middle Ages* 1 (Rome and Cambridge, Mass., 1986), p. 127-128. Bloch (*Monte Cassino*...1, p. 131-132) thinks that the titles in Peter the Deacon's catalogue either refer to the same treatise, or that one of the titles refers to the *Isagoge* of Hunain ibn Ishaq, aka 'Johannitius'. He points out that Constantine's authorship of the *Tegni* included in the *Articella* is supported at least by one manuscript. In a later *addendum*, Bloch (*Monte Cassino*...1, p. 1101) remarks that another work, i.e. *De spermate*, has been identified with the *Microtegni* by Passalacqua ('Microtegni...'). Bloch rejects Constantine's role as its translator by saying that the *Ars parva* seems 'by far the more likely candidate for being connected with Constantine'. Assigning the *Tegni* (*Ars parva*) to Constantine does not, however, rule out the possibility that he also translated the work circulating under the titles *De spermate* or *Microtegni*. Constantine's role as the translator is accepted, for example, by Passalacqua ('Microtegni...') and Jacquart D. & Thomasset C. (*Sexuality and Medicine in the Middle Ages*, tr. by M. Adamson, (Cambridge, 1988)). Monica H. Green has pointed out that *De spermate* often appears in codices which also contain genuine Constantiniana (letter to P. Pahta). See also Burnett, 'The Chapter...'.

14. Pietro d'Abano was apparently more familiar with Galen's authentic writings than most of his contemporaries, and he translated some of them. His translations include, for example, *De tremore, palpitatione, convulsione et rigore*, which is on a subject too specialised to have been known to ordinary physicians (for references, see Kudlien, 'The Seven Cells...', p. 416, n. 5). Pietro's remarks on the pseudo-origin of *De spermate* are later mentioned, for example, in *Liber anathomiae corporis humani et singulorum membrorum illius* by Gabriele Zerbi, published in 1502. It is noteworthy that a contemporary cursive hand in the French early fourteenth-century manuscript Vat. Pal. lat. 1094 f. IV (list of contents) has written regarding *De xij. portis* which in the body of the manuscript is combined with *De spermate,* with an explicit referring to *Microtegni*: 'no(n) credo set e(st) alius lib(er) bo(nus)'.

support the ascription. An early printed version of the text is included among the works of spurious origin, for example, in *Galeni opera omnia*, Vol. 8 (Basle, 1542). In addition to that, Thorndike and Kibre[15] refer to an edition printed in Venice, 1508; Diels[16] and Lemay[17] refer to an edition printed in Lyon, 1528; and Jacquart and Thomasset[18] mention a Venetian Giuntina edition from 1597[19].

3. *De spermate* and the concept of a text

The study of the Middle English translation of *De spermate* raised the question of defining the text, a problematic task in the case of many medieval works. The text identified as the Middle English *De spermate* in the *Catalogue of Manuscripts in Trinity College, Cambridge*[20] clearly consists of two different sections. The first section, on f. 28r-36v of the Trinity manuscript (lines 1-669), focuses on the theory of human generation and embryology. The second section, beginning on f. 37r and finishing abruptly in the middle of a sentence on f. 40v (lines 698-973), deals with topics in astrologically-based physiology, aetiology, and therapy, without any reference to embryology or matters of human generation. Between the two clearly different sections there is a short account on the nature of the twelve zodiacal signs, which links the two sections together. It is connected with the embryological passage in referring in a general manner to cases where celestial influences can cause inconsistencies in the resemblance between parents and the child. It is also connected with the passage on physiology, aetiology, and therapy, which frequently stresses the importance of astrological knowledge for successful medical practice. There is no sign in the hand of the original scribe of a break or change from one text to another at any point. A later hand has added a marginal comment 'iiij elements' on f. 37r, where the discussion of the general physiological theory of the four elements, four seasons,

15. Thorndike L. & Kibre P., *A Catalogue of Incipits of Mediaeval Scientific Writings in Latin* (Cambridge, Mass., 1963).

16. Diels, *Die Handschriften....*

17. Lemay H.R., *Women's Secrets: A Translation of Pseudo-Albertus Magnus's* De secretis mulierum *with Commentaries* (Albany, New York, 1992).

18. Jacquart & Thomasset, *Sexuality and Medicine....*

19. The early modern printed version included in the Basle edition of 1542 differs from all manuscript copies checked so far. The corrections may originate in the corrector, since they seem to classicise rather conspicuously the language. See also Passalacqua, 'Microtegni...'.

20. James M.R., *The Western Manuscripts in the Library of Trinity College, Cambridge: A Descriptive Catalogue* 2 (Cambridge, 1900-1904), p. 338.

and four changes caused by the planets begins, indicating that a later user of the manuscript has also noticed the change of topic. This marginal comment is regarded as a marker of a break-up point in the text by Mooney in her *Index of Middle English Prose* handlist describing the Trinity collections[21].

The thirteen manuscripts of the Latin text currently preserved in British and US libraries may serve to illustrate the variation in what was considered a text called *De spermate* in the Middle Ages. What appears to be a full version of both the embryological treatise and the astrological treatise on aetiology and therapy is contained in nine copies. Of these, Oxford, Balliol College 231, a Galenic encyclopedia known in the Balliol collections as *Galenus latinus*[22], divides the material into two separate treatises, indicating the break with an explicit and a new rubric. The first treatise contains the discussion on generation and embryology. The second one begins with the description of the signs of the zodiac, and contains the full version of the text on astrologically-based physiology, aetiology, and therapy. The division is also reflected in the titles used in the incipits and explicits (see below). The other four Latin manuscript copies are incomplete. Oxford, Bodleian Library, Ashmole 399 and Oxford, Pembroke College 21 are short fragments containing the beginning of the embryological treatise. A fragment in a New York Academy of Medicine manuscript contains most of the embryology. British Library, Cotton Galba E.iv contains the entire embryological tract and approximately half of the second part, breaking off at almost the same point as the Middle English translation.

Three different titles of the text occur frequently in the manuscripts, sometimes two or all three in a single manuscript. Of twenty Latin copies checked so far, a title, rubric, or explicit referring to the treatise by name is found in nine. Ten manuscripts use the title *Liber spermatis* or *Liber de spermate*. The variant title *Microtegni* occurs in twelve manuscripts. Nine manuscripts also use the title *Liber de XII portis* or *Liber de XII signis*, with reference to the zodiac. Two manuscripts, Cambridge, Peterhouse, 33 and Erfurt, UFEG F. 249, use four titles, adding *De humana natura* to the list. Balliol College 231, which divides the treatise into two, refers to the first treatise in the rubric as *Liber microtegni Galieni l. de spermate* and in the explicit as *Liber de*

21. Mooney L., *The Index of Middle English Prose. Handlist XI: Trinity College, Cambridge* (Cambridge, 1995).

22. Mynors R.A.B., *Catalogue of the Manuscripts of Balliol College Oxford* (Oxford, 1963), p. 244-247.

spermate. The second treatise, which begins with the account of the signs of the zodiac, is referred to as *Liber de XII signis uel elementoris* in the rubric, and as *Liber de XII portis uel migrotegni Galieni* in the explicit. In addition to the more frequent titles, others are also used. Of the Vatican manuscripts, two (Pal. lat. 1234 and Urb. lat. 246) use the title *de corporis et anime armonia conseruanda* while also giving *Microtegni*. One (Pal. lat. 1298) also uses the title *Macrotegni* (see below). Finally, one of the Erfurt manuscripts (Qu. 178) refers to the text as *Libellus de generacione embrionis*.

Although it was common for medieval treatises to circulate under several titles, in this case some of the many different titles can be connected with the fusion of two texts. The division in Balliol College 231 seems a logical one in assigning the beginning of the treatise on the twelve signs to a place where the discussion also begins in the text. This manuscript also reserves the title *Microtegni* for the combination of the two treatises.

4. On the text history of *De spermate*

Of the 42 Latin copies of *De spermate*, twelve are held by libraries in England, and one, owned in the fifteenth century by the Benedictine Abbey of St Augustine in Canterbury, is now part of the collections of the New York Academy of Medicine. Seven of the Latin copies in England are associated with Oxford, one of them written in 1451 by Hermann Zurke of Greifswald, who also worked for Gilbert Kymer, Chancellor of Oxford, Dean of Salisbury, and was the personal physician to Humphrey, Duke of Gloucester[23]. Two Latin copies survive in college library collections in Cambridge; one of them is associated with Roger Marchall, a fifteenth-century Cambridge-trained Doctor of Medicine and royal physician to Edward IV[24].

Twenty-nine copies of the text are now known to survive in libraries in France, Germany, Poland, Italy, Switzerland, and the Vatican. The three manuscripts now at the *Universitäts- und Forschungsbibliothek Erfurt – Gotha*, belong to the Amploniana collection, i.e. the library bequeathed to the University of Erfurt by Dr. Amplonius Rating de Bercka (d. c. 1435) and preserved in its original form to

23. Voigts L., 'Scientific and Medical Books', in: J. Griffiths & D. Pearsall (eds.), *Book Production and Publishing in Britain, 1375-1475* (Cambridge, 1989), p. 385.

24. Voigts L., 'A Doctor and his Books: The Manuscripts of Roger Marchall (d. 1477)', in: R. Beadle & A.J. Piper (eds.), *New Science out of Old Books: Studies in Manuscripts and Early Printed Books in Honour of A.I. Doyle* (Aldershot, 1995), p. 249-314.

this day[25]. Of the manuscripts preserved in Italy, three housed at the Malatestiana library of Cesena and having belonged to Giovanni di Marco, doctor to Novello Malatesta, Lord of Cesena in the fifteenth century, have received attention, mainly from art historians[26]. We shall examine here more closely the five manuscripts housed at the Vatican, i.e. Pal. lat. 1094, 1234, 1298, Vat. lat. 2383 and Vat. Urb. lat. 246, as well as the three manuscripts which belonged to Dr. Amplonius.

4.1. The Vatican manuscripts

The Palatine manuscript 1094 is French and may be dated to the early 1300s. It consists of seven codicological units, 636 numbered folios, and presents no fewer than seven hands, both textualis (for the body of text) and cursive (for marginal notes), all French. The different parts were bound together and provided with a table of contents, foliotation and running titles in France sometime in the fourteenth century. The first codicological unit has a note on f. 402v in a fourteenth-century cursive French hand: 'istud dixit nobis magister Gerardus medicus regis Francie', the reference being probably to Gérard de Saint-Dizier, doctor to John II, King of France, in 1352[27]. *De spermate* is contained near the end of the first codicological unit, mainly regular senions, on f. 391va-398rb. The manuscript as a whole is a collection of Galen's authentic and spurious works. In the fourteenth-century table of contents on f. IV, *De spermate* is carefully distinguished from *De xij portis*. In the text itself the latter covers ff. 395ra-398rb): 'Explicit lib(er) de .xij. portis u(e)l microtegni Gal(eni)'. Sometime before 1436 this manuscript was incorporated into the collection of the Elector Palatine Louis III (r. 1417-1436), who bequeathed it, together with 153 volumes of theology, astronomy, Canon and Roman law as well as medicine, to the Heidelberg Church of the

25. See Speer A. (ed.), *Die Bibliotheca Amploniana: Ihre Bedeutung im Spannungsfeld von Aristotelismus, Nominalismus und Humanismus*, Miscellanea Mediaevalia 23 (Berlin-New York, 1995), p. 16-31.

26. L. Vandi ('Dalla biblioteca…', p. 326) analyses the remarkable miniatures of the Cesena manuscripts, including D 25.2, probably written by Giovanni da Parma (d. 1289) and illustrated very soon afterwards by artists of the school of Bologna. The other two Cesena manuscripts containing the *De spermate* are D 25.1 and S 5.4 (which also has Niccolò da Reggio's translation of the authentic *De semine*). See also Nutton V., 'De motibus liquidis and the medieval Galen', *Galenos* 1 (2007), p. 163-173.

27. See Schuba L., *Die medizinischen Handschriften der Codices Palatini in der Vatikanischen Bibliothek: Universitätsbibliothek Heidelberg* (Wiesbaden, 1981), p. XXIV, suggesting either this person or Gérard de Lacombe, doctor to Charles VI, d. 1408. The palaeographical evidence would indicate the former. See also Wickersheimer E., *Dictionnaire biographique des médecins en France au moyen âge* 1-2 (Paris, 1936), p. 184-185.

Holy Ghost at his death that same year. The University of Heidelberg took over the administration of the Holy Ghost library in 1438[28].

Another Palatine manuscript, Latin 1234, is almost one hundred years younger, but is also French[29]. It is a composite paper and parchment manuscript with nine codicological units, eight of which are datable and localisable on palaeographical grounds as well as on the basis of watermarks to the very first decades of the fifteenth century and the area of Southern France. The ninth unit, containing *De spermate*, seems to be from the third quarter of the fourteenth century and localisable to Italy, Southern France or even Franconia. The first two codicological units have distinct references to Montpellier[30]. The manuscript ended up in Germany sometime before the beginning of the sixteenth century, since the eighth codicological unit presents, on f. 322, a distinctly German hand. The volume is a collection of medical texts ranging from anonymous recipes to Constantine the African, *De remediorum ex animalibus materia (Epitoma libri medicinae Sexti Placidi)* over Bernard de Gordon's glosses to Galen's *De marasmo* and Mondino de Luzzi's *De dosi medicine* and *Anatomia corporis humani*. Galen is present with Gerald of Cremona's translation of *De secretis*. A fragment of *De spermate*, entitled *Migrotegni*, covers f. 268rab (ninth codicological unit). A most interesting point is no doubt the identification of Constantine the African as the translator: 'Incipit migrotegni Galieni translatum a Constantino Montis Cassinensis monacho de corporis et anime armonia conseruanda'.

The third Palatine manuscript, Latin 1298, was written by five different scribes, four of them Transalpine, apparently all working at the beginning of the 1460s. A very competent hybrida hand annotated the whole volume[31]. The Italian-type hand dated his part on f. 106va: *146iiij die xx.a me(n)s(is) Junij*. One of the scribes is called Paulus de Olomucz (i.e.

28. Schuba, *Die medizinischen Handschriften...*, p. XXI-XXII.

29. Schuba (*Die medizinischen Handschriften...*, p. 250) gives the origin as German, probably basing his hypothesis on the frequency of transmission in the German manuscripts listed in Thorndike L. & Kibre P., *A Catalogue of Incipits of Mediaeval Scientific Writings in Latin* (Cambridge, Mass., 1963). The hands as well as the watermarks quite distinctly point to France, the paper showing watermarks from South France – North Italy – Switzerland and the Rhine area.

30. F. 25vb, 44ra, 98vb, connections to Bernardus de Gordonio, *Lilium medicinae*, in the fever definitions on f. 12rvab, 13rb, 21va, Lilium 2, 9, 3.4 and 10 (Schuba, *Die medizinischen Handschriften..., ad loc*). Another Montpellier author is Stephanus Arlandi (Étienne Arland), whose *Viridarium* ends on f. 223rb: 'Explicit viridarium super Antidotarium Nycolai editum a magistro Stephano Arlandi in medicina in Montepessulano'.

31. Schuba (*Die medizinischen Handschriften...*, p. 387) gives a transcription of the note on f. 206ra: 'Explicit liber Raby Moysi... (quem ego magister Cunradus medicine [? OM] doctor deinceps Patauio per) Paulum de Olomucz de Morauia transscribere permisi.

Olomouf) de Moravia. The manuscript was written in Padua, a centre for Transalpine students of law and medicine. It is another medical reference work, with texts by Mondino de Luzzi (*Anatomia corporis humani*), Moses Maimonides (*Aphorismi, De regimine sanitatis, De venenis, De hemorrhoidibus*), Galen (*De secretis, De sectis, De assuetudinibus seu de consuetudinibus, De rigore, tremore, iectigatione et spasmo seu de tremore, palpitatione, convulsione et rigore, De facile acquisibilibus*), Rasis (*De secretis in medicina*), Arnald of Villanova (*Medicationis parabolae*), a commentary on Avicenna and Mesue the Elder's (Johannes Damascenus) *Aphorismi* as well as *Macrotegni siue liber de sperma seu de xiii* (*sic*) *portis* and *Praecepta Hippocratis*. The paper used throughout the volume is, understandably enough, Italian[32].

The fourth manuscript preserved at the Vatican library, Urb. lat. 246, takes us to the court of Federico di Montefeltro, Duke of Urbino, the wealthy and versatile bibliophile noted especially for his patronage of the bottega of Vespasiano da Bisticci, which specialised in humanist luxury manuscripts. The volume containing *De spermate* shows that Federico also collected medical texts and that his luxury manuscripts could be made up from older manuscripts decorated according to the standard Urbino scheme. In this case the four codicological units, all Italian parchment manuscripts from the thirteenth and fourteenth centuries, have been bound together and provided with the usual Federico list of contents (as in most manuscripts produced by Vespasiano) in a medallion on f. IIv: 'In hoc codice. continetur. liber-mesue de sinplici.medicina. et almansoris. et sinonime. simonis. Ienuensis'. Folio 1 has a humanist floral decoration with Federico's arms. Before Federico, the *Synonyma* of Simon of Genua, with additions by Mundinus de Foro Julij, copied in the third codicological unit, were owned and corrected by Dr. med. Jacobus de Omnibus Sanctis, of Padua on 27 June 1400 at Vesper time (f. 239vb). The *De spermate* fragment is included in the second unit, an early fourteenth-century Italian manuscript: 'Incipit lib(er) galieni d(e) armo(n)ia cor(por)is (et) a(n)i(m)e q(ui) d(icitu)r miçio tegni' (f. 192va-193rb). The

Amen. Quos ambos deus sublimis hec preter remedia feliciter sanos conseruat', as well as that on f. 32va (diplomatic transcription by OM): 'Et dix(i)t m(a)g(iste)r mathe(us) nob(is) se bis i(n) vita (Schuba: vno, mistake) vidisse sud(or)e(m) saguineu(m) (Schuba: *sanguinem*, mistake) (et) ia(m) a(n)no lxiiij.o vid(i)t q(uen)d(am) h(a)b(e)nte(m) i(n)-camisia sua ex sudo(r)e gutas ple(nas) sang(uinis) pu(r)i'. See Merisalo O., 'Regionalism and Interregionalism in the Production of Mss.: The Case of *Vat. Pal. Lat.* 1298', *Miscellanea Bibliothecae Apostolicae Vaticanae* XIII (2006), p. 493-517.

32. An interesting codicological detail is the chalking of the written space in most of the manuscript, which, according to Malachi Beit-Arié, is an Oriental practice (oral communication on 18 February 2004).

title resembles that given in Pal. lat. 1234 above. The Urbino list of contents is more than approximative: the volume also contains texts on the analysis of urine, paediatrics, herbs, the *Aphorismi* of Johannes Damascenus (cf. Pal. lat. 1298 above), the *Pronostica Democriti* and, finally, Mondino de Luzzi's *Anatomia* (cf. Pal. lat. 1234 and 1298).

The last of the Vatican manuscripts, Vat. lat. 2383, has been studied by Passalacqua. This Italian manuscript, written in two rotunda text hands from c. 1300, consists of two codicological units, the second of which differs considerably in size from the first one. The first unit has several historiated initials; several Transalpine Gothic cursive hands appear in the margins as well as in the list of contents on f. 146v. The second codicological unit has an ex libris, from c. 1300, on f. 172v: 'lib(er) de (com)(plexi)onib(us). nicolun[? nicolao?] d(e) u(er)cell(is). p(retio).s(olidorum).xx.' The volume consists almost exclusively of texts by Galen (and pseudo-Galen). *De spermate* is contained in the first codicological unit, ff. 44b-50vb. The work bears the title *De spermate* both in the incipit and the explicit.

4.2. The Amplonius manuscripts

Erfurt, UB, Fol. 249 is a large volume, containing two codicological units. The first one (quaternions) coincides with a small, regular, angular Gothic textualis in bright brown ink, datable to the late thirteenth century. The second unit (quinions) is characterised by a larger-format, late-thirteenth century Gothic textualis with a stronger lateral compression, in light brown ink as well as a third hand, a well-developed Gothic cursive from the end of the thirteenth or beginning of the fourteenth century, not dissimilar to Northern French documentary hands from the same period. The decoration is identical in both codicological units, consisting of Gothic arabesqued initials in red and blue. Schum[33] read on f. I the fifteenth-century ex libris 'Iste liber est Wigboldi... cten de Osenbrughe canonici ecclesiarum Monasteriensis et Osnabrugensis et valet XII fl.' This note is preceded by the still visible 'l(ibras).x. p(ar)I', i.e. 'parisienses' (read by Schum as 'parisiencium'). Several volumes previously owned by Canon Wigbold Stucten of Münster and Osnabrück were acquired by Dr. Amplonius[34], who had himself gone to school in

33. Schum W., *Beschreibendes Verzeichnis der Amplonianischen Handschriften-Sammlung zu Erfurt* (Berlin, 1887).

34. The volumes in question contained moral philosophy, natural philosophy and medicine; see Kadenbach J.,'Die Bibliothek des Amplonius Rating von Bercka. Entstehung, Wachstum, Profil', in: Speer (ed.), *Die Bibliotheca Amploniana...*, p. 25.

Osnabrück[35]. The volume is another Galenic encyclopedia, with, to round it off, Averroes' *De theriaca*. The pseudo-Galenic *De spermate* is contained in the second codicological unit, in the section written by the second hand, over two quinions (XXVIII and XXIX), as Schum's item no. 17. f. 253-259 (= 254rb-260rb, counting error): 'Incipit Mic(r)otegnj. Galienj. idest lib(er) de sp(er)mate seu de duodeci(m) portis. seu de hu(m)ana natura', starting with 'Sperma hom(in)is descendit ex om(n)i corp(or)is humore' and ending with 'per na(tura)m sui corporis. Explicit lib(er) de xij. portis u(e)l microteg(n)i G. siue de sp(er)mate (=. Seu de hu(m)ana natura.' Although the text resembles that preserved in the Balliol manuscript and in Amplonian Qu. 178 (see below), it finishes with a series of variants unknown in the rest of the tradition, thus pointing to the existence of different redactions.

Another large manuscript, Erfurt, UB, Fol. 278, consists of five codicological units, all datable to the late thirteenth and early fourteenth centuries. There are four hands, a pearl-formed Gothic textualis corresponding to the first codicological unit (quaternions), a South-French or Italian Gothic textualis covering units two and three (senions), a laterally compressed Transalpine textualis from the early fourteenth century corresponding to unit four, as well as another early fourteenth-century textualis appearing in unit five. The volume is a Galenic encyclopedia, with *De spermate* in the first codicological unit. The text, identified on f. 78v as *microtegni seu de spermate seu de XII portis*, begins with the words 'Sperma hominis descendit ex omni corporis' and ends on f. 82v with 'ex qua acuicione et calo<re>'.

Manuscript Qu. 178 consists of four codicological units (f. 1r-52r; f. 53r-117v; f. 118r-160v; f.161r-175r), written in seven different Transalpine Gothic hands. Six hands are datable to the beginning of the fourteenth century: hand 1 appears in the first and second codicological units, hand 2 in the first unit, hand 3 in the second and the third, hand 4 in the first and the third, hand 5 in the second, and hand 6 in the third and fourth. Hand 7, datable to the end of the fourteenth and beginning of the fifteenth century, occurs in the rubrics of f. 46-49. This large volume is a medical encyclopaedia, containing, in the first codicological unit, the *Isagoge* of Johannitius accompanied by notes by John of St. Amand and Bartholomew of Bruges, the *Tegni* (*Ars parva*) of Galen with notes by Haly, Thaddaeus of Bologna, Bartholomew of Bruges and John of St. Amand, and the *Aphorisms* of Arnald of

35. Kadenbach, 'Die Bibliothek...', p. 18.

Villanova. The second codicological unit contains texts by Johannes Damascenus, with extracts from Isidore, texts by Hippocrates, with the commentary by Galen and *expositiones* by Thaddaeus of Bologna, Bartholomew of Bruges and others, the *Pronosticationes* of Hippocrates, with a commentary by Galen and notes by Thaddaeus and John of St. Amand, *Liber praescientiae* by Hippocrates, translated by Gerald of Cremona, and Hippocrates' *Liber de secretis* and *Liber regimenti acutorum* (with Galen's commentary). The third unit presents the *Parabolae* of Arnald of Villanova, *Liber colliget* by Averroes, the *Almansor* by Rasis, Aegidius' *De pulsibus*, *De urinis* and *De simplicibus aromatibus* with commentaries and concordances, and an anonymous *De ponderibus*. This is followed by a Galenic section: *De complexionibus*, *De inaequali discrasia*, and *De simplicibus medicinis*. The fourth codicological unit presents Albertus Magnus' commentary on Aristotle's *Meteora*, Henri of Mondeville's *Anatomia chyrurgiae*, the pseudo-Galenic *De spermate* as well as Arnald of Villanova's *Tractatus de membris corporis humani* and Walter of Montpellier's *De urinis bonis*. *De spermate* bears the title *Libellus de generacione embrionis* (misattributed by Schum to Arnald of Villanova and completely omitted in the 1412 catalogue). The text begins with 'Sperma homi(ni)s desce(n)dit ex om(n)i hu(m)ore', and ends with 'nascu(n)t(ur) ex cal(er)a [sic!] nig(r)a', which corresponds to Galba f. 235a.

5. Preliminary remarks on the Latin tradition

The following observations are based on a preliminary and partial comparison of the twenty manuscripts we have examined so far. A full collation of all the evidence is in preparation.

On the basis of the sampled manuscripts, the following may be stated:

1) there would appear to be different redactions of the text;

2) all manuscripts contain a number of mechanically explainable mistakes, such as *saut du même au même*;

3) so far, no distinct regional variants have emerged;

4) among significant variants, the opening line may be used as an illustration: for example, Cotton Galba E. IV and Malat. D 25.1 read: 'Sperma hominis descendit ex humore totius corporis', whereas Oxford, Balliol College, 231; Oxford, Ashmole 399; New York American Medicine; Malat., D 25.2; Pal. lat. 1094; and Urb. lat. 246 read: 'Sperma hominis descendit ex omni humore corporis'.

Conclusion

The date and place of the composition and translation of the pseudo-Galenic *De spermate* still remain unknown; it may date from late antiquity and may have originated in Alexandria. It certainly circulated in a Latin translation, possibly by Constantine the African himself, from the twelfth century onwards. The text, now extant in at least 42 manuscripts, gained considerable popularity in the thirteenth and fourteenth centuries, and was still being copied in the second half of the fifteenth and early sixteenth century. The Italian material seems to indicate popularity in Padua, the French material in Montpellier; Oxford and Cambridge had it; even Federico di Montefeltro, Duke of Urbino acquired it; and the doctor of Novello Malatesta, Lord of Cesena, wanted it. Some of the manuscripts examined for this paper were exported from France and Italy to the Empire; and Dr. Amplonius Rating de Bercka bequeathed his copy to the University of Erfurt in the first half of the fifteenth century. Galen's popularity in the Renaissance certainly profited *De spermate*: it is included in the printed works of Galen in 1508, 1528, 1542 and still in 1597. The accompanying texts in the manuscripts so far examined are primarily medical. In a few cases the volumes are collections of Galenic texts, more often they contain other central texts of late medieval medicine. The number of copies and the accompanying texts bear witness to the importance and authoritative character of *De spermate*, a central work of medical theory in a key area of scholastic interest in twelfth to fifteenth century Europe[36].

University of Jyväskylä and University of Helsinki

36. For transcriptions and other material regarding *De spermate*, see http://www.jyu.fi/Members/merisalo/despermate.htm. For a discussion of several *De spermate* manuscripts, see Merisalo O. 'Transition and Continuity in Medical Manuscripts (Thirteenth-Fifteenth Centuries)', in: J. Hamesse & J. Meirinhos (eds.), *Continuities and Disruptions* (forthcoming).

Pieter BEULLENS

ARISTOTLE, HIS TRANSLATORS, AND THE FORMATION OF ICHTHYOLOGIC NOMENCLATURE

1. Introduction*

According to the book of *Genesis*, on the fifth day of creation God made the creatures of the water and the air according to their species, and He saw that it was good[1].

> Creavitque Deus cete grandia et omnem animam viventem atque mota-bilem, quam produxerant aquae in species suas, et omne volatile secun-dum genus suum. Et vidit Deus quod esset bonum.

However, when He brought all living beings to man in order to have them named, there is no trace of the water animals[2].

> Formatis igitur Dominus Deus de humo cunctis animantibus terrae et universis volatilibus caeli, adduxit ea ad Adam, ut videret quid vocaret ea; omne enim, quod vocavit Adam animae viventis, ipsum est nomen eius.

Must it be understood that, as a result of this divine negligence, fish appear as an unspecified multitude and that not a single species is called by its proper name in the entire Bible[3]?

* This article presents some results of the research carried out within the frame-work of the project 'Translating biological treatises in the Middle Ages: a comparative study of translators and their strategies in the learned language and the vernacular' (OT/01/06), financed by the Research Fund of the Catholic University Leuven (K.U.Leuven).

1. *Gen.* 1:21.
2. *Gen.* 2:19.
3. See e.g. 'Immensi maris litora quantum inter se distant; hoc natantium et reptilium innumeras continet tum maximas tum minimas animantes.' (*Ps.* 103:25). Conrad Gesner quotes the passage in Latin, Greek, and Hebrew as the motto for the ichthyologic section of his *Historiae animalium*, and he takes care to note with painstaking philological accuracy that in the Hebrew version the psalm is found as number 104, see Gesner C., *Historiæ Animalium Liber IIII. qui est de Piscium & Aquatilium animantium natura* (Tiguri, 1558).

In general, the keenest interest most humans ever display in fish is in its edible form. Therefore, the main difficulty for the study of the fishes' way of life and the variety of species obviously is the problematic access to their natural habitat. In addition to this fact, several other aspects hamper an adequate inventory. The territory of many species is restricted to certain areas, while others migrate and thus live in different places at different times, alternately in seawater and in freshwater. Some kinds have different names according to the stages in their life cycle, or depending on their sex. In contrast, there exist local semantic variants for identical species, and the fisher, the fishmonger, and the consumer do not necessarily use the same vocabulary.

It is for this reason somewhat surprising to see that Aristotle in his zoological works cites more than one hundred fish species by name. One can hardly imagine the famous philosopher gathering information while hanging upside down from a fishing boat and looking through a bucket at the marvelous world of sea creatures below him. Most scholars accept that he collected most of his knowledge by talking to fishermen and by studying their catches.

In fact, some medieval legends record that his pupil, Alexander the Great, was also eager to investigate nature. In the romance by the so-called archpriest Leo, the king tells in his own words that he studied the fishes from a glass diving clock[4]:

> 'Faciamus doleum olovitreum et ligetur catenis et regant eum fortissimi milites.' Hoc audito praecepi cito talia facere et tali modo perquisivi profundum maris. Vidi ibi diversas figuras piscium atque ex diversis coloribus; vidi ibi et alias bestias habentes imagines terrenarum bestiarum ambulantes per fundum maris quasi quadrupedia. Veniebant usque ad me et fugiebant. Vidi ibi et alias ammirabiles causas, quas recitare non possum.

And Alexander Neckam, who praises the king's scientific enthusiasm, regrets that he did not put the natures of fishes to writing[5]!

4. 'Let us make a glass barrel and have it tied up with chains and let the strongest soldiers control it.' 'After I heard this, I ordered to do these things quickly and in that way I investigated the depth of the sea. There, I saw different forms of fishes and with different colors. There, I also saw other beasts that have the appearance of earth creatures and walk over the bottom of the sea as if they were quadrupeds. They came to me and then fled. There, I also saw other wonderful things that I cannot mention.' Pfister F. (ed.), *Der Alexanderroman des Archipresbyters Leo* (Heidelberg, 1913), p. 126-127 (III,27^{II},6).

5. 'The Macedon was such a sedulous investigator of nature, that he observed the natures and habits of the fishes in a glass cask. ... Alas! he did not put the natures of the fishes to writing.' Wright Th. (ed.), *Alexander Neckam. De naturis rerum libri duo* (London, 1863), p. 142 (II,21).

> Tam sedulus autem naturarum indagator extitit Macedo, ut in vase vitreo naturas et consuetudines piscium deprehenderet... Pro dolor! naturas piscium scripto non commendavit.

Still, Aristotle probably did most of the research in this field on his own. His extraordinary knowledge of fishes and their behavior even formed the core of a scholarly discussion throughout the twentieth century concerning the time of the redaction of his zoological works. Starting from a suggestion by Thompson in the preface to his 1910 Oxford Translation of the *History of Animals*, Lee in a classic article defended the opinion that Aristotle collected the bulk of the material he used for his zoological works during his stay at Lesbos around 345[6]. Although Solmsen[7] later countered most of Lee's arguments, 'the fishes of Lesbos' have earned their place in many biographical introductions to Aristotle[8].

2. Translations and interpretations

Even if Aristotle's achievement was formidable, Georges Cuvier in his *Histoire des sciences naturelles*, posthumously published in 1841, sharply formulated its principal shortcoming[9].

> Comme tous les naturalistes anciens, Aristote semble avoir cru que les noms, par lesquels on désignait de son temps les animaux, ne changeraient jamais, et il se borne presque toujours à nommer les espèces, sans en faire la description. Il en résulte qu'il est extrêmement difficile, dans beaucoup de cas, de reconnaître les animaux qu'Aristote dénomme.

Precisely this aspect often challenges the creativity of modern translators. How, then, did their medieval and renaissance counterparts tackle the problems posed by this very specific vocabulary? We will look at the

6. Thompson D.W, *Historia Animalium*, The Complete Works of Aristotle IV (Oxford, 1910), p. VII; Lee H.D.P., 'Place Names and the Data of Aristotle's Biological Works', *Classical Quarterly* 42 (1948), p. 61-67.

7. Solmsen F., 'The Fishes of Lesbos and their Alleged Significance for the Development of Aristotle', *Hermes* 106 (1978), p. 467-484. Reaction by Lee H.D.P., 'The Fishes of Lesbos Again', in: A. Gotthelf (ed.), *Aristotle on Nature and Living Things* (Pittsburgh – Bristol, 1985), p. 3-8.

8. E.g. Barnes J., *Aristotle* (Oxford, 1996[2]), p. 8-9.

9. 'As all ancient naturalists, Aristotle seems to have believed that the names by which in his time animals were identified would never change. He nearly always limits himself to naming the species without describing them. As a result, it is in many cases extremely difficult to identify the animals that Aristotle mentions.' Cuvier G., *Histoire des sciences naturelles depuis leur origine jusqu'à nos jours chez tous les peuples connus*, M. de Saint-Agy (ed.), première partie, comprenant les siècles antérieurs au 16e de notre ère, tome premier (Paris, 1841), p. 166.

ichthyologic terminology in the Latin translations of Aristotle's *De historia animalium* by William of Moerbeke, George of Trebizond, and Theodore Gaza. As it will turn out, their approaches in this matter offer a privileged insight in their translation strategies in general.

There are surprisingly few species, indeed, for which the three translators have identical interpretations. Their versions tend to be uniform for some edible kinds, such as the eel (ἔγχελυς, 'anguilla'), or in cases where etymology leads the way to an adequate translation, as for the gilthead (χρύσοφρυς, 'aurata'). Apart from these instances and a few others where analogy with land-dwelling animals suggested an obvious translation, their strategies differ substantially.

2.1. William of Moerbeke

William of Moerbeke played a pioneering role in the thirteenth century[10]. As he probably had very few research tools at his disposal, he often preferred to give a transcription of the word he found in his Greek model. 'Guglielmo voleva far legger il greco in lettere latine', that is, according to Lorenzo Minio-Paluello, Moerbeke's conception of the highest possible degree of accuracy[11]. In the case of marine fishes, his strategy corresponds with the practice in antiquity, as up to 80% of the Latin words in that field were borrowed from the Greek[12]. Although this figure does not distinguish the rate in prose texts from the use in poetry, where a greater leniency towards Greek elements exists, it is clear that the Romans already struggled with the matter.

Yet, as an additional assistance to his readers, Moerbeke occasionally incorporated terms from the vernacular to translate or gloss Greek words for which he had a concept but no fitting Latin expression. In doing so, he followed the authority of Hildegard of Bingen, Albert the Great, and even Alan of Lille, who before him had introduced vernacular 'fishes'

10. See the critical edition *De historia animalium. Translatio Guillelmi de Morbeka.* Pars prima: Lib. I-V, ed. P. Beullens & F. Bossier, Aristoteles Latinus XVII 2.I.1. (Leiden – Boston – Köln, 2000). The second volume is in print.

11. 'William wanted to allow the Greek to be read in Latin characters'. Lorenzo Minio-Paluello compares Moerbeke's strategy with Boethius's in these terms in 'Boezio, Giacomo Veneto, Guglielmo di Moerbeke, Jacques Lefèvre d'Etaples e gli Elenchi Sofistici', *Rivista di filosofia neo-scolastica* 44 (1952), p. 408; reprinted in Minio-Paluello L., *Opuscula. The Latin Aristotle* (Amsterdam, 1972), p. 174.

12. Battisti C., 'Sui grecismi dell'ittionimia latina. Accatti ed adattamenti di ittionimi greci nella tradizione latina', *Bollettino dell'atlante linguistico mediterraneo* 2-3 (1960-61), p. 61-95; 4 (1962), p. 37-52.

into the Latin idiom[13]. More than half of all vernacular expressions used in Moerbeke's *De historia animalium* are names of fish species. I suspect that the translator acquired his ichthyologic knowledge from the typical menu in the abbey, as the ecclesiastical calendar forbade the consumption of mammal meat on many days[14].

The catalogue of vernacular words leaves us with an interesting glimpse into his linguistic background[15]. The origin of the words 'raie' (skate), 'plaiz' or 'plais' (turbot), 'tenche marine' (wrasse) and 'treille' (in various spellings, red mullet) must be the north of France and Picardy. Moreover, the word 'carpre' (carp), which Moerbeke may have picked up either in Picardy or in Flanders, provides a particularly fascinating piece of evidence that was overlooked in an interesting study by Hoffmann. In it, records of archaeological finds are correlated with literary testimonies in order to show that in antiquity the carp was native only to rivers running into the Black Sea, slowly spreading towards the north and west of Europe during the Middle Ages[16]. Moerbeke's use of a vernacular word demonstrates that the species had become indigenous to the northern regions by the mid-thirteenth century.

If we reasonably accept that Moerbeke's ichthyologic vocabulary gradually grew after he had traveled to Italy and Greece, it may even offer some additional information about his whereabouts later on in his life. The suggestion that he visited Southern Italy or Sicily may be corroborated by the use of the word 'spinula' as a translation for λάβραξ (sea bass). The term is attested in Calabria, the Salentine peninsula, and Sicily, while variants 'spinola' and 'spigola' are used in Naples and Rome respectively[17]. It is tempting to link this information – with due caution – to the history of the famous Greek manuscript Vindobonensis phil. gr. 100 (*J*). It was in Moerbeke's possession for several years and may have

13. Valente V., 'Ittionimia medievale', *Bollettino dell'atlante linguistico mediterraneo* 10-12 (1968-70), p. 271-282.

14. For the role of fish in medieval life in general, see Zug Tucci H., 'Il mondo medievale dei pesci tra realtà e immaginazione', in: *L'uomo di fronte al mondo animale nell'alto medioevo*, tomo primo. Settimane di studio del centro italiano di studi sull'alto medioevo 31 (Spoleto, 1985), p. 291-372.

15. *De historia animalium*, ed. Beullens & Bossier, p. xiii-xv.

16. Hoffmann R.C., 'Remains and Verbal Evidence of Carp (*Cyprinus carpio*) in Medieval Europe', in: W. Van Neer (ed.), *Fish Exploitation in the Past*. Proceedings of the 7th meeting of the ICAZ Fish Remains Working Group. Annales du Musée Royal de l'Afrique Centrale, Sciences Zoologiques nº 274 (Tervuren, 1994), p. 139-150.

17. Rossi A., 'I nomi dei pesci, dei crostacei e dei molluschi nei trattati cinquecenteschi in volgare di culinaria, dietetica e medicina', *Studi di lessicografia italiana* 6 (1984), p. 67-232, s.v. spinola (p. 128); Vinja V., 'Analyse du contenu des ichtyonymes. Les noms de *Labrax lupus* et de *Chrysophrys aurata*', *Studia Romanica et Anglica Zagrabiensia* 25-26 (1968), p. 5-22.

been the model for his translation of *De historia animalium*. On a leaf in this manuscript, Moerbeke copied a list of works by Hippocrates from the Vaticanus gr. 276. That manuscript was in Southern Italy or Sicily in the mid-thirteenth century, where it had served as model for some of Bartholomew of Messina's translations[18]. It might well be that Moerbeke visited that region and picked up the term *spirula* there and then.

2.2. George of Trebizond

Whatever sources Moerbeke used or strategies he employed, they remain implicit and one has to infer them from his work in which all self-referentiality is missing, at least in his Aristotle translations.

The situation is quite the opposite for two other Graeco-Latin versions of the same text that were made in the fifteenth century. There, the self-awareness of the translators found its expression in the form of elaborate prefaces, in which they discussed in detail the difficulties and strategies of their work. George of Trebizond, who finished his translation in 1450, at the twilight of the Middle Ages, spent an entire paragraph on the way he coped with Greek fishes[19].

> Cum igitur vir ille animalium innumeras pene species nominet propriaque ipsorum exponat accidentia, quid traducens feci cum ut plurimum de quibus diceret ignorarem? Collegi omnia que legi apud Latinos de animalibus, pluraque in dies querebam, et ab accidentibus certo multa conspicatus sum argumento, ut, verbi gratia, glanydas dicit Aristoteles

18. *Metaphysica. Lib. I-XIV. Recensio et Translatio Guillelmi de Moerbeka*, vol. 1: *Praefatio*, ed. Vuillemin-Diem, Aristoteles Latinus XXV 3.1. (Leiden – New York – Köln, 1995), p.184.

19. 'Since that man (*viz.* Aristotle) names almost uncountable animal species and describes their own characteristics, what did I do while translating, as I frequently did not know about which he spoke? I collected everything I read in the writings of the Latin authors and I daily looked for more. I perceived many things with conclusive evidence from the characteristics, as, e.g., Aristotle says of the male fishes of the *glanis* species, that if they come across their young brood, they immediately halt for the defense of their brood and do not go from there until the young are of age to care for themselves. I found in the Latin authors that 'siluri' exactly do this. I therefore did not hesitate and put 'silurus' in all places where in the Greek the 'glanis' fish is named. Furthermore, we modeled many terms according to the meaning of the words and we did not do it without model, as several Latin authors call a 'belone' 'acus' and 'erythini' 'rubeoli', modeling Greek words in Latin. As in some cases neither procedure could conveniently be applied, we put the Greek words themselves, often slightly modified to fit the Latin form.' The preface is edited by Dittmeyer L., 'Untersuchungen über einige Handschriften und lateinische Übersetzungen der aristotelischen Tiergeschichte', in: *Programm des k. neuen Gymnasiums zu Würzburg für das Studienjahr 1901/1902* (Würzburg, 1902), p. 16-18 and by Monfasani J., *Collectanea Trapezuntiana: Texts, Documents, and Bibliographies of George of Trebizond* (Binghamton, N.Y., 1984), p. 298-300, doc. XCIII.

pisces masculos, si tenello fetui occurrerint suo, residere ad defensionem fetus illico nec inde abire donec se parvuli tutari iam pro etate queant. Hoc ipsum siluros facere apud Latinos inveni. Non dubitavi ergo ubicunque glanydes pisces Grece dicuntur siluros ponere. Multa preterea secundum verborum potestatem expressimus, neque id nobis absque exemplo factum est cum non nulli Latinorum auctorum belonem acum et rubeolos erythinos appellent, verba Greca Latine exprimentes. In non nullis etiam cum neutrum fieri commode posset, Greca ipsa posuimus paululum plerumque ad formam Latinam commutata.

According to this preface, Trebizond used three different strategies in his search for an adequate ichthyologic vocabulary. First, he searched the ancient Latin authors for corresponding characteristics. The most obvious place to look for zoological details in Latin antiquity was the elder Pliny, who entirely devoted books IX and XXXII of his *Naturalis historia* to the life of water-dwelling animals and the cures made from them. In book IX, the description of the 'silurus' ('Silurus mas solus omnium edita custodit ova, saepe et quinquagenis diebus, ne absumantur ab aliis[20].') indeed matches Aristotle's description of the breeding-habits of the 'glanis' in book VI of his *De historia animalium* in some facts[21].

Οὗ δ' ἂν πλεῖστον ἐκτέκωσιν, φοφυλακεῖ ὁ ἄρρην ... προσεδρεύει ὁ ἄρρην καὶ τετταράκοντα καὶ πεντήκοντα ἡμέρας, ὅπως μὴ κατεσθίηται ὁ γόνος ὑπὸ τῶν παρατυχόντων ἰχθύων.

Aristotle takes up the same report with more details in book VIII[22].

Τῶν δὲ ποταμίων ὁ γλάνις ὁ ἄρρην περὶ τὰ τέκνα ποιεῖται ἐπιμέλειαν πολλήν· ἡ μὲν γὰρ θήλεια τεκοῦσα ἀπαλλάττεται, ὁ δ' ἄρρην οὗ ἂν πλεῖστον συστῇ τοῦ κυήματος φοφυλακεῖ παραμένων, οὐδεμίαν ὠφέλειαν ἄλλην παρεχόμενος πλὴν ἐρύκων τἆλλα ἰχθύδια μὴ διαρπάσωσι τὸν γόνον· καὶ τοῦτο ποιεῖ

20. 'With the catfish alone of all species the male guards the eggs, often for as long as fifty days at a time, to prevent their being eaten by other fish.' (Pliny, *Naturalis historia*, IX, 165; translation: H. Rackham).

21. 'Wherever the bulk of the eggs were laid, the male stays on guard; ... the male keeps watch for as much as forty or fifty days, to prevent the brood being devoured by any small fishes that may happen to come by.' (Aristotle, *De historia animalium*, VI, 568b13-17; translation: A.L. Peck).

22. 'Among the river fishes the male glanis takes great care of the young. For the female separates from them after giving birth, but the male stays by and guards the eggs wherever the most part of the spawn has set; he confines his defensive efforts to warding off the other small fishes from stealing the embryos; and this he does for as many as forty or fifty days, until the embryos have grown and can make their escape from the other fishes.' (Aristotle, *De historia animalium*, VIII[IX], 621a20-27; translation: D.M. Balme).

ἡμέρας καὶ τετταράκοντα καὶ πεντήκοντα ἕως ἂν αὐξηθεὶς ὁ γόνος δύνηται διαφεύγειν ἀπὸ τῶν ἄλλων ἰχθύων.

It is worth mentioning that Pliny used the word 'glanis' as well, although in a different context[23], but already Albert the Great accepted the identification of both names[24].

Trebizond's second choice was to create new words with analogous etymologies. Pliny did that, equaling 'acus' and 'belone' (pipefish)[25], but 'rubeolus' for ἐρυθ(ρ)ῖνος was an independent creation by Trebizond, inspired by the red color suggested in its name[26].

Finally, when other methods failed, Trebizond left room to transcribe Greek terms in Latin characters and adapt the inflection. Although this practice is close to what Moerbeke did before him, parallels with the work of Theodore Gaza yield significant results concerning their mutual view on translating.

2.3. Theodore Gaza

After all, Gaza's preface to his translation of *De animalibus*, written in the early 1470's, was clearly intended as a break with the past. He sneered at his predecessors for their lack of knowledge of the Latin language and ignorance of the Aristotelian science[27].

> Videbam permulta errasse interpretes tum imperitia linguę tum aristotelicę disciplinę inscicia.

He specifically blamed them for using Greek instead of Latin terms, for inappropriately putting the wrong names to the wrong concepts, and for incompetently inventing new ones[28].

> Aut enim gręce illi dixerunt quę latine audire homines latini desiderant, aut rerum aliarum nomina aliis impropie tribuerunt, aut nova ipsi inepte finxerunt.

23. Pliny, *Naturalis historia*, IX, 145.

24. 'Piscis fluviatilis, quem glanin quidem Graeci, nos autem silurum vocamus.' *Albertus Magnus. De animalibus libri XXVI*, vol. I., ed. H. Stadler, Beiträge zur Geschichte der Philosophie des Mittelalters 15 (Münster, 1916), p. 262.

25. *Naturalis historia*, IX, 166: 'acus sive belone'; but also: 'belonae quos aculeatos vocamus' (*Naturalis historia*, XXXII, 145).

26. See Ovid, *Halieutica*, 104: 'caeruleaque rubens erythinus in unda'.

27. 'I saw that many translators strayed by lack of knowledge of the language as well as ignorance of the Aristotelian science.' Quoted from the *editio princeps*, Venice, Iohannes de Colonia and Iohannes Manthen, 1476 (*Gesamtkatalog der Wiegendrucke*. Teil I, zweite Auflage (Stuttgart/New York, 1968), n. 2350), fol. a 2 v.

28. 'For they either said in Greek what Latin people wanted to hear in Latin, or they inappropriately put the wrong names to the wrong concepts, or incompetently invented new ones themselves.' *Editio princeps*, fol. a 2 v.

As he had obtained no help from their work, he claimed, he had gone through a host of ancient Latin authors in search of the adequate information[29].

> Me plurimum elaborasse in his libris interpretandis fateor, cum nihil a primis interpretibus illis iuvari possem, sed omnia ex codicibus veterum autorum petere necesse haberem. Lectione longa, notationeque varia, Plinium, Cornelium, Columellam, Varronem, Catonem, M. Tullium, Apuleium, Gel<l>ium, Senecam, complures alios linguę latinę autores evolvere diligentius oportuit.

These authors were the sources of his translations into Latin, which was a foreign language for him. If he introduced new or Greek words, he wanted them to sound familiar and close to the Latin practice, as for those words that either were generally used in Latin, as 'delphinus' or 'camelus', or for which there was no Latin equivalent readily available, as for the freshwater fish 'balerus'. In some cases, one should not even despise the vernacular[30].

> Nominat usu veterum probatissimorum autorum genera animalium. Si quid novi imponit, ita inserit ut familiare cognatumque id quoque videri possit. Si quid gręce appellat, id non durum occurrit; aut enim usitatum et ut ita dixerim vagum latinis est, quale est nomen delphini, cameli, elephantis, crocodili, ichneumonis, aspidis, salamandrę et reliquorum generis eiusdem, aut ita sonat ut non minus latinum esse quam gręcum videatur... Quis latacem ambiguam illam beluam aut balerum fluviatilem piscem ita nominans peregrinum sonare videbitur? Nec vero contemnendum vulgus interdum est.

Although Gaza claimed that he did not want new words to sound harsh to Latin ears, some of his choices were remarkable and innovative. For many of them, Pliny must have been the source of inspiration. Gaza introduced the word 'limaria' for the well-known πηλαμύς, probably hinting

29. 'I admit that I have laboured hard on the translation of these books, since I could not get help from those prior translators, but had to get everything from the manuscripts of antique authors. By long reading and varied annotating I very carefully read through Pliny, Cornelius, Columella, Varro, Cato, Cicero, Apuleius, Gellius, Seneca, and several other authors of the Latin language.' *Editio princeps*, fol. a 2° r. Apparently the 'Cornelius' Gaza is referring to must be the medical author Cornelius Celsus.

30. 'He names the animal species according to the use of antique and canonized authors. If he assigns a new name, he inserts it so that it could also look familiar and related. If he names something in Greek, it does not sound harsh. For it must either be commonly used and, so to speak, widely spread among the Latins, as are the words 'delphinus', 'camelus', 'elephas', 'crocodilus', 'ichneumon', 'aspis', 'salamandra', and others similar, or sound so that it seems to be no less Latin than Greek... Who would think that by naming that doubtful animal 'latax' or the river fish 'balerus', one sounds foreign? Sometimes not even the vernacular should be despised.' *Editio princeps*, fol. a 4° v.

at an etymology already suggested by Pliny[31]. Possibly also under Plinian influence Gaza's first rendition for ξιφίας was 'mucro', before he settled for the more obvious 'gladius'[32]. More extraordinary still was his translation of ζύγαινα by 'libella', a word found in Pliny, but in a different context altogether. It designates a T-shaped carpenter's tool, similar to the head of the fish and to the insect's larva, which eventually drew its name from it[33]. The transcriptions of the Greek names of these fishes are also present in Pliny, except for the last one, which Saint Ambrose used in his *Hexameron*, but Gaza apparently considered them inappropriate in a Latin translation[34].

There are other cases where Gaza clearly challenged the authority of the Latin authors. Following Trebizond, he identified the ἔλλοψ with the Latin 'accipenser', although Pliny explicitly states that, according to Ovid, the 'helops' is unknown in Mediterranean waters, which shows that those who think it is the same as the 'acipenser' are mistaken[35]. Gaza held even more reservations against the Greek manuscripts of Aristotle's text, which, he felt, were corrupted during their transmission. Therefore, he did not hesitate to emend the text whenever he judged it necessary. As he probably was unaware of the existence of the fishing-frog, he consistently considered the word βάτραχος as a corruption of βάτος in those passages that clearly implied a fish and not an amphibian, and thus translated it as 'raia'[36].

One of Gaza's main objections to the work of his predecessors was their incompetent invention of new names. Yet, it is precisely in this area that he definitely outdid Trebizond[37]. While the translation for ἀθερίνη at its first occurrence was introduced with a periphrastic formula as 'atherina quam aristam appello' and then became 'arista' and 'aristula', the

31. 'Limosae vero aut e luto pelamydes incipiunt vocari.' (Pliny, *Naturalis historia*, IX, 47).

32. '...xiphian, id est gladium, rostro mucronato esse' (Pliny, *Naturalis historia*, XXXII, 15).

33. Kemnes N.A., 'Über die Herkunft des Namens Libella für die Odonaten sowie die ältere Geschichte dieses Namens', *Lychnos. Lärdomshistoriska samfundets årsbok* (1942), p. 76-86.

34. Thompson D.W., *A Glossary of Greek Fishes* (London, 1947), p. 178-180 and 74-75.

35. Pliny, *Naturalis historia*, XXXII, 153 (quoting Ovid, *Halieutica*, 96): 'Helopem dicit esse nostris incognitum undis, ex quo apparet falli eos, qui eundem acipenserem existimaverint.'

36. There are at least eight examples of this practice in *De historia animalium*: 505a6; 564a18; 565b29; 570b30; 606b16; 620b12; 13; 27.

37. Citations of Gaza's translation are from the *editio princeps*, Trebizond's version is quoted according to ms. Firenze, Bibl. Laur., 84,9.

renditions 'vitta' for ταινία and 'fidicula' for κίθαρος were daring, although they came tremendously close to Trebizond's 'coronula' and 'fidicula'. A further comparison, though, demonstrates that the latter was more conservative in accepting these innovations: χαλκίς first became 'esrarius' and 'fabrilis', but then Trebizond returned to a transcription of the Greek, while Gaza used 'erica', 'spuria', and just once the Greek 'chalcis'. There is a similar evolution in the rendering of χάννη, where Trebizond first launched 'oscita' and then settled for a transcription, while Gaza stuck to the innovative 'hyatula'. As a final example, the translation pedigree of the ὀρφός is very instructive. Moerbeke transliterated the term, but used vernacular renderings as well, demonstrating that he had a particular fish in mind[38]. Trebizond handled the word with observable difficulty. He first omitted the word from his translation, leaving a mark in the margin, later transliterated it on three occasions and once made a marginal note of the Greek word[39]. Gaza introduced the term 'cernua', which may well be an implementation of his warning not to despise the vernacular: in some Italian regions, the ὀρφός was known as 'cerna'[40].

3. Philology and Ichthyology

Surely, Aristotle's translators fought a Herculean struggle with the ichthyologic vocabulary, and that may well be representative for many other fields of science[41]. Yet, was there any influence of the Aristotelian terminology on the vocabulary of scientific discourse from the late fifteenth century onwards?

In the centuries before, Moerbeke's translation was copied, but the few commentaries that were made and survive focus on the scholastic analysis of the text's structure, as if the identification of concrete animal species were not of interest. The following passage from the work of Gerard of Breuil (Gerardus de Brolio), the only scholastic commentator to have used Moerbeke's translation, may be indicative of this attitude[42].

38. *De historia animalium*, ed. Beullens & Bossier, p. xv.

39. Monfasani, *Collectanea Trapezuntiana*, p. 624, ad 591a11.

40. Folena G., 'Per la storia della ittionimia volgare. Tra cucina e scienza naturale', *Bollettino dell'atlante linguistico mediterraneo* 5-6 (1963-64), p. 61-137. Thompson, *A Glossary*, p. 187-188.

41. Compare with the article by Craig Martin, 'Scientific Terminology and the Effects of Humanism: Renaissance Translations of *Meteorologica* IV and the Commentary Tradition' in this volume.

42. 'Partus autem neque omnes. First he treats the mating and breeding times of the egg-laying fishes, then he tells about them according to the species of the different fishes

> 'Partus autem neque omnes. Primo tangit tempora partus et gestationis piscium ovantium in generali, secundo narrat ea specialiter in diversis piscibus ibi *Gestant autem quos vocant*. Et illa pars habet tot partes quot sunt modi piscium in quibus inducit ibi et patent omnia nisi quantum barbaries nominum impedit.'

The publication of Gaza's version in 1476 brought greater attention to zoological studies, though the approach was quite different from our modern viewpoints. It mainly consisted of the inventory and harmonization of the ancient sources that were made available in the first decades of printing. In his encyclopedia *Commentariorum urbanorum libri XXXVIII*, published in 1504, Raphael of Volterra (Raphael Volaterranus) tellingly ranks the chapters about the natural world under the heading 'Philologia'. Hermolao Barbaro (Hermolaus Barbarus) and Paolo Giovio (Paulus Jovius) are two fascinating representatives of this philological method.

3.1. Hermolao Barbaro

Hermolao Barbaro was probably the first to make extensive use of Aristotle's zoological writings for philological purposes. In his *Castigationes Plinianae et in Pomponium Melam*, published in 1492, he suggested more than 5,000 emendations of the then extant editions of Pliny's *Naturalis historia*[43]. He therefore not only used old and authoritative manuscripts of Pliny's text, but also the testimonies of other writers, including Aristotle. Barbaro referred to both the original Greek text, which was not yet available in printed form by then, and Gaza's translation. Being a pupil of Gaza's, he gave much weight to his testimony for the constitution of Pliny's text. As he essentially inverted the method adopted by Gaza for his translation of Aristotle, he came ever so close to circular reasoning. Beatus Rhenanus, who had noticed this methodological problem, criticized him on that point in his preface to the work of a successor of

from the paragraph *Gestant autem quos vocant*. And that passage has as many parts as there are sorts of fishes that he considers, and everything is understandable insofar as the strangeness of the names does not impede it.' Commentary by Gerard of Breuil ad 570a25-31 sqq., quoted according to ms. Milan, Bibl. Ambr., Z.252.Sup, fol. 77v.

43. First edition: Romae, E. Silber, 1492-3. Critical edition: Pozzi G. (ed.), *Hermolai Barbari Castigationes Plinianae et in Pomponium Melam*. Thesaurus Mundi. Bibliotheca Scriptorum Latinorum Mediæ et Recentioris Ætatis 11;14;18-19, 4 vol. (Patavii, 1973-1979). The publication of the critical edition would have greatly pleased Georges Cuvier, who noted about the work: 'Il est d'ailleurs imprimé en caractères gothiques assez difficiles à lire'. (Cuvier G., *Histoire des sciences…*, deuxième partie, comprenant les 16e et 17e siècles, tome deuxième, p. 166).

Barbaro's, Franciscus Massarius (Francesco Massari), who wrote *Castigationes* on Pliny's book IX[44].

> Placet porro mihi quod in castigando ad vetusta exemplaria confugit. Nam ille nunquam pro meritis satis laudatus Hermolaus Barbarus, ut in annotando omneis adhuc vicit qui Plinio manus sunt admoliti, ita in castigando dimissis antiquis manu scriptis libris, ex autoribus saepe lectionem corruptam restituere conatus, frustra laborat.

Barbaro also quite rightly assessed the value of vernacular languages for the identification of fishes. In a discussion of the term 'psitta', which he equated with 'solea', he referred to the word *lingulaca* used by Plautus (*Cas.* 497-98) and Varro (*Ling.* V 77) for a flatfish and added that on the coast of Germany and 'Belgium' the sole was called 'tongen', which means 'tongue'[45].

> Est ergo psitta soleis aspectu proxima. Soleas Graeci, quantum capere coniectura possum, buglosson hoc est bovis linguam vocant. Nam quem Plautus atque Varro lingulacam vocant, aut solea est aut ei non absimilis. Moveor quod Graeci fere semper psittam cum mentione buglossi iungunt, ut et hac parte Plinius: 'Psittam – inquit – et soleam'... Accedit quod ad Oceanum Germanicum et Belgicum non aliter vocari soleam piscem animadverti quam significante linguam vocabulo, id est tongen.

3.2. Paulo Giovio

Paulo Giovio's interest was not a philological one in the first place. He admitted that he got his inspiration for the treatise *De Romanis piscibus libellus* from 1524 from his presence at the papal banquets[46]. Yet, in the

44. 'Furthermore, it pleases me that he has recourse to antique copies for correction. For this man Hermolaus Barbarus, who never gets enough praise according to his merits, in the field of commentary beat all who have yet put their hand on Pliny, while on matters of correction he labours in vain, as he dismissed the antique manuscripts and often tried to restore a corrupt reading from authorities.' Dedicatory letter by Beatus Rhenanus in Massarius F., *In nonum Plinii De naturali historia Castigationes et Annotationes* (Basileae, 1537), fol. AA2v.

45. 'So the 'psitta' closely resembles the 'soleae'. The Greeks call the 'solea', as far as I can grasp the reasoning, 'buglossos', that is a cow's tongue. For the fish that Plautus and Varro call 'lingulaca' is either the 'solea' or a very similar fish. I am struck that the Greeks almost always combine 'psitta' with the mentioning of 'buglossos', as also Pliny on our side says: 'Psitta and solea'... Moreover, I noticed that on the coasts of the German and Belgian Ocean the fish 'solea' gets no other name than the word that means 'tongue', that is 'tongen''. Pozzi, *Hermolai Barbari Castigationes*, p. 607-8.

46. See Zimmermann T.C. Price, 'Renaissance Symposia' in: S. Bertelli & G. Ramakus (eds.), *Essays presented to Myron P. Gilmore*, vol. I: *History* (Florence, 1978), p. 363-374. First edition: Iovius P., *De Romanis Piscibus libellus* (Romae, 1524). Critical edition: Travi E. & Penco M. (eds.), *Dialogi et descriptiones*. Pauli Iovii Opera 9 (Roma, 1984),

work he primarily tried to identify the fishes that were served at these occasions, using written sources and displaying an impressive literacy. He not only quoted from ancient and contemporary writers, including Raphael of Volterra and Hermolao Barbaro, but also from less evident medieval sources as Albert the Great, Avicenna, and Peter of Abano (Petrus Aponensis). Among the ancient authors, Aristotle evidently took pride of place, and quotations always came from the translation by Theodore Gaza with only minor points of criticism.

Giovio underpinned his search for the correct identification of Roman fishes by listing the species names in Roman, Ligurian, Etruscan, Venetian, French, and Spanish vernaculars. Later scientific writers often borrowed precisely these elements from his work. Interestingly, Giovio also conjectured about the particular value of the Spanish language, which, according to him, preserved many uncorrupted elements from the Latin language[47].

> Sed id me plurimum movet, quod spigola hodie in Hispania lupus appelletur, quae una terrarum latinam linguam magna ex parte incorruptam conservavit.

Although Giovio produced an elegant little work, Cardinal Louis de Bourbon, to whom he had dedicated the work, was not impressed and rewarded the author with 'a fabulous benefice on the island Thule, beyond the Orcades', which convinced Giovio to return to the historical works that he had begun[48].

p. 1-64 (the edition of *De piscibus* is by E. Travi). I have not seen the German translation and commentary by T. Baumann (Mannhein Univ. diss., 1993). The treatment of the work in Bäumer Ä., *Zoologie der Renaissance – Renaissance der Zoologie*, Geschichte der Biologie, Band 2 (Frankfurt am Main – Bern – New York – Paris, 1991), p. 347-349, is marred by several serious mistakes and has to be used with the utmost caution: Giovio's preface is not dated April 4th, 1523, but March 29th, 1524 ('.IIII. CALENDAS APRILIS. M.D.XX.IIII.'); he states that he will not imitate Pliny ('Neque ipsum Plinium civem meum penitus imitabor'), while Bäumer writes that he will take him as his example ('Allerdings nimmt sich Giovio nicht die Griechen zum Vorbild, sondern Plinius'); the 'capo' is not treated in ch. 6, but in ch. 16; Giovio does not identify the 'sturio' with the 'lupus' of the ancients, but attributes this identification to Raphael of Volterra; Giovio does not say he treated the identification of the 'lupus' in ch. 4 unclearly, but extensively ('diffuse monstravimus'), which makes Bäumer's point about the early typesetting of ch. 4 and Giovio's changed opinion about the identity of the 'lupus' quite ludicrous.

47. 'I am deeply struck by the fact that the 'spigola' in present day Spain is called 'lupus'. It is the only country in the world to have preserved the Latin language for a large part uncorrupted.' Travi, *Dialogi et descriptiones*, p. 18.

48. He writes this bitter remark in a letter to Galeazzo Florimonte from 1551: 'But, to tell you the truth, after the labour on fishes I appealed to the cardinal de Bourbon, to whom I dedicated the book and who rewarded it with a fabulous benefice on the island Thule, beyond the Orcades. I regretted that I had uselessly browsed what could be browsed and that I was condemned to the costs and I returned, according to my talent, to the work I had started on the *Istoria*'; Ferrero G.G. (ed.), *Lettere*, vol. II., Pauli Iovii Opera 2 (Roma, 1958), lettera 379, p. 204-206.

> Ma, a dirvi il vero, dopo che la fatica de' pesci m'andò busa col Cardinal di Borbone, al quale dedicai il libro, rimunerandomi esso con un beneficio fabuloso situato nell'isola Thile, oltre l'Orcadi, pentendomi d'aver scartabellato lo scartabellabile indarno et esserne condennato nelle spese, mi ritornai secondo il mio genio sopra il cominciato lavoro dell'Istoria.

Still the public thought otherwise and Giovio's treatise was reprinted in Rome and in Basle in 1527, in Antwerp in 1528, in Basle in 1531, and in Strasbourg in 1534[49]. In the latter edition, it formed an ichthyologic 'reader' in combination with the Latin translation of Oppian's *Halieutica* by Lorenzo Lippi (Laurentius Lippius) and books IX and XXXII of Pliny's *Naturalis historia*.

3.3. Dictionaries and wordlists

Shortly after these works had appeared, other works including ichthyologic vocabulary found their way to the public. Their presentation as dictionaries and wordlists emphasized the philological approach to the subject. In 1544, Charles Estienne (Carolus Stephanus) published a dictionary of biology called *De Latinis et Graecis nominibus arborum, fruticum, herbarum, piscium, et avium liber ... cum Gallica eorum nominum appellatione*. Aristotle topped the list of authorities and, evidently, Gaza's translation was the source of the Latin equivalents. There were other similar displays of interest in vernacular biological vocabulary: Pierre Gilles (Petrus Gyllius) added a list of fish names from Marseille to his 1533 Latin translation and compilation of Aelian's *De natura animalium* and Léger Duchesne (Leodegarius a Quercu) drew up a list of French biological vocabulary similar to that of Charles Estienne[50].

The public interest in these publications was undeniable. Estienne's work got four editions in the space of ten years, while Gilles' translation and wordlist were reprinted in the first year after its *editio princeps*.

4. From the great naturalists to Linnaeus

In the decade after 1550, five naturalists introduced drastic changes into the study of fishes. Belon, Rondelet, Salviani, Gesner, and Aldrovandi still relied on written sources, but they additionally examined fish

49. Travi, *Dialogi et descriptiones*, p. 7-8.

50. It was printed as an appendix to a wordlist on plants by Jean Brohon. *Editio princeps*: Cadomi, in officina M. Angier ac D. Bouvet, 1541. Later edition: Parisiis, ex officina Nicolai Buffet, 154?.

specimens for their research, which they often acquired during their travels[51]. As a further means of identification, their descriptions were documented with drawings of specimens[52]. Yet, the philological approach remained the cornerstone of their studies, as Salviani's work in particular demonstrated. It opened with a series of synoptic tables, listing the species' names in Greek, Latin, and vernacular, its attributes, and the relevant quotations in Aristotle, Oppian, Pliny, Athenaeus, Aelian, and various authors. According to the preface, references to Aristotle were to the books and chapters from Gaza's Latin version. One can measure Gaza's importance for Salviani's work from the index, which listed 33 references for the item 'Theodori Gazae errores'. Salviani conveniently left out the instances where he approved of Gaza's version.

Gaza's enormous authority also affected Conrad Gesner. In the preface to the ichthyologic volume of his *Historiae animalium*, he justified his choice to coin new words in German if the appropriate ones did not exist, by referring to Gaza's method, who was as innovative to Latin as his was then for the German language[53].

> Proderit autem hic labor meus imprimis grammaticis, cum illis qui pueros docent, tum qui scribendo interpretantur et illi qui forte hoc opus in Epitomen contractum Germanice reddet. Ficto quidem apposite vocabulo, simul animalis forma naturaque uno saepius nomine interpretantur. Licuit hoc Gazae, dum Graeca Latine interpretatur; et laudatus est a viris doctis, etiamsi in more Latinis id antea non fuisset.

In the early eighteenth century Swedish ichthyologist Peter Artedi made the final step towards Linnaeus' binary nomenclature. He, ironically, drowned in the Amsterdam canals in 1735 at the age of thirty. Linnaeus recovered his notes with great difficulty and had them published in Leyden three years later[54]. The first of these five small and rare volumes

51. See the overview by Gudger E.W., 'The five great naturalists of the sixteenth century: Belon, Rondelet, Salviani, Gesner and Aldrovandi: a chapter in the history of Ichthyology', *Isis* 22 (1934), p. 21-40. Aldrovandi's works were printed around 1600, some of them posthumously, but his research stretches to the mid-16th century.

52. Kolb K., *Graveurs, artistes et hommes de science. Essai sur les Traités de Poissons de la Renaissance* (Paris, 1996).

53. 'This labor of mine may in particular be useful for grammarians, those that teach children as well as those who write commentaries, and for the possible translator into German of this work in an epitomized form. By appropriately inventing a word, the form and the nature of an animal were most often represented in one name. Gaza was allowed this procedure when he translated Greek into Latin. And he was praised for it by the scholars, even though it was not a Latin practice before.' Gesner, *Historiae Animalium Liber IIII.*, fol. b 3 r.

54. Artedi P., *Ichthyologia sive opera omnia de piscibus*, 5 vol. (Lugduni Batavorum, 1738).

comprised a history of ichthyologic literature, starting in the eleventh postdiluvian century with the poet Linus[55]. In this historical overview Artedi cited Aristotle according to the chapter division of the 1504 [1503] Aldine edition of Gaza's translation, but Gaza himself got only three lines of biography and his intrinsic value for ichthyologic studies was restricted by the label 'quasi interpres Aristotelis'[56]. Gaza's influence had faded: in Artedi's fourth volume, which listed all available names of fish species in ten languages, most of Gaza's fanciful creations were relegated to a place low on the list.

Linnaeus binary ichthyologic nomenclature, as it was established in the tenth edition of his *Systema naturae*[57], was greatly influenced by Artedi's work, as he testified in the preface: 'Synonyma ab Artedio petenda'. Linnaeus preserved several traditional names, sometimes combining Greek and Latin elements for binary combinations as in the case of the 'Xiphias gladius' or the 'Silurus glanis'. Besides, there also exists a 'Perca cernua'. If the hypothesis is right that Gaza imported the word 'cernua' from the vernacular into Latin, one of his neologisms may still live on in contemporary ichthyology.

Conclusion

Gaza's translation enormously promoted the development of ichthyologic vocabulary in early modern times. His authority often came close to that of Pliny and Aristotle himself. Gaza created a new technical vocabulary. This accomplishment probably caused the total eclips of his predecessors, even in the rare cases where his text received criticism and a comparison with information from William of Moerbeke and George of Trebizond would have been useful. There is only one instance where Moerbeke's choice is compared to Gaza's: Paolo Giovio mentioned that 'the old as well as silly translator of Galen's work' never translated 'galeus' with 'mustella', as opposed to what Gaza would have done, who

55. Artedi had a perfectly acceptable reason for his starting at that point in time, as he explains in his introduction. In fact, there are neither remains nor writings of the antediluvian fathers in the field of science. This may not be surprising after all, as some trustworthy authors assert that the Egyptians invented writing after the Flood. (Artedi, *Ichthyologia*, vol. I, p. 1).

56. Artedi, *Ichthyologia*, vol. I, p. 3-5; 62.

57. Linnaeus C., *Systema naturae per regna tria naturae, secundum classes, ordines, genera, species, cum characteribus, differentiis, synonymis, locis*, 2 vol. (Holmiae, 1758[10]). A photographic reprint of the first volume is available (London, 1956).

always rendered Greek names by Latin words[58]. Giovio certainly was not aware of the fact that the anonymous translator of Galen's *De virtute alimentorum* whom he was referring to had also translated Aristotle's zoological works.

The quotation illustrates two important features that need further research. First, the importance of printing in the spread of scientific knowledge needs additional study. Barbaro used the Greek text of Aristotle's *De historia animalium* in manuscript form, but all works cited by Giovio in his treatise apparently were already available in printing in the early sixteenth century. Moerbeke's Galen translation was printed in 1490 and was not yet superseded by humanistic versions[59], while Gaza's version of the Aristotelian zoology was the only one available without competition since its 1476 *editio princeps*.

Another aspect that deserves fuller investigation is the influence of dietetic works on the development of ichthyologic vocabulary, as the interest in fishes was related to their function as nourishment. In the sixteenth century, the taste of fishes is commonly used as an element of identification, and even Linnaeus mentions those species that are prized as foodstuff in a preliminary note to his *Systema naturae*. Obviously, this aspect does not only hold for fishes: in the same letter in which Giovio complained about the lack of appreciation for his earlier work, he hinted at another project and inquired about the correct Latin translation of several birds, including the 'beccaccia, tanto grata a Francesi'[60]. As it seems, Giovio never wrote a work on ornithology. Nevertheless, he jokingly promised anyone who could help him with this linguistic problem the reward of a pair of socks. The winner may thus have obtained an indispensable garment for the attendance at the banquets 'al fresco', where humanist philologists gathered, 'mingling philosophy and Falerno, periods and perch'[61].

K.U.Leuven

58. 'Neque enim ille vetus pariter ac ineptus Galeni interpres unquam pro galeo mustellam interpretatus est. Quod profecto Theodorus fecisset, qui semper nominibus Graecis, pro virili, latina vocabula reddere consuevit.' Travi, *Dialogi et descriptiones*, p. 49.

59. Durling R., 'A chronological census of Renaissance editions and translations of Galen', *Journal of the Warburg and Courtauld Institute* 24 (1961), p. 230-305, in particular p. 283, no. 13.

60. See note 48.

61. Zimmermann, 'Renaissance Symposia', p. 363.

Iolanda VENTURA

TRANSLATING, COMMENTING, RE-TRANSLATING: SOME CONSIDERATIONS ON THE LATIN TRANSLATIONS OF THE PSEUDO-ARISTOTELIAN *PROBLEMATA* AND THEIR READERS

Introduction: the Latin *Problemata*

In the last decades, both medieval and Renaissance translations have been a topic of particular interest to many scholars. Several studies have been dedicated to Greek-Latin, Arabic-Latin, and Latin-vernacular translations; the aim of these studies was the analysis of the specific vocabulary employed by the different translators, the strategies they developed in order to reproduce and adapt the internal structure of the original text in the new language, and the relationship between text and translation. Last but not least, some studies attempted an evaluation of the success achieved by translations, and their diffusion among medieval and Renaissance libraries[1].

Moreover, the role played by translations in the development of specific branches of knowledge, such as medicine, astronomy, astrology, or for the progress of natural science and philosophy has been acknowledged and stressed[2]. Translations contributed to the enrichment and the updating of the cultural background and the library the readers had at their disposal, as well as to the changing of the image and evaluation of a specific branch of knowledge.

Specific studies have targeted the definition of the personality and activity of representative medieval translators, such as Gerard of

1. Contamine G. (ed.), *Traductions et traducteurs au Moyen Âge* (Paris, 1989); Hamesse J. & Fattori M. (eds.), *Rencontres de cultures dans la philosophie médiévale. Traductions et traducteurs de l'Antiquité tardive au XIVe siècle*, Publications de l'Institut d'Études Médiévales. Textes, Études, Congrès 1 (Louvain-la-Neuve, 1990); Hamesse J. (ed.), *Les traducteurs au travail. Leurs manuscrits et leur méthodes,* FIDEM. Textes et Études du Moyen Âge 18 (Turnhout, 2001).

2. See, for instance, Beyers R. e.a. (eds.), *Tradition et traduction. Les textes philosophiques et scientifiques grecs au Moyen Age latin. Hommage à Fernand Bossier*, Ancient and Medieval Philosophy Series 1, 25 (Leuven, 1999).

Cremona[3], Michael Scot[4], or William of Moerbeke[5], who ranked among the most significant contributors to the transmission of philosophic and scientific culture during the Middle Ages. On the other hand, the translating activity of some Renaissance translators like Leonardo Bruni[6], Lorenzo Valla[7], George of Trebizond[8], and Theodore Gaza was considered for its contribution to the creation and development of a more sophisticated philosophic language and to the birth of the humanistic culture.

Particular attention has also been paid to the vernacular translations, for several reasons[9]. First of all, a vernacular translation of a philosophic, theological, or scientific work shows the interest of a group of readers not only for a specific text, but for the branch of knowledge it belongs to; otherwise, their ignorance of Latin would have hindered their access to them. Second, the task of the vernacular translator cannot just be identified as the transfer of the content of a work into another language; on the contrary, he must provide his readers – virtually excluded from the cultural milieus within which the text is diffused – with information concerning the network of glosses, commentaries, and explanations developed all around it[10]. With this aim, translators like

3. Burnett C., 'The Coherence of the Arabic-Latin Translation Program in Toledo in the Twelfth Century', *Science in Context* 14/1-2 (2001), p. 249-288 (with further bibliography).

4. See Burnett C., 'Michael Scot and the Transmission of Scientific Culture from Toledo to Bologna via the Court of Frederick II Hohenstaufen', *Micrologus 2: Le scienze alla corte di Federico II* (1994), p. 101-126.

5. Brams J. & Vanhamel W. (eds.), *Guillaume de Moerbeke. Recueil d'études à l'occasion du 700e anniversaire de sa mort (1286)*, Ancient and Medieval Philosophy Series 1, 8 (Leuven, 1989).

6. Hankins J., 'Notes on Leonardo Bruni's Translation of the Nicomachean Ethics and Its Reception in the Fifteenth Century', in: Hamesse (ed.), *Les traducteurs au travail...*, p. 427-447.

7. Regoliosi M., *Nel cantiere del Valla. Elaborazione e montaggio delle "Elegantie"* (Roma, 1993).

8. Monfasani J. (ed.), *Collectanea Trapezuntiana. Texts, Documents, and Bibliographies of George of Trebizond*, Medieval and Renaissance Texts and Studies 25 (Binghamton, 1984); Lardet P., 'Georges de Trébizonde traducteur et scholiaste de la *Rhétorique* d'Aristote', in: Hamesse (ed.), *Les traducteurs au travail...*, p. 311-348.

9. Beer J. (ed.), *Translation Theory and Practice in the Middle Ages*, Studies in Medieval Culture 38 (Kalamazoo Michigan, 1997); Bray N. & Sturlese L. (eds.), *Filosofia in volgare nel Medioevo*, FIDEM. Textes et Études du Moyen Âge 21 (Louvain-la-Neuve, 2004).

10. Boucher C., 'De la subtilité en français: vulgarisation et savoir dans les traductions d'*auctoritates* des XIIIe-XIVe siècles', in: R. Voaden e.a. (eds.), *The Medieval Translator. Vol. 8: The Theory and Practice of Translations in the Middle Ages* (Turnhout, 2003), p. 89-99.

Evrart of Conty and Nicole Oresme needed to develop systems and strategies of inclusion and organization of data derived, among others, from glosses and commentaries, in order to provide their readers with a complete information concerning the translated text, its content, and its interpretation[11]. The insertion of data derived from glosses and commentaries into the vernacular translation also has the aim of offering a simplified access to the text and more complete information about its content.

Finally, a specific task of translators to both Latin and the vernacular is the creation and development of a specific vocabulary with the aim of transferring and reproducing concepts and meanings the original language could express, but the new one could not. In this sense, translators also contribute to the progress of a specific language by enriching the vernacular vocabulary of a particular branch of knowledge.

A medieval and Renaissance translation can be considered as the result of the activity of a translator working to provide his readers with a new book, and as a 'product' read and used by its public. In this sense, the translation can be successful or not, can be read or abandoned on the shelf of a library, contributing or not to the scientific, philosophic, and cultural development of a certain period.

Moving from this evaluation of the nature, the role, and the purpose of a translation, I shall explore in this article the aspect of the 'life' of a translated text considered as a cultural product diffused and read through many centuries, and observed in its relationship to its readers. In order to define the ways the readers approached and read the translated text, and interpreted and evaluated the translation they had at their disposal, I will consider the most common form of discussion of a work, viz. the commentaries and the lecture notes or glosses handed over by some manuscripts. With this aim, I will focus on a specific text, that is the Pseudo-Aristotelian *Problemata*, because this work represents, for several reasons, a good example of the advantages and difficulties the public would experience dealing with a translated text. Firstly, by working with the Pseudo-Aristotelian work, we face two sorts of translation,

11. On Evrart de Conty, see the bibliography in De Leemans P. & Goyens M. (eds.), *Aristotle's* Problemata *in Different Times and Tongues*, Mediaevalia Lovaniensia Series 1 Studia XXXIX (Leuven, 2006), p. 310-315.

On Nicole Oresme, cf. Souffrin P. & Segonds A.Ph. (eds.), *Nicolas Oresme. Tradition et innovation chez un intellectuel du XIVe siècle* (Paris, 1988); Caroti S., 'Nicole Oresme. Dalla «quaestio» alla «Glose»: la presenza del dibattito universitario nella glosse del «Livre du ciel et du monde»', in: Bray & Sturlese (eds.), *Filosofia in volgare...*, p. 155-190.

a medieval and a Renaissance one, each of which is characterized by a different approach to the Greek text, by specific strategies of transferring its content and reproducing its form, and by particular translator's aims. Secondly, the specific structure and content of the *Problemata*, a text where scientific data are gathered together and embedded in a logical framework, challenged both translators and readers, and allowed them to deal with the work in different ways. Finally, the intrinsic difficulty of the text, which probably scared many of its potential readers and kept them away from it, also lead the commentators to develop specific strategies to create an access to it and to focus on particular problems.

However, before starting a discussion of some selected commentaries devoted to the Pseudo-Aristotelian *Problemata* from the fourteenth to the seventeenth centuries, it might be useful to briefly sketch the history of the various attempts to translate the work and of its reception throughout the Middle Ages and the Renaissance[12].

The Pseudo-Aristotelian *Problemata Physica* are a collection of questions and answers divided into thirty-eight books whose actual form dates back to the third century BC. The work deals with natural sciences such as medicine, dietetics, physiology, physiognomy, meteorology, and with more theoretical branches of knowledge like mathematics, music, philology and rhetoric[13]. Handed over in different redactions in some manuscripts of the High Middle Ages, known and used by many authors, from Macrobius to the anonymous author of the *Lorscher Arzneibuch*[14], and included in some collections of natural questions, the genuine *Problemata* were first translated from Greek into Latin between 1258 and 1266 by Bartholomew of Messina, who was active at the court of King Manfred of Sicily, son of Frederic II. We do not know anything

12. On the *Problemata*, see *Aristoteles. Problemata Physica*, tr. H. Flashar (Berlin, 1983), p. 295-384 (esp. p. 372-382 on the Latin tradition); *Aristote. Problèmes*, ed. and tr. P. Louis (Paris, 1991-1994), Vol. 1, p. VII-LIV; *Aristotele. Profumi e miasmi*, ed. G. Marenghi (Napoli, 1991). See also the selected bibliography in De Leemans & Goyens (eds.), *Aristotle's Problemata*...,p. 295-317.

13. According to the latest studies, the original bulk of the *Problemata* is represented by a collection of questions written by Aristotle. The text preserved in the manuscripts, however, is not the original one, but the result of a later elaboration carried out in the Aristotelian school. For an overview on the debate concerning the attribution of the *Problemata* to Aristotle, see Aristotele, *Problemi musicali*, ed. G. Marenghi (Firenze, 1957), p. 13-18.

14. On the *Lorscher Arzneibuch*, cf. Stoll U., *Das 'Lorscher Arzneibuch'. Ein medizinisches Kompendium des 8. Jahrhunderts (Codex Bambergensis Medicinalis 1). Text, Übersetzung und Fachglossar*, Sudhoffs Archiv, Beihefte 28 (Stuttgart, 1992).

about the life of this translator, who also produced Latin versions of some other Pseudo-Aristotelian texts, among them the *De coloribus*, the *De mirabilibus auscultationibus*, and the *Liber de pomo*[15]. One aspect of his translating activity, however, has been stressed by several scholars: in his translations, he constantly makes use of a strict, rigid *verbum de verbo* way of transferring the Greek text into Latin. In this sense, he continues with a method of translating established before him, and respected by translators like Gerard of Cremona or Burgundio of Pisa. It was in his translation that the Pseudo-Aristotelian *Problemata* were read and diffused from the thirteenth to the fifteenth century. Only in the fifteenth century, Theodore Gaza made a new translation, which was published for the first time in its definitive form in 1475[16]. This humanist translation achieved a certain success; it was printed repeatedly, while the medieval one was published seven times, and only together with Gaza's version and Peter of Abano's commentary. A further Renaissance translation made by George of Trebizond did not succeed in affirming itself against Gaza's, and was never printed; only ten manuscripts preserve it[17].

Bartholomew's translation is preserved in ca. seventy manuscripts, among them we find both copies of the complete text and extracts from single books; the work is either copied alone, or in combination with other Aristotelian or Pseudo-Aristotelian works. Some of the books, particularly the ones dealing with medicine and physiology, were also transmitted separately, and included in miscellaneous manuscripts together with other medical works[18].

Despite the number of manuscripts preserving the Pseudo-Aristotelian *Problemata*, the work can hardly be defined as a success. The diffusion of the text is relatively small, compared with the success of other Aristotelian works like the *Physica*, the *Meteora*, or the Pseudo-Aristotelian *Physiognomica* and *Secretum secretorum*. The translation

15. On Bartholomew, see the bibliography in De Leemans & Goyens (eds.), *Aristotle's Problemata...*, p. 305-307.

16. On Gaza's translation of the *Problemata*, see below, p. 133-135.

17. Monfasani J., 'The Pseudo-Aristotelian *Problemata* and Aristotle's *De animalibus* in the Renaissance', in: A. Grafton & N. Siraisi (eds.), *Natural Particulars. Nature and Disciplines in Renaissance Europe*, Dibner Institute Studies in the History of Science and Technology, 1 (Cambridge-London, 1999), p. 205-247. See also his contribution in: De Leemans & Goyens (eds.), *Aristotle's Problemata...*, p. 275-294.

18. Manuscripts in: Lacombe G. e.a., *Aristoteles Latinus. Codices. Pars Prior* (Rome, 1939); Id., *Pars Posterior* (Cambridge, 1955); Minio-Paluello L., *Aristoteles Latinus. Codices. Supplementa altera* (Bruges-Paris, 1961).

of the *Problemata* was, in fact, completed when the corpus of scientific works written by Aristotle and dealing with zoology, botany, meteorology, had already come to light, due to the work of translators like Gerard of Cremona and Michael Scot. At the same time, William of Moerbeke's translations which were completed during the second half of the thirteenth century deeply influenced the approach to the Aristotelian texts, sometimes substituted former previous translations, and reached the contemporary learned libraries. In this context, the interest the Pseudo-Aristotelian *Problemata* could arouse was, with few exceptions, minimal[19]. Moreover, the content of the Pseudo-Aristotelian work, which can be defined as a sort of philosophical encyclopaedia mixing medicine, zoology, meteorology, etc., might have been perceived as a complex hodgepodge of data which could be found (sometimes in a better developed form) in other of Aristotle's works, or in Hippocrates' or Galen's writings. Finally, the suspicion that the *Problemata* did not belong to the authentic Aristotelian works might have influenced their reception in a negative way[20]. Clues of the debate concerning the attribution of the work to Aristotle can also be found in some commentaries witnessing the doubts of the readers[21].

The complexity of the Latin form, which prejudiced the interpretation of the content, might have also contributed to the fiasco experienced by Bartholomew's translation of the *Problemata*. The rigid

19. One of the few exceptions may be represented by the theory of melancholy outlined in *Problemata* XXX,1. Cf. Klibansky R., Panofsky E. & Saxl F., *Saturn and Melancholy. Studies in the History of Natural Philosophy, Religion and Art* (London, 1964; ital. translation Torino, 1983), p. 19-39, and Rius i Gatell R., 'Sobre la melancolia: M. Ficino', in: *Filosofia y Ciencia en el Renacimiento* (Santiago de Compostela, 1988), p. 49-55. See also Cadden J., 'Preliminary Observations on the Place of the *Problemata* in Medieval Learning', in: De Leemans & Goyens (eds.), *Aristotle's Problemata*..., p. 1-19.

20. Cf. Williams S.J., 'Defining the *Corpus Aristotelicum*: Scholastic Awareness of Aristotelian *Spuria* in the High Middle Ages', *Journal of the Warburg and the Courtauld Institutes* 58 (1995), p. 29-51, esp. p. 45. On the question of authenticity of some Aristotelian works, see Kraye J., 'Aristotle's God and the Authenticity of *De mundo*: An Early Modern Controversy', *Journal of the History of Philosophy* 28 (1990), p. 339-358 (repr. in Kraye J., *Classical Traditions in Renaissance Philosophy*, Variorum Collected Studies Series 743 [Aldershot, 2002], nr. XI).

21. Peter of Abano and, later, Ludovico Settala, agreed that the *Problemata* are based on an authentic Aristotelian bulk, integrated by later authors. The authenticity of the work is also affirmed, among others, by the anonymous glossator of the manuscript Vaticano, B.A.V., Vat. Lat. 2112, who affirms (f. 1r): 'Iste liber, non obstante dicto aliquorum in oppositionem, compilatus fuit ab Aristotile, cum de ipso in secundo Metheororum ac in plerisque locis parvorum naturalium et librorum aliorum facit multociens mentionem [...]'.

verbum de verbo system he used often fails to reproduce the elliptic, complex, and logically structured Greek text, and does not contribute to a clarification of its obscurities as a translation based on paraphrases would have done. Bartholomew's technique of translation was strongly criticized by some Humanists in the Renaissance and by some modern scholars as a sign of incompetence or of ignorance of the text; moreover, it was also considered as too dependent on the Greek original. However, this characteristic should not be overestimated or condemned as a reason for the failure of the translation; in the Middle Ages, it was perceived in another way. The *verbum de verbo* translation strategy did not astonish or disturb medieval readers; on the contrary, a literal translation was a sign of the translator's respect for the text and made the Latin text more reliable for lectures, commentaries, excerptions[22]. King Manfred, for instance, advertised the new translation of Aristotle's works to the students in Paris by pointing to the *servata virginitate* of the text, and consequently its untouched authority[23]. Nevertheless, in the case of the *Problemata*, the perfect reproduction of the syntax and the structure of the Greek text, and the choice to use Latin words representing the perfect equivalent of the correspondent Greek terminology created some problems in the comprehension of the text. The readers of the *Problemata* had to develop strategies in order to explain the real meaning of the single words by creating a system of glosses all around it, and to reconstruct the logical structure of sentences or paragraphs which was easily perceivable in Greek, but completely disfigured by the Latin translation.

Clues to the difficulties caused by the translation of the *Problemata* among their medieval readers can be found in some more or less illustrious

22. Burnett C., 'Translating from Arabic into Latin in the Middle Ages: Theory, Practice, and Criticism', in: S.G. Lofts & P.W. Rosemann (eds.), *Éditer, traduire, interpréter: essais de méthodologie philosophique*, Philosophes Médiévaux 36 (Louvain-la-Neuve, 1997), p. 55-78.

23. 'Dum librorum ergo volumina quorum multifarie multisque modis distincta cyrographa diviciarum nostrarum armaria locupletant, sedula meditatione revolvimus et accurate contemplatione pensamus, compilationes varie, ab Aristotile aliisque philosophis sub Grecis Arabisque vocabulis antiquitus edite, in sermocinalibus et mathematicis disciplinis, nostris aliquando sensibus occurrerunt, quas adhuc originalium dictionum ordinatione consertas et vetustarum vestium, quas eis etas prima contexerat, operimento connectas vel hominis defectus aut operas ad Latine lingue noticiam non perduxit. Volentes igitur ut reverenda tantorum operum senilis auctoritas apud nos non absque multorum commodes vocis organo traduce[re] iuvenescat, ea per viros electos et utriusque lingue prolatione peritos instanter duximus verborum fideliter servata virginitate transferri' (quoted from Burnett, 'Translating from Arabic into Latin...', p. 68, n. 32).

witnesses, which seem to stress the two most difficult aspects of the translation, viz. the internal structure and the vocabulary. Roger Bacon, for instance, heavily criticized the translation, which makes the content incomprehensible and consequently unusable for scientific studies or philosophical debates. This severe judgement characterizes the approach of the English philosopher to many others translations: according to him, not one of the previous or of the contemporary translators had been able to translate the text he aimed at rendering into Latin, or at reproducing its content in a comprehensible way. Roger Bacon, in fact, objected to the procedure of the translation in itself, because the acquisition of science ought to be achieved by reading texts in their original language[24]. However, in the case of the *Problemata*, the English philosopher seems to go further, and to point to specific sections, whose internal incoherence or obscurity does not allow the comprehension of Aristotle's opinion. In his *Perspectiva* I, distinction 8, chapter 2, for instance, he remarks:

> ... Aristoteles in libro problematum debuit nos certificare hoc [scil. quod nos audiamus undique et olfaciamus, sed videmus per oppositionem visibilis respectu visus], nam ibi tangit istud alia sua problemata secreta; sed vel mala translatio vel falsitas exemplaris Graeci vel aliqua alia causa nos impedit in hac parte[25].

Some pages later, in *Perspectiva* II, distinction 3, chapter 4, while dealing with the different perception of the moonlight, he again accuses the text he reads of hiding the real meaning behind an obscure translation:

> ... Cuius causam Aristoteles secundo Celi et mundi et Averrois imperfectis et transitoriis sermonibus occultant; atque Liber problematum Aristotelis Latinus sua translationis obscuritate dubios nos relinquit[26].

In this way, he points to a specific difficulty the readers have to face in dealing with the text: the translation fails to render the explanation of natural phenomena, its obscurity does not allow the use of this source to

24. Lemay R., 'Roger Bacon's Attitude toward the Latin Translations and Translators of the Twelfth and Thirteenth Centuries', in: J. Hackett (ed.), *Roger Bacon and the Sciences. Commemorative Essays*, Studien und Texte zur Geistesgeschichte des Mittelalters 57 (Leiden-New York-Köln, 1997), p. 25-47.

25. Edition: *Roger Bacon and the Origins of Perspectiva in the Middle Ages. A Critical Edition and English Translation of Bacon's Perspectiva with Introduction and Notes by* D.C. Lindberg (Oxford, 1996), here p. 112 and n. 257 p. 362.

26. Ibidem, p. 212 and n. 448 p. 374.

understand, to interpret, and to certify natural processes. Here, his accusations are specifically directed against the structure and the form of the Latin questions, which cannot be used to analyze nature.

Roger Bacon's judgement perfectly describes the difficulties a medieval reader might have experienced while approaching the *Problemata*. However, it may be interesting to analyze the ways in which some more or less famous and learned commentaries expressed the complicated relationship to the obscure language and the complex structure of the Pseudo-Aristotelian work, and found ways to make the text more accessible. In the next pages, I will firstly examine two examples of the ways some medieval readers attempted at commenting and explaining the *Problemata* with the aim of solving the difficulties created by the Latin vocabulary employed by Bartholomew of Messina, and of clarifying the obscurities of its structure. A successive section will leave the Middle Ages and focus on the Renaissance translation accomplished by Theodore Gaza, aiming at analyzing the way in which two early modern commentators deal with it and evaluate the scientific validity of its content and the translator's ability in rendering the Greek text into Latin.

1. The *Problemata* and their Medieval Readers: the commentary of Peter of Abano and the Anonymous Glosses of the MS Erfurt, UFEG, Coll. Amploniana, 4° 237

The Latin tradition of the philosophical commentaries has been analyzed extensively in the recent years[27]. The tradition of the commentaries and glosses devoted to Aristotelian texts like the *De generatione et corruptione*[28], the *Meteora*[29], and the writings *de*

27. Cf., for instance, Del Punta F., 'The Genre of Commentaries in the Middle Ages and its Relation to the Nature and Originality of Medieval Thought', in: J.A. Aertsen & A. Speer (eds.), *Was ist Philosophie im Mittelalter?* Miscellanea Mediaevalia 26 (Berlin-New York, 1998), p. 138-151; Goulet-Cazé M.-O. (ed.), *Le commentaire entre tradition et innovation,* Bibliothèque d'Histoire de la Philosophie (Paris, 2000); Fioravanti G., Leonardi C. & Perfetti S. (eds.), *Il commento filosofico nell'Occidente latino (secoli XIII-XV)*, Rencontres de Philosophie Médiévale 10 (Turnhout, 2002).

28. Thijssen J.M.M.H. & Braakhuis H.A.G. (eds.), *The Commentary Tradition on Aristotle's De generatione et corruptione. Ancient, Medieval, and Early Modern*, Studia Artistarum 9 (Turnhout, 1999).

29. On the commentaries on the *Meteora*, see Martin C., 'Francisco Vallés and the Renaissance Reinterpretations of Aristotle's *Meteorologica* IV as a Medical Text', *Early Science and Medicine* 7/1 (2002), p. 1-30. See also his contribution in this volume. On the

animalibus[30] has become object of particular interest. Bartholomew's translation of the *Problemata*, however, did not enjoy the same success as other Aristotelian or Pseudo-Aristotelian texts. By turning the pages of Charles Lohr's catalogue of the medieval commentaries devoted to Aristotle's writings, we notice that we do not face a rich tradition of expositions; only a few names appear, and it is rather improbable that further researches will add many others to the list[31]. We know, for instance, of some anonymous commentaries preserved in the manuscripts Den Haag, Koninklijke Bibliotheek, MS 70 E 9[32], Gent, Universiteitsbibliotheek, MS 178[33], and Brugge, Stadsbibliotheek, 481[34]. A *Scriptum bonum super tribus libris De problematibus Aristotelis Montipessulanense* was also available in the library of Amplonius Rating de Berka; unfortunately, the codex is now lost[35]. The same is the case with the commentary written by Nicolaus Trevet, a Dominican friar known for his interest in science and classical literature[36].

The first, and most famous, commentator of the *Problemata* was Peter of Abano, a Paduan physician and astrologer who extensively commented on the whole text translated by Bartholomew, and probably also translated

French translation of the work, cf. Ducos J., *La météorologie en français au Moyen Âge* (XIIIe-XIVe siècles) (Paris, 1988); Ducos J., 'L'œuvre de Mahieu le Vilain: traduction et commentaire des "Metéorologiques"', in: Hamesse (ed.), *Les traducteurs au travail…*, p. 285-309. See also her contribution in this volume.

30. On the commentary tradition on *De animalibus*, see Perfetti S., *Aristotle's Zoology and its Renaissance Commentators (1521-1601)* (Leuven, 2000); Perfetti S., 'How and When the Medieval Commentary died out: The Case of Aristotle's Zoological Writings', in: Fioravanti e.a. (eds.), *Il commento filosofico…*, p. 429-443; on the commentaries on *De partibus animalium*, see Perfetti S., '*Docebo vos dubitare*. Il commento inedito di Pietro Pomponazzi al *De partibus animalium* (Bologna 1521-24)', *Documenti e Studi sulla Tradizione Filosofica Medievale* 10 (1999), p. 439-466.

31. Lohr C.H., 'Medieval Latin Aristotle Commentaries', *Traditio* 23 (1967), p. 313-314; 24 (1968), p. 149-245; 26 (1970), p. 135-216; 27 (1971), p. 251-351; 28 (1972), p. 281-396; 29 (1973), p. 93-137; 30 (1974), p. 119-144.

32. *Inventaris van de handschriften van de Koninklijke Bibliotheek. Deel 1: kastnummers 66-71* ('s-Gravenhage, 1988), p. 43.

33. Pattin A., *Repertorium commentariorum Medii Aevii in Aristotelem Latinum quae in Bibliothecis Belgicis asservantur* (Leuven-Leiden, 1978), p. 118.

34. Pattin, *Repertorium commentariorum…*, p. 16-17.

35. Lohr, 'Medieval Latin…', *Traditio* 27 (1971), p. 351.

36. Lohr, 'Medieval Latin…', *Traditio* 28 (1972), p. 306-307. See Ehrle Fr., 'Nikolaus Trivet, sein Leben, seine *Quodlibet* und *Quaestiones ordinariae*', in: *Abhandlungen zur Geschichte der Philosophie des Mittelalters. Festgabe Clemens Baeumker zum 70. Geburtstag (16. September 1923), dargebracht von seinen Freunden und Schülern* (Münster i.W., 1923), p. 1-63, here p. 17.

the *Problemata* written by Alexander of Aphrodisia[37]. The commentary devoted to the *Problemata*, the *Expositio Problematum Aristotelis*, was written by him between the end of the thirteenth century and 1310[38]. The exposition is preserved in numerous manuscripts, some of which belong to the so-called redaction of Jean of Jandun, which is nothing but a copy of Peter's text, with only slight differences, most noticeably in Section XV[39]. Peter's commentary also influenced a third commentary attributed to Walter Burley and preserved, among others, in the manuscript Paris, BNF, lat. 16089[40]. Traces of this last commentary are also found in a specific redaction of the *Problemata* preserved in the manuscripts London, Gray's Inn Library, MS 2, Cambridge, Peterhouse, MS 220, and Oxford, Magdalene College, MS 65. Here, the text of the *Problemata* is not divided into books, but rearranged thematically as a collection of lemmas put in alphabetical order[41]. Finally, Peter of Abano's commentary

37. On Peter of Abano, see the selected bibliography in De Leemans & Goyens (eds.), *Aristotle's Problemata...*, p. 307-310. See esp. Paschetto E., *Pietro d'Abano medico e filosofo* (Firenze, 1984), and Olivieri L., *Pietro d'Abano e il pensiero neolatino: filosofia, scienza e ricerca dell'Aristotele greco tra i secoli XIII e XIV* (Padova, 1988).

38. Peter of Abano, *Expositio Problematum Aristotelis*, in: *Problemata Aristotelis cum dpulici (!) translatione* (Paris, 1520). On this text, see Siraisi N.G., 'The *Expositio Problematum Aristotelis* of Peter of Abano', *Isis* 61 (1970), p. 321-339 and, more recently, Federici Vescovini G., 'L'*Expositio Succinta Problematum Aristotelis* de Pierre d'Abano', in: De Leemans & Goyens (eds.), *Aristotle's Problemata...*, p. 55-70.

39. Kuksewicz Z., 'Les *Problemata* de Pietro d'Abano et leur "rédaction" par Jean de Jandun', *Medioevo* 11 (1985), p. 113-137.

40. Cadden J., '"Nothing Natural Is Shameful": Vestiges of a Debate about Sex and Science in a Group of Late-Medieval Manuscripts', *Speculum* 76/1 (2001), p. 66-89; Van der Lugt M., 'Aristotle's *Problems* in the West: a Contribution to the Study of the Medieval Latin Tradition', in: De Leemans & Goyens (eds.), *Aristotle's Problemata...*, p. 71-111; Biller P., 'Black Women in Medieval Scientific Thought', *Micrologus 13: La pelle umana/ The Human Skin* (2005), p. 477-492.

41. Cf. Cambridge, Peterhouse, MS 220, here f. 95r-95v: 'Rigescere. Quare rigentes sunt lividi sicut senes et ad formationem corporis motici? Responsio primo propter recessum sanguinis ab exterioribus ad interiora sicut in dampnatis. Secundo propter congelationem sanguinis et venarum ex frigore. Nota quod color lividus est quasi color plumbeus in pergameno. Frigor est tremor corporis proveniens ex frigore cum timore. Tertio propter desperationem et dolorem anime, quia exteriora sunt signa interiorum: VIII, 1. Quare rigentes non possunt dormire? Responsio. Primo propter defectum sanguinis et caloris sicut melancolici. Secundo propter ymaginem, tristitiam et timorem. Tertio propter difficultatem et gravitatem anhelitus. Tertio (!) propter tremorem spirituum et dolorem. Quinto propter perturbacionem cordis et cerebri. Sexto propter grossitudinem, gravitatem et frigiditatem humorum. Sompnus est ligatio sensuum proveniens ex consensu vaporum ad cerebrum, unde post prandium est optimum dormire: VIII, 2. Quare in rigore acutiores sunt et qui egritaverunt tristes et irati? Responsio quia stipticius facit infrigidari – IX, 3 –, ubi frigoris per poros apertos post exercitium et laborem sicut mingendo. Secundo propter eventationem et raritatem corporis. Tertio propter tremorem spirituum et nervorum ex labore: VIII, 4.'

was also summarized and transmitted in form of short and extremely simplified explanations; two witnesses of these later re-elaborations of Peter's text are represented by the anonymous commentary preserved in the manuscript Den Haag, Koninklijke Bibliotheek, MS 70 E 9, and by the *Expositio super libro Problematum Aristotelis* attributed to Thomas of Mantevilla, an otherwise unknown author who probably lived in the fourteenth century and belonged to the Dominican Order. This work is preserved in but one manuscript, Napoli, Biblioteca Nazionale, VII.AA.39, f. 47r-116r[42].

Peter of Abano's commentary was first printed in 1474 in Mantua, together with Theodore Gaza's translation of the *Problemata*[43]. This commentary can be defined as the work of a scientist, not a philosopher, and shows the erudition of its author. The encyclopaedic nature and purpose of this commentary can be recognized in Peter's intention of commenting on the content of the text with the help of other texts, and of showing how the natural phenomena displayed in the *Problemata* can be found in other texts. In this sense, he tends to transform his commentary into a kind of encyclopaedia, by inserting the Pseudo-Aristotelian questions into a net of sources dealing with the same topics[44]. Medical and scientific texts are used to gloss the description of natural phenomena involving the world of nature or the human body included in the *Problemata*, and in a certain sense to 'confirm' it. This procedure is particularly necessary, because the Pseudo-Aristotelian work cannot be easily read and understood if the reader is not acquainted with other texts dealing with the same topics discussed in the *Problemata*, or is not versed in all branches of natural sciences and philosophy. As Peter says:

> ... in hoc libro inveniuntur fere totius philosophie per modum cuiusdam allegationis sermones compilati. Ex quo quidem deducitur quod liber hic non potest plene intelligi, nisi ab illo qui philosophiam secundum omnem eius partem inspexerit, propter quod fortasse in eius expositione pigritarunt glossatores[45].

An encyclopaedic background is required of any commentator of the *Problemata*; the same background is gathered together by Peter and put at disposal of his readers through his exposition.

42. Cf. Lohr, 'Medieval Latin ...', *Traditio* 29 (1973), p. 183.

43. Petrus de Abano, *Expositio Problematum Aristotelis quam Petrus edidit paduanus* (Mantua, 1474).

44. Siraisi, 'The *Expositio Problematum*...'.

For our purposes, it is interesting to ask if Peter approaches the Latin text of the *Problemata* as a translation, and if he deals with questions related to the status of the text, or with specific philological problems. At first glance, Peter of Abano does not seem to deal with the translation at all; he does not seem to perceive that the *Problemata* are a translated text. Only in a few occasions like *Probl.*, II, 30 he notices that the original Pseudo-Aristotelian text should be adapted to the European language and mentality, although this adaptation does not correspond neither to the text the Latin culture had at disposal nor probably to the original one. Here, he affirms:

> Notandum est, ut sermo Aristotelis magis nostris climatibus adaptetur, quod per corpus esse nudum potest intelligi corpus modicis vestimentis coopertum, ut lineis aut laneis et levibus. [...] Amplius maxime apud nos videtur cuidam honestati attinere, quod fortassis non est apud calidissimos barbaros[46].

However, more than with the differences between oriental and occidental mentalities, here he seems to be concerned with the problem of the right way in which some concepts can be properly rendered in another milieu. Moreover, a closer look on Peter's commentary suggests that his discussion of the text shows a deep interest in the correct choice of terminology and the right interpretation and exposition of Aristotle's opinion from the side of both the translator and the commentator. First, as far as his methodology and his activity as a commentator is concerned, Peter seems to be more interested in glossing single terms of the *Problemata*, in defining their meaning in a general way and within the context, in discussing the language used in the text, than in explaining the logical process in which the natural phenomenon is described. Particular interest is devoted to technical terms, or to rarely used expressions, such as 'causon' (I, 20)[47], 'farmacia' (I, 42)[48], or the verb 'spumescere' (III, 11)[49], which in a medical context has a particular meaning. Generally,

45. Petrus de Abano, *Expositio Problematum Aristotelis* (Paris, 1520), f. 1rb.

46. *Ibidem*, f. 34va.

47. *Ibidem*, f. 10vb. For the text of the explanation, see below.

48. *Ibidem*, f. 19ra: 'Notandum quod farmacia sumitur pro omni medicina cuiuscumque sit virtutis; hoc apud Galenum sepe. Interdum vero sumitur pro veneno, communiter tamen accipitur pro medicina solutiva solum, et sic accipitur hic.'

49. *Ibidem*, p. 43va: 'Dicitur autem spumescere quoniam principalis dea spermatis ut Venus fuit a spuma generata, sicut dicit Aristoteles in 2° de generatione animalium; fabulant enim poete ipsam natam fuisse ex abscisione testiculorum Saturni et proiectorum in mare, ex quorum agitatione ex undis maris ad scopulos facta inde spuma erit generata. Vel dicitur spumescere transumptive, eo quod per modum spumam generatur in

Peter attempts to find a simpler and more general term which can at the same time reproduce the exact meaning of the word used by Bartholomew (so to say, cover the same semantic area), and be understandable to all his readers, not just the specialists. At the same time, however, he attempts to delimit very carefully a simple Latin term which seems too general to him, or, to set the borders within which it can be used. In the case mentioned above of *causon*, for instance, after having connected the word with the Latin *febris*, he devotes some lines to explain the reasons for the equivalence, and to describe the cases in which a fever can be defined as *causon*:

> Notandum quod causon potest accipi pro omni febre in qua est fervor magnus et adustio, secundum quod nomen causonis incendium importat, ita quod nomine causon comprehendatur causon verus tertiana, continua et sinocus; vel potest accipi causon solum pro causone vero, cuius materia est colerica, aut etiam flegmatica salsa [...][50].

A similar need for specification of the terminology may be noticed in Peter's substitution in the commentary of the term used by Bartholomew in his translation with a more precise one. One interesting case of this kind of substitution has been recently noticed by I. Verdenne in Peter's explanation of the fact that swimmers and divers fill up their ears with a sponge. Here, viz. in commenting *Probl.*, XXXII, 2, Peter comments the sentence 'Aut quia mare superveniens non frangat aures' as follows: '[...] Soluit dicens causam esse ne aqua maris occurrens aurium pulpe immediate rumpat ipsam', where the word 'aures' has been replaced by the more technical, but also more precise 'aurium pulpa'[51].

The interest for the language is particularly evident in the medical books, where the various content related to the medical questions shows a wide background of particular words, and leads the commentator to focus on the choice and the use of words displayed by Bartholomew. Shortly, a net of 'glossed words' seems to characterize the commentary, and to demonstrate the attention paid by the commentator to the vocabulary of the work.

oribus equorum cum stimulantur; sic per motum et confricationem sperma generatur in testiculis, sicut dicit Haly in Tegni, capitulo signorum.'

50. *Ibidem*, f. 10vb.

51. *Ibidem*, f. 249vb. See Verdenne I., 'L'homme sous la mer. La figure du plongeur dans le monde gréco-romain et l'occident médiéval', in: D. James-Raoul & C. Thomasset (eds.), *Dans l'eau, sous l'eau. Le monde aquatique au Moyen Âge*, Cultures et Civilisations Médiévales 25 (Paris, 2002), p. 273-319, here p. 299-302.

However, the reason for devoting attention to the language of the *Problemata* is not only derived from the text itself and in the necessity for commenting it; the *Problemata* is not an 'isolated work', but a text to be related to other texts. The fact that the content of the *Problemata* should be interpreted with the help of other works dealing with similar topics also induces the commentator to verify the correctness of the terminology of the Pseudo-Aristotelian works by comparing it with the vocabulary used in the reference texts. In other terms, glossing, explaining, commenting on the description of a natural phenomenon in the *Problemata* (for instance, a characteristic of the human body, a disease, etc.) and the words used for this aim, also subsumes the explanation of the way in which a tradition of sources (Hippocrates, Galen, etc.) describes the same phenomenon, and a comparison between these texts. Consequently, the correct definition of the expressions used by Pseudo-Aristotle and the comparison with the way of describing natural phenomena shown by parallel sources is the first, necessary step toward the construction of a meta-commentary, an encyclopaedic gloss to the *Problemata* based on the constant reference to other texts.

Moreover, Peter's discussion of the Pseudo-Aristotelian text and of its form is also reflected by his evaluations of the text and his proposals to improve it. On some occasions, for example, the commentator pronounces a severe judgement on the text he is explaining. In *Probl.*, I, 56, for example, he observes that 'Aristotle' – whom he identifies with the translated text – did not care to search for the right words ('...parva cura fuit ei de nominibus...'); the same condemning sentence is pronounced in the comment on *Probl.*, III, 8[52]. Furthermore, in some sections the commentary is used as an occasion to develop a kind of 'philological analysis' with the aim of improving the text; a procedure which reminds us of the approach to classical works we know from the humanists. In *Probl.*, I, 10, for example, Peter begins a section of his discussion by noticing that the form in which the question is preserved might appear to be corrupted, if not compared with a similar point of Hippocrates' *De aere et aquis*[53]. In

52. *Ibidem*, I, 56, f. 25vb: 'Et scias quod Aristoteles hic et ita in sequenti problemate accipit ignem aliquando pro calore innato et spiritu, et interdum pro calore extraneo febrili; parva enim fuit cura ei de nominibus. Intellige itaque hoc problema quia pravum valde'; III, 8, f. 41vb: 'Et nota quod Aristoteles, de nominibus minime sollicitans, non intelligit per ebrios vehementer ebrios; illi enim non possent iudicare, sicut dictum fuit in 2° problemate, et sic neque saporem sentire, sed per ebrios intelligit illos quos parum appellabat ebrios, et per parum ebrios intelligit hic sobrios aut qui modicum ultra hoc biberunt.'

53. *Ibidem*, f. 6rb: 'Et notandum quod istud problema invenitur corruptum apud illos qui inspexerunt tantum librum Afforismorum et non librum De aere et aquis.'

Probl., III, 2 he suggests that the verb 'potant' in most manuscripts should be substituted by the better variant 'peccant'[54]. The same happens in *Probl.*, III, 6, a question concerning the fact that drunken people often tremble. Here Peter observes that the presence of a neutral instead of a masculine ('mixtum' vs. 'mixtus') could have helped to offer a better way of commenting and solving the problem; it is probable, he further observes, that 'mixtum' was the original form[55]. Finally, in *Probl.*, II, 20, Peter points out that a different reading can lead to two different solutions of the question[56]. Comparison of different *lectiones*, suggestions made in order to furnish a better text, judgements of it: all these ways of dealing with the text seem to indicate not only an interest for language and linguistic problems as such, but also for the special language used by Bartholomew in his translation; Peter's commentary can therefore be interpreted as an active discussion with the translation he had on his table.

Nevertheless, some questions remain open and perhaps need some further research. We do not know, for instance, if Peter of Abano's activity as a commentator might have been related to a previous (or contemporary) work as a translator. Peter, as he declares in the Prologue of his commentary to the *Problemata*, moved to Constantinople to learn Greek; there, he informes his readers, he translated another collection of natural questions. A direct confrontation with the problems connected with the translating activity might have influenced his approach to the text of the *Problemata*. Unfortunately, there is no manuscript evidence of translations made by Peter of Abano. Furthermore, we do not know on what kind of text he commented. It may sound like a strange question, but perhaps it is not. In fact, if we consider the number of instances in which Peter glosses the text, makes reference to other forms or *lectiones*, we should at least ask ourselves whether Peter could have used a glossed manuscript, or that he glossed a manuscript, maybe creating a new form of the text[57]. It would

54. *Ibidem*, f. 38vb: 'Notandum est quod littera communiter est "potant", quam credo corruptam. Debet enim dicere "peccant", et ita habetur libro 23, [capitulo] ubi reperitur hoc problema. Quedam enim sic: propter quod parum ebrius insanit magis? Item apparet ex solutione quam dat hic, sed tamen et potant stare potest, ut patebit statim.'

55. *Ibidem*, f. 40va: 'Et scias quod si littera esset mixtum et non mixtus, faciliorem nobis prestaret intellectum, et fortassis fuit mixtum primitus scriptura.'

56. *Ibidem*, f. 31vb: 'Et scias quod quidam libri habent "aut" <...> non habent, et tunc videtur duplex solutio assignari; ponitur tamen littera que dicit 'aut' communior et melior. '

57. See, for instance, the commentary to *Probl.*, I, 22 (*ibidem*, f. 11rb): 'Est autem per ranas frynoidas secundum quamdam glossam marginis intelligendum ranas quarum facies habent signa rubea; huiusmodi vero ranas immo potius ranunculos multotiens vidi, et sunt parvi minores quidem ranis communibus valde, et in parte superiori sunt coloris karopos

be intriguing to search for this manuscript, or at least for the family of manuscripts Peter's copy belonged to.

A final consideration concerns the purpose of linguistic problems within Peter's commentary. As it has been pointed out already by O. Weijers, the commentary tradition of the Middle Ages also includes some techniques which are closely related to the form and the language of the texts, such as the *expositio litterae*, the *notanda*, the *dubia*, everyone of which has its place and function within philosophical commentaries[58]. Which role do these techniques play in Peter's commentary? Do they influence his approach to the text and his way of commenting it? Do they influence his relationship with a translation? It could be interesting to examine in which way Peter uses the traditional techniques of commenting on philosophical texts to comment on the *Problemata*, and on which level they permeate the commentary to the Pseudo-Aristotelian text and the evaluation of its language. Such researches may also help to define better the influence of the tradition of the philosophical commentaries on Peter of Abano's explanation and to outline the relationship between natural science and philosophy within his text.

Some difficulties in the reading the Pseudo-Aristotelian *Problemata* are not only witnessed by the main commentators, but also by common readers or simple copyists. Their doubts, their need for clarification, explanation, and interpretation are usually reflected by the apparatus of glosses and marginalia accompanying the manuscripts. As it has already been pointed out, the glosses 'represent the first halting attempts at interpretations by readers new to difficult texts, and enable us to see the beginnings of the assimilation of Aristotle'[59]. Moreover, glosses demonstrate how different sorts of readers approached the Aristotelian works: glosses reflect, for instance, the contents of lecture courses, or simply the doubts of ordinary users of the Aristotelian texts. As far as the *Problemata* concerns, we notice that the collection was not often supplied with glosses, and that the apparatus of marginalia

vel cineritii seu cusi, sub ventre vero habent colores citrinos aureos; quos non raro contingit in Italia videre post pluviam immediate estu precedente vehementi saltantes per loca prius arida pulverulenta, et habent parvam caudam subtilem.'

58. Weijers O., 'La structure des commentaries philosophiques à la Faculté des Arts: quelques observations', in: Fioravanti e.a. (eds.), *Il commento filosofico...*, p. 17-41.

59. Dod B.G., 'Aristoteles Latinus', in: N. Kretzmann, A. Kenny, J. Pinborg & E. Stump (eds.), *The Cambridge History of Later Medieval Philosophy. From the Rediscovery of Aristotle to the Disintegration of Scholasticism, 1100-1600* (Cambridge, 1982), p. 45-79, here p. 73.

does not accompany the text throughout the manuscripts. According to the list of manuscripts provided by the *Aristoteles Latinus*, the following codices preserve, among others, a set of marginal or interlinear glosses: Klosterneuburg, Stiftsbibliothek, 677 [A.L.[1] 44], Praha, Národní Knihovna, IV. D. 6 [A.L.[1] 201], Saint-Omer, Bibliothèque Municipale, 598 [A.L.[1] 426], and Città del Vaticano, B.A.V., Ottob. Lat. 2083 [A.L.[2] 1764]. At the moment, it is not possible for me to offer any complete evaluation, neither of the nature and the purpose of these glosses, nor of the personality of the glossators and his interests; furthermore, it should be analyzed in which milieu the different readers lived, to which type of manuscripts (private copy, University exemplar, private collections) the specific codex belonged, and which sections of the *Problemata* especially attracted the interest of the users. Only complete comparative studies will provide us with more information about these unknown readers.

As far as the content of the glosses is concerned, some unknown readers seem to concentrate on the words chosen by Bartholomew and on their use, by explaining or emending them, or by attaching a more known or understandable synonym. By consulting the manuscripts, it is possible to distinguish between two different functions of glosses, a 'commenting' one, mostly marginal, and an interlinear 'explaining' one. A special attention deserve the emendations, which show the philological skills of the reader and his ability to reconstruct the text, but also the relationship he had with the book he was consulting.

In order to exemplify the nature and the function of the glosses in showing the relationship of the readers with the translated text, I will focus on a concrete example, namely the manuscript Erfurt, Universitätsbibliothek Erfurt/Gotha, Collectio Amploniana, 4° 237 [A.L.[1] 897]. Here, the text of the *Problemata* is preserved at f. 1r-104r, and begins with *Probl.*, II, 6. The manuscript shows some marginal glosses commenting the content of the work with the help of parallel sources, while the interlinear ones are concentrated on single words aiming at providing explanations by adding synonyms to rare or obscure words which translated the Greek terms perfectly but were not immediately intelligible to the medieval readers; in some few cases, the glosses suggest a more appropriate synonym. These glosses are often accompanied with *scilicet* or *id est*. Among these glosses, we can excerpt the following examples: by commenting *Probl.*, II, 40, the anonymous reader annotated the word 'efferimus' with the gloss 'id est emittimus', while the term 'haustione' in *Probl.*, II, 33 is accompanied by a more comprehensible 'id est expurgatione'. Moreover, some other problematic expressions are simplified by accurate annotations, such as in the case of *Probl.*, III, 20 the adjective 'distenso' received a note 'id est velociter per aerem emisso', or, in *Probl.*, VII, 6 the ambiguous 'stuporem' is explained by the gloss 'id est congelationem', which points to the appropriate meaning within the context. However, another kind of annotations, which we may define as 'syntactic', also attracted my attention: here, we are not observing the difficulties of a reader facing an unusual or inappropriate vocabulary, but a complex sentence structure. Bartholomew's *verbum de verbo* translation probably created some problems, especially to the readers who were not specialized in the logically constructed language of the Pseudo-Aristotelian *Problemata*. In some cases, verbs or adjectives were not accompanied by nouns, since the subject was mentioned in the beginning of the sentence; in these places, the unknown hand of the Erfurt manuscript made a note between the lines, recalling the missing subject. While the latter type of gloss can be justified with the need of reconstructing the logical structure of the questions included in the *Problemata*, the former one, which focuses upon the vocabulary employed by the translator, leads us to ask if the presence of specific words in the translation corresponds to the need from the side of the reader of glossing it. Moreover, we should ask if the insertion of a gloss can be referred to the specific cultural background of a particular reader, or if the glossators point to the same words, creating a sort of corpus of typical glosses reflecting the same problems. Moreover, it may be possible that the readers feel the necessity of glossing a particular word both because they are well acquainted with the context they are dealing with (for instance, with medicine), and, consequently, because they know other works explaining the same phenomenon described in the Pseudo-Aristotelian *Problemata* with different (or maybe better) terms. We have already seen that Peter of Abano considers the *Problemata* within a network of other texts, e.g. explaining the medical part with the help of Hippocrates or Galen; in this way, he creates a sort of encyclopaedic commentary. Do the unknown glossators of the *Problemata* do the same, without developing a complete commentary? And which sections are particularly affected by this sort of marginal explanations? I argue that this kind of analysis would provide us with more details about the personality and the culture of the anonymous readers of the *Problemata*.

2. Reading the *Problemata* in Early Modern Time: the Commentaries of Ludovico Settala and Giulio Guastavini

The approach to the *Problemata*, as well as the way of reading and commenting other Aristotelian works dealing with natural science, consistently changed during the Renaissance. The philosophic culture of the fifteenth and the sixteenth centuries is witnessed by a deep renewal in the library of the Aristotelian text[60]. Firstly, the increase of the knowledge of Greek allowed more people to read the Aristotelian works in the original language. Secondly, the patronage system promoted a rich production of translations, especially in Italy. Finally, Renaissance authors and translators did not limit themselves to translation, but also discussed the value and purpose of translations by trying to determine the best way of translating a work, and to define the relationship between text and translation, or between original language and new one[61]. New translations slowly supplanted the medieval ones, not only among the book readers, but also in the University milieu[62]. Particularly important for the reception and the diffusion of the 'Renaissance Aristotle' were the changes experienced in the biological works of the philosopher, which were retranslated during the fifteenth century. Among them, the *Problemata* represent a very special case.

In 1473, Theodore Gaza's first translation of the *Problemata* was published in Mantua; the second revised edition was posthumously published in 1475[63]. This work, completed in its first redaction around 1454, was a

60. On Aristotle in the Renaissance, see, for instance, Schmitt C.B., *Aristotle and the Renaissance*, Martin Classical Lectures 83 (Cambridge Mass., 1983), p. 64-89 and correspondent notes; Schmitt C.B., *The Aristotelian Tradition and Renaissance University*, Variorum Collected Studies Series 203 (London, 1984); Kessler E., 'The Transformation of Aristotelianism during the Renaissance', and Murdoch J.E., 'From the Medieval to the Renaissance Aristotle', both in: J. Henry & S. Hutton (eds.), *New Perspectives on Renaissance Thought. Essays in the History of Science, Education, and Philosophy in memory of Charles B. Schmitt* (London, 1990), respectively p. 137-150 and p. 163-176; Pade M. (ed.), *Renaissance Readings of the Corpus Aristotelicum*, Renæssancestudier 9 (Copenhagen, 2000); Bianchi L, *Studi sull'aristotelismo nel Rinascimento*, Subsidia Mediaevalia Patavina 5 (Padova, 2003).

61. Regoliosi M., '"Mercatura optimarum artium". La traduzione secondo Lorenzo Valla', in: Hamesse (ed.), *Les traducteurs au travail...*, p. 449-470.

62. Schmitt C.B., 'Philosophy and Science in Sixteenth-Century Italian Universities', in: Schmitt, *The Aristotelian Tradition...*, nr. XV. On Aristotle's use in Italian Universities, see Kraye J., 'La filosofia nelle università italiane del XVI secolo', in: C. Vasoli & P.C. Pissavino (eds.), *Le filosofie del Rinascimento* (Milano, 2002), p. 350-373.

63. Monfasani, 'The Pseudo-Aristotelian *Problemata*...'.

product of the new humanistic theory of translation. According to the authors following this theory of translation firstly recommended by Leonardo Bruni, and later used by Joannes Argyropoulos, Joachim Périon and others, despicable medieval *de verbo ad verbum* translations like those made by William of Moerbeke had to be replaced by new ones which should present specific characteristics. Among them, the use of a perfect Latin form was particularly important; here both words and sentence structures were not transliterated from Greek, but chosen according to the ideal of Ciceronian Latin; that means, everything expressed in Greek should be adequately expressed in Latin. Moreover, the 'actualization' of the text through the Latin translation required the translator to become acquainted with the cultural context in which the text had been created and to acquire all competences necessary for a complete discussion of it. In short, a translator had to become an expert of Latin grammar, syntax and vocabulary, a philosopher specialized in the text he aimed at translating, a skilled philologist able to discuss and to correct a potentially corrupted Greek text[64].

Theodore Gaza put Bruni's theory into practice; one may say he took it to its limits. Refusing to produce a Latin double of the original Greek work, Theodore worked at the same time at the Pseudo-Aristotelian text he had at disposal and at the translation in progress, trying to provide his readers with a perfect Ciceronian Latin translation and to re-establish a perfect Greek text of the *Problemata* which was not preserved in the available manuscripts. To achieve this goal, he refused *a priori* the medieval model of the *verbum de verbo* translation, decided to render the Pseudo-Aristotelian text in Latin through pretty looking (but mostly inexact in their content) paraphrases and elegant circumlocutions, changed structure and content of questions and books, and tried to offer with his translation a new edition of the Greek text, reconstructed through a collation of various manuscripts or the divination of the supposed correct readings.

The result of his work was a Latin text derived from a new, artificial version of the Greek original. Aim of this contribution, however, is not to analyze or to assess Gaza's translation or to retrace its history. On

64. On Renaissance translations of Aristotle's works, see Copenhaver B., 'Translation, Terminology and Style in Philosophical Discourse', in: C.B. Schmitt, Q. Skinner, E. Kessler, and J. Kraye (eds.), *The Cambridge History of Renaissance Philosophy* (Cambridge, 1988), p. 77-110, and Cortesi M., 'La tecnica del tradurre presso gli Umanisti', in: C. Leonardi & B. Munk Olsen (eds.), *The Classical Tradition in the Middle Ages and the Renaissance*, Biblioteca di Medioevo Latino 15 (Spoleto, 1995), p. 143-168.

these topics, recent studies of scholars like J. Monfasani and S. Perfetti have already shown how Gaza produced his translations[65]. They have also sketched the reception of his translations of the *Problemata* and of Aristotle's *De animalibus* among Renaissance commentators, by stressing how the new Latin Aristotelian texts had a big success but were strongly criticized, or at least severely judged, both by commentators versed in philosophy and natural science and by philologists and experts of Latin grammar and syntax. Among these critics, we find Angelo Poliziano, who attacked Gaza in his *Miscellaneorum centuria*[66]. Therefore, by observing the reception of Gaza's works, it seems that the big success of the 'products' derived from the new humanist ideal of translation did not correspond to a positive evaluation of the ideals permeating them.

Several reasons for this apparent contradiction between official success and negative attitude towards the translation can be found: the nature of the translation itself, certainly, but also the milieu(s) in which the translated text circulated, and the development in the contemporary culture and in the access to the text. However, I would like to point particularly to one aspect. Actually, the success and the failure of Gaza's translations, also of the Latin version of the *Problemata*, is closely connected with the progressive emergence of the Greek text, with its final affirmation, and with the changes in the relationships between Greek and Latin in respect of the reception of the Aristotelian works. Let us try to sketch the situation. Between the second half of the fifteenth and the end of the seventeenth century, a radical change in the ways of approaching Aristotelian texts took place. If Theodore Gaza had tried to re-establish a new, correct Greek text by working on the Latin translation, or maybe to 'recall' it by tracing back the Greek through the translation, the further cultural development first led to an equivalence of the importance of Greek and Latin within the bilingual edition published throughout the sixteenth century, and finally to a final affirmation of the Greek language among the editions of the Aristotelian texts. In short terms, the dictatorship of Latin was transformed into a Latin-Greek diarchy until an overthrow pushed

65. Monfasani, 'The Pseudo-Aristotelian *Problemata*...'; Perfetti S., '*Cultius atque integrius*. Teodoro Gaza, traduttore umanistico del *De partibus animalium*', *Rinascimento* IIª serie, 35 (1995), p. 253-286; Perfetti S., 'Metamorfosi di una traduzione: Agostino Nifo revisore dei 'De animalibus' gaziani', *Medioevo* 22 (1996), p. 259-30.

66. Angelus Politianus, *Miscellaneorum centuria*, I, 90, in: *Opera omnia* (Basel, 1533, repr. Torino, 1971), vol. I, p. 301-303. See Monfasani, 'The Pseudo-Aristotelian *Problemata*...', p. 211 and note 49.

Latin down from its throne. In fact, during the second half of the sixteenth century, many Greek-Latin editions of the Aristotelian works were published and the former humanist translations corrected, commented, and annotated[67]. After 1600, the number of Latin and Greek-Latin editions became rare, and slowly started to disappear. This phenomenon characterized the whole corpus of Aristotelian works, and the Pseudo-Aristotelian *Problemata* as well. As far as we know, Gaza's translation had a monopoly on the Aristotelian translations until the end of the seventeenth century; a last attempt of re-translating the work into Latin was made by the Jesuit Silvestro Mauro, who published his translation in the fifth volume of his edition of Aristotle's works in 1668[68]. However, at this time the attention was more concentrated on the Greek original version.

In this context, where the clash between Greek and Latin was ending with the affirmation of the original version to which the Renaissance culture goes back, some commentaries on the *Problemata* were written[69]. The texts we can consider are not numerous. The work did not enjoy a big success, nor did it attract particular attention among commentators. Beside the intrinsic difficulty of the text, which discouraged every explanation, the influence exerted on the readers by Peter of Abano probably played a certain role, although the commentary was not reprinted after the sixteenth century[70]. Among the printed commentaries, we only find the ones written by Ludovico Settala and Giulio Guastavini, to which I will come back later, the *Urbanae disputationes in primam Aristotelis Problematum sectionem*, published in Rome in 1630 by Ioannes Manelfi[71], and the *Problemata XXXIV ex X primis sectionibus Problematum Aristotelicorum decerpta*, a selection of problems accompanied by short explanations written by Laurent Forer SJ, and printed in Dillingen in

67. On Aristotle's editions of the Renaissance, see Kraye J., 'The Printing History of Aristotle in the Fifteenth Century: a Bibliographical Approach to Renaissance Philosophy', *Renaissance Studies* 9 (1995), p. 189-211 (repr. in Kraye, *Classical Traditions...*, nr. XV).

68. *Aristotelis Opera, quae extant omnia, brevi paraphrasi, ac litterae perpetuo inhaerente expositione illustratae a Sylvestro Mauro* (Romae, 1688), 6 volumes. The *Problemata* are included in the fifth volume.

69. On the commentary in the Renaissance, see Buck A. & Herding O. (eds.), *Der Kommentar in der Renaissance*, Deutsche Forschungsgemeinschaft, Kommission für Humanismusforschung 1 (Bonn, 1975). See also Lohr C., 'Renaissance Latin Translations of the Greek Commentaries on Aristotle', in: J. Kraye & W.F. Stone (eds.), *Humanism and Early Modern Philosophy* (London-New York, 2000), p. 24-40.

70. The last print of Peter's commentary was, in fact, issued in 1520.

71. Ioannes Manelfi also wrote some works on medical matter (among them, a treatise *De febribus*) and edited, in 1654, Niccolò Leoniceno's Latin translation of Hippocrates' *Aphorismi*.

1615. As far as the unprinted commentaries are concerned, by turning, for instance, the pages of the repertory of Latin commentaries published by Charles Lohr in 1988[72], we find mentions of the following commentaries: 1) the *Problemata Aristotelis explicata et determinata* written by Basilius Alamanni SJ (lost)[73]; 2) the *Problematum partes XI* (translation and notes) by Nicasius Ellebodius, preserved in two manuscripts in the Biblioteca Ambrosiana in Milan[74]; 3) two commentaries devoted to respectively *Probl.*, IV, 2 (*In Aristotelis problema II sectionis IV de melancholicis*) and XXV, 12 (*In Aristotelis problema XII sectionis XXV*) written by Octavianus Ferrarius, witnessed by three manuscripts also preserved in the Ambrosiana[75]; 4) the *Adnotationes in Aristotelis Problemata* of Benedetto Manzoli, handed over by another codex of the Ambrosiana[76]; 5) a *Commentarium in Aristotelis Problemata*, also preserved in the Ambrosiana, and composed by Dominicus Montesaurus[77]; 6) the commentary written by Robertus Titius, viz. the Italian erudite Roberto Titi (1551-1609), who also composed a work on controversies called *Locorum controversorum libri X* (published in 1588); this commentary is handed over by the ms. Oxford, Bodleian Library, MS D'Orville 54[78]. It is interesting to note that items 1-5 were all written in the North of Italy (more precisely, in Milan) between the second half of the sixteenth century and the first half of the seventeenth; most of them were the result of the work done by Italian Jesuits. Some of these works are preserved in manuscripts formerly owned by Gian Vincenzo Pinelli[79]. On the other

72. Lohr C.H., *Latin Aristotle Commentaries. II: Renaissance Authors*, Unione Accademica Nazionale; Subsidia al «Corpus Philosophorum Medii Aevi» 6 (Firenze, 1988).

73. *Ibidem*, p. 7-8.

74. *Ibidem*, p. 133-134. On Nicasius Ellebodius, see Wagner D., *Zur Biographie des Nicasius Ellebodius († 1577) und zu seinen „Notae" zu den aristotelischen Magna Moralia*, Sitzungsberichte der Heidelberger Akademie der Wissenschaften, Philosophisch-Historische Klasse, Jg. 1973 Abh. 5 (Heidelberg, 1973).

75. *Ibidem*, p. 147. Octavianus Ferrarius was also professor at the University of Pavia. Among his students, we find Ludovico Settala, author of a commentary on the *Problemata* (see below), who praised Ferrarius by calling him *praeceptor incomparabilis*. However, I was not able to find any mention of Ferrarius' commentary in Settala's explanation of the Pseudo-Aristotelian work.

76. Not mentioned in Lohr's catalogue.

77. Lohr, *Latin Aristotle Commentaries...*, p. 271.

78. *Ibidem*, p. 458. For a brief description of this manuscript, see *A Summary Catalogue of Western Manuscripts in the Bodleian Library of Oxford which have not hitherto been catalogued in the Quarto Series, with References to the Oriental and Other Manuscripts*, by F. Madan M.A. (Oxford, 1897), vol. IV, p. 51-52.

79. On Gian Vincenzo Pinelli and his library, see Rivolta A., *Contributo ad uno studio sulla biblioteca di Gian Vincenzo Pinelli* (Fribourg, 1914).

hand, there is no evidence of an interest for the *Problemata* in the Coimbra commentaries of Aristotle's writings.

Titi's commentary shows a particular situation; the text handed over by the D'Orville manuscript actually consists of two parts, the first (f. 1r-152v) being a collection of short lecture notes related to the sections from 1 to 15 (up to Problem 4) of the *Problemata*, while the second (f. 153r-196r) presents a more developed explanation (possibly, a first draft of a commentary Titi had in mind to write) of Problems 1 to 55 of the first book. At first glance, it is possible to notice that specific questions related to translation, or to exact or wrong ways of rendering a Greek word or expression into Latin, form the bulk of Titi's notes and the main interest of his approach to the text, while a general explanation of the structure and the content of the Pseudo-Aristotelian questions is left aside. Not only does he examine Gaza's translation closely; he often criticizes it. In the case of *Probl.*, I, 7, he observes that Gaza's Latin version 'non omnino aperit mentem consiliumque philosophi, immo innubuit magis'[80]! Moreover, he often attempts a comparison between Theodore's choices as a translator and the decision taken by a *vetus interpres*, whose name he does not know but whose text he could probably take from one of the editions of the sixteenth century including both the medieval and the humanist translation[81]. Peter of Abano also attracts some criticism; he is sometimes accused of having perverted the sense of the *Problemata*. Nevertheless, Titi admits that Peter is not always responsible for these misunderstandings, since the Latin text he had at his disposal was corrupted and unreliable[82]. Titi's commentary does not consist only of criticism, he also attempts a discussion of the text he reads. He believes that the best way to comment the *Problemata* and to restore the sense of the single questions is to compare them with some parallel sources, such as Hippocrates or Galen. This comparison, however, does not concern the Latin versions of the works, but the original Greek text, which should contribute towards a reconstruction and interpretation of the *Problemata*. Only after identifying the general meaning of a Greek expression and determining its value within the context of the original text, it is possible to discern which is the most exact (or the best) way of translating it, and thus choose the most reliable translation.

80. MS D'Orville 54, f. 4r.
81. Cf. for instance the commentary on *Probl.*, VIII, 6 on f. 79v-80r.
82. Cf. the comment on *Probl.*, II, 29 on f. 29v.

As far as the diffusion and the use of the *Problemata* are concerned, we do not know of any witnesses of lectures devoted to the work in the early modern universities, with the exception of Oxford, whose statutes of 1549 include the Pseudo-Aristotelian work in the lectures programs[83]. The privileged public of the *Problemata* probably consisted of erudite physicians; and to this category belong the most important commentators. The most complete Early Modern commentary to the *Problemata* was written, for instance, by a Milanese physician, Ludovico Settala (1552-1633)[84]. Among his works, we also find a commentary devoted to Hippocrates' *De aeribus, aquis et locis*, the *Animadvertiones et castigationes medicae*, and some political writings. The *In Aristotelis Problemata commentaria* were published in a complete form in Lyon in 1632; the first volume containing the first seven books had already been published in 1602 in Frankfurt, while the second (Books 8 to 15) appeared in 1607[85]. The work was never reprinted. The aim of Settala's work, as he declared in the *Prolegomena*, was to give a complete and reliable explanation of the *Problemata*, which had been neglected for two reasons, viz. 'interpretis vitio et commentatoris defectu'[86]. Largely dependent on Settala's conclusions, and inspired by them, is the commentary written by Giulio Guastavini, a Genovese professor at the University of Pisa, and published in 1609; the exposition, however, in this work is limited to Books 1 to 10[87]. No further edition of this commentary is known.

Both Settala's and Guastavini's commentaries share the same structure. Every question is printed in its Greek original form, followed by a Latin translation; the commentary closes every single section. The exposition is structured in the same form, pointing to the same aspects, using almost the same sources: among others, the writings of Hippocrates and

83. Cf. Blair A., 'The *Problemata* as a Natural Philosophical Genre', in: Grafton & Siraisi (ed.), *Natural Particulars...*, p. 171-204, here p. 179 and n. 43, referring to Schmitt C.B., *John Case and Aristotelianism in Renaissance England*, McGill-Queens' Studies in the History of Ideas 5 (Klingston - Montreal, 1983), p. 41.

84. On Settala, see Rota Ghibaudi S., *Ricerche su Ludovico Settala. Biografia, bibliografia, iconografia e documenti*, Biblioteca Bibliografica Italica 18 (Firenze, 1959).

85. Ludovicus Septalius, *In Aristotelis Problemata commentaria* (Lyon, 1632). See Lohr, *Latin Aristotle Commentaries...*, p. 418-419.

86. Septalius, *In Aristotelis Problemata commentaria...*, *Prologus* (no page numbers given): 'Et qui legunt, quia Latine fere legunt, faciunt nae intelligendo, ut nihil intelligant. Id autem et interpretis vitio, et commentatoris defectu.'

87. Julius Guastavinius, *Commentarii in priores X Problematum sectiones* (Lyon, 1608). See Lohr, *Latin Aristotle Commentaries...*, p. 177.

Galen, that is, the books belonging to the necessary background of every commentator of the *Problemata*, as Peter of Abano had already demonstrated. Another point in common between Settala and Guastavini is the form of the exposition. This similarity can be easily explained, since most of the Renaissance commentaries dealing with the biological works of the *corpus aristotelicum* are composed in the same way; they keep the same focus on the content already present in their medieval counterparts, yet they derive from the humanistic experience a more developed sensibility for philological matters and for problems related with the status of the text, its constitution, and its vocabulary[88].

However, if we analyze the two texts closer, we discover some differences in the approach to the text, in the roles played by the Greek original and by the Latin translation, in the applied sources. Firstly, Settala in his commentary seems to be more interested in the scientific content displayed by the *Problemata* than in the text itself, while Guastavini lingers with special care on philological problems, especially focusing on the Greek text, neglecting the Latin translation[89]. Both authors heavily criticize Gaza's translation, but Settala accuses the humanist translator of failing to render the Pseudo-Aristotelian description of natural phenomena into Latin because of his defective scientific and philosophic culture, while Guastavini more reproaches him for his low respect for the Greek text and for the unnecessary (if not wrong) conjectures and corrections. Last but not least, Settala is better acquainted with the medieval tradition of the *Problemata* than Guastavini: he knows – and mentions – a *vetus interpres*, as well as Roberto Titi did. Neither Settala nor Titi mentions Bartholomew's name. Settala also quotes Peter of Abano, although mostly

88. Perfetti, *Aristotle's Zoology...*, p. 4-7.

89. Cf., for instance, the commentary on *Probl.*, I, 4 in Guastavinius, *Commentarii...*, p. 7: 'In codice Giustiniani post vocem εμέτους legitur particula negandi, δέ, quae problematis sensum priori ex toto contrarium reddit; quam si recipiamus ita locum exponemus: In mutationibus temporum non esse utendum vomitibus, quoniam cum varia in mutatione, et ob mutationem sint genita excrementa: alia enim temporis conditio alia gignit excrementa, non possunt omnia per vomitum expelli, quo tenuiora tantum, magisque prompta, et ad superiores partes tendentia evacuantur: reliqua igitur dimovebuntur solum, et corpori turbationem afferent: Praestat igitur corpora non movere, valetudinis tamen habere rationem, quod proximo problemate dixit. Porro Theodorus aut aliam scripturam secutus est, aut illa de suo addidit. Unde fit, ut cibum alii parum concoquant, alii parum cupiant. Iustinianus in margine pro εν τη μεταβολη habet διά τάς μεταβολάς. Admonivisse contenti non immutavimus, quod ad sensum minime interesset.' Settala, on the other side, simply notices the presence of the doubtful δέ; however, he does not discuss the problem, but concentrates on the scientific content of the question (cf. Septalius, *In Aristotelis Problemata commentaria...*, p. 10).

to criticize him[90]; most of these quotations are not reproduced by Guastavini, although he could have drawn them from Settala[91]. Another aspect, however, deserves our attention.

In their prologues, both commentators explain the way in which they have written their commentaries. Settala tells us that he started his commentary with checking the Latin translation, which he recognized as failing, or better as more elegant than precise or valid from a scientific point of view; for this reason, he decided to retranslate the work (at least the first seven books of the *Problemata*) and to include his own translation into the commentary[92]. Therefore, he informs his reader, his Latin translation (not Gaza's one) will be, together with a corrected Greek text and with the commentary, the key to accessing the work, to making its meaning clear. Guastavini, on the other hand, describes his work as a commentator as follows: he started his explanation by discussing the text of the Basel edition in a copy formerly owned by a scientist, Alessandro Giustiniani, comparing it with Giustiniani's annotions. This erudite had

90. Peter of Abano was object of criticism also by A. Luiz (Antonius Ludovicus), who published in Lissabon in 1540 a *Liber de erroribus Petri Aponensis in Problematibus Aristotelis exponendis*. Cf. on this text Lohr, *Latin Aristotle Commentaries…*, p. 227.

91. See, for instance, the commentary on *Probl.*, I, 2, in Guastavini, *Commentarii…*, p. 3, and in Septalius, *In Aristotelis Problemata commentaria…*, p. 2.

92. Septalius, *In Aristotelis Problemata commentaria…*, *Prologus* (no page numbers given): 'Hactenus enim amoenissimus hic, et fertilissimus quaestionum philosophicarum ager [viz. the Problemata], velut sterile herediolum, pene incultus, et intactus remansit. Et qui Latine vertit, elegans quam fidelis esse maluit, quorum illaudabile quidem, hoc vero necessarium maxime esse haud aliter, quam in imaginibus artificium, et similitudinem, peritus quisque novit. Hoc enim utrumque si assequi pictor poterit, cum summa laude exprimet: si non poterit, similitudinem artificio praefert. Sic Apelles laudatus, et aestimatus fuit, cum illum aulicum derisorem carbone de foco sumpto ita in pariete expressit, ut paucis et nigris lineis ipsum pene hominem perduxerit. Sed longe ab Apelle interpres noster abfuit, qui dum ex omni lato verborum colores ambitiose temperavit, et periodorum umbras e Romuleo luco petitas, Aristotelem alium alibi fecit, deque sententia sua abduxit, atque transformavit. Utinam in hac tantum problematum parte, et membro: sed universum prope modum corpus laborat, et in vitio est. Hoc etiam magis, quod esse docti sciunt, et hactenus patiuntur. Quid igitur est? In parte ergo laborem sum aggressus, et emolitus sum, Problemata illustrando, quorum septem primas sectiones (foelix a nomine meo omen sit) Latinas velud carbone feci, simplici, sed fideli stilo, et pene (quantum linguae ratio passa fuit) ad verbum. Tu, o bone Lector vide, confer, iudica. Nam et meam interpretationem et Graecum prototupon simul habes. Wechelianum inquam a doctissimo Sylburgio emendatum tantum optimum, et purissimum, alibi tamen a correctione quoque nostra adiutum (?). Latinas itaque Problematum primas has septem sectiones feci, sed post alium: et ut vestem, et ornamentum novo corpori darem, Commentariis illustravi: sed illustravi primus illa forma, et illo studio, que probari aut excusari aut aequo et prudenti lectore debeant. Nam ad formam quod attinet, brevem arctamque illam sprevi aut neglexi: ac lubens me dedi in plenum commentandi cursum, explicando, disserendo, disputando.'

corrected and glossed the text, suggesting many improvements for both the Greek text and the Latin translation. Giustiniani's glosses represented the first inspirations to Guastavini's work as a commentator. Some of the ameliorations Guastavini derived from Giustiniani were later confirmed by the text furnished by Casaubon's edition, whose text Guastavini later used as basis for his commentary[93].

From their prologues, but also from the methodology they used to deal with both the Latin and the Greek texts, we can draw some provisory conclusions and compare the two ways of facing the original work and its translation. First of all, Guastavini is more interested in the Greek text than in the Latin one; he concentrates on the original Greek version, which becomes the real object of his commentary. Gaza's version is heavily criticized, but kept; it is included into his book but should be evaluated according to the original Greek version, the Greek giving, so to say, the tools and the possibility to judge the Latin translation and discover its 'dark sides'. On the other hand, Gaza's failing Latin translation will not provide clues to the reconstruction of the Greek text. By putting the Latin translation on a lower level, Guastavini openly denies the ultimate goal of Gaza's version, which is not the starting point for the reconstruction of the original text anymore. Moreover, Guastavini contests both the textual form of the translation and the scientific validity of its content; for these reasons, he chooses not to rely on it. Settala is slightly different: although he recognizes that the real object of the commentary is the Greek text, he considers it important to correct and improve the Latin translation, to make it as important for the reader and

93. Guastavini, *Commentarii...*, p. 5*-6*: '[...] Nactus eram problematum exemplar satis bonum, quod fuerat Alexandri Iustiniani Chii Medici, et Patricii Genuensis: Is liber erat ex editione Basileiensi: sed ad cuius oram, vir ille eruditissimus, ex manuscriptis codicibus multas adscripserat lectionum varietates, et castigationes; in quas cum accurate inspicerem, Theodorique simul eorum problematum translationem, illa occasione considerarem, hanc permultis locis non fidelem, eas autem lectiones saepius meliores, quam vulgatas, dignoscere mihi visus sum. Quare si illae in publicum ederentur, ad Theodori autem translationem aliqua adnotarentur, rem fore studiosis haud inutilem, putavi. Rem igitur aggressus (id fuit ante annos plus minus tredecim) in commentarios, non ita breves, laborem evadere sensi. Absolveram autem iam tertiam sectionem, cum in Aristotelis operum editionem ex Bibliotheca doctissimi Casauboni incidi, in quibus lectionum illas Iustinianaei codicis quae ad problemata essent, varietates extare vidi pene omnes, et vero laetatus sum ea auctoritate etiam fuisse illas confirmatas, cum tamen nonnullas adhuc esse viderem, satis ut mihi videbantur concinnae, quae in illa editione desiderarentur; progrediebar itaque eo in labore satis naviter, et quicquid ab aegrorum curatione dabatur otii, id totum in illo studio continebam. [...]'

as reliable for a real *accessus* to the work as the genuine Greek Pseudo-Aristotelian *Problemata*. In other terms, he places both versions of the text closer to each other, although not on the same level. Moreover, in feeling free not only to correct the Latin text, but also to retranslate the Greek, he seems to have a more pragmatic relationship with the Latin version, which should be transformed and improved according to the needs of the commentator and his public[94].

In short, Settala's and Guastavini's works show different choices in determining the role played by the original text and its translation, and in defining their relationship. Their choices are not casual, but should be considered as a consequence of the changes experienced by the contemporary philosophical culture and by its transformation from bilingual into monolingual. During the seventeenth century Latin does not represent an alternative to Greek anymore. Consequently, a Latin translation does not constitute a tool to access the Greek text more easily, since more people are now able to read the text in its original form. Moreover, the translation cannot constitute the basis for an explanation, since it is perceived not only as unnecessary, but also as something perverting the sense of the text, making the reader reject it. It is exactly in this period that Settala and Guastavini write their commentaries and are confronted with the progressive diminution of the importance of Latin as the language of the philosophical culture, and the parallel, definitive affirmation of Greek. Facing this phenomenon, they take two different positions, Settala by 'preserving' the role of the Latin text, Guastavini by 'denying' or 'limiting' it.

94. See, for instance, the different translations of *Probl.*, I, 4 included in Settala's and Guastavini's commentaries. Septalius, *In Aristotelis Problemata*..., p. 10-11: 'Quamobrem morbi ex bile in aestate generati solitis (febricitant enim maxime in aestate) acutis ob bilem existentibus, in hyeme generantur hii magis? An quia cum febre iuncti, acuti sint, violentos esse oportet? Violentia autem est contra naturam. Humescentibus enim quibusdam locis, calida inflammatio generatur, inflammatio autem, cum caloris sit excessus, febres facit. In aestate quidem sicci et calidi sunt morbi, in hyeme vero humidi et calidi, proinde acuti: celeriter enim interimunt: non enim concoquere valeant abundantiam excrementi'; Guastavini, *Commentarii*..., p. 8-9: 'Quamobrem cum vitia bilis per aestatem contrahi debeant, morbi acuti, qui certe ex bili consistunt, per hyemem potius urgeant? An quia febribus illi comitati, acuti propterea sunt, quia violenti; violentia vero contra naturam est. Locis enim quibusdam in corpore humescentibus fervens oritur inflammatio, quae cum non nisi caloris exsuperantia sit, febres utique accendere potest. Aestivo igitur tempore arida, calidaque vitia male homines habent, hyberno autem humida, calidaque: proinde acuta: celeriter namque interimunt, cum propter humoris superexcrescentis abundantiam concoqui nequeant.'

Conclusions

After this small journey through the translations of the Pseudo-Aristotelian *Problemata* and the commentaries devoted to this work, it is possible to draw some provisory conclusions. Firstly, the *Problemata* does not belong to the most famous and diffused works of the corpus of Aristotle's writings. Compared to other Aristotelian and Pseudo-Aristotelian works, its readers are fewer, but also interested and sometimes specialized ones. All commentaries considered here are, with the exception of the anonymous Erfurt glosses, written by physicians; that demonstrates the meaning of this collection of natural questions for both the medieval and early modern medical culture. Secondly, the nature of the two translations played a considerable role in the diffusion of the work, and in the attitude of the readers towards it. While the medieval commentaries and glosses show that the Latin translation had to be not only commented on, but also reordered both in vocabulary and structure, the Renaissance expositions reflect the complicated relationship of the commentators with an unsatisfying Latin translation and their more and more evident need to approach the Greek original text. However, both the medieval and the early modern commentaries reflect the readers' disappointment with the available translations, judged as inadequate, badly documented, or simply incomprehensible.

Nevertheless, we have also seen that the meaning of a translation and of its positive and negative features is not only witnessed by the major or 'official' commentaries, but also by the reactions of the common readers. As far as the question of the approach to the text shown by readers and users of both manuscripts and printed versions of the *Problemata* is concerned, this study could only offer a very small and necessarily superficial overview. Much work is still to be done on this field in order to define the ways in which unknown readers used the translations of the *Problemata* and struggled with the difficulties presented by the work. We know, for instance, very little about the glosses included in the manuscripts, and nothing about the notes that accompanied the printed books.

As far as the difficulties stressed by both commentators and anonymous glossators are concerned, we notice that they are situated both in the sometimes inadequate vocabulary and in the complicated sentence structure. Facing these difficulties of the text, medieval and Renaissance commentators react by using the same methodologies. Nevertheless, their final results are different. The task of making the complex internal logic of the *Problemata* more clear and explaining the questions is usually accomplished by introducing the comparison with parallel sources; in the case of medicine, for instance, these reference texts are represented by the works of Hippocrates and Galen. On the other hand, the ways in which the commentators react against the inadequacies of the vocabulary of the translators are different. Bartholomew of Messina, for instance, is not openly reproached by Peter of Abano; nevertheless, the commentator notices that the sense in which some words should be interpreted needs clarifications, or at least specifications, in order to define the exact meaning of some unusual, generic, or hardly intelligible expressions.

The way in which early modern commentators tackle the problem of the vocabulary employed by Theodore Gaza is completely different. The translator is often reproached for using the wrong word, or for providing a misleading translation perverting the sense. The reason for this attitude towards Gaza's translation can be found in the nature of the translation itself, whose goal was to offer a perfect text from a philological and literary point of view, but not the correct defining and rendering of the scientific meaning, but also in the commentators' possibility to use the original Greek text for both the *Problemata* and the sources employed for the explanation, and to refer to them. The scepticism about the results achieved by the translator and the possibility to appeal to the Greek text to explain complex sections or to correct manifest misunderstandings, however, are only a part of the explanation for the loss of domain Latin went through compared to Greek during this period. The fact is that the Latin version is not the object of the commentary, since it is not perceived as the real text which has to be commented. The Latin text, we may say, does not only cease to represent the centre around which the commentary devoted to a work originally written in Greek has to revolve; Latin texts have ceased to be the key to access the content of the work itself. From the seventeenth century onwards, the editions of Latin translations of philosophical works become more and more rare, and the access to authors such as Aristotle is accomplished through the consultation of the original Greek text. Consequently, also the *Aristoteles latinus* and the *Aristoteles graeco-latinus* are to be substituted by the *Aristoteles graecus* on the shelves of the libraries of both commentators and learned readers.

Université Catholique de Louvain

Craig MARTIN

SCIENTIFIC TERMINOLOGY AND THE EFFECTS OF HUMANISM: RENAISSANCE TRANSLATIONS OF *METEOROLOGICA* IV AND THE COMMENTARY TRADITION

0. Introduction

A recent study on scientific translations has called these products a result of 'a zone of cultural and linguistic collision'[1]. The crossing of linguistics and culture is particularly apparent for translations of Greek science into Latin. We have available two cultures separated by centuries that attempted to bring Aristotle's corpus into the same language. The differences in their results can be best explained by differing cultural understandings of ideals of language. What were seen as successes in linguistic style were also rejected by others, particularly those who dedicated themselves to the study and explication, who saw medieval translations as no worse than those of their contemporaries.

Renaissance Latin translators of Aristotle sought to revolutionize scientific terminology by eliminating transliterations and what they considered to be medieval corruptions. While these translators often exaggerated the revolutionary nature of their work, their new terminology created a new textual tradition sharply divided from that of the Middle Ages. Among commentators on Aristotle, however, the success of humanist translations of *Meteorologica* IV was limited. Commentators and university lecturers most often preferred to use the vulgate version of *Meteorologica* IV, translated by William of Moerbeke in the thirteenth century, although some, while retaining the medieval translation, altered the specialized vocabulary of the book so that it matched humanist innovations[2]. The Latin translations of *Meteorologica* IV give particular insight into Renaissance attempts to revise Aristotelian scientific vocabulary,

1. Montgomery S., *Science in Translation* (Chicago, 2000), p. 2.
2. This limited success was not universal for Aristotelian natural philosophy. For example, Gaza's translation of the *De partibus animalium* was the standard for Renaissance commentators, see: Perfetti S., *Aristotle's Zoology and its Renaissance Commentators (1521-1601)* (Leuven, 2000), p. 4-5.

because this book contains a multitude of technical terms for natural and artificial processes. Furthermore, the vocabulary of *Meteorologica* IV challenged translators and commentators, since even Aristotle was conscious that he was developing his own specialized vocabulary in this work and noted the inadequacy of existing Greek terminology.

By the 1260's William of Moerbeke replaced a number of the existing translations of Aristotle, many of which had been composed from Arabic, with new versions made from Greek sources. *Meteorologica* IV, however, stood out from much of the Latin Aristotle at this time, including *Meteorologica* I-III, because what had been the most widely circulated Latin version, the so-called *vetus translatio,* was produced from the Greek. Nevertheless, William of Moerbeke's translation of *Meteorologica* IV contrasts with that of Renaissance translators in its wide acceptance. Despite its similarities with Henry Aristippus' *vetus translatio*, which it replaced, commentators and scholars adopted his version soon after its completion and continued to do so throughout the early modern period. For *Meteorologica* IV, William of Moerbeke's work did little to update the terminology, it nevertheless eliminated the most glaring Arabic aspect of many of the copies of the older translation, namely the appendage of a portion of Avicenna's *Kitab al-shifa* as the final chapters of the book.

The broad acceptance of William of Moerbeke's work offers a point of comparison with respect to the role of new translations in the medieval and in Renaissance commentary traditions. While medieval scholars welcomed William of Moerbeke's translation, Renaissance commentators on *Meteorologica* IV, especially Italian ones, despite their professed concerns for textual analysis, had little regard for the humanist translations of the fifteenth and sixteenth centuries.

Renaissance translations of the *Meteorologica* arrived relatively late, some fifty years after Leonardo Bruni's *Nicomachean Ethics*. George Trapezuntius complained in a letter of 1465 that no one recently had translated this book because it was widely considered unworthy of eloquence. He also erroneously claimed that the only extant translation was made from the Arabic[3]. Soon after, Mattia Palmieri remedied his laments by translating

3. Monfasani J., *Collectanea Trapezuntiana: Texts, Documents, and Bibliographies of George of Trebizond* (Binghamton, NY, 1984), p. 107: 'Cur autem pessimi homines, si nomen suum transferendo preclarum facere cupiunt, cur, inquam non transferunt que ab antiquioribus minus Latine traducta sunt? Dicent, fortassis non extare digna eloquentia sua opera philosophi non verba Latine. At Metaurorum quatuor libri perversi magis sunt quam versi nec a Greco, sed ab Arabico traducti mendose sunt.' The two most common medieval translations, William of Moerbeke's and Henry of Aristippus' of the fourth book, were both made from the Greek.

the *Meteorologica* in the late 1460's. His was the first translation of this work in approximately two centuries[4]. A half century later, a wave of new translations of the *Meteorologica* emerged from printing houses.

Typically these translations were part of complete works of Aristotle, although two commentators, Francesco Vimercati (1512-1571) and Francisco Vallés (1524-1592), produced new translations that accompanied their own expositions[5]. Occasionally the *Meteorologica*, and sometimes *Meteorologica* IV alone, appeared as separate volumes[6]. François Vatable's (first printed in 1518) and Joachim Périon's (first printed in 1552) translations were printed most frequently. According to F.E. Cranz, Vatable's work was printed 36 times, and Périon's 26. Pietro Alcionio and Juan Ginés de Sepúlveda created Latin translations that were first printed in 1521 and 1532 respectively[7]. All of these works shared a general concern with *latinitas*. They were also fairly consistent in their adoption of novel technical terms.

Attempts to transform Aristotle's scientific vocabulary can be traced back to the early part of the 1400's and the efforts of Leonardo Bruni and Theodore of Gaza. Rejecting the belief that Latin lacked a sufficient vocabulary, Bruni attacked the *verbum de verbo* method of medieval translators and their adoption of Graecisms, believing that if Latin was as rich as Greek, suitable Latin terms could be found for even the most opaque terminology[8]. Gaza abhorred the medieval translations of Aristotle that

4. Palmieri M., *Meteorologica*, Ms. Milano, Biblioteca Ambrosiana, L 40 sup (xv), fol. 3v-119v. This manuscript can be dated to the mid or late 1460's by its dedication to Marco Barbo, Bishop of Vicenza. He held this position from 17 September 1464 to 17 March 1470, see: Eubel C. (ed.), *Hierarchia catholica medii aevi* (Münster, 1914), vol. 2, p. 267. It is misidentified as a commentary in: Lohr C.H., 'Medieval Aristotle Commentaries', *Traditio* 27 (1971), p. 342. Eugenio Garin was familiar with an unidentified copy of this manuscript, see: Garin E., 'Le traduzioni umanistiche di Aristotele nel secolo XV', *Atti e memorie dell'Accademia fiorentina di scienze morali 'La Colombaria,'* 16, n.s. 2 (1947-50 [1951]), p. 100; Garin E., 'Mattia Palmieri traduttore di Aristotele', in: *La cultura filosofica del Rinascimento italiano* (Florence, 1961), p. 308-312.

5. Vimercati F., *Commentarii in IV libros Meteorologicorum* (Paris, 1556); Vallés F., *In IV. librum Meteorologicorum commentaria* (Padova, 1591 [first edition Alcalá, 1558]).

6. For an example of this work as a separate volume, see: *Aristotelis Stagiritae Meteorologicorum liber quartus*, tr. Joachim Périon (Wittenberg, 1585). This, however, is not the first edition.

7. Cranz F., *A Bibliography of Aristotle Editions 1501-1600*, 2nd ed. (Baden-Baden, 1984), p. 182. A word of caution should be expressed about relying on this work. Despite its usefulness, it is incomplete and contains several inaccuracies with regard to translations of *Mete.* IV.

8. For Bruni see: Hankins J., 'The new language', in: G. Griffiths, J. Hankins & D. Thompson (eds.), *The Humanism of Leonardo Bruni* (Binghamton, New York, 1987), p. 197-212. See also: Schmitt C., *Aristotle and the Renaissance* (Cambridge, Mass., 1983),

simply transliterated technical terms, and attempted to render these technical words into a better classical Latin. The aim of their translations was not merely to render Aristotle's prose into a more aesthetically pleasing corpus. Rather, they attempted to make medieval commentators obsolete. Once a new vocabulary was accepted for Aristotelian science, it would only be with great difficulty that one could understand medieval comments based on a substantially different text. If successful, humanist translations could undermine the work of previous centuries and give Aristotelian science a *tabula rasa* on which interpreters, unhindered by the weight of scholastic jargon and disputations, could put forth new interpretations of the text and consequently of nature.

These goals provoked a certain degree of controversy and were in fact rejected by contemporaries of Gaza such as Cardinal Bessarion and George Trapezuntius, both of whom supported at least some aspects of the medieval intellectual tradition[9]. Nevertheless, Renaissance translators of *Meteorologica* IV, following Gaza's lead, eschewed transliterating Greek words. They did not, however, succeed in creating a lasting revolution. Commentators and translators more than occasionally noted that Latin lacked an appropriate vocabulary for translating the technical terminology of *Meteorologica* IV. Furthermore, the conservatism of universities, Jesuit teaching, and the revival of Thomism in the second-half of the sixteenth century ensured that medieval commentators and translations continued to be vital. The vulgate text remained extremely popular throughout the sixteenth century, and the adoption of a new set of technical terms made medieval commentators and translations neither incomprehensible nor obsolete. The persistence of William of Moerbeke's translation was particularly strong in Italian universities where it remained the standard text for Italian commentators[10]. The resilience of

p. 64-88 for Bruni's positions in addition to a survey of Renaissance translations of all the works of the Aristotelian corpus. On Renaissance translations of philosophical works in general see: Kraye J., 'Philologists and philosophers', in: J. Kraye (ed.), *Cambridge Companion to Renaissance Humanism* (Cambridge, 1996), p. 142-160; Copenhaver B.P., 'Translation, terminology, and style in philosophical discourse', in: C. Schmitt (ed.), *Cambridge History of Renaissance Philosophy* (Cambridge, 1988), p. 75-110; Garin, 'Le traduzioni...', p. 55-104.

9. Perfetti S., '"Cultius atque integrius" Teodoro Gaza, Traduttore umanistico del *De partibus animalium*', *Rinascimento* 35, series 2, supplement (1995), p. 253-286; Monfasani J., 'The Pseudo-Aristotelian *Problemata* and Aristotle's *De animalibus* in the Renaissance', in: A. Grafton & N. Siraisi (eds.), *Natural Particulars: Nature and the Disciplines in Renaissance Europe* (Cambridge, Mass., 1999), p. 205-247.

10. The text of William of Moerbeke's translation is embedded in the following sixteenth- and seventeenth-century commentaries: Nifo A., *Commentaria in libro De mixtis qui a veteribus quartus Meteororum liber inscribitur* (Venice, 1560); Boccadiferro L.,

his translation was such that even the only Renaissance commentary on *Meteorologica* IV written in the vernacular follows his Latin translation[11].

In *Meteorologica* IV Aristotle outlined a number of concepts that previously lacked a systematic nomenclature. The first three chapters of *Meteorologica* IV set out to explain πέψις, its opposite, ἀπεψία, and their species. Πέψις refers to a substance's perfection, typically caused by the actualization of the substance's innate heat. Modern scholars typically translate it into English as 'concoction' or leave it transliterated as 'pepsis'. Aristotle defines three species of πέψις: πέπανσις or ripening, ἕψησις or boiling, and ὄπτησις or roasting. For each of these three species there is a corresponding imperfection. These are named ὠμότης, μόλυνσις and στάτευσις, respectively. The opposite of ripening, or ὠμότης, can accurately be translated as rawness; the precise meanings of the other two opposites, however, are trickier. The difficulty in pinning down clear definitions and translations for these terms is evident in the divergence of two of the more influential modern English translations of *Meteorologica* IV. Ingemar Düring translates μόλυνσις as 'parboiling', H.D.P. Lee as both 'half-cooked', and 'scalding'[12]. Στάτευσις posed an exceptional difficulty, even when compared to its close relative μόλυνσις. Despite supplying a name for it, Aristotle called this process 'very difficult to name' (ἀνωνυμώτερον)[13]. It means something close to 'scorching'.

Many a critic of the last two-thousand years has accused Aristotle of obscurity. In the case of terminology of *Meteorologica* IV, this obscurity arises from the attempt to craft a specialized vocabulary out of everyday terms, from the self-consciousness that he was developing

Lectiones in librum IV Meteorologicorum (Venice, 1563); Scipione Chiaramonti S., *In librum IV Meteorum commentaria* (1644); Cabeo N., *In libros Meteorologicorum Aristotelis commentaria et quaestiones* (Rome, 1644); Pallavicini A., *Explicatio paraphrastica in quatuor Libros Meteororum Aristotelis* (Genoa, 1613); Eck J,. *Libri Meteororum IV, adiectis commentariis* (Augsburg, 1519). The following commentaries cite his translation: Pázmány P., *Tractatus in libros Meteororum*, in: *Opera omnia* III (Budapest, 1897), p. 415-453; Pomponazzi P., *Dubitationes in quartum librum Meteorologicorum* (Venice, 1563); Piccolomini F., *Lectiones in quartum Meteorologicorum*, Ms. Milano, Biblioteca Ambrosiana, D 396 inf. (XVI); Zabarella J., *Commentarii in Meteora* (Frankfurt, 1602); Javelli G., *Quaestiones super IV librum Meteororum,* in: *Opera Omnia* (Lyon, 1580); Cremonini C., *Expositio in IV libros Meteororum,* Ms. Padova, Biblioteca Universitaria, 1210, p. 561ff. Remarkably, Francesco Patrizi used William of Moerbeke's text to translate the Greek passages that he quoted in: Patrizi F., *Discussiones Peripateticae* (Venice, 1581), p. 116-119.

11. de Vieri F., *Trattato delle metheore* (Florence, 1582).

12. Düring I., *Aristotle's Chemical Treatise* (Göteborg, 1944), p. 39; Aristotle, *Meteorologica,* tr. H.D.P. Lee, (Cambridge, Mass., 1952), p. 293, 311.

13. 381b16.

a new scientific vocabulary out of ordinary language. After listing the species of concoction and inconcoction, Aristotle wrote:

> It is necessary to understand that these names [concoction, etc.] do not strictly belong to the subjects at-hand, for there are no terms universally accepted for these processes. So it is necessary to consider these terms as applied to these forms in general and according to common usage[14].

There are, however, distinctions between the normal usage of these eight terms and how Aristotle intended to apply them. He wrote that: 'roasting and boiling occur via artifice; there are, however, natural processes that are the same in respect to form, for their affections are similar, but we do not have names for them'[15]. Therefore, it was necessary to extend the usage of these widely understood terms and apply them to natural processes that do not have specific names.

These eight terms for concoction, inconcoction, and their species represent only a fraction of the scientific terminology of *Meteorologica* IV. For example, this book uses πόροι (pores) to describe the inner workings of material change. Renaissance translators departed from their medieval predecessors, opting for *meatus* instead of the transliterated term. Similarly, *putredo* replaced *putrefactio* for the translation of σῆψις, or rotting. Additionally, *Meteorologica* IV.8-9 describes 18 secondary passive qualities of solid bodies, each of which has an opposite. A number of these terms are obscure, and have not played a significant role in the history of natural philosophy. Nevertheless, the obscurity of the names of these qualities presented translators with opportunities to fashion their interpretations[16]. Aristotle's description of some of these qualities is not without interest for the history of natural philosophy. For example, in *Meteorologica* IV.9 he defined the term φλογιστόν (phlogiston), which eventually played a large role in eighteenth-century chemistry. He divided combustible substances (καυστά) into two kinds: those that emit flames (φλογιστά) and those that do not (ἀφλογιστά)[17]. The persistence of the Greek term φλογιστόν even into the late eighteenth century is indicative of the limited success of Gaza's ideals, since revivals of philhellenism eventually swept away the linguistic reforms of the fifteenth and sixteenth centuries and transliterated Greek words remain implanted in scientific and medical vocabularies.

14. 379b14-18. My translation.
15. 381b3-5. My translation.
16. For example see the discussions of the translation of these terms in: Vallés, *In IV. librum...*, fol. 56v-86r.
17. 387b18-21.

1. Medieval Translations

Medieval translations of *Meteorologica* IV were not as homogeneous as one might expect (see Appendix I). There were in fact four Latin translations that circulated in the thirteenth and fourteenth centuries[18]. A fragment of a fifth exists as well, although it is unclear whether this fragment represents part of a completed work or is merely the start of an unfinished project[19]. Henry Aristippus made the first complete Latin translation of *Meteorologica* IV from the Greek text in the late twelfth century[20]. His work formed the final chapter of what has become to be known as the *vetus translatio*. Gerard of Cremona was responsible for the translation of the first three books of the *vetus translatio* of the *Meteorologica*, all of which were the product of a *verbum de verbo* method from Ibn al-Bitriq's Arabic paraphrase that differed significantly from the Greek text. Gerard began translating *Meteorologica* IV, but stopped just before the end of the first chapter. Why he stopped is unknown, but it was possibly the result of learning of the existence of Henry's translation from the Greek[21].

William of Moerbeke's translation, which was also made from the Greek, made the *vetus translatio* obsolete by the 1260's. His translation was widely available in the Middle Ages, and even in the sixteenth century his translation remained among the most widely circulated translations. William of Moerbeke also translated Alexander of Aphrodisias' commentary on the *Meteorologica*. Fragments of Aristotle's text are embedded in that commentary and do not vary greatly from William of Moerbeke's other translation[22]. In addition to these works, an anonymous translation of Ibn al-Bitriq's Arabic paraphrase, almost always accompanying Averroes' commentary on the *Meteorologica,* circulated beginning in the thirteenth

18. Fobes F., 'Mediaeval versions of Aristotle's *Meteorology*', *Harvard Studies in Classical Philology* 10 (1915), p. 297-314; Minio-Paluello L., 'Henri Aristippe, Guillaume de Moerbeke et les traductions latines médiévales des *Météorologiques* et du *De Generatione et Corruptione* d'Aristote', in: Id., *Opuscula* (Amsterdam, 1972), p. 57-86 (= reprint from *Revue Philosophique de Louvain* 45 (1947), p. 206-235); Lacombe G., *Aristoteles Latinus Codices: Pars prior* (Rome, 1939), p. 56-57.

19. Schoonheim P., *Aristotle's* Meteorology *in the Arabico-Latin Tradition* (Leiden, 2000), p. 144-150.

20. He apparently only translated the fourth book of this work.

21. Schoonheim, *Aristotle's* Meteorology..., p. xxxiv.

22. Alexander of Aphrodisias, *Commentaire sur Les Météores d'Aristote*, ed. A.J. Smet (Leuven, 1968); Vuillemin-Diem G., 'Zu Wilhelm von Moerberkes Übersetzung der aristotelischen Meteorologie,' in: R. Beyers, J. Brams, D. Sacre & K. Verrycken (eds.), *Tradition et traduction* (Leuven, 1999), p. 115-166.

century and was printed numerous times in the Renaissance, most famously in the Giunta editions[23]. This translation is the most corrupt of all medieval versions, lacking many substantial passages altogether.

The broad acceptance of William of Moerbeke's translation of *Meteorologica* IV was accompanied by a redefinition of the book's contents. Led by the Arabic translation and by Gerard of Cremona's translation from the Arabic of *Meteorologica* I.1, which erroneously mentioned an Aristotelian work on minerals, copyists of the *vetus translatio* often appended three chapters of Avicenna's *Kitab al-shifa* to *Meteorologica* IV translated by Alfred of Sarashel[24]. These three chapters became to be known as the *De mineralibus*. In the first half of the thirteenth century Latin scholars generally accepted them as part of the Aristotelian corpus. For example, Adam of Buckfield's commentary on the *Meteorologica,* written probably around the middle of the thirteenth century, included comments on the *De mineralibus* without distinguishing it from the rest of *Meteorologica* IV[25]. In Roger Bacon's commentary on the *Physics* we find citations and even quotations of the *De mineralibus* that he identified as coming from *Meteorologica* IV[26].

William of Moerbeke's work did not contain the errors found in the Arabic version of *Meteorologica* I.1, in addition to cutting the *De mineralibus* from the work. By the middle of the thirteenth century, a number of scholars, including Vincent of Beauvais, questioned the authorship of the *De mineralibus*. When William of Moerbeke's translation became available in the 1260's scholarly opinion on this work changed as a whole[27]. For example, Bacon reversed his opinion on the authenticity of

23. Averroes, *In quartum librum Meteorologicorum* (Venice, 1562-1574).

24. Gerard of Cremona's translation from Yahya ibn al-Bitriq's Arabic translation of Aristotle's *Meteorologica* reads: 'Quando ergo narraverimus de illis, dicemus mineras et animalia narratione universali et particulari, et communicabimus sermonem.' Cf. Schoonheim, *Aristotle's* Meteorology..., p. 4-5. Aristotle's text, however, speaks of plants and animals, not minerals (339a7). They were perhaps also persuaded by opinions such as al-Farabi's that included a work on minerals in his classification of Aristotelian science. See: al-Farabi, *Catálogo de las Ciencias*, ed. and tr. Á. Palencia (Madrid, 1953), p. 60-61 for the Spanish translation of al-Farabi; see p. 162-163 for Gerard of Cremona's Latin translation of al-Farabi.

25. Adam of Buckfield, Ms. Bologna, Biblioteca Universitaria, 1180 (2344) (XIII), fol. 202v-204v.

26. Roger Bacon, *Quaestiones supra libros quatuor Physicorum Aristotelis*, ed. F. Delorme (Oxford, 1928), p. 119.10: 'scribitur .4. *Metheororum* quod multa facit natura que non potest ars facere, etsi multum laboret' is a paraphrase of Avicenna's: 'ars est debilior quam natura et non consequitur eam quamvis multum laboret.' See also: 128.35-129.5.

27. Williams S., 'Defining the Corpus Aristotelicum: Scholastic Awareness of Aristotelian Spuria in the High Middle Ages', *Journal of the Warburg and Courtauld Institutes* 58 (1995), p. 38-40.

the *De mineralibus*. In his *De erroribus medicorum,* Bacon faulted physicians for their reliance on poor translations and their incomplete knowledge of Aristotle's works in general. For example, he claimed that these ignorant doctors relied on 'a brief little chapter that is added onto the end of the *Meteorologica*, and is not derived from Aristotle's text,' that is the *De mineralibus*[28].

That the *De mineralibus* was widely rejected as inauthentic by the end of the thirteenth century is born out by the medieval commentary tradition. There is no extant medieval or Renaissance commentary explicitly on the *De mineralibus,* although a small number of commentaries on *Meteorologica* IV, all from the thirteenth century or early fourteenth century, address Avicenna's work as part of *Meteorologica* IV. Boethius of Dacia cited his own *Quaestiones* on the *De mineralibus* in his *Quaestiones super Topica*, although his work on minerals has never surfaced[29]. It is possible that this self-reference is to his final two questions of his *Quaestiones super quartum Meteorologicorum*, which addressed topics found in the *De mineralibus*[30]. However, being familiar with William of Moerbeke's *translatio nova,* he recognized Avicenna as the author[31]. Giles of Rome, who died in 1316, supposedly wrote a commentary on the *De mineralibus*, although it has yet to be discovered[32]. Beyond these there is no evidence that scholars treated the *De mineralibus* in commentaries

28. Roger Bacon, *Opera hactenus inedita Rogeri Baconi*, Fasc. IX, ed. A. Little & E. Withington (Oxford, 1928), p. 159: 'Liber de corporibus inanimatis totaliter deficit, quia pauca capitula, que addita sunt in fine Metheororum, non sunt de textu Aristotelis, sicut ex alia translatione notum est.' William R. Newman notes that by the 1260's Bacon considered the *De mineralibus* to be in the domain of fools: Newman W., *The Summa Perfectionis of Pseudo-Geber* (Leiden, 1991), p. 20-25. E.J. Holmyard and D.C. Mandeville identify the date of Bacon's reversal as 1266: *Avicennae De congelatione et conglutinatione lapidum*, ed. E.J. Holmyard & D.C. Mandeville (Paris, 1927), p. 10.

29. Boethius of Dacia, *Quaestiones super librum Topicorum*, ed. N. Green-Pedersen & J. Pinborg (Hauniae, 1976), p. 256; cf. Pinborg J., 'Die Handschrift Roma Angelica 549 und Boethius de Dacia', *Classica et Mediaevalia* 28 (1969), p. 379.

30. Boethius of Dacia, *Quaestiones super quartum Meteorologicorum*, ed. G. Fioravanti (Hauniae, 1979). *Quaestio* 119 asks: 'Utrum terra pura possit fieri lapis?' which corresponds to the opening line of the *De mineralibus,* which reads: 'Terra pura lapis non fit quia continuacionem non facit sed commutacionem' (cf. *Avicennae De congelatione...*, ed. Holmyard & Mandeville, p. 45).

31. Boethius of Dacia, *Quaestiones super quartum...*, p. 86: 'ut dicit Avicenna...'. For his reference to William of Moerbeke's translation see: Boethius of Dacia, *Quaestiones de Generatione et Corruptione,* ed. G. Sajó (Hauniae, 1972), p. 123: 'sed motus caeli regulat virtutes omnium elementorum, ut dicitur in secunda translatione Meteororum, in principio; ergo etc.'

32. Müntz E. & Fabre P., *La Bibliothèque du Vatican au XVe siècle* (Paris, 1887), p. 190.

or *Quaestiones*. Thus, in cleansing *Meteorologica* IV of Arabic works, William of Moerbeke enjoyed much success in the scholarly world.

Besides redefining the accepted contents of *Meteorologica* IV, William of Moerbeke's work had a smaller impact on the language of the translation itself. Because Henry Aristippus had also translated from the Greek, the *translatio nova* does not differ from the *vetus translatio* as significantly for *Meteorologica* IV as it did for other texts. An examination of these works (see Appendix 1) shows the similarities in language and syntax. Furthermore, the *translatio nova* did not change the rendering of this book's technical vocabulary (see Table 1).

Medieval translations uniformly render πέψις and ἀπεψία as *digestio* and *indigestio*. Both William of Moerbeke and Henry Aristippus adopted transliterations for the six species of πέψις and their incompletions. On the other hand the Averroistic version, paving the way for Renaissance revisions, translated πέπανσις as *maturatio*, ὄπτησις as *assatio*, and ὠμότης as *cruditas*. Even though these translations primarily used transliteration, medieval commentators strove to find purer Latin terms for these words. Since the commentator's role is to explain words via other words, it should not be surprising that some of the earliest written efforts to clarify the Latin text of Aristotle put forth translations of these words, which remained transliterated in the translations. Two of the most influential medieval commentaries on the *Meteorologica*, those of Albertus Magnus and Ps.-Thomas Aquinas, defined at least some of these opaque Greek terms in Latin, thereby establishing synonymous Latin terms for these processes[33]. The terms they used were identical, or at least similar, to those that were adopted by many of the humanist translators of the Renaissance, thereby suggesting that the continuity between medieval and Renaissance Aristotelianism was stronger than sixteenth-century scholars imagined it to be.

Albertus' commentary, which is based on his reading of the *vetus translatio*, explained the text in a straight forward manner. He counted

33. Albertus Magnus, *Liber quartus meteororum,* ed. A. Borgnet, *Opera Omnia* Vol. 4 (Paris, 1890), p. 705-808; Ps.-Thomas Aquinas, *In libros Aristotelis Meteorologicorum expositio*, ed. R. Spiazzi (Rome, 1952), p. 645-685. While Thomas is no longer accepted as the author of this work, it was considered authentic throughout the Renaissance. On the problem of its authorship see: Dondaine A. & Bataillon L., 'Le commentaire de saint Thomas sur les Météores', *Archivum Fratrum Praedicatorum* 36 (1966), p. 81-152; Vuillemin-Diem G., 'Anonymus Normannus (Mahieu le Vilain): Super Meteora II.9-III. Zur Identifizierung des Autors, zur Eigenart des Textes, mit einer Edition von zwei Kapiteln der noch unveröffentlichten Schrift,' *Recherches de Théologie et Philosophie médiévales* 71, 1 (2004), p. 1-130.

translation among the tools of exposition and notably used this tool to explain some of the species of πέψις and their opposites. For example, in his definition of ὄπτησις he stated that: 'a digestion of this kind is in the manner of roasting, in which the heat acts with the dry: and it is called *optesis,* which in Latin is called *assatio,* just as *hepsesis* is called *elixatio*'[34]. Albertus alternated between the transliterated Greek terms and the Latin translations, and used the words *epsesis* and *elixatio* interchangeably. He failed, however, to provide translations of all of the species of concoction and inconction. He translated ὠμότης as *crudum*[35], and defined στάτευσις at times as *semiassum* and at others as *imperfectum perassatum,* but left πέπανσις and μόλυνσις undefined. These gaps probably do not necessarily suggest an inability to translate these words. If asked, he probably would have translated πέπανσις as *maturatio* or some variation of that word, judging from the frequency of the verb 'maturescere' in his explanation of πέπανσις. While μόλυνσις is more difficult to translate than πέπανσις, it would have been easy to render it in the same way he did στάτευσις and call it *semielixum* or *imperfectum perelixatum.* Rather these gaps suggest that translation was only one of the keys to understanding Aristotle's text, and Albertus' use of translation was meant to clarify the text and was not necessarily a critique of the existing standards of translation.

2. Renaissance Commentaries and Translations

While Albertus and [Ps.-]Thomas discussed translation in *Meteorologica* IV to a limited degree, and Roger Bacon wrote polemics against those unskilled in translation and textual exegesis[36], medieval treatments of *Meteorologica* IV rarely emphasized the details of Aristotle's Greek and commented on the methods of rendering it into Latin far less frequently than their Renaissance counterparts.

34. Albertus Magnus, *Liber...*, p. 726: 'et hujusmodi digestio est per modum assationis, in qua calor agit cum sicco: et dicitur esse *optesis*, quod Latine sonat *assatio*, sicut epsesis *elixatio* interpretatur.'

35. Albertus Magnus, *Liber...*, p. 732: 'Cruditas autem est *homothes* in Graeco.... dicitur Latine *crudum...*'; p. 746: 'Dicatur autem, si placet statheusis, quod sonat semiassum, vel imperfectum perassatum.'

36. On Bacon's view of contemporary translations and translators see: Lemay R., 'Roger Bacon's Attitude toward the Latin Translations and Translators of the Twelfth and Thirteenth Centuries', in: J. Hackett (ed.), *Roger Bacon and the Sciences* (Leiden, 1997), p. 25-48; Grabmann M., *Forschungen über die lateinischen Aristotelesübersetzungen des XIII. Jahrhunderts* (Münster, 1916), p. 56-73.

Following the lead of Bruni and Gaza, sixteenth-century translators of *Meteorologica* IV uniformly avoided using transliterated terms. This is also true for the only known fifteenth-century translation composed by Mattia Palmieri[37]. While Palmieri retained the terms *digestio* and *indigestio,* sixteenth-century translators, and a good number of commentators, typically rejected these terms in favor of *concoctio* and *inconcoctio* (see Table 2).

A revolution in terminology could not take place over night, if it was to take place at all. Replacing a transliterated Greek word with a more commonly known Latin word did not necessarily clarify the text. Indeed, the multiplicity of new terms used to translate the word brought forth debate, and for the first time in the Latin tradition, commentators debated the merits of various translations of *Meteorologica* IV. Commentators, however, for the most part did not use humanist translations even though they accepted the changes of terminology. To my knowledge only four early modern commentaries on *Meteorologica* IV contain Renaissance translations; two commentators, Francesco Vimercati and Francisco Vallés, used their own, and Johannes Hawenreuter and Christoval Nuñez used Alcionio's. Vimercati and Hawenreuter provided the Greek text, thereby allowing readers to bypass the issue of translation altogether[38]. Sixteenth-century editions of Thomas Aquinas typically included both the *nova translatio* and a Renaissance translation, usually, if not always, Vatable's[39]; the Giunta editions of Averroes included the *nova translatio* accompanied by the Averroistic version[40]. There was particular need to supplement the Averroist translation because of its numerous and large lacunae. The remaining commentaries that included a text used the vulgate translation. The vulgate, however, was not etched in stone, and many

37. Palmieri, *Meteorologica*, fol. 94r-94v: 'Est & enim calidi quidem digestio: digestionis vero maturitas: elissatio ad hunc assatio: frigiditatis autem indigestio, cuius partes sunt cruditas, delissatio, atque subassatio.'

38. Vimercati, *Commentarii...*; Vallés, *In IV librum Meteorologicorum*; Hawenreuter J., *Commentarii in Meteorologicorum libros IV* (Frankfurt, 1605); Nuñez C., *Opus... in quo commentantur tria priora capita Aristotelis ex Meteororum libro quarto* (Madrid, 1613). See also: Charpentier J., *Descriptiones universae naturae* (Paris, 1562), fol. 60v-74r, which gives glosses on Périon's translation.

39. Cranz F., 'The Publishing History of the Aristotle Commentaries of Thomas Aquinas', *Traditio* 34 (1978), p. 157-192.

40. Cranz F., 'Editions of the Latin Aristotle Accompanied by the Commentaries of Averroes', in: E. Mahoney (ed.), *Philosophy and Humanism* (New York, 1976), p. 116-128; Schmitt C., 'Renaissance Averroism studied through the Venetian editions of Aristotle-Averroes (with particular reference to the Giunta edition of 1550-1552)', in: *L'averroismo in Italia,* Atti dei Convegni Lincei 40 (Rome, 1979), p. 121-142.

early modern commentaries emended the medieval translation so that it reflected the recently crafted vocabulary. Thus, the translation remained for the most part the same, only the objectionable transliterations were replaced.

Commentators participated in propagating these changes in the Latin vocabulary of *Meteorologica* IV. Early attempts used equivalences to establish the new terms. Lefèvre's paraphrase on *Meteorologica* IV is perhaps the best example[41]. He gave notes before each chapter of the paraphrase allowing the reader to review the vocabulary before examining the text. For example his notes to *Meteorologica* IV.2 began by giving equivalents to the six species of concoction and inconcoction and continued to rephrase other longer and complex phrases[42]. Agostino Nifo's commentary, written in 1523 and first printed in 1531, went one step further and provided a chart that contains the Greek words, their transliterations, and the new translations[43]. The existence of charts, such as Nifo's, shows that the meaning of the text and the words have not changed at all. The new translation is directly equivalent to the old one, and does not necessarily pave the way for a clearer interpretation of Aristotle. Thus, these new terms could substitute the old words in the vulgate without actually improving the translation or reforming natural philosophy. Commentators, however, were aware that Renaissance translations remedied errors that had crept into the medieval tradition. For example, Nifo points out that the medieval versions translate μύλαι as 'lead' (*plumbum*) rather than

41. On Lefèvre's Aristotelianism see: Kessler E., 'The Lefèvre enterprise', in: C. Blackwell & S. Kusukawa (eds.), *Philosophy in the Sixteenth and Seventeenth Centuries: Conversations with Aristotle* (Brookfield, Vermont, 1999), p. 1-22; Rice Jr. E., 'Humanist Aristotelianism in France: Jacques Lefèvre and his circle', in: A. Levi (ed.), *Humanism in France* (Manchester, 1970), p. 132-149; Lohr C.H., *Aristotle Renaissance Commentaries* II. *Renaissance Authors* (Florence, 1988), p. 138-142; Schmitt C., 'The rise of the philosophical text-book', in: *Cambridge History of Renaissance Philosophy*, p. 795-796.

42. Here are the entire notes to *Meteorologica* IV.2: Lefèvre J. *Totius Philosophiae naturalis paraphrases* (Paris, 1521), fol. 183r: 'Pepansis / maturatio. Epsesis / elixatio. Optesis / assatio. Omotes / cruditas / & (quae pepansi opponitur) immaturatio. Sunt praeterea duae incoctiones: inelixatio opposita epsesi / & inassatio quae opponitur optesi. Proprium / naturale. Humidum & siccum naturale / uniuscuiusque mixtorum natura materia dicuntur: quod natura ex elementorum contemperamentis / secundum materiam omnia gignantur mixta. Et cum materiae sit pati / sitque compositum: ex calidorum / frigidorum / humidorum /& siccorum contemperamentis pariter coalitum / quorum duo calidum & frigidum potissimum sunt ad agendum / & duo humidum & siccum ad patiendum: hinc potius humidum & siccum / uniuscuiusque mixti / natura materia dicuntur / contemperamentaque passiva. Apostema / recollectio. graecum est. nos huiusmodi in uno loco humorum recollectionem / agglobationemque: panos & vomicas nominamus.'

43. Nifo, *Commentaria*... [1531 version], p. 548.

the more accurate 'mill-stones' (*molares lapides*)[44]. These errors, however, were limited in number.

By the middle of the sixteenth century a new type of commentary had come into vogue that specifically dedicated itself to the problems of translation. Joachim Périon's *Observationes*, first printed in 1552, gave notes at the end of his translation that explained the rationale of his choice of words thereby providing a linguistic foundation for students of Aristotle[45]. Not surprisingly, his two notes on *Meteorologica* IV were dedicated to the explanation of the species of πέψις and the names of some of the eighteen passive qualities. Echoing Lucretius' pleas of excuse for the inadequacy of the Latin language and implicitly rejecting Bruni's defense of Latin, Périon noted that there are names for many things in Greek that do not exist in Latin. As a result he was compelled to use compound terms to describe ἕψησις and ὄπτησις. Cicero, and classical Latin in general, guided Périon in his choice of usage and terms. For example, he justified his choice of terms as *non plena assatio* and *non plena elixatio* by claiming that Cicero often called what is imperfect 'not full'[46]. Following what he considers to be a more classical Latin, Périon rejected *maturatio* in favor of *maturitas*. In general he avoided transliterations, but nevertheless used *phymata* to translate φύματα, which as a technical medical term means boils, arguing that the Latin word has a proper classical provenance because it was adopted as a technical term by Aulus Cornelius Celsus, the Roman physician of the first century A.D.[47].

Both humanist translations and Renaissance versions of William of Moerbeke's translation rejected the medieval use of *digestio* to translate

44. Nifo, *Commentaria...*, fol. 135v: 'Quod vero ad verba attinet, μυλίαι, non plumbum, ut in antiquis translationibus legitur, sed molares lapides intelligendum est, ut Alexander exposuit.' Nifo took the term 'molares lapides' from William of Moerbeke's translation of Alexander. Alexander of Aphrodisias, *Commentaire...*, p. 327.

45. Périon J., *Pars tertia operum Aristotelis Stagiritae, libros naturalis & divinae philosophiae ordine continentes* (Basel, 1563). See also: Gesner K., *Physicarum meditationum, annotationum, et scholarium libri V* (Zürich, 1586), p. 27-54, 140-161, for a commentary on *Meteorologica* IV largely dedicated to explaining Aristotle's vocabulary. On Périon's explanations of Aristotle, see: Stegman A., 'Les observations sur Aristote du Bénédictin J. Périon', in: J.-C. Margolin & M. de Gandillac (eds.), *Platon et Aristote à la Renaissance* (Paris, 1976), p. 376-389.

46. Périon, *Pars...*, p. 156: 'Et certe Cicero saepe id quod imperfectum est, non plenum appellat.'

47. Périon, *Pars...*,p. 156: 'φύματα appello phymata, cum Celso, qui haec a tuberculis distinguit'. Périon states that Celsus distinguishes *phyma* from *tuberculum* presumably referring to *De medicina* 5, 28, 9. Périon's reasoning for translating φύματα as *phymata*, however, may be on shaky ground, as elsewhere Celsus specifically identifies *phumata* as *tubercula*. See *De medicina* 6, 18, 2: 'Tubercula etiam, quae φύματα Graeci vocant...'.

πέψις, opting for *concoctio* instead[48]. It may not be readily apparent why Renaissance translators and commentators chose *concoctio* over *digestio*. Neither term transliterates the Greek, and both appear in classical Latin. For example, the rhetorician Quintillian employed both words[49]. However, the first appearance of *digestio* in written Latin came after the Augustan era, giving it a shakier provenance than its rivals *concoctio* and *coctio*. Controversy over the replacement of *digestio* with *concoctio* began when George Trapezuntius attacked Gaza's translation of the *Problemata*, in his view Gaza was mistaken for equating these terms, when they in fact refer to separate processes[50].

While sixteenth-century Aristotelians no longer used transliterated forms of the varieties of πέψις, *digestio* remained part of the scientific vocabulary. Ambivalence, and at times confusion, reigned in regards to the replacement of *digestio* with *concoctio*. This ambivalence, perhaps, was not new and a parallel is found in Celsus' *De medicina* where it is suggested that *digestio* and *concoctio* are synonymous[51]. Similarly, Ermolao Barbaro's *Compendium scientiae naturalis*, composed in 1484 but first printed in 1545, labeled the chapter that summarizes the first part of *Meteorologica* IV as 'De digestione et concoctione'. While he joined these two words with a copula in the section's title, he made clear in the summary of the chapter that his simultaneous use of both terms is pleonastic. Thus he defined the goal of the chapter as 'Quid sit concoctio sive digestio quotque eius species sint aperiemus'[52]. Lodovico Boccadiferro's *Lectiones in quartum Meteorologicorum* written in the 1530's or 40's at the University of Bologna, used the two terms interchangeably except where he applied them simultaneously, typically separating them by 'vel' or 'seu'[53]. These terms, however, eventually came to be

48. See, for example, the translations embedded in the commentaries of: Chiaramonti, *In quartum Metheorum…*; Cabeo, *In libros Meteorologicorum…*; Boccadiferro, *Lectiones….*

49. *Institutio oratoria,* 8, 4, 16; 11, 3, 19. Quintillian used the noun *digestio* and the verb *coquere*.

50. Mohler L., *Kardinal Bessarion als Theologe, Humanist, und Staatsmann: Funde und Forschungen* (Aalen, 1967; Reprint: Paderborn, 1923-42), p. 286-289.

51. 'Sive concoctio, sit illa, sive tantum digestio'. *De medicina,*1 praefatio, 63.

52. Barbaro E., *Compendium naturalis scientiae* (Venice, 1545), fol. 51v.

53. Boccadiferro, *Lectiones…*, p. 15: 'omnis digestio fit a calido, & digestio est commixtio perfecta humidi cum sicco, & omnis digestio est a calido'; p. 9: 'coctio est quaedam operatio calidi, & incoctio est operatio frigidi…'; p. 65: 'Dicebat Aristoteles in principio huius libri, quod quatuor erant qualitates primae, duae activae, & duae passivae: & dixit, quod intendebat declarare, quae sunt actiones activarum: & dixit, quod tres erant operationes activarum in genere, scilicet generatio, & putredo, & digestio seu coctio.' For Boccadiferro's dates and writings see: Lohr, *Aristotle…*, p. 57; Lohr C.H., 'The Aristotle

distinguished in medical literature. More or less following George Trapezuntius' critique of Gaza, Santorio Santorio (1561-1636), a professor of medicine at the University of Padova, basing his views on Galen's *De methodo medendi,* claimed that concoction refers to the transformation of food into *chylum* or blood, while *digestio* is the transmission of these newly created substances throughout the body[54]. Santorio's views, which are not based on Aristotle's text, illustrate the incompleteness of the attempt to reform Aristotelian vocabulary. According to Santorio there is nothing stylistically problematic with *digestio* but it merely refers to a different process than *concoctio.*

The inclusion of Vatable's translation with William of Moerbeke's in editions of Thomas Aquinas occasionally caused confusion rather than clarification. For example, Francisco Fernandez Bexarano included in his *Super IV libros Meteororum quaestiones,* published in 1643, a discussion over the distinction between *digestio* and *concoctio*[55]. Apparently having consulted an edition of Thomas' and Pseudo-Thomas' commentary that contains both medieval and Renaissance translation, Fernandez Bexarano found support for the existence of both concepts within Aristotle. Quoting Vatable's translation that uses *concoctio*, he cited [Ps.-]Thomas' comments that refer to *digestio*, leading him to the conclusion that *digestio* is the perfection of the principle of generation in animals, while *concoctio* refers to perfection of inanimate substances, the most notable example being the perfection of must when it turns into wine[56].

commentaries of Ludovicus Buccaferrea', *Nouvelles de la république des lettres* 1 (1984), p. 107-118; Lines D., 'Natural philosophy in Renaissance Italy: The University of Bologna and the Beginnings of Specialization', *Early Science and Medicine* 6 (2001), p. 308.

54. Santorio S., *Commentaria in primam fen primi libri Canonis Avicennae* (Venice, 1646 [first edition 1625]), p. 507: 'Praeterea notandum, per concoctionem intelligere conversionem cibi in aliam substantiam, videlicet in chylum, vel chyli in sanguinem, & non digestionem: differt enim concoctio a digestione, ut docet Galenus 12. methodi 3. ubi ait, coctionem fieri per somnum: digestionem vero per vigiliam: digestio vero nihil aliud significat, quam delationem seu transmissionem alimenti ab una ad aliam partem sine conversione cibi ex una in aliam substantiam: conversio enim solum fit per somnum, quando videlicet calor dispersus per sensiteria revehitur ad imas partes, & cum innato calore viscerum unitur: in vigilia calor innatus seiunctus ab influente non est tantae virtutis, quantae possit convertere chylum in sanguinem, vel sanguinem in ossa.' Santorio's view is a reasonable interpretation of the passage cited. Galen writes: πέπτει μὲν γὰρ ὁ ὕπνος, διαφορεῖ δὲ ἡ ἐγρήγορσις. Cf. *Claudii Galeni Opera Omnia,* ed. C. Kühn (Leipzig, 1825), vol. 10, p. 824.

55. Fernandez Bexarano F., *Super IV libros Meteororum quaestiones* (Lyon, 1643), p. 449-452: 'Utrum coctio sit idem, quod digestio?'

56. Fernandez Bexarano, *Super...*, p. 451 cites Vatable's translation: 'Arist. cap. 3. ubi ait: *multaque alia ex hisce quae cocta sunt, eadem quidem forma, sed translatione matura dicuntur, ut tubercula.*' Cf. Thomas Aquinas, *In libros Meteorologicorum praeclarissima commentaria* (Venice, 1561), p. 222.

Boccadiferro's commentary, despite not showing a marked preference for *concoctio* over *digestio*, nevertheless marked a new role for commentaries that could only have arisen with the availability of multiple new translations. For the first time we find commentators comparing, criticizing, and offering amendments to translations of *Meteorologica* IV. In particular, the choice of Latin words for these obscure processes was a bone of contention, leading Boccadiferro to judge the relative merits of these new translations. His inability to resolve these issues prompted him to argue that in some cases it was futile to find the appropriate word in Latin. Although Boccadiferro's commentary is accompanied by a modified vulgate translation, he was familiar with both Vatable's and Alcionio's translations, which he did not find entirely satisfactory. Even though the modified vulgate that accompanies his text translated στάτευσις as *tostio*[57], Boccadiferro argued that this is not the appropriate word. Citing Alcionio's translation of 'tastio [sic] et frixio', he argued that these words do not correspond to Aristotle's intent, because the appropriate word does not exist in Latin. Rather he offered his own Latin phrase, which is not found in any other translation, that explains the term via negation: *diminuta assatio*[58]. Boccadiferro's critiques of the new translations highlight some of the difficulties in rendering many of the new words of *Meteorologica* IV into Latin. In particular, the eighteen qualities outlined in *Meteorologica* IV.9 provided difficulties for the translator. Boccadiferro praised Vatable's translation of τμητόν into *scissile*, which barely alters William of Moerbeke's translation of *scissibile*[59]. The descriptions of these properties and the examples of them, however, posed difficulties for the translator. For example, the translation of the Greek word σταῖς, which Lee translates as dough, into *pasta* was unsatisfying and again Boccadiferro was pessimistic about Latin's ability to render Greek words with precision[60].

57. It should be noted that this volume was printed posthumously, and Aristotle's text may not have been identical to the one Boccadiferro used.

58. Boccadiferro, *Lectiones...*, p. 67: 'Est autem statheusis diminuta assatio, sed non bene translatum est hoc verbum a latinis: Vatablus transtulit ex ... [Lacunae are in the text, presumably where Greek words belong.] Alcionius autem tastio aut frixio; sed nec ... nec tostio est proprium vocabulum ipsius statheusis, quia hoc verbum ex ... est quid commune ad intensam assationem & ad remissam assationem: neque etiam tostio est conveniens, ut dicit Alcionius, quia est tostio vehemens assatio; neque est frixio, quia ut infra dicemus frixio est species coctionis seu digestionis: & ideo isti non bene transtulerunt hoc vocabulum statheusis: & ideo ego credo, quod non habemus vocabulum & verbum appropriatum Latinum, sed debet circumscribi; & ideo debet dici diminuta assatio & imperfecta, & principium ipsius assationis, quia non habemus proprium verbum.'

59. Boccadiferro, *Lectiones ...*, p. 216: '& transtulit Vatablus scissile & inscissile, & bene iudicio meo transtulit.'

60. Boccadiferro, *Lectiones...*, p. 214: 'ita traduxit pasta, quia dicit, quod Latini carent hoc nomine proprio: & ideo latini transtulerunt, farina aqua subacta.'

Critiques, however, did not end at the simple choice of words; Boccadiferro criticized the translators for not being true to the text and interpolating interpretation that is either incorrect or beyond the literal sense of Aristotle's text[61]. This method of translation, however, had defenders as well as critics, as is illustrated by the works of Vallés and Vimercati.

As one might expect, the two commentators who translated *Meteorologica* IV used their commentaries to justify their translations. Vimercati, rejecting methods of translating 'word to word' and even those that 'only render the sense' of the text, wanted his translation to 'explain its sense'[62]. Thus his commentary on the *Meteorologica* is in part a justification of his translation. Accordingly his introduction promises that the commentary will note and correct the negligence and errors of other translations[63]. Despite his self-professed care in translating, Vimercati translated the technical terms of *Meteorologica* IV with hesistancy. He appeared almost apologetic for his translations of μόλυνσις and στάτευσις as *imperfecta elixatio* and *imperfecta assatio* respectively, writing that 'we translate these words in this way because more appropriate words do not exist'[64]. As a result he left these two terms in the Greek in his discussions of the species of concoction and inconcoction, in addition to providing Latin translations. Στάτευσις remained particularly problematic, and we are

61. See Boccadiferro, *Lectiones...*, p. 219: 'Exhabilia autem &c. hic sumitur, qualiter expositio Vatabli, & Alcyonii non est ad verba Aristotelis'; p. 230: 'quia Alexander dicit, sicut lignum, idest lignum viride; & sic Alcyonius transtulit lignum viride, licet hoc non sit in textu'; p. 68: '& male transtulit Vatablus, & male exposuit: sed intelligitur factum, idest actu tale ex potentia tale; & prius erat nutrimentum potentia caro, & postea factum est actu tale in fine coctionis legi verba (...) enim sic debet legi lettera.'

62. Vimercati, *Commentarii...*, preface, fol. A iiii[r]: 'In conversione id contendimus, ut nec verbum verbo, veteris interpretis more, nec verborum ulla habita ratione, quod plerique alii faciunt, sensum solum rediremus, sed media via ingressi, ita sensum explicaremus...' On Vimercati's method of translation used in his commentary on the *Physics* see: Schmitt, *Aristotle...*, p. 79-81. For Vimercati's biography and works see: Gilbert N., 'Francesco Vimercato of Milan: a bio-bibliography', *Studies in the Renaissance* 12 (1965), p. 188-217.

63. Vimercati, *Commentarii...*, preface, fol. A iiii[r]: 'Complura loca a caeteris interpretibus perperam & oscitanter conversa, & emendata, fideliter, ni fallimur, reddidimus: illorumque errores interdum in Commentariis notavimus, alios ex conversione nostra, & verbis Graecis, quae singulis particulis ascribenda curavimus, diligentiae vestrae animadvertendos reliquimus, ne illos reprehendendi gratia novam nos conversionem edidisse videremur.'

64. Vimercati, *Commentarii...*, vol. 4, p. 23: 'Concoctionis veluti species sunt, maturitas, elixatio, assatio: inconcoctionis, cruditas, μόλυνσις, id est, imperfecta elixatio, & στάτευσις, id est, imperfecta assatio. Ita enim vocabula haec reddimus, cum magis propria non suppetant.' He repeats this claim with respect to μόλυνσις on vol. 4, p. 44: 'Eo itaque vocabulo appellantur, quod aliud commodius non habeatur, quo affectio illa exprimatur.'

presented with three alternative translations: *imperfecta assatio*, *prava assatio,* and *tostio*. Vimercati traced his inability to find the precise Latin words back to Aristotle himself, because he claimed that these processes lack proper names[65]. In these cases, the meagerness of Latin mimics the meagerness of Aristotle's Greek.

In contrast to Vimercati's promotion of his translation that sought to give a better interpretation of the text, Vallés promoted his translation by openly attacking the interpolations of Alcionio. In Vallés' view, Alcionio overstepped his role as translator by rearranging the text and adding words that are not in the Greek version[66]. While in some instances Alcionio's editorializing expressed the essence of Aristotle's words, in other instances it alters the sense of the passage or is just plain wrong according to Vallés. For example, Alcionio, apparently unsatisfied with the sense of the text, translated ἐξίοντος τοῦ ὑγροῦ as 'per egressum caloris'[67]. Vallés, who rendered these words with the much more literal phrase 'exeunte humido', noted that Alcionio's translation 'reads not without a large corruption of the sense of the words'[68]. Less egregious alterations did not escape Vallés scrutiny. For example, he noted that Alcionio has inserted the word *calefacta* to his translation of the sentence that reads: ὕδωρ γὰρ οὐ παχύνεται μόνον τῶν ὑγρῶν ('water is the only wet substance that does not thicken'), so that his translation reads 'aqua enim rerum humidarum sola calefacta non crassescit'[69]. Vallés' translation more closely follows the Greek text, reading: 'aqua enim sola humidarum non crassescit'. In his eyes, Alcionio's addition interpreted rather than translated the text. The addition of the word *calefacta* suggests that while water does not thicken when heated, it does solidify under the influence of the cold[70]. Thus, Vallés, following a method of translation contrary to Vimercati's, criticized

65. Vimercati, *Commentarii...*, vol. 4, p. 48: 'Id cum fit, concoctionis quamdam inaequalitatem sequi necesse est, quam inconcoctionem Aristoteles nomine magis vacare asserit, quam inconcoctionem elixationi oppositam, quam dixit μόλυνσιν nominari. Similem tamen esse στάτευσις, hoc est pravae assationi, aut etiam fortassis tostioni.'

66. Vallés, *In IV. librum...*, fol. 9v: 'quod fortassis intelligens Alcyonius, qui mihi videtur pro suo arbitratu verba Arist. perunque transferre, reddidit, & quodvis similare; sensum exprimens, tamen addens dictionem illam similare, quae in graeco codice non est. quare ego malo illam explicando suppleri, quam leggendo addi.'

67. 383a20; *Aristoteles Stagiritae Opera* [Alcionio's translation] (Lyon, 1578) vol. 1, p. 769: 'Et certe mollia, eaque non liquida per egressum caloris durescunt...'

68. Vallés, *In IV. librum...*, fol. 43r: '(ubi, non sine magna sensus verborum corruptione legit Alcyonius, per egressum caloris)'.

69. 380a34; for Alcionio's translation see: *Aristoteles Stagiritae Opera* (Lyon, 1578), vol. 1, p. 761.

70. If this was Alcionio's intent it does not appear justified as Aristotle repeats this claim when explaining that water solidifies but does not thicken. Cf. 383a6-13.

Alcionio for not faithfully translating the actual words and thereby giving the text an altered meaning[71]. According to Vallés, the weakness of Alcionio's translation comes not only from his loose method, by also a lack of philological sophistication: Alcionio did not follow satisfactory manuscripts[72]. The errors of translation that Vallés called an impediment to natural philosophy, in his *Controversiae medicinae*, were not confined to medieval translations and mar the works of *recentiores* as well[73].

Pomponazzi, Boccadiferro's teacher, even though he shared concerns over Alcionio's translation, dismissed excessive preoccupation with translation as being beneath the concerns of philosophy. Nevertheless, in his *Dubitationes in quartum Meteorologicorum,* he noted that *tostio* is not a good translation of στάτευσις because it does not denote a superabundance of heat. Pomponazzi, however, cut short his discussion, arguing that this topic is not suitable to philosophers but rather 'pertains to the Grammarians. Therefore I leave it to them'[74]. Pomponazzi's disdain for discussions of grammar and vocabulary is indicative of the limited success of the humanist translation movement. Among most university lecturers and commentators on Aristotle, problems of understanding natural philosophy were not limited to rendering texts into Latin. Even those dedicated to translation and grammatical issues, such as Périon, Vimercati, and Boccadiferro, defied Bruni's denial of the poverty of Latin, and confessed the impossibility of finding accurate Latin terms that correspond to Aristotle's Greek. Thus, uncovering Aristotle's intent and its relation to the truth involved more than translation, and William of Moerbeke's translation was for the most part a sufficient textual foundation for explorations of natural philosophy that utilized logic, dialectic, and the consideration of experience, rather than considerations of philology and prose style[75].

Oakland University

71. Vallés, *In IV librum...*, fol. 25r: 'ut illinc constet, male ac superflue, Alcyonium interposuisse dictionem calefacta, ita legentem aqua humidarum rerum sola calefacta non crassescit. praeterquam quod verba autoris non reddidit fideliter, sensu etiam alio transtulit.'

72. Vallés, *In IV librum...*, fol. 21r: 'Locus primus huius dictionis invenitur in aliquibus codicibus mutilis, quos sequutus est Alcyonius, non sine magna sententiae corruptione, ita dicens. Natura quam designamus (dimissis prioribus verbis) ut forma & essentia est; concoctionis enim finis nonnullis in rebus ad substratam quandam formam est. Quae verba post ea, quae antea scripserat, non video, quid pulchrum possint significare. melius itaque multo habetur in codice Aldino, quem sum secutus.'

73. Vallés F., *Controversiae medicae* (Alcalá, 1556), p. 2.

74. Pomponazzi, *Dubitationes...*, fol. 22v: 'Hoc pertinet ad Gramaticos. ideo illis relinquo.'

75. I am grateful for comments and corrections received from Gudrun Vuillemin-Diem, Katharine Park, William Newman, John Murdoch, John Monfasani, and an anonymous reader.

Table 1: Medieval Terminology

Translation	πέπανσις	ἕψησις	ὄπτησις	ὠμότης	μόλυνσις	στάτευσις
Vetus	pepansis	epsesis	optesis	omotes	molinsis	Stateusis
Nova	pepansis	hepsesis	optesis	omotes	molynsis	Stateusis
Alexander	pepansis	hepsesis	optesis	omotes	molynsis	Stateusis
Averroes	maturatio	[wanting]	assatio	cruditas	[wanting]	[wanting]

Table 2: Renaissance Terminology

Author	πέπανσις	ἕψησις	ὄπτησις	ὠμότης	μόλυνσις	στάτευσις	Date of 1st ed.	# of edd.
Palmieri	maturitas	elissatio	assatio	cruditas	elissatio	subassatio	ms.	c. 1460
Vatable	maturatio	elixatio	assatio	cruditas	inquinatio	excaldatio	1518	36
Périon*	maturatio	elixatio	assatio	cruditas	non plena elixatio	non plena assatio	1552	26
Alcionio	maturitas	elixatio	assatio	immaturitas	leviscoctio	frixio aut tostio	1521	10
Vallés	maturitas	elixatio	assatio	immaturitas	levis elixatio	levis assatio	1558	3
Vimercati*	maturatio	elixatio	assatio	cruditas	imperfecta elixatio	imperfecta assatio	1556	4
Sepúlveda	maturitas	elixatio	assatio	cruditas	inquinatio	concalfactio	1532	1
Camozzi	maturatio	elixatio	assatio	cruditas	pollutio	statheusis	1551	1
Piccolomini	maturatio	elixatio	assatio	cruditas	inquinatio	tostio	1540	4
Gesner*	maturatio	elixatio	assatio	acerbitas	imperfecta elixatio	imperfecta assatio	1586	1

* Périon includes Greek words for all except πέπανσις and ὠμότης. Vimercati included Greek words for μόλυνσις and στάτευσις in addition to the Latin terms. Gesner included all of the Greek terms, his work is a paraphrase rather than a translation.

All Renaissance translators translated πέψις and ἀπεψία as *concoctio* and *inconcoctio* respectively, save Palmieri, who retained *digestio* and *indigestio*, and Alcionio, who rendered them as *concoctio* and *cruditas*. In Piccolomini's translation of Alexander that retains *digestio* and *indigestio*. Camozzi's translations of Olympiodorus used *digestio* as well as *concoctio*.

Appendix: Incipits to translations of *Meteorologica* IV

a. Aristotle's Text (ed. F.H. Fobes (Cambridge, Mass., 1919)):

[378b10-20] Ἐπεὶ δὲ τέτταρα αἴτια διώρισται τῶν στοιχείων, τούτων δὲ κατὰ συζυγίας καὶ τὰ στοιχεῖα τέτταρα συμβέβηκεν εἶναι, ὧν τὰ μὲν δύο ποιητικά, τὸ θερμὸν καὶ τὸ ψυχρόν, τὰ δὲ δύο παθητικά, τὸ ξηρὸν καὶ τὸ ὑγρόν· ἡ πίστις τούτων ἐκ τῆς ἐπαγωγῆς· φαίνεται γὰρ ἐν πᾶσιν ἡ μὲν θερμότης καὶ ψυχρότης ὁρίζουσαι καὶ συμφύουσαι καὶ μεταβάλλουσαι τά θ' ὁμογενῆ καὶ τὰ μὴ ὁμογενῆ καὶ ὑγραίνουσαι καὶ ξηραίνουσαι καὶ σκληρύνουσαι καὶ μαλάττουσαι, τὰ δὲ ξηρὰ καὶ ὑγρὰ ὁριζόμενα καὶ τἆλλα τὰ εἰρημένα πάθη πάσχοντα αὐτά τε καθ' αὑτὰ καὶ ὅσα κοινὰ ἐξ ἀμφοῖν σώματα συνέστηκεν.

b. Medieval

Corpus Vetustius

Ms. Venezia, Biblioteca Marciana, Lat. VI. 47 (3464), fol. 184r:

Quoniam autem quatuor cause determinate sunt elementorum harum autem iuxta coniugationes et elementa quatuor contingit esse. Quarum sane due sunt active calidum et frigidum; due passive aridum et humidum. fides utique horum ex inductione. apparet namque in universis caliditas et frigiditas terminantes et permutantes unigenea et non unigenea. et humectantes et arefacientes nec non indurantes ac mollificantes. arida namque et humida formata. et quas dixit tollerantia passiones ipsaque secundum se ipsam et quecumque communia ex ambobus corpora constant.

Fragmentum Parisinum of Gerard of Cremona (apud Schoonheim, p. 144):

Capita primitiva elementorum quattuor sunt quattuor, sicut elementa composita, ex quibus sunt duo elementa agentia et duo elementa patientia. Verum duo elementa agentia sunt caliditas et frigiditas, et duo quidem elementa patientia sunt umiditas et siccitas. Illius vero demostratio est quod caliditas et frigiditas sunt distinguentes res et componentes eas et mutantes generata convenientia in genere et indurantes et umectantes. Umiditas autem et siccitas patientes sunt per se ipsas, et patiuntur propter eas omnia corpora composita ex eis.

Nova translatio from [Ps.-] Thomas Aquinas, *In libros Aristotelis Meteorologicorum expositio*, ed. R.M. Spiazzi (Rome, 1952), p. 645:

Quoniam autem quatuor causae determinatae sunt elementorum: harum autem secundum coniugationes et elementa quatuor accidit esse: quarum duae quidem factivae, calidum et frigidum; duae autem passivae, siccum et humidum. Fides autem horum est ex inductione. Videntur enim in omnibus caliditas quidem et frigiditas terminantes et copulantes et permutantes, et homogenea et non homogenea, et humectantes et exsiccantes, et indurantes et mollificantes: sicca autem et humida terminata, et alias dictas passiones patientia, ipsaque secundum se et quaecumque communia ex ambobus corpora constant.

Translation associated with Averroes, *In quartum librum Meteorologicorum* (Venice, 1562-1574), fol. 467v:

Postquam divisum est quod principia elementorum, quae sunt secundum modum formae, sunt quatuor, sicut est numerus elementorum, ex quibus componuntur: & duo sunt activa, & sunt calor, & frigus: & duo passiva, & sunt humidum, & siccum. Et signum huius est, quod calor, & frigus sunt ambo, quae componuntur res ad se invicem & admiscent, quousque res nova generetur: & universaliter istae duae potentiae, & virtutes sunt, quae mutant creaturas convenientes in genere unam in aliam. Sed siccitas, & humiditas sunt passiva in se ipsis istarum duarum qualitatum, & per illas duas patiuntur omnia composita. Et signum huius est. quod etiam Antiqui definiunt cum istis distinctionibus illas, & nominaverunt illas nominibus istorum: & dixerunt quod calor, & frigus erant virtutes....

Alexander of Aphrodisias, translated by William of Moerbeke. From: *Commentaire sur Les Météores d'Aristote*, ed. A.J. Smet (Leuven, 1968), p. 281:

Quoniam quattuor sunt causae elementorum determinatae... [282] videntur in omnibus caliditas et frigiditas terminantes et copulantes et permutantes...[283] congenea...

c. Renaissance

Mattia Palmieri, *Meteorologica,* Ms. Milano, Biblioteca Ambrosiana, L 40 sup (xv), fol. 92r:

Cum vero quatuor iam elementorum causae quae secundum coniunctionem, atque elementa. Quatuor esse continget, diffinitae sint: quarum duae quidem activae, calidum nempe & frigidum: quemadmodum ex inductione ostenditur. Nam in cunctis quidem caliditas, atque frigiditas ea quae aut eiusdem generis sunt, aut partibus differunt circumscribere coniungere et permutari videntur humectari quaque illis et exiccari: dura fieri ac mollia reddi apparent. Quae autem sicca humidaque determinata sunt:

tum et alia, quae eiusmodi pati dicuntur ipsa per se: tum quaecumque ex ambobus communia corpora constant.

Joachim Périon, *Aristotelis Stagiritae Meteorologicorum liber quartus* (Wittenberg, 1585), [A2]:

[The version amended by Nicolaus Grouchy is essentially identical. Cf. Aristotelis Stagiritae tripartitae philosophiae opera omnia absolutissima, ex optimis quibusque, maxime novis interpretibus collecta, aliquot libris aucta, recognita, argumentis valde bonis in singulos libros & capita, multis.... (Basel, 1563)]

Quoniam quatuor causae elementorum expositae sunt, ex earumque copulationibus quatuor etiam elementa consecuta sunt, quae quidem earum faciendi, calor & frigus, reliquae duae patiendi vim habent: idque inductione doceri potest. Videntur enim in omnibus rebus calor & frigus tum determinare, tum copulare, tum mutare, tum humida reddere, tum exiccare, tum durare, tum mollire, tam ea, quae eiusdem generis sunt, quam ea quae diversi: siccitatem vero & humorem determinari, aliasque affectione, s [sic] quae expositae sunt accipere [A3] tum ipsas per se, tum corpora, quae ex utraque earum constant....

Pietro Alcionio, *Aristotelis Stagiritae Opera* (Lyon, 1578), vol. 1, p. 761:

Cum quatuor elementorum causae allatae iam sint, pro earum certe paribus accidit quatuor etiam enumerari elementa. illarum quidem duae agentes habentur, calor, & frigus: duae patibiles, siccitas, & humor. Horum autem omnium fidem ipsa inductio facere potest: nam calor & frigus in rebus naturae omnibus videntur figurare, suisque finibus describere, coagmentare: & quae tum eodem similique genere, tum dissimili sunt, mutare, humectare, arefacere, obdurare, & emollire, siccitas autem & humor tum illa separatim, tum quae communiter ex illis compacta sunt, videntur finibus describi, & alias a calore & frigore contrahere affectiones....

Giovanni de Camozzi, *In Meteora Aristotelis Commentarii* (Venice, 1551), fol. 76r:

Quoniam quatuor elementorum causae decretae sunt... [76v] Quia duae sunt causae efficientes calidum videlicet & frigidum. Duae autem materiales, siccum videlicet & humidum...

François Vatable in *Operum Aristotelis Stagiritae philosophorum omnium longe principis, nova editio, Graece & Latine*, ed. Isaac Casaubon (Lyon, 1590), vol. 2, p. 359:

Cum autem elementorum causas quatuor esse a nobis definitum sit, & iuxta coniugationes harum ipsas quoque elementa esse quatuor acciderit, e quibus duae quidem, calor inquam & frigus, agere solent: duae vero ariditas, videlicet & humiditas, pati: (cuius rei fides ex inductione sumi potest.) nam in omnibus calor & frigus, tam quae unius, quam quae diversi sunt generis definire, copulare, transmutare, humectare, arefacere, indurare, & mollificare videntur: arida vero & humida, tum ipsa per se, tum communia corpora quaecunque constant ex utrisque, definiri, & caeteros quos diximus affectus pati. praeterea & hoc quoque rationibus ipsis quibus eorum naturas definimus, perspicuum existit.

Francesco Vimercati, *In quatuor libros Aristotelis Meteorologicorum commentarii et eorundem librorum e graeco in latinum per eundem conversio* (Paris, 1556), fol. 163r:

Quoniam autem elementorum causae quatuor sunt constitutae, ex quarum coniugationibus evenit, ut quatuor quoque existerent elementa, duae illarum efficiendi vim habent, calor & frigus, duae vero patiendi, siccitas & humor. id quod inductione probatur. Calor enim & frigus, tum ea, quae eiusdem, tum quae diversi generis sunt, determinare, coniungere, immutare, humectare, exiccare, durare, emollire, in omnibus videntur. Humida autem & sicca, tum ipsa per se, tum quae communiter ex utrisque concreta sunt corpora terminari, & aliis, quae dictae sunt, qualitatibus affici.

Francisco Vallés, *In quartum librum Meteorologicorum Aristotelis commentaria* (Torino, 1583), p. 4:

Quandoquidem quatuor elementorum causae expositae iam sunt, atque pro earum coniugiis quatuor etiam elementa esse contingit: illarum quidem duae agentes habentur calidum & frigidum: duae patibiles, siccum & humidum. Horum autem omnium fides ex inductione est. nam calor & frigus in omnibus videntur finire, coniungere, tum quae eodem genere, tum quae diverso sunt, mutare, humectare, exiccare, durare & mollire: sicca autem & humida, cum illa per se, tum quae ex illis ambobus compacta sunt corpora, finibus describi, & alia pathemata dicta pati. Rationibus etiam quibus naturas illorum definire consuevimus, fit manifestum.

Juan Ginés de Sepúlveda, *Opera Aristotelis latina facta* (Paris, 1532), fol. 29r:

Elementorum causa quatuor esse, supra est a nobis definitum. harum enim copulis quatuor existunt elementa, quarum duae sunt affectivae calidum, & frigidum: duae passivae siccum & humidum. quarum rerum fides non solum inductione fit, cum calor & frigus in cunctis rebus finire, concernere, mutare tum cognata, tum quae diversi generis sunt: ad haec humectare desiccare, durare, atque mollire notentur: sicca vero & humida tum ipsa per se tum corpora communia, quae ex ambobus conflantur, finiri, & caeteras quae memoratae sunt affectiones pati, sed etiam rationibus, quibus ipsorum naturas definimus.

Joëlle DUCOS

TRADUIRE LA SCIENCE EN LANGUE VERNACULAIRE: DU TEXTE AU MOT

0. Introduction

La traduction médiévale, ses pratiques et sa théorie sont, depuis plusieurs années, un objet d'intérêt pour les chercheurs[1]. Par ailleurs, les traductions scientifiques latines sont de mieux en mieux connues, en particulier grâce à un travail important d'édition et de nombreux travaux sur les procédures, la transmission des textes, les erreurs, les compléments et les écoles de traduction. C'est qu'elles constituent l'un des vecteurs privilégiés du savoir au moyen âge, qu'il s'agisse de traductions arabo-latines ou gréco-latines. Elles s'insèrent dans les pratiques universitaires de la communication scientifique, le commentaire et l'interrogation en étant les fondements. Les entreprises de traduction vernaculaire s'installent avec un décalage par rapport au latin et se développent dans différents pays entre le XIIIe et le XIVe siècle: des traductions scientifiques en langue d'oc, en catalan, en italien apparaissent et constituent un fonds important. En langue d'oïl, on traduit la science véritablement à partir du XIIIe siècle et considérablement au XIVe siècle. Cette activité de traduction n'est pas en marge des autres types: elle est vivante, importante, et si l'on sait l'importance des milieux princiers ou royaux dans leur conception, elle dépasse largement l'enjeu d'œuvres de commande. Une théorie s'exprime, principalement avec Nicole Oresme qui pose la question du français comme langue de savoir à côté du latin et du grec, ce qu'a pu mettre en évidence S. Lusignan[2]. Les études sur les prologues des traducteurs permettent de comprendre le dessein de certains textes. Toutefois la théorie fondatrice de ces textes n'apparaît que par le biais des pratiques plutôt que par ce qu'en dit le traducteur: la question de la littéralité, qui est toujours invoquée, ne correspond pas toujours à la technique de

1. Voir en particulier la revue *The Medieval Translator*, les travaux de C. Buridant, R. Copeland, R. Ellis, S. Lusignan et G. Di Stefano par exemple et les nombreux colloques sur les pratiques de traduction tant latines qu'en langue vernaculaire depuis une dizaine d'années.

2. Lusignan S., *Parler vulgairement au Moyen Age* (Paris-Montréal, 1986), p. 154-166.

traduction, laissant penser qu'elle n'est que l'expression d'un souci de fidélité. Il est clair aussi que la finalité des textes de traductions varie beaucoup selon la nature du texte traduit, peut-être plus que celle du lectorat supposé ou exprimé. Mais par ailleurs, les traducteurs, qu'ils soient anonymes ou nommés, se heurtent à une difficulté toujours exprimée qui est la transmission du savoir par une équivalence lexicale.

La question de la vulgarisation française se pose autour de ces deux axes. C'est d'abord le genre de texte à traduire ou du texte produit: traduire suppose en effet une invention de formes textuelles et non une simple équivalence. Le corpus de textes traduits en langue d'oïl marque une diversité des choix et des pratiques. La question de la fidélité doit alors être posée, non pas en termes d'une équivalence ou d'une prise de position de départ, mais plutôt dans une pratique qui est éclairante, non seulement sur la vulgarisation, mais aussi sur les savoirs latins. Mais la question de l'équivalence au niveau du mot ne doit pas être oubliée, car elle aussi est significative de l'enjeu des textes et donne des pistes sur l'utilisation effective des textes et ainsi de leur destinataire.

1. Diversité des pratiques et des genres

Parler d'une traduction scientifique française au moyen âge n'est guère révélateur de la réalité de l'activité à partir du XIIIe siècle, car c'est supposer une unité qui n'existe pas. L'aire géographique peut amener des spécificités à la fois dans le type de texte traduit et dans la pratique[3]. Mais la multiplicité des formes et des types de textes traduits donne à penser que ce mouvement est à la fois important et d'une grande diversité. Quoi de commun en effet entre les grandes traductions aristotéliciennes, les recettes médicales, les pronostics, les traités de chirurgie, les traités de peste, les traités d'astronomie ou encore les encyclopédies? S'il est clair que la production scientifique française est largement constituée par des traductions, leur fonction et leur forme sont profondément différentes, les unes pratiquant l'abrégé comme l'adaptation de Guy de Chauliac, les autres au contraire privilégiant la glose et le commentaire. Cet ensemble, dont l'unité vient d'un contenu scientifique et d'un enjeu, vulgariser le savoir, s'inscrit par ailleurs dans un mouvement général de traduction

3. Voir en particulier dans l'aire anglo-normande l'importance des traductions scientifiques. Cf. Hunt T., *Popular Medicine in 13th century England* (Cambridge, 1990), *The Medieval Surgery* (Woodbridge, 1992), *Anglo-Norman Medicine* (Woodbridge, 1994-1997), *Three Receptaria from Medieval England: the Language of Medicine in the 14th Century* (Oxford, 2001).

dont il partage certaines caractéristiques. Ajoutons que l'étendue chronologique entre le XIIIe siècle et le XVe siècle contribue également à la diversité: elle n'est pas seulement liée à l'évolution linguistique ou dialectale, mais participe aussi d'une conception plus large du rôle du français dans la transmission des savoirs et d'une fonction de la langue vernaculaire dans la communication scientifique.

Cette hétérogénéité des textes, en longueur, en date, en thème, en discipline, laisse à penser que chaque traduction sécrète ses propres règles et que finalement il n'est guère possible de marquer des tendances ou des pratiques communes. Pourtant des points communs se dessinent: des conceptions de traduction, des options peuvent être dégagées. En 1995, Charles Bruckner, à l'occasion d'un colloque[4], avait tenté une typologie des traductions au XIVe siècle à partir de critères internes et externes dans un corpus varié, constitué des traductions les plus célèbres du XIVe siècle comme les traductions oresmiennes, la traduction des *Evangiles* et des *Epistres* de Jean de Vignay, la traduction du *Policratique* de Jean de Salisbury par Denis Foulechat; celle de Jean Daudin de l'*Epitre consolatoire* de Vincent de Beauvais. Parmi ces critères, intervient le prologue, dont la présence est généralement conjointe avec celle de gloses à l'intérieur du texte, ce qui, selon C. Bruckner, signale une traduction plus soignée, ou du moins marquant une mise à distance des procédés de traduction, ce qui révèle une réflexion du traducteur sur ses propres pratiques. On peut noter cependant qu'à l'intérieur de ce corpus, figure une traduction de Nicole Oresme, celle des *Economiques,* qui est sans prologue, ce qui montre la difficulté d'établir une loi générale: les trois autres traductions aristotéliciennes de Nicole Oresme ont un prologue et le traducteur est sans doute l'un de ceux qui a le plus clairement exprimé sa théorie de la traduction. Quoi qu'il en soit, ce premier critère paraît majoritairement valable, le prologue permettant au traducteur de présenter, outre le dédicataire, son œuvre et généralement la raison et la conception qui ont présidé à l'écriture de la traduction. Dans le domaine des traductions scientifiques, on peut noter en effet que les textes les plus longs, les plus importants, tant dans la pensée scientifique que dans l'évolution de la traduction française, sont majoritairement accompagnés d'un prologue[5].

4. Brucker C., 'Pour une typologie des traductions en France au XIVe siècle', in: C. Brucker (éd.), *Traduction et adaptation en France, Actes du colloque de l'université de Nancy, 23-25 mars 1995* (Paris, 1997), p. 63-79.

5. Voir les traductions de Nicole Oresme, Mahieu le Vilain, Evrart de Conty, Jean Corbechon, Martin de Saint-Gilles, Guy de Chauliac, mais aussi celles qui sont anonymes, comme celle du traducteur de Léopold d'Autriche.

Le prologue signale ainsi le caractère prestigieux de la traduction et peut-être même son statut d'autorité dans le domaine du savoir scientifique.

Toutefois, le métier et les pratiques du traducteur ne sont pas toujours révélés par ce texte inaugural. Si on lit celui qui introduit la *Compilation de la science des estoilles* de Léopold d'Autriche[6], on peut constater la disparition du traducteur dans le prologue: c'est que ce dernier ne fait que traduire l'original latin. La présence d'un prologue témoigne donc plutôt dans ce cas de l'absence de la voix du traducteur et non d'un texte d'autorité. Inversement, dans d'autres textes, il peut être attribué à l'auteur, alors que le traducteur s'y exprime dans un entrelacs d'extraits tirés de l'original et d'énoncés qui lui sont propres[7]. La question du prologue n'est donc qu'un critère externe de classement, qui se révèle bien plus complexe dans la pratique.

Il peut exister un écart entre les propos du traducteur dans son introduction et sa pratique. Tel est le cas de Mahieu le Vilain qui, dans le prologue du *Livre des Météores*, affirme *translater le devant dit livre de mot a mot, si comme je pourray miex en langue franchoise*[8]. L'examen précis de l'ensemble du texte prouve au contraire un écart avec la version latine de Guillaume de Moerbeke. Outre les passages volontairement expurgés ou ceux qui sont ajoutés, le texte présenté comme celui d'Aristote est en fait un ensemble composite, constitué de la traduction de la *nova*, mais aussi de la paraphrase de Thomas d'Aquin avec également des extraits du commentaire d'Alexandre d'Aphrodisias traduit par Guillaume de Moerbeke[9]. On ne saurait parler véritablement de traduction littérale, mais bien plutôt d'une paraphrase en français. Ainsi, quand le traducteur s'arrête de traduire ou quand il fait une longue digression, il marque clairement qu'il y a bien rupture par rapport à la lettre: la littéralité doit être comprise, non pas comme une traduction mot à mot, mais comme une paraphrase qui suit l'ordre d'Aristote[10]. Par ailleurs, il faut noter que, par

6. Léopold d'Autriche, *Li compilacions de la science des estoilles, Book I-III*, éd. F.J. Carmody (Berkeley – Los Angeles, 1947).

7. Voir en particulier le cas du *Livre de Albubeth*, œuvre d'origine arabe traduite en français: Jacquart D., 'De l'arabe au moyen français, en passant par le latin: "le livre de Albubeth"', in: M. Folkerts & R. Lorch (éds), *Sic itur ad astra, Studien zur Geschichte der Mathematik und Naturwissenschaften* (Wiesbaden, 2000), p. 286-287.

8. Mahieu le Vilain, *Livre des Méteores*, éd. R. Edgren (Uppsala, 1945), p. 1.

9. Ducos J., 'L'œuvre de Mahieu le Vilain: traduction et commentaire des *Météorologiques*', in: J. Hamesse (éd.), *Les traducteurs au travail, leurs manuscrits et leurs méthodes* (Turnhout, 2001), p. 285-309.

10. À la fin d'une digression sur les comètes: 'Et nous ne parlon pas en cest livre de choses qui aviennent par miracle, mais de choses qui aviennent par nature' (Mahieu le Vilain, *Livre des Méteores*, p. 43); après un *excursus* sur la France et la *nachele saint*

comparaison à ce que fait le traducteur anonyme de *La compilation* de Léopold d'Autriche, Mahieu le Vilain ne reproduit pas la première personne présente dans Aristote, mais la réorganise par des phrases introductrices comme *or dit le philosophe, or veut le philosophe enquerre et savoir, or veut le philosophe dire et savoir* ou présente un apparat rhétorique qui distingue entre la parole des adversaires, ce qui relève du doute ou de l'argument. La traduction est ainsi mise en valeur de la dialectique et de la rhétorique aristotélicienne et non pas seulement équivalence dans une langue vernaculaire. Le prologue ne peut donc être véritablement compris qu'à la lueur de la pratique: traduire mot à mot signifie en effet une fidélité à la logique et à l'organisation du discours aristotélicien et non une pratique littérale au sens contemporain du terme.

La présence du prologue n'est pas non plus toujours un indice clair de la stratégie employée par le traducteur dans son travail. À titre d'exemple, celui de Jean Corbechon est nettement en rupture par rapport à sa pratique de traduction[11]. Le prologue de la traduction française est constitué de deux parties. La première est de Jean Corbechon où il se présente en tant que tel et situe son entreprise par rapport à la question de la *translatio studii*, Charles V, roi lettré, étant considéré comme l'héritier de Jules César, de Ptolémée et de Charlemagne. La deuxième est la traduction du prologue de Barthélemy l'Anglais repris dans son intégralité. Ce double prologue laisse supposer que la voix du traducteur soit présente de manière importante dans la traduction qui suit. Or, à part quelques rares cas d'intrusion du traducteur, le texte traduit ne présente que peu d'additions et, bien au contraire, abrège parfois, supprime des références. La présence d'un prologue n'est donc pas significative d'un type de traduction, ni par ce qu'en dit le traducteur, ni par son unique présence.

Un seul critère ne permet donc pas de définir le type d'œuvres, et plusieurs sont nécessaires. Il paraît assez évident que la source du texte détermine aussi le genre qu'emploie le traducteur et qu'il faut croiser prologue, absence ou présence de gloses pour tenter d'élaborer quelques critères. Si l'on compare en effet les textes précédemment cités, on peut

Pierre: 'Or lesseray je ce a dire et revendray a nostre propos' (Mahieu le Vilain, *Livre des Méteores*, I, 9, p. 70); pas de description géométrique du halo: 'et que ce couviengne il estre, pourroit l'en prouver par description en figure de geometrie, mez la figure est trop forte a entendre et de poi de proffit, pour quoy orendroit je m'en soufferay' (Mahieu le Vilain, *Livre des Méteores*, p. 142).

11. Pour l'édition de ce prologue voir Ribémont B., 'Jean Corbechon, un traducteur encyclopédiste au XIVe siècle', *Cahiers de Recherche Médiévale (XIIIe-XVe s.)* VI (1999), p. 95-97.

constater que les textes aristotéliciens ou les *Aphorismes* d'Hippocrate, par exemple, comportent un commentaire important, alors que la traduction de Jean Corbechon et celle de l'œuvre de Léopold d'Autriche se présentent sans trace explicite du traducteur. Une distinction claire s'établit ainsi entre les textes anciens et ceux qui ont été écrits par des auteurs médiévaux: le texte ancien, d'origine grecque ou autre, connu par une version latine, donne lieu à une traduction assortie d'un commentaire.

Pourtant là encore, il s'agit de nuancer. En effet certains textes anciens traduits en français ne donnent pas lieu non plus à un commentaire. Par exemple, il existe toute une tradition de pronostics traduits en langue vernaculaire, sans aucun commentaire. C'est le cas des pronostics sur les tonnerres qui relèvent de la tradition ancienne du calendrier brontoscopique, et qui sont la trace de la traduction byzantine de Nigidius Figulus, traduite vraisemblablement au IXe siècle[12]. L'ancienneté n'est donc pas le seul critère, mais cet exemple prouve aussi que la finalité et le caractère vivant de l'œuvre dans la vie universitaire conditionnent le type de traduction. En effet, si ce type de pronostics demeure dans la tradition astronomique, comme en témoigne la compilation de Léopold d'Autriche, en revanche, ils ne sont plus mentionnés dans les commentaires scolastiques, ni à titre de compléments ni comme matière à discussion. Leur caractère ancien n'est donc pas un critère suffisant pour amener un commentaire dans la traduction et ils sont assimilés aux multiples textes à visée pratique comme les recettes, les pronostications, traités sur les fièvres dont ils partagent deux caractéristiques, l'absence d'auteur explicite et le caractère très abrégé de l'écriture qui s'apparente à celle des adages.

De fait, il faut distinguer entre texte court, fragmentaire ou non, qui constitue un volume important de la traduction en langue vernaculaire, et les traités longs, signalés également par un auteur qui fait partie des autorités reconnues par l'université. Il ne faut pas oublier que, dans ce que nous connaissons des traductions scientifiques françaises, une large part est constituée de fragments[13], d'abrégés, la traduction intégrale étant

12. Draelants I., 'Astrologie et divination dans le ms 77 Bibl. Univ. Liège, f°62 r°-70v°', *Troisième congrès de l'association des cercles francophones d'histoire et d'archéologie de Belgique* (Namur, 1990), p. 319-335. Ducos J., 'Astrométéorologie et vulgarisation: le livre VI de *Li compilacions de le science des estoilles* de Léopold d'Autriche', in: D. Jacquart, D. James-Raoul & O. Soutet (éds), *Par les mots et les textes* (Paris, 2005), p. 251-253.

13. Voir par exemple Tovar C. de, 'Les versions françaises de la *Chirurgia parva* de Lanfranc de Milan. Etude de la tradition manuscrite', *Revue d'Histoire des textes* XII-XIII (1982-1983), p. 196, signalant un fragment de manuscrit qui est en fait la traduction la plus ancienne de la *Chirurgia parva*.

finalement l'exception. Cette dernière est d'ailleurs souvent réussie et marque un soin extrême pour une œuvre qui n'est pas une entreprise éphémère, mais au contraire une sorte de monument du savoir. L'importance du fragment, de l'extrait dans la traduction française est cependant frappante. La traduction anonyme de la *Compilation* de Léopold d'Autriche, conservée dans un unique manuscrit, présente des lacunes, des passages non traduits signalés par un folio blanc: ce travail n'est donc pas achevé, mais paraît plutôt comme une traduction un peu hâtive. Il est vrai cependant que le manuscrit présente d'autres textes incomplets, ce qui laisse à penser que le copiste n'a pas terminé son travail. Il reste cependant que la traduction de fragments d'œuvres est une pratique bien connue. Ainsi C. Thomasset a pu opposer la transmission de l'œuvre de Henri de Mondeville à celle de Guy de Chauliac: si la première est transmise intégralement, la deuxième est 'soumise à un démembrement systématique', selon les termes de C. Thomasset[14], chaque chapitre étant traduit séparément et doté en quelque sorte d'une matière autonome. L'état le plus abrégé du texte vernaculaire est celui de la définition, le texte étant réduit à une série de noms à définir, suivant un ordre généralement alphabétique. L'on peut penser que l'une des pratiques de la traduction scientifique est alors assez proche du florilège et fonctionne comme aide-mémoire d'œuvres plus importantes et on pourrait imaginer une pratique vivante de la traduction en langue vernaculaire dans l'activité scientifique, dont nous n'avons plus gardé que ces quelques traces dans les manuscrits.

Rien de comparable à première vue entre ces fragments, ces ébauches et les grandes œuvres traduites. Pourtant, un examen plus précis des textes marque une continuité: en effet l'encyclopédie de Barthélemy l'Anglais traduite par Jean Corbechon fonctionne elle aussi comme un ensemble d'éléments brefs réunis autour de thématiques. Chaque chapitre paraît ainsi comme l'agglomérat de fragments de savoir attribués ou non à une autorité. À la limite, l'encyclopédie n'est qu'une somme de fragments, ce qui explique finalement l'absence de glose, la présence de l'auteur étant limitée à un choix des citations et à un ordre. Le traducteur n'est alors qu'un intermédiaire qui communique un savoir composite dans une autre langue, mais n'a pas à expliciter une pensée ou une logique dans la mesure où le texte-source n'est pas issu d'une seule voix, mais de plusieurs, réunies par la grâce d'un compilateur. Le rapprochement avec la traduction

14. Thomasset C., 'La médecine', *GRMLA*, VIII, 1 (Heidelberg, 1998), p. 315.

de Léopold d'Autriche est alors inévitable puisque Léopold se présente comme un compilateur et non un auteur: 'Li nons de l'aucteur ne soit mie demandés, car pluseurs ont estét aucteur et ne mie uns seus: je suis leurs loiaus compileres et diligens'[15].

Cette continuité de pratiques entre le traducteur de Léopold d'Autriche, Jean Corbechon et les textes courts et anonymes marque la différence nette entre l'autorité et la compilation. Mais elle est aussi révélatrice sur la pratique encyclopédique qui fonctionne comme un ensemble de fragments, une collection en quelque sorte, plutôt que comme une œuvre unitaire. C'est en quoi la traduction et la vulgarisation encyclopédique sont en continuité, avec le même rapport au savoir, qu'il soit garanti par une autorité ou qu'il soit présenté comme anonyme.

Bien différente est la traduction aristotélicienne telle que l'ont pratiquée Mahieu le Vilain, Evrart de Conty et Nicole Oresme. Le texte est en effet accompagné d'un commentaire qui le soutient, l'éclaire, le *desclaire*, selon l'expression d'Evrart de Conty. Ce commentaire prend des formes différentes, depuis la simple glose lexicale destinée à éclaircir le sens d'un néologisme jusqu'au développement sur une thématique exprimée ou évoquée par Aristote en passant par des gloses, que l'on peut qualifier de philologiques et qui invoquent la corruption du texte et l'obscurité du texte latin. Dans cette pratique, Mahieu le Vilain, qui écrit à la fin du XIIIe siècle, apparaît comme fondateur, puisque c'est lui qui a inauguré ce mouvement de traduction, créant en quelque sorte un genre qui est celui de la traduction aristotélicienne ou d'autorité. Il ne s'agit pas de traduire seulement, mais bien d'initier à un savoir qui est celui de la scolastique exprimé dans les commentaires. Aussi les développements interviennent-ils pour les questions en débats à l'université ou peuvent-ils remplacer les questions litigieuses comme celle de l'éternité du monde. C'est aussi une rhétorique transposée en français, parce qu'il s'agit de celle du savoir universitaire: le texte d'Aristote est composé de questions, de *doutes* auxquels Aristote répond par des arguments. Le traducteur amplifie les réponses par des explications supplémentaires empruntées aux commentaires. La relation entre traduction et pratiques universitaires est ainsi visible et elle apparaît aussi bien comme mode de pensée, savoir vulgarisé que comme matrice d'écriture du texte, qui ne peut se comprendre véritablement sans être situé en continuité avec les pratiques d'écriture scolastique. Il ne s'agit pas d'une imitation, mais d'une

15. Léopold d'Autriche, *La compilacion...*, p. 56: il s'agit d'une traduction du texte latin: le *je* désigne Léopold d'Autriche.

imprégnation qui amène la création d'un genre dans les traductions françaises[16].

Evrart de Conty et Nicole Oresme sont dans la continuité de la traduction de Mahieu le Vilain, dont deux manuscrits étaient présents dans la Librairie de Charles V. Leur traduction comporte également plusieurs niveaux de commentaires: une mise en forme rhétorique et dialectique de l'œuvre initiale, une glose lexicale et un commentaire qui complète, mais aussi discute les propositions d'Aristote. La nature des traités toutefois transforme l'articulation du commentaire: Nicole Oresme segmente le texte en propositions qu'il commente plus ou moins longuement, alors qu'Evrart de Conty, en raison du caractère éclaté des *Problèmes*, sépare son commentaire de la traduction-paraphrase. L'amplification du texte-source est ainsi considérable, et le commentaire apparaît à la fois comme une aide à la lecture mais aussi comme la présence claire d'une pensée et d'une écriture propre au traducteur[17]. L'écart par rapport aux traductions littérales ou fragmentaires est ainsi marqué, puisqu'il est clair qu'il ne s'agit pas seulement de communiquer un texte, mais aussi une pensée aussi bien aristotélicienne que scolastique. Paradoxalement, le caractère abrégé demeure: l'adverbe *briefment* privilégié par Evrart de Conty le prouve[18]. Le commentaire procède de la même manière que l'encyclopédie ou les compilations. Il s'agit de donner un aperçu condensé du savoir, qu'il soit déclaré comme vrai ou en débat, et on ne peut que noter combien l'écriture aristotélicienne est aussi transformée en français: dans la pratique de Nicole Oresme et de Mahieu le Vilain, l'œuvre éclate en propositions glosées, l'articulation de l'ensemble soulignée par des récapitulatifs réguliers paraissant finalement moins importante que ces sortes d'exposés en réduction du savoir.

Ces échos entre les différentes pratiques laissent alors à penser que les critères habituels de classification des traductions ne fonctionnent que partiellement dans ces textes. La traduction littérale, voire abrégée, signifie sans doute un savoir-faire ou un savoir moins prestigieux que pour les traductions d'autorité, mais elle ne suppose pas nécessairement

16. La personne du traducteur et sa formation ont certainement joué un rôle dans l'intégration de la rhétorique scolastique: il est clair que Mahieu le Vilain est issu de l'université parisienne.

17. Voir pour Evrart de Conty et son commentaire scientifique, les analyses de D. Jacquart qui en a souligné à la fois la filiation par rapport à l'université mais aussi l'originalité, *La médecine médiévale dans le cadre parisien* (Paris, 1998) et 'Médecine et morale: les cinq sens chez Evrard de Conty', *Micrologus* X (2002), p. 365-378.

18. Guichard-Tesson F., 'Le métier de traducteur et de commentateur au XIVe siècle d'après Evrart de Conty', *Le Moyen Français* 24-25 (1989), p. 131-167.

un lectorat laïc ou ignorant: il paraît assez évident que le mode fragmentaire et discontinu est un mode de vulgarisation en latin comme en français et que les pratiques françaises s'inscrivent dans cette filiation, tout en développant d'autres modes d'écriture. En conséquence, le lectorat de telles œuvres ne peut, pour comprendre le fonctionnement de l'écriture française, qu'être familier avec des pratiques universitaires ou monastiques de vulgarisation. De même, l'association entre fidélité et littéralité ou fidélité et sens ne se pose pas dans ces termes pour la traduction scientifique: la présence ou l'absence d'une glose, la pratique de la paraphrase plutôt que d'une traduction proche de la lettre n'est pas significative d'une absence ou d'un souci de fidélité par rapport au texte original. Elle correspond plutôt à des enjeux différents entre des textes appartenant à la littérature secondaire en latin et fonctionnant comme des outils, et d'autres qui relèvent de l'autorité et invitent à l'explicitation, la mise en débat éventuelle ou à la création, dans une écriture qui se sépare, dans ses thématiques comme dans son mode d'exposition, de celle d'Aristote.

2. Traduction et lexique

La fidélité et la littéralité affirmées par les traducteurs supposent aussi un important travail sur les équivalents lexicaux. Elles sont liées au transfert sémantique qui intervient au niveau du mot dans la traduction. La préoccupation générale des traducteurs vient justement de cette question du langage: l'écart entre le latin et le français est souvent présenté comme une insuffisance de la langue vernaculaire par rapport à d'autres. Telle est l'opinion de Mahieu le Vilain affirmant que *l'en ne puet si proprement translater science en franchois comme en latin*[19]. La difficulté de l'équivalence se pose avec d'autant plus d'acuité pour la langue scientifique, car elle intègre des emprunts, mais aussi des usages lexicaux qui lui sont propres. Vulgariser la science, c'est ainsi permettre au lecteur de s'approprier un langage spécialisé. Mais ce peut être aussi garder un lexique usuel, en en transformant le signifié. Si les pratiques de traduction varient du point de vue du genre, elles varient aussi au niveau du mot en fonction de la chronologie, mais aussi du type de texte. En effet, si l'on confronte différents genres de traduction, chaque époque a des tendances différentes: c'est ainsi que Jean de Meun[20], comme Mahieu le Vilain à

19. Mahieu le Vilain, *Livre de meteores*, p. 1.

20. Billotte D. & Hicks R., 'Jean de Meun lexicographe: usage de la réduplication synonymique dans deux traductions', in: C. Brucker (éd.), *Traduction et adaptation en*

la fin du XIIIe siècle, pratique peu l'emprunt direct et encore moins le binôme synonymique, considéré pourtant comme l'un des moyens privilégiés de la traduction pour introduire un néologisme, le nouveau mot s'appuyant sur un autre usuel. Inversement, au XIVe siècle, est pratiqué couramment l'emprunt, qui est le procédé le plus courant, aussi bien dans les textes courts que dans les textes longs[21]. Quant au binôme synonymique, il reste minoritaire: à titre d'exemple, Jean Corbechon, qui traduit un texte où ce procédé est courant en latin, ne l'utilise guère et préfère d'autres procédures. Certains textes intègrent facilement des mots latins ou d'autre origine, en particulier quand il s'agit de textes très spécialisés ou techniques[22]. D'autres ne le font que sur certaines thématiques, comme c'est le cas pour les textes météorologiques où les translittérations apparaissent principalement à propos des météores ignés, des vents. La traduction apparaît ainsi comme un précieux témoignage sur la langue et ses possibilités, aussi bien d'un point de vue lexical que syntaxique, car les traducteurs scientifiques l'exercent dans la mesure où ils tentent d'adapter une expression latine à des structures françaises. Il est très rare que la traduction soit l'exacte reproduction d'une syntaxe ou d'un lexique latin, alors que la pratique latine peut parfois être plus proche de la langue d'origine. La traduction française n'est en effet pas à usage strictement savant, mais se double d'un souci de vulgarisation. Aussi garde-t-elle une structure propre à la langue vulgaire, même si, comme ont pu le démontrer Claude Buridant et Michèle Goyens, elle contribue à son évolution[23].

Il reste que l'équivalence du mot paraît l'enjeu le plus important dans la traduction de l'avis même des traducteurs. Cette équivalence peut se faire sur plusieurs modes: en dehors de la translittération, l'emprunt avec lexicalisation française, le néologisme sur un radical français, le calque

France, Actes du colloque de l'université de Nancy, 23-25 mars 1995 (Paris, 1997), p. 141-157; Billotte D., *Le vocabulaire de la traduction par Jean de Meun de la Consolatio Philosophiae* (Paris, 2000).

21. Pour un aperçu de la diversité des pratiques et des usages à partir du XIVe siècle, voir Jacquart D. & Thomasset C. (éds), *Lexique de la langue scientifique (Astrologie, Mathématiques, Médecine), Matériaux pour le Dictionnaire de Moyen Français* (Paris, 1997).

22. C'est en particulier le cas pour les textes astronomiques.

23. Buridant C., 'La traduction intralinguale au Moyen Age et à la renaissance (les modernisations, remaniements et réécritures de textes médiévaux et leur intérêt pour l'histoire de la langue française et les caractéristiques typologiques de son évolution)', *Perspectives médiévales* (2000), p. 29-50; Goyens M. & Van Hoecke W., 'La traduction comme témoin de l'évolution linguistique', in: R. Lorenzo (éd.), *Actas do XIX Congreso Internacional de Linguistica e Filoloxia Romanicas, Universidade de Santiago de Compostela, 1989* (Coruña, 1992), p. 13-32; Goyens M. & Van Hoecke, W. 'La traduction comme source pour l'étude d'anciens états de langue', *Le Moyen Français* 44-45 (2000), p. 243-264.

sémantique, où à un équivalent courant est ajoutée une signification scientifique et spécifique, ou encore l'équivalent dans le langage courant soit dans la même catégorie grammaticale soit par transfert: c'est ainsi que le nom *antiperistasis* peut être traduit par différentes lexies comme un syntagme comme *contraire repercussion*, ou des verbes comme *assembler et rebouter* ou *aüner et revertir*, ou une phrase où apparaissent généralement l'adjectif *contraire* et le verbe *ressortir*[24]. Inversement, si *corruptio* peut être traduit dans Jean Corbechon par *corruption,* le nom français peut correspondre à deux verbes, *destruere et consumere*. La variété des équivalents des adjectifs en *-ivus* fréquemment employés par Barthélemy l'Anglais relève du même ordre: adjectifs, noms, verbes, relatives, les équivalents se multiplient.

La comparaison entre les textes aboutit alors à une multiplicité des équivalences où progressivement un accord semble se faire sur des désignations: c'est ainsi qu'*impression* qui se distinguait clairement de *passion* en suggérant une approche astronomique du phénomène et non physique, devient progressivement, dans toutes sortes de textes médicaux, astrologiques et même littéraires, le nom spécifique des phénomènes météorologiques, ce qui reproduit l'évolution également du latin. Mais cet accord sur le signifiant ne détermine pas nécessairement une identité de signifié: *vapeur, fumee, exalation* sont toujours présents, mais ont la même diversité sémantique qu'en latin selon les auteurs. De même les adjectifs *gros, espés, subtil*, qui désignent les états de la matière, auxquels il faut ajouter *tenve* et *condempsié*, restent flous, certains auteurs privilégiant *gros* dans un contexte humide à la différence d'*espés* qui apparaît plutôt dans un contexte de sècheresse[25]. Les dictionnaires donnent ainsi une image assez faussée de la réalité lexicale car, outre un écart parfois avec les éditions les plus récentes des textes, ils ne permettent pas de mesurer la variété des procédures: ainsi la traduction d'un mot par une phrase descriptive ne figure jamais dans un lexique. C'est le cas pour les tropiques dans la traduction de Mahieu le Vilain: la lexie est transformée en description du phénomène (*point la ou le soleil est au plus bas en yver*; *point la ou le soleil est plus haut en esté*). L'absence de variation dans l'équivalence laisse penser qu'il s'agit bien de lexies figées qui reposent sur l'étymologie de *solstitium*[26]. Dans ce cas peut-on

24. Ducos J., *La météorologie en français* (Paris, 1998), p. 229-232.
25. Sur toutes ces questions, voir Ducos, *La météorologie…*, p. 214-229.
26. Cf. Le Boeuffle A., *Astronomie Astrologie: lexique latin* (Paris, 1987), p. 244, où il donne la définition de Varron: *solstitium, quod sol eo die sistere*. L'usage de périphrases descriptives se trouve également dans Aratos et Cicéron.

parler de traduction lointaine ou inexacte? C'est une adaptation sémantique par d'autres moyens linguistiques, ce qui ne signifie pas l'absence de fidélité mais l'appropriation par le traducteur de toutes les possibilités lexicales dans une création qui peut rester un hapax ou donner lieu à une habitude lexicale.

La traduction paraît alors un chantier lexical où toutes les tentatives sont faites pour rendre un sens clair, ce qui est particulièrement visible dans les textes longs où les gloses lexicales sont fréquentes et permettent d'initier à ce lexique alors que les textes courts restent allusifs et demandent le plus souvent une connaissance préalable du lexique. Le texte court est souvent opaque pour le non-initié, à cause de l'abondance de lexèmes spécifiques sans traduction, alors que ce n'est pas le cas pour les textes longs. C'est pourquoi on a l'impression d'une technique plus approfondie chez les traducteurs d'autorité ou de textes longs comme l'encyclopédie de Barthélemy l'Anglais. Elle vient de l'écart par rapport aux habitudes lexicales du texte d'origine et du souci d'adapter un lexique français ou d'initier un lecteur par le biais des gloses, des reprises qui marquent des tentatives pour faire coïncider un sens donné dans une autre langue avec un mot. C'est l'inadéquation consciente du langage qui se marque ainsi et tente d'être corrigée par les multiples remarques lexicales des traducteurs qui cherchent une équivalence et une exactitude la plus parfaite possible. Cette recherche permanente ne se mesure guère si l'on se contente d'une analyse statistique où est relevée la fréquence des équivalents. À titre d'exemple, la traduction de Mahieu le Vilain est particulièrement significative de cette création et ces réajustements permanents. C'est ainsi que *vapeur* et *fumee* apparaissent d'abord dans la traduction pour être remplacées par *buee seche* et *buee moiste*; dans le dernier livre, à propos de l'optique, un autre lexème apparaît, la *roration*, qui désigne un état intermédiaire entre la vapeur et la goutte d'eau. De fait, si la traduction paraît comme un lieu d'expériences lexicales, elle est aussi celle d'une initiation du lecteur qui doit progressivement acquérir le lexique du traducteur. Ainsi, dans la traduction de Mahieu le Vilain, les gloses lexicales et encyclopédiques sont plus rares dans la deuxième partie de l'œuvre. C'est donc l'œuvre globale qui permet la construction progressive du lexique et de ses relations sémantiques aussi bien dans la pratique de la traduction que dans la lecture. La vulgarisation passe ainsi par cet inventaire linguistique qui procède par tâtonnements et structuration en cours d'écriture.

L'inventaire peut être parfois donné *a priori* et le traducteur choisit un mode de traduction: tel est le cas d'Evrart de Conty qui dessine une rose des vents qui sert de dictionnaire entre usages philosophiques, maritimes et de la *gent commune*[27]. Mais cette introduction disparaît ensuite puisque les désignations du vent sont des lexèmes latins (*subsolanus, eurum, boreas, zephyrus*...) à l'aide de syntagme composé du substantif *vent* et d'un adjectif dérivé du latin (*vent boreal, austrin*...). Dans ce cas, la traduction n'est qu'introduction au texte latin, le schéma introductif servant comme aide-mémoire et remplaçant des gloses lexicales. La vulgarisation est ainsi initiation au savoir latin, mais n'est pas conçue véritablement comme ayant une existence autonome à la différence d'un texte comme celui de Mahieu le Vilain. De la même manière, malgré la théorie qu'expose Nicole Oresme d'une traduction qui exerce la langue, il est frappant de constater la présence de citations latines ou de mots latins ainsi que l'abondance d'emprunts dont l'équivalence n'est pas toujours donnée ou repose sur une équivalence latine[28]. Cette pratique éclaire la conception qu'a Oresme du rôle du français dans la pensée scientifique: la langue vernaculaire n'est pas un moyen de vulgariser, elle est aussi un outil de pensée, de rang égal avec le latin. Employer le latin dans un commentaire de traduction n'est donc pas un obstacle, car la traduction ne s'adresse pas à des ignorants, mais à des savants qui utilisent aussi bien latin que langue vernaculaire. Cette procédure peut surprendre: de fait elle se retrouve aussi dans les textes courts et signifie clairement un enjeu différent de celui de Mahieu le Vilain, Jean Corbechon ou, à certains égards, Evrart de Conty. Dans cette perspective en effet, la traduction est un témoignage de bilinguisme où latin et français sont équivalents. Dans les textes courts, latin et français se mêlent sans nuire au sens. On peut supposer, dans ce dernier cas, que cette alternance reflète une pratique occasionnelle du latin, peut-être technique ou de certains milieux professionnels. En revanche, Nicole Oresme choisit délibérément une expression française, opposée à un latin relégué dans le passé. La pratique lexicale est ainsi significative du rôle des langues vernaculaires dans la transmission des savoirs: elles apparaissent comme complément ou prolongement du latin, plutôt que comme concurrent inférieur.

27. Ducos J., 'Traduction et lexique scientifique: le cas des *Problemes* d'Aristote traduits par Evrart de Conty', in: C. Brucker (éd.), *Traduction et adaptation en France, Actes du colloque de l'université de Nancy, 23-25 mars 1995* (Paris, 1997), p. 237-248.

28. Voir en particulier dans *Le livre du ciel et du monde*, éds A.D. Menut & A.J. Denomy (Londres, 1968).

3. Conclusion

Du texte au mot, la traduction scientifique marque ainsi sa variété, mais aussi sa place aussi bien dans une théorie générale de la traduction que dans la transmission des savoirs. Les pratiques dans leur diversité mettent à jour des enjeux, des finalités différentes, mais aussi des zones de convergence: la brièveté, la fragmentation, la condensation, même dans les œuvres les plus longues, en sont les modes d'écriture préférés. Les traductions aristotéliciennes et d'autorité intègrent cette dimension dans un cadre rhétorique et dialectique qu'elles soulignent dans le texte à traduire et à commenter. La fidélité ne se mesure pas à la présence ou à l'absence de gloses, à la littéralité ou à l'adaptation, mais bien plutôt à la communication d'un savoir qui passe par des procédures variées d'écriture et de création lexicale. L'écart entre les traductions prestigieuses et les fragments ou les textes anonymes, en particulier de traités techniques, n'est donc pas non plus à mettre sur le compte d'un métier de traducteur plus grand ou plus précis. C'est avant tout le destinataire ou plus clairement la destination du texte qui l'explique. Les textes courts n'ont comme finalité que de donner à lire un savoir sans souci explicite de vulgarisation conçue comme aide: support d'une pratique comme le laisserait supposer la nature de beaucoup de ces textes? Aide-mémoire d'un savoir vivant ou survivant? Leur présence atteste en tout cas d'une pratique plus généralisée que ne le laissent supposer les grandes traductions. La finalité de ces dernières est plus complexe et le destinataire désigné dans le prologue peut masquer un autre lectorat, plus savant ou plus accoutumé à une pratique latine. La traduction scientifique en français, telle que nous pouvons l'appréhender malgré les lacunes de notre information et l'absence d'inventaire complet, est ainsi, dans sa diversité et sa complexité, un témoignage d'une vie du savoir et d'une communication scientifique où la langue vernaculaire est bien davantage qu'un support secondaire.

Université Paris IV-Sorbonne

Alessandro VITALE-BROVARONE

TRADUIRE DES MOTS ET TRANSPORTER DES CHOSES: QUELQUES RÉFLEXIONS SUR LA LITTÉRATURE SAVANTE ET L'EXPÉRIENCE MARCHANDE DANS LA FORMATION DU LEXIQUE

Introduction

L'activité de traduction est un fait de médiation linguistique qui met en contact deux systèmes linguistiques, en ayant recours à un rôle que nous pourrions dire professionnel, celui du traducteur. Les acteurs de ce fait de transmission de culture sont clairs, ainsi que leurs actes et les résultats de leur intervention, qui sont homogènes, par exemple de texte écrit à texte écrit. Les diagrammes auxquels on était habitué à l'époque du structuralisme peuvent facilement nous aider a représenter ce fait:

d'une macrostructure au niveau de la langue:

1 Langue A → Traduction → Langue B

à une microstructure au niveau de la parole

2 Texte dans la langue A → Traducteur → Texte dans la langue B

Mais comme dans toute simplification, ce schéma nous fait perdre des détails, et souvent les détails sont bien intéressants. Prenons d'abord la première proposition: celle-ci se fonde sur une hypothèse qui doit être vérifiée avant d'être acceptée: l'existence de deux systèmes linguistiques fermés, sans communication entre eux. Dans la majorité des cas, en effet, les systèmes linguistiques ne sont pas des ensembles fermés, et l'acte même de la traduction modifie les systèmes, en introduisant des mots ou des idées, et souvent les deux en même temps. L'acte de traduction affecte le système d'arrivée, et il perçoit le système de départ selon son paradigme préconstitué, qu'on peut évaluer seulement en franchissant la barrière abstraite qui se trouverait entre langue et parole: l'acte de traduction ne se réalise que dans la personne et dans les choix du traducteur, qui sont marqués par des données historiques au sens fort du terme. Il n'y a pas de traduction en dehors des circonstances concrètes. Mais un autre élément perturbateur intervient, quand nous considérons le fait que souvent l'acte de la traduction n'affecte pas seulement les textes écrits, mais aussi les textes oraux, et cela met en jeu directement une série de faits linguistiques dont les étapes concrètes se perdent dans le silence des sources, et arrivent à des formes plus ou moins définitives que nous avons l'habitude de définir comme des emprunts linguistiques.

La notion même d'emprunt linguistique fausse un peu la réalité historique: il n'y a pas entre les langues un rapport abstrait qui détermine le passage d'un mot d'une langue à une autre, mais une série de traductions qui guident des habitudes linguistiques, grâce à des personnages qui interviennent et mettent en jeu leurs connaissances linguistiques, qu'ils soient internes au système de départ ou au système d'arrivée, ou aux deux, ou à aucun. Il s'agit de personnages qu'on appelait de façons diverses: *interpretes, biglossi, dragomans* et autres.

Ces termes définissent de façon différente la nature de ces personnages. Le terme *biglossus* est employé par Pierre Damien[1]; il indique celui qui possède deux langues, et qui est capable de passer d'une langue à une autre; le terme *interpres* est lié au mot *pretium* et indique directement le commerçant, celui qui met, ou se met, d'accord sur le prix d'une marchandise[2]; le *dragoman* prend son nom d'une forme sémitique, d'où l'arabe *targuman*, puis le grec *dragoumani*, et les dérivés romans (*drogman, truchemant*; prov. *dragomanz*; it. *dragomanno*; cat. *torcimany*), pour indiquer, au départ, un officier turc qui s'occupait d'ambassades et faisait fonction d'interprète. À l'exception de *biglossus*, qui fait allusion à une qualité individuelle, les termes réfèrent à une activité professionnelle, liée au commerce ou à l'échange professionnel de messages verbaux. *Dragoman* a aussi eu une certaine fortune poétique, comme dans le cas de *Drogomanz* faisant fonction de *senhal* chez Peire Vidal:

> Drogoman senher, s'eu agues bon destrier
> en fol plag foran intrat tuit mei guerrier[3]

et des vers des *Canti carnascialeschi*:

1. *Vita Odilonis, Patrologiae Latinae cursus completus, in quo...* Accurante J.P. Migne, Tome 144, col. 944: *Laurentius, Amalfitanae sedis archiepiscopus, qui potens in litteris ac biglossus, Graece noverat et Latine.*

2. Ernout A. & Meillet A., *Dictionnaire étymologique de la langue latine. Histoire des mots* (Paris, 1967), p. 320.

3. Peire Vidal, *Poesie,* a cura di D'A.S. Avalle, Documenti di Filologia (Milano – Napoli, 1960), p. 219-229.

Dragomanni siam, donne, levantini,
che qui dalla Velona,
della buttagra assai perfetta e buona
abbiam per voi portata, o fiorentini[4]

Le rôle de médiateur linguistique est donc étroitement lié au commerce, à côté d'autres types d'échange, ce qui ne saurait guère être sous-estimé dans l'évaluation des faits que nous appelons en linguistique des emprunts, avec une formule qui réduit trop facilement un fait historique à une définition mécanique. Nous prendrons en considération quelques mots et quelques situations linguistiques, textuelles en particulier, qui nous aident à dresser une typologie de phénomènes, souvent caractérisés par une allure paradoxale qui révèle la présence active des médiateurs linguistiques dont nous venons de parler. Pour ce faire, nous analyserons d'abord les quatre mots suivants: *marzapane, maïs, marguerite* et *tamerix*, que nous pouvons appeler des 'mots sans mémoire'.

1. Les 'mots sans mémoire'

1.1. *Marzapane*

La première recette à ma connaissance qui décrit la préparation de ce qu'on appelle *marzapane* en italien, *marturana* en sicilien, *massepain* en français et *marzipan* en anglais et en allemand, est la suivante:

Marciapanis
Recipe amigdalorum dulcium novellarum et mondatarum libras.I.. Pistantur bene in uno mortario de lapide, deinde addatur zucharum libras 1 et dim., olei amigdolarum dulcium libras dim., aqua roxa parum misse, et ponatur in furno ad coquendum.

Ce passage est tiré d'un réceptaire datant de 1476, qui va bientôt être publié (ms. G.IV.27, fol. 27r a-b, de la Bibliothèque nationale de Turin).

Bien que la recette soit claire et facile, et possible à réaliser quasi partout[5], le massepain a encouragé l'étymologie populaire: la décomposition en 'pain' et 'masse' en français, ou le 'pain de saint Marc' en

4. *Nuovi canti carnascialeschi del Rinascimento*, a cura di C.S. Singleton (Modena, 1946), p. 166.

5. Les amandes sont en effet vendues même loin du lieu de production: 'Anchora se vende le mandolle in Pullia a soma e la soma si è tonbolli 8 et a Venexia se vende le mandolle a stero...'. Voir *Zibaldone da Canal. Manoscritto mercantile del sec. XV*, éd. A. Stussi, Fonti per la Storia di Venezia, sez. V (Venezia, 1967), p. 21.

anglais et en allemand. D'autre part, l'étymologie savante a correctement vu un problème de métonymie, passant par le vénitien *matapan*, ou l'arabe *martaban*, ou encore *martuban*[6], en constatant une coïncidence entre le nom de la pâte et le nom de l'étui dans lequel elle était vendue, ou le poids par lequel elle était mesurée. Mais même là une verité apparente a servi d'écran pour cacher la réalité historique. En effet, c'est la ville indienne de Martaban qui donne son nom à l'étui en faïence qui provenait de ce lieu. Le principe de la raison suffisante a caché la réalité historique, qui a fini par être découverte par G.R. Cardona et G.B. Pellegrini[7]: les acteurs historiques, ceux qui préparaient la pâte sous forme de pain, ceux qui l'utilisaient comme gâteau cérémonial pour la fête de saint Marc, les érudits qui y voyaient un signe du commerce méditerranéen (le *Zibaldone da Canal* parle du *marçapan* comme d'une mesure variable et trompeuse utilisée à Laiazzo)[8], ont d'une certaine façon caché le commerce en bateau entre l'Inde et la Péninsule arabique. Vraie ou fausse, l'étymologie populaire, à travers les interprètes, a forgé d'autres réalités.

1.2. *Maïs*

Le terme *maïs*, originaire d'Amérique Centrale et utilisé par Christophe Colomb dans ses *Journaux*, fut remplacé dans les différents pays où il a été importé par les mots *grain, blé d'Inde* ou bien *de Turquie*, et c'est sous cette forme que Leonhart Fuchs le nomme dans son *Historia stirpium* de 1542, accompagné par une somptueuse illustration. Or, Fuchs, tout en donnant des indications de culture assez exactes, paraît avoir oublié son origine, à peine une quarantaine d'années après son introduction en Europe:

> DE TURCICO FRUMENTO. Hoc frumentum, ut alia multa, ex eorum est genere quae aliunde ad nos translata sunt. E Graecia autem et Asia in

6. Ainsi Bloch O. & von Wartburg W., *Dictionnaire étymologique de la langue française* (Paris, 1932) *sub voce* 'massepain'.

7. Cardona G.R., 'Marzapane', *Lingua Nostra* 30 (1969), p. 34-37; Pellegrini G.B., '*Martabana – marzapane*', *Lingua Nostra* 37 (1976), p. 92.

8. *Zibaldone da Canal; Manoscritto…*, p. 61-62: 'In llaiaça se vende formento e orço ad una mexura ch'è nome marçapan e per la volontade de li Armini nexun non può dire lo vero de un mexe a l'olltro como torna nexuna mexura cum questa, perché ella crexe e menoma a soa volluntade e perçiò li marchadanti.de reçeve assè fiade dano'. ('À Laiazzo le blé et l'orge sont vendus à une mesure qui porte le nom de *marçapan*, et par la volonté des Arméniens personne ne peut savoir avec certitude d'un mois à l'autre comment varie une mesure comme celle-ci, parce qu'ils la font augmenter ou diminuer selon leur volonté, et par conséquent les marchands en reçoivent bien de fois des dommages').

Germania venit, unde Turcicum frumentum appellatum est: Asiam enim universam hodie immanissimus Turca occupat. Germani etiam loca unde affertur respicientes Turkisch korn nominant

... Seritur mense aprili, nec nisi sub finem Augusti et initia Septembris ad maturitatem pervenit[9].

La légende sur son origine se développe très tôt, et cache l'histoire: un deuxième filtre idéologique et interprétatif s'interpose. Il est à noter que le filtre est de nature savante, et passe par les livres, et notons aussi que même dans une histoire qui paraît être toute paysanne – celle de la rapide diffusion du maïs –, c'est l'intermédiaire savant qui joue le rôle principal. Cet exemple est le premier parmi d'autres qui met en valeur l'interférence de la culture savante dans la transmission des noms de plantes: un vrai fait de traduction détermine l'histoire de la langue parlée.

1.3. *Marguerite*

Nous savons que le mot *marguerite* renvoie, en grec comme en latin, à la perle. Le premier passage est un peu plus complexe que ne le suggèrent les dictionnaires étymologiques, Carnoy en premier lieu[10]. Il ne remonte probablement pas à l'arabe *marwa*, 'montagne', 'pierre à feu', mais à un mot iranien, qui ne doit pas être lié au babylonien *mu-gun (-gun)-nui* ou à des formes similaires, mais à un type *mrga-ahr/ni- ita*, lié à un nom signifiant 'oiseau', 'huître en forme d'oiseau' (all. *Vogelmuschel*)[11]. Ce mot proviendrait non pas du persan, mais à son tour d'un mot non iranien appartenant à une langue parlée par les pêcheurs de perles du Bahrein. Nous sommes donc en présence d'un parcours complexe: de la langue des pêcheurs à l'iranien, à l'arabe et à l'éthiopien, au grec, et du grec encore vers l'arabe, et puis encore à nos langues[12].

D'ici au mot indiquant la marguerite, le passage n'est pas long: les prés couverts de marguerites fermées paraissent couverts de perles; ou bien le mot, à partir du sens 'perle', devient un nom de fille, et du nom de fille un nom de fleur[13]. L'image est belle, forte d'une évocation

9. Leonhart Fuchs, *Historia stirpium* (Basel 1542), p. 824.

10. Carnoy A., *Dictionnaire étymologique des noms grecs de plantes* (Louvain, 1959), s.v.

11. Gershevitch I., 'Margarites the Pearl', *Studia Iranica* 7 (1989).

12. Pennacchietti F., 'Arabo *Marwan* e *margan* "perla": ipotesi di una comune origine iranica', *Afroasiatica neapolitana* 6 (1997), p. 305-315.

13. Wartburg W. von, *Französisches Etymologisches Wörterbuch. Eine Darstellung des galloromanischen Sprachschatzes* (Bonn – Zbinden, 1928ss), s.v., p. 327-328.

quasi-poétique. Mais comment croire qu'une image poétique, très anodine, ait eu une si large diffusion (en plus, *marguerite* étant avec ce sens un terme très tardif, XIIIe siècle)? La réponse se trouve peut-être dans le rôle intermédiaire des livres scientifiques, qui ont souvent exercé leur influence sur la terminologie botanique courante. Or, les livres n'emploient que très rarement *marguerite*, préférant *chrysantemon*, ou *leucanthemon,* ou encore *bellis*. Il faudrait alors imaginer un autre parcours, qui n'a pas encore été pris en considération et que nous formulons ici sous forme d'hypothèse. On pourrait en effet partir d'un autre nom donné à des plantes qui ressemblent à la marguerite – et dans certains cas donné à la marguerite elle-même – en partant de son emploi en pharmacologie pour les maladies utérines, à savoir MATRICARIA (à partir de MATRIX, qui donne justement MATRICARIA). En partant d'une forme dont la deuxième syllabe est réduite, avec le suffixe -ITA / -ITTA, on peut aboutir à *marguerite*. Là encore, si notre hypothèse est vraie, le mot a coupé son lien étymologique établi par le commerce des simples[14] et a acquis un nouveau sens pseudo-poétique ('jeune fille belle comme une fleur', 'fleur belle comme une jeune fille'), en se liant à un autre terme ('perle'), qui à son tour avait oublié ses liens avec les acteurs originaires, les pêcheurs du Bahrein, et les traducteurs / commerçants iraniens.

1.4. *Tamerix*

Le quatrième mot de cette série est probablement le plus connu: *tamerix,* qui remonte à un terme grec, puis au latin *myricae*. Le préfixe *ta-*, comme on l'a observé, le rapproche d'un petit nombre de mots indiquant des plantes portant le même préfixe. Or, ce préfixe est typique du berbère, qui emprunta une longue série de mots latins, parfois en les restituant avec des caractères adjoints[15]. Le mot *tamerix* cache une histoire complexe, qui est l'histoire des rapports entre la latinité et la côte septentrionale de l'Afrique. Nous pourrions même formuler des hypothèses sur l'époque à laquelle ces rapports ont eu lieu. En effet, la présence à Rome de gens originaires de l'Afrique du nord à l'époque des guerres puniques

14. Voir aussi le rôle des *herbolarii* dans la médiation entre les couches linguistiques.
15. Sur les mots berbères ayant une origine latine, voir Schuchardt H., *Die romanischen Lehnwörter im Berberischen*, Kais. Akademie der Wissenschaften in Wien. Philosophisch-historische Klasse. Sitzungsberichte, 188. Band, 4. Abhandlung (Wien, 1918). On peut se référer aussi à Wagner M.L., *Restos de Latinidad en el Norte de África*, Biblioteca geral da Universidade. Cursos e conferências de Extensão Universitária XLV-XLVI (Coimbra, 1936).

et peu après, peut bien expliquer l'entrée précoce de ces termes ayant une marque berbère[16].

2. Les gens du métier et les traductions

2.1. Dino Dini

Le rapport entre les acteurs de la vie pratique et les traductions scientifiques est normalement peu considéré, à l'avantage de formulations simplifiées telles que 'les traductions sont faites par les traducteurs'. Ainsi, tout comme les langues ne sont jamais des systèmes purs, mais des systèmes mixtes, les gens sont souvent des individus complexes. Nous pouvons considérer le cas d'un vétérinaire florentin du XIVe siècle, Dino Dini. Septième et dernier héritier d'une longue tradition familiale, il écrivit son traité de la *Mascalcia* à partir de 1352, où il affirme:

> Onde volendo il padre che apresso di sé uno di figlioli rimanesse all'arte chonvenne a me [...] venire all'arte onde dinanzi da me furono di mia gente l'uno presso all'altro ciascuno malischalcho feci io il settimo. E veggio molti, chi per millantarsi, chi per bugie, quale per improptitudine torre alcuna volta l'arte di mano a'ssimiglianti di me: onde per questo isdegno mi sono posto in quore d'essere l'ultimo di miei di questa arte[17].

Si on a d'abord l'impression d'une œuvre entièrement fondée sur l'expérience directe ou sur la tradition propre à l'art de la médecine vétérinaire, Dino fait preuve néanmoins de ce caractère culturel mixte qui est le propre aussi des marchands, comme les études de Christian Bec l'ont bien montré[18]. Toutefois, cette qualification-là réduit aussi la complexité de l'histoire, puisque, comme nous le verrons, les artisans possèdent à leur tour une culture complexe. Dans le bref passage omis dans la citation qui

16. Le mouvement de va et vient est bien connu. On pourra voir Steiger A. & Keller H.-E., 'Lat. *Mantelum.* Ein Beitrag zur Geschichte des mediterranen Kulturlehngutes', *Vox Romanica* 15 (1956), p. 103-154, et en particulier p. 120-125. Parfois les mouvements des mots peuvent être suivis à partir des données historiques aussi bien que de la forme linguistique. Cfr. aussi Ineichen G., 'Parole migratorie e la terminologia farmacologica', in: *Yadnama. In memoria di Alessandro Bausani,* vol. II (Roma, 1991), p. 145-148. On verra à ce propos la discussion entre R.S. Lopez et G.B. Pellegrini, dans *L'Occidente e l'Islam nell'Alto Medioevo, 2-8 aprile 1964*, Atti delle Settimane di Studio del Centro Italiano di Studi sull'Alto Medioevo, XII (Spoleto, 1965), p. 843-844.

17. Firenze, Bibl. Nazionale, II. VIII. 120, fol. 73r

18. Bec C., *Les marchands écrivains: affaires et humanisme à Florence, 1375-1434* (Paris – La Haye, 1967).

précède, Dino raconte comment il a été initié à la profession vétérinaire par son père, qui lui a fait interrompre ses études de grammaire. Un peu plus tard il reprend l'étude de Vegèce, que – nous dit-il – personne n'avait compris mieux que sa famille:

> e più da mio patre e da me, e più ch'io stetti diciotto mesi chon sofficientissimo maestro a schuola solamente a essermi dichiarato il testo della gramaticha del detto libro i nomi medicinali e sufficientissimi medici di fisicha.

La tradition de la science appliquée ne peut pas être considérée comme entièrement séparée des connaissances théoriques et apprises par voie écrite, ni l'œuvre qui naît de l'expérience comme entièrement détachée de la traduction. S'il y avait aussi des praticiens purs, le mépris de Dino pour eux est fort:

> ma nientemeno gli artefici di quest'arte sonno dello studiare bene schusati, inperò che la magior parte sonno figliuoli di lavoratori di terra levati dalla marra e da guardare le pecore, per la quale ragione non possono essere veri artefici: imperò che sonno sanza lettere sì che non possono studiare, e però sonno sdegnati molti valenti huomini di questa arte, perché ànno veduto, per generale pecchato d'ignoranza, dare presgio a ssì fatta generatione d'artefici, per uno pocheno di lor praticha: ed essendo presi la maggior parte di loro da soperchi di vino, dicendo alchuno: "S'el tale non fosse ubriaco, de' chom'è buono maestro".

En effet, dans le traité de Dino, il y a une reprise constante du texte de Végèce, sous une forme qu'on pourrait nommer traduction, mais qui est en même temps confirmation, expérience, continuation. On pourrait difficilement imaginer une attitude comme celle de Dino Dini en dehors du contexte marchand et d'une civilisation marchande comme celle de Florence; les citoyens qui lui amènent les chevaux pour les soigner, les cas difficiles qu'il raconte, la concurrence des incompétents: c'est justement la civilisation des Arti de la ville qui se manifeste, et qui s'enrichit du rapport, respectueux et en même temps vainqueur, avec l'Antiquité classique.

Son témoignage est particulièrement important pour des raisons qui en font un cas peu commun: il n'est pas un traducteur purement livresque; il n'est pas un praticien pur; son lexique n'est, par conséquent, ni dominé par l'influence latine, ni par un éphémère jargon de métier; son lexique est aussi – nous pouvons bien le supposer – le lexique d'une famille qui a pratiqué le métier de vétérinaire au cours de plusieurs générations, et qui a donc exercé une influence continue et ininterrompue sur la langue d'une ville aussi importante que Florence, influence importante pour son époque, mais aussi pour le futur de la langue italienne.

2.2. Simon de Gênes

À peu près cinquante ans plus tôt, un autre cas de médiation linguistique s'impose à notre attention, celui de Simon de Gênes. Nous savons très peu à propos de sa vie et de sa personnalité[19]. Il traduisit de l'arabe le texte de Jean Sérapion, selon le parcours habituel, en ayant recours à la collaboration d'Abraham de Tortosa. Mais peut-être son œuvre la plus intéressante dans le domaine du lexique et de la traduction est sa *Clavis sanationis*, un répertoire alphabétique de simples (avec quelques composés), par lequel il veut étudier la terminologie botanique pour essayer d'y discerner les synonymies et arriver à une sorte de normalisation. Il utilise une vaste série de sources, dont il donne la liste: 11 sources grecques, 8 arabes, 13 latines, et plus ou moins 36 autres que nous rencontrons au cours de la lecture du texte. Il connaît les trois langues, ce qui ne l'empêche pas de s'adresser à d'autres experts:

> Nec his solum contentus sed ad diversas mundi partes per sedulos viros indagare ab advenis sciscitari non piguit, usque adeo quod per montes arduos nemorosas convalles campos ripasque sepe lustrando aliquando comitem me feci cuiusdam anicule cretensis admodum sciole non modo in dignoscendum herbis et nominibus grecis exponendis, verum etiam in ipsis herbarum virtutibus secundum Dyascoridis sententiam explicandis[20].

aussi en comparant les langues dans leur variété interne:

> Querens a quadam saracena de Alef ostensa matricaria qualiter vocaretur, respondit uchuem. Et iterum a quadam alia alterius regionis, respondit cahavem, aliquando achavem, aliquando alachuem[21].

Lors de la constitution d'un corpus très vaste, l'assemblage des matériaux implique une extrême variété d'actes de traduction. Nous avons pris en considération plusieurs actes de traduction qui supposaient des milieux et des circonstances divers: entre activité pratique et activité livresque dans le cas de Dino Dini, avec le recours à des compétences ultérieures dans le cas de Simon de Gênes; grâce à une série complexe de rapports entre des civilisations et des mondes divers, comme dans le cas des traductions

19. Voir en particulier Paravicini Bagliani A., *Medicina e scienze della natura alla corte dei papi nel Duecento* (Spoleto, 1991), p. 247; Jacquart D., 'La coexistence du grec et de l'arabe dans le vocabulaire médical du latin médiéval: l'effort de Simon de Gênes', in M. Groult (éd.), *Transfert du vocabulaire dans les sciences* (Paris, 1988), p. 278-290.

20. Simon Ianuensis, *Clavis sanationis*, c. 2r; nous citons de l'incunable imprimé à Venise, par Guielmum de Tridino, 1486.

21. Simon Ianuensis, *Clavis*... c. 2r.

qui sont supposées par l'histoire des mots que nous avons examinés ci-dessus, et qui souvent cachent, derrière une fausse simplicité, des problèmes historiographiques de grande envergure.

2.3. Dominique Gundisalvi

Nous avons vu que le fait de réduire l'acte de la traduction à un passage simple d'une langue à une autre au moyen d'un traducteur, simplifie la réalité et trahit la complexité historique. Le fait que souvent la traduction ne passe pas par un seul intermédiaire, mais par plusieurs personnes a été observé dans le cas de Simon de Gênes et ses rapports avec Abraham de Tortosa, ainsi que les *vetulae* auxquelles il s'adresse: ce fait n'est pas rare dans le cas de la traduction de textes écrits dans des langues très éloignées, comme l'arabe ou le grec par rapport au latin. Mais il y a aussi le cas d'un intermédiaire vulgaire dans le passage d'une langue de culture à une autre, comme par exemple dans le cas de la traduction de la *Source de vie* de Ibn Gabirol: dans cet ouvrage, Dominique Gundisalvi, archidiacre de Ségovie, est aidé par Jean d'Espagne, qui dit dans son prologue à la traduction d'Avicenne: 'Me verba vulgariter proferente, et domino archidiacono singula in latinum convertente', ou encore, dans les vers à la fin du *Fons vite*: 'Transtulit Nispanis interpres lingua Iohannis / Hunc ex Arabico non absque iuvante Domingo'[22].

2.4. Laurent de Premierfait

La distance entre les langues peut aussi être bien plus faible, mais tout de même demander la présence d'un intermédiaire, comme dans le cas de Laurent de Premierfait et de Giovanni Boccaccio; le cas est exactement le contraire de celui que nous venons de prendre en considération: le passage entre deux langues vulgaires se fait cette fois à travers une langue de culture. Bien que le texte ne soit pas lié aux sciences, nous le citons pour sa richesse de détails:

> Et pour ce que suis Françoiz par naissance et conversacion, je ne say plainement langaige florentin, qui est le plus preciz et plus esleuz qui soit en Italie, je ay convenu avec ung frere de l'ordre des cordeliers nommé maistre Anthoine de Aresche, homme tresbien saichant vulgar

22. Cité par Brunner F., 'Etudes sur le sens et la structure des systèmes réalistes', *Cahiers de Civilisation médiévale* 1 (1958), p. 295 et 296.

> florentin et langaige latin. Cestui frere Anthoine bien instruit en deux langaiges, maternel et latin, pour juste salaire translata premierement ledit Livre des Cent Nouvelles de florentin en langaige latin, et je Laurens, assistant avec lui, ay secondement converty en françois le langage latin receu dudit frere Anthoine, ou au moins mal que j'ay peu ou en gardant la verité des paroles et sentences, mesmement selon les deux langages, forsque j'ay estendu le trop bref en plus long et le obscur en plus cler langaige afin de legierement entendre les matieres du livre[23].

Le système des éléments en jeu n'est pas nécessairement complet: nous avons vu le cas d'un traducteur *biglossus*, ouvert sur les deux langues et capable de transférer le texte d'un système linguistique à un autre, et le cas d'un passage intermédiaire par une troisième langue (Laurent de Premierfait grâce à Antonio d'Arezzo[24]), ou par une étape d'interprétation linguistique (Dominicus Gundisalvi grâce à son intermédiaire) ou sémantique (Simon de Gênes grâce aux dames *sciolae*). Parfois il n'y a pas, *stricto sensu*, de texte de départ: c'est le cas d'un texte de départ purement virtuel, celui envisagé par Marco Polo, qui dans les prisons de Gênes raconte, avec précision, ses expériences de voyage, probablement en vénitien, à Rustichello de Pise, lui aussi prisonnier; Rustichello à son tour met le texte par écrit en français, grâce à son expérience de romancier[25]:

> Le quel [Marco Polo] puis, demourant en la chartre de Jene, fist retraire toutes cestes chouses a messire Rusticiaus de Pise, que en celle meisme chartre estoit, au temps qu'il avoit MCCXCVIII anz que Jesucrit nesqui[26].

3. Les glossaires

Nous arrivons maintenant à un autre témoignage de l'acte de traduction, qui en est à la fois une prémisse et une conséquence: les glossaires. Souvent sous forme de listes à deux entrées, les glossaires ne constituent pas, ou pas seulement, des instruments pour la traduction ou pour la compréhension, mais aussi une tentative de mettre de l'ordre dans une terminologie qui est devenue incertaine après un jeu combinatoire complexe de signifiants et de signifiés. Un ensemble d'éléments mal classés et mal

23. Boccace, *Decameron*, traduction (1411-1414) de Laurent de Premierfait, éd. Di Stefano G., Bibliothèque du Moyen Français 3 (Montréal, 1998), p. 5.

24. Le texte nous fait hésiter quant à l'existence d'un texte latin écrit, ou bien d'une simple traduction orale.

25. On lui doit une vaste compilation arthurienne.

26. Marco Polo, *Il Milione*. Prima edizione integrale a cura di L.F. Benedetto, Comitato Geografico Nazionale Italiano 3 (Firenze, 1898), p. 3-4.

définis – dans le cas que nous allons prendre en considération, il s'agit de plantes, d'animaux, et en général de composants et composés ayant un intérêt pharmacologique –, auxquels s'appliquent des mots sur lesquels on peut exercer un contrôle très incertain. On y voit une tentative de mettre de l'ordre dans la terminologie employée par un ou plusieurs textes qu'on veut traduire ou comprendre, ou bien par un ou plusieurs textes qu'on a traduits. D'ailleurs, souvent cette tentative engendre un désordre d'un degré supérieur[27]. Voyons le cas d'un glossaire contenant des mots souvent exotiques: les langues qu'on a pu identifier pour le moment sont le latin, le grec, l'arabe et l'hébreu, le turc, le persan et l'arménien, mais aussi le roumain. Ce glossaire est contenu dans le ms. Vicenza, Biblioteca Bertoliana, G. 3. 9. 1:

> Cuscute id est podagra lini et vulgarem in Cremonia est gringus et est herba nascens in pratis et linis
> Condita dicitur que in sole conficiuntur ut çucchero
> Carabe id est gumbi quam trahit ad paleam
> Curacis id est urtica
> Canderos est romanum triticum:
> Cubugi id est perdicis

Le texte ne diffère pas fort d'autres glossaires; on peut le prendre en considération pour observer des données de contexte, qui ouvrent une série de problèmes. Le contexte général est pharmacologique, et la présence massive de termes arabes est justifiée non seulement par les sources – les textes arabes traduits gardent dans la plupart des cas la terminologie arabe, à côté de la traduction: ainsi, par exemple, Avicenne et Sérapion –, mais aussi par des raisons commerciales.

Prenons en considération le dernier concept traité, 'perdrix'. D'abord, a-t-il un sens? Dans les textes, le mot *cubugi* (à l'origine *qbg*) se trouve toujours associé à sa traduction, et la présence du terme dans le glossaire ne semble pas remonter à une nécessité de traduction, ni de compréhension d'un texte qui est déjà traduit. D'autre part, la présence ne s'expliquerait guère non plus par une nécessité commerciale: la présence de la perdrix dans la diététique, qui forme une part légitime de la médecine et

27. La meilleure description de ce désordre se trouve, à mon avis, dans le court texte cité par T. Hunt, en ouverture de son *Plant Names of Medieval England* (Cambridge, 1989): ~'A complete language of meaningless nonsense, almost impossible to retain, and certainly worse than useless when remembered – a vast vocabulary of names many of which signify that which is false, and most of which mean nothing at all', à son tour emprunté à Britten J. & Holland R., *A Dictionary of English Plant-Names* (London, 1886) et à Richter H.C. dans *Hardwicke's Science Gossip* (Nov. 1., 1868), p. 251.

par conséquent de la pharmacologie, ferait en tout cas référence au marché local et ne pourrait être l'objet d'un commerce de grande distance. La source livresque paraît donc prévaloir, non pas dans une perspective pratique de traduction, mais dans le cadre d'un souci de mise en ordre d'une terminologie acquise par la traduction[28].

Le premier concept traité dans le passage cité, apparemment inoffensif, pose tout au contraire une question complexe. En effet, il est précisé que la *cuscuta* s'appelle *gring* à Crémone[29]. Mais d'où vient la nécessité de donner cette précision? La *cuscuta* est une mauvaise herbe, envahissante, qui se trouve partout sur les bords de rues, dans les champs, sur les débris. Est-il indispensable de référer à la *cuscuta* de Crémone? Serait-elle le signe d'une localisation possible du glossaire? En effet, on trouve dans le glossaire d'autres références à des localités de l'Italie du Nord, mais pas à Crémone.

Le glossaire en question, ainsi qu'une vaste quantité d'autres glossaires, serait le collecteur d'une série complexe d'actes de médiation linguistique, de traduction actuelle ou virtuelle, qui s'accumulent, en mêlant les très longues distances (Proche- et Moyen-Orient), et les distances courtes (Nord de l'Italie), tout comme les sources livresques et les sources pratiques.

4. Le contexte social

Avant d'en venir à notre dernier exemple, ajoutons une considération sur le rôle social des acteurs des médiations linguistiques. Si nous avons pu apprécier le travail des *dragomanz*, les traducteurs professionnels, les collaborateurs à la médiation entre systèmes linguistiques, nous avons également pu relever des médiations entre différentes couches internes à un seul milieu linguistique. Rappelons le mépris de Dino Dini pour les praticiens, reprenant à sa façon le contraste entre théorie et empirisme, que l'on retrouve aussi dans la tradition médicale classique à partir d'Hippocrate. Dans la tradition botanique et pharmacologique, on constate, par analogie, un contraste entre les botanistes savants et les fournisseurs d'herbes, les *herbolarii*, qui, eux aussi, jouaient un rôle de médiateurs entre la tradition savante et la tradition populaire, bien que le terme ne

28. Ce n'est pas plus qu'un souci: les glossaires créent en effet des problèmes plutôt que d'en résoudre.

29. Cfr. Penzig O., *Flora popolare italiana. Raccolta dei nomi dialettali delle principali piante indigene e coltivate in Italia* (Genova, 1924), I, p. 150-151.

soit pas très précis: ceux qui, dans un milieu non lettré, possèdent des connaissances techniques, forment une élite savante[30].

Le témoignage explicite de Simon de Gênes pourra nous aider encore une fois: 'Oculus consulis vocatur a quibusdam planta que antipater dicitur; alii oculum Christi vocant, ut placet herbulari<i>s fingere nomina'[31].

La présence des *herbolarii* dans la médiation linguistique verticale se trouve partout dans la tradition botanique. Nous pourrions citer l'exemple de l'œuvre de Leonhard Fuchs (1501-1566), l'un des pères de la botanique moderne. En effet, Fuchs va même un peu plus loin, en identifiant d'autres intermédiaires, partant d'expériences professionnelles diverses, dont nous disposerions à des niveaux différents; Fuchs ne fait en effet pas de distinctions entre rangs sociaux, mais entre *Sondersprache*, en voyant peut-être très clair dans le problème. Qu'on ait la patience de suivre les listes présentées dans l'appendice, rédigées sur la base de l'édition de Bâle, in officina Isengriniana, 1542, et qui essaient de reconstruire le cadre qui est dispersé dans l'œuvre[32]. Nous pouvons en résumer ici les données. Mis à part la riche série de dénominations vulgaires dont Fuchs est le précieux témoin, il attribue une terminologie latine à une série de groupes professionnels. Au premier rang, les *officinarii*, les apothicaires, qui ont des usages terminologiques qui parfois ne coïncident pas avec ceux des sources classiques; on ne pourrait pas confondre leurs habitudes avec celles qui gouvernent la langue ordinaire; leur langue savante est celle que nous voyons figée dans les inscriptions sur les pots de pharmacie, et doit être rangée aux habitudes pratiques d'un groupe professionnel. Au deuxième rang le *vulgus*, qui possède une terminologie que Fuchs enregistre, parfois peut-être en adaptant la forme à la langue latine;

30. Ce fait explique la perméabilité linguistique entre la tradition écrite et la tradition orale. On a constaté plusieurs fois que des termes, aussi d'origine grecque ou arabe, sont entrés dans la langue parlée; c'est le cas de *mišmiš* ou *mišmišì*, le terme arabe courant pour *abricot*, qui se rencontre à San Remo et Porto Maurizio à l'extrême ouest de la Ligurie, sous les formes *miscimì* et *miscimim* (Penzig, *Flora popolare*..., I, p. 381; nous remercions M. Fabrizio Pennacchietti de nous avoir fourni cette référence), pour lesquels il n'y a pas de correspondants en français, ou pour les différentes formes du terme *akakia* dans les dialectes, aussi avec accent grec. Cf. Vitale-Brovarone A., 'La terminologie botanique', in: R. Bidler & G. Di Stefano (éds), *Autour de Jacques Monfrin. Néologie et formation verbale. Actes du Colloque International, Université McGill, Montréal 7-9 octobre 1996*, *Le Moyen Français* 39-41 (1996-7), p. 649-680.

31. Simon Ianuensis, *Clavis*..., f. 67rA.

32. Un travail de 'tesi di laurea' (avant la réforme des cycles des études) a été consacré à la récolte des termes vulgaires et latins dans les éditions de l'œuvre de Fuchs: Guffanti A., *La terminologia botanica nelle edizioni cinquecentesche* De historia stirpium *di Leonhard Fuchs* (Università di Torino, Facoltà di Scienze della Formazione, 2000-2001), sous ma direction.

la terminologie du *vulgus* est parfois partagée par les *herbarii*. Au troisième rang les *herbarii*, ce groupe peu connu qui a le rôle de fournisseur des *officinarii*: leur terminologie est propre à eux, mais souvent elle recoupe la savante, celle des *officinarii* ou celle du *vulgus*. Enfin deux autres groupes, rarement évoqués, le *vulgus medicorum* et les *iurisconsulti*; ces derniers sont nommés à propos du coton en tant qu'objet de commerce, nommé selon l'usage qui reste dans la langue allemande, *lana ligni*; les premiers sont les médecins qui ont une modeste connaissance des simples et qui par leur ignorance sont à l'origine d'une mauvaise renommée qui nuit beaucoup au bon nom des médecins sérieux (tous les traités de botanique pharmacologique soulignent cette mauvaise habitude avec un ton très critique).

Le point le plus important qui reste à étudier, et là on n'en est qu'au début, est celui des *herbarii*, qui mettent en rapport la production et la récolte d'herbes non cultivées avec le monde proprement savant. Les textes examinant leur rôle existent et méritent une attention qui pour le moment n'a pas eu le juste développement: la longue série de traités de botanique, de Ruelle à Brunfels, à Bauhin, à Mattioli, à Aldrovandi, qui nous fournissent de riches indications; mais aussi Antonio Musa Brasavola (1501-1554), qui dans ses traités développe ses thèmes sous forme de dialogue, et en particulier dans son *Examen omnium simplicium*, publié plusieurs fois au cours du XVIe siècle et où un *herbolarius* participe directement au dialogue, non pas sous la forme charlatanesque des personnages des *Herberies*, où bien de l'*Erbolato* de Ludovico Ariosto, mais en exposant connaissances et points de vue. Ce qui nous manque encore est une connaissance précise de ce type d'acteurs sociaux, qui nous permettrait de connaître certains aspects des traductions 'diastratiques', et nous donnerait des indications très précieuses pour comprendre le dernier cas que nous voudrions présenter, dans la tentative de bouleverser la simplicité linéaire du schéma que nous avons présenté au début, schéma souvent sous-jacent à plusieurs études sur la traduction. Nous pouvons avoir, comme nous le verrons, un point d'arrivée qui n'est pas nécessairement une langue.

Pour développer un exemple à cet égard, nous ne donnerons pas le texte de départ, rédigé dans le bon français de Philippe de Clèves, qui se trouve en appendice à son traité de l'art de la guerre[33]. Nous citons le texte qui se trouve dans le manuscrit Z. I. 28 des Archives de l'État de Turin,

33. Le texte est édité, mais sans tenir compte du manuscrit turinois, par Paviot J., *Philippe de Cleves, seigneur de Ravestein, L'instruction de toutes manieres de guerroyer* (Paris, 1997).

Biblioteca antica. Celui-ci commence dans un français plus ou moins correct, et pourrait être vu comme une simple copie; et puis il glisse vers une langue différente, qui devait paraître bien claire à son auteur, resté inconnu (pourtant il devait faire partie de l'équipe des bombardiers du duc de Savoie). Nous ne savons rien sur son origine, ni sur sa formation, mais nous pouvons constater que sa langue possède des éléments, toujours en conflit, français, italiens, latins, ligures, provençaux et même sardes. Le lieu où ce mélange peut être imaginé fait penser à Nice. Mais en réalité l'armée est souvent en mouvement, et un lieu fixe ne rend pas compte de la réalité historique. Il faut plutôt penser à l'armée comme lieu du concours et du nivellement de types linguistiques souvent disparates. Le phénomène est connu, et il a été considéré à la base de langues bien formalisées comme le persan et l'urdu. Mais voyons le résultat, en partant du français qui reste français dans le prologue:

> Pour donner doctrine a tous roys, ducz, barons, chevaliers, communiaulté des villes et cités et aultres seigneuries et gens quelconcques qui se doubtent de leurs ennemys estre assiegez, constrainctz compellez ou aultrement efforcez en les chasteaulx, villes ou forteresses et citez, soit besoing et necessité non pas seullement serviteurs familliers, mais aussi bien entenduz de la guerre qui ont en eulx grand loyaulté, sens, discretion et preudhommie envers leurs seigneurs

et la langue qui se développe au sein du texte:

> Dei conditiones et virtutem contenentia que ounium (= it. *ognun*) patroune de dictou arte debet abere. Cap.lo II
>
> A tuti patrone del bitou (sic, pour *ditou*) alte convenit et apartenet habere conditiones et virtutem e sciencia que ultimamente debet abere amare et servire Dieu e pensare a suo hounore et sempre l'abere enam secioche (= 'devant ses yeux') e primo debet apencare a l'anima soua et debet habere aviso que Dieu è iudici vero et que provede li iusti per les callos reiounes debet advisare de servire lou segnour Dieu

et un peu plus loin:

> Item la carte nom debet providere nom venir en fastidio (corr. sur *fastido*!) ne pillar collero may debet istare in allegressa piou de liessa e se fa veder in bom portamente infra le gente de guerra quelque d'une cossa qui pensa en soui intellecta et ce contenire distramente habere la faccia bella e i doumando bon esperantia d'abere victoria car en faciendo bouna facia facit ali altra gente perdre bounou esperanca

Une transcription se transforme en traduction au cours du texte, mais la langue d'arrivée est un objet instable, mixte: c'est bien la traduction qui crée une langue. Cette langue devait être pour une raison ou pour une autre préférable au français, et certainement plus claire, du moins à l'auteur. Nous entrevoyons une couche d'artisans, mis en contact avec des modèles linguistiques mouvants et complexes, où aucun modèle ne prévaut sur les autres: un fait de contact interlinguistique qui met sous nos yeux une réalité linguistique qui doit correspondre à un fait historique, à un état de langue certainement vrai, qui pourtant a souvent été caché par les modèles linguistiques figés auxquels nous sommes habitués. C'est justement grâce à des cas rares et particuliers que nous comprenons mieux à quel point l'adhésion à des langues conventionnelles est le fruit d'un choix, qui prévaut largement dans les témoignages écrits, mais qui n'a pas été le seul choix possible. La langue des couches sociales peu lettrées, qu'on peut observer dans des traductions comme celle-ci, a été cachée par le silence discret des sources, mais subsiste parfois dans des cas que la modalité de la transmission et de la conservation des textes a rendus rares.

Conclusions

Il est clair que nous ne pouvons pas limiter le phénomène de la traduction à un fait formel de passage d'un texte écrit à un autre texte écrit.

Dans plusieurs cas l'emprunt linguistique n'est pas simplement une étiquette sur un objet, mais il est ce qui reste d'un fait (ou de plusieurs faits répétés) de traduction, orale plutôt qu'écrite, mais souvent due à des individus qui possèdent forcément une langue spéciale: il ne s'agit pas d'un rapport abstrait entre deux langues, mais d'un rapport entre personnes appartenant à des groupes différents. Nous recoupons ici l'observation de David Langslow qu'il faut prêter une grande attention aux aspects sociolinguistiques, aussi quand on considère des phénomènes du passé.

Les mots qui sont le fruit de cette traduction embryonnaire perdent souvent le contact avec leurs origines, ou même cachent celles-ci. Notre effort ne doit pas simplement être celui de retrouver les origines, mais de comprendre les éléments qui ont participé aux différents moments de la médiation linguistique. Les commerçants de l'Océan Indien, les potiers, les paysans qui cultivaient les amandes, les vendeurs arabes, les confisiers européens, le clergé, les pêcheurs de perles, les souverains achéménides, les Indiens Hopi, les cultivateurs allemands, les Berbères qui ont reçu des mots provenant d'autres parties de la Méditerranée, les ont enrichis et les ont restitués, les 'rhizotomes', les soldats et les bombardiers.

Les professionnels ont aussi participé aux faits de traduction: ils ont eu une double position de médiateurs, entre les langues et entre la théorie et la pratique. Ils nous montrent aussi comment ils peuvent traduire et adapter les textes, et comment leur position nous garantit leur influence directe sur les autres membres de la communauté linguistique: dans ce cas la traduction ne se limite plus à être un rapport entre deux livres, mais elle rentre à plein titre dans la liste des acteurs de l'histoire linguistique.

La traduction suppose un dialogue implicite entre le traducteur, ses sources et son destinataire. Tout ce que nous appelons 'désir de clarté', 'simplification' etc. n'a pas une valeur en soi, mais seulement dans le contexte dans lequel la traduction s'insère: il s'agit d'un des éléments d'un flux de communication large et étendu dans le temps. Pouvons-nous arracher les textes mathématiques, qui ont fait l'objet d'une autre communication, des boutiques d''abaco'? Pouvons-nous imaginer la diffusion de textes et d'idées sans les marchands? Avons-nous le droit de dresser une terminologie botanique en accueillant toutes les formes sans tenir compte des temps et des lieux?

Pouvons-nous être sûrs que la traduction conduit un texte, écrit ou oral, d'une langue à une autre? Le dernier texte que nous avons pris en considération constitue en effet un cas extrême, qui pourtant nous permet des considérations qu'on peut appliquer, dûment réduites, aussi aux traductions ordinaires, ou plus ordinaires. La traduction, tout comme elle crée des textes, crée aussi des langues, langues plus ou moins spéciales dans leur lexique, dans leur style, et qui contribuent à la modification des langues et des cultures sur le plan général.

Università di Torino

Appendice

Termes attribués par Leonhart Fuchs à des groupes particuliers:

Termes des *officinarii*

5 De althaea (= *Althaea officinalis* L.): eviscum

6 De anetho (= *Anethum graveolens* L.): anethum

8 De anthemide (= *Chamomilla recutita* (L.) Rauschert): camomilla

11 De arnoglossa, Plantago maior (= *Plantago maior* L.): lanceolata Plantago minor (= *Plantago media* L.): lanceolata

11 De arnoglossa, Plantago minor (= *Plantago media* L.): lanceola

13 De Arthemisia (= *Arthemisia vulgaris* L.): artemisia (generaliter)

15 De acantho (= *Acanthus spinosus* L.): branca ursina

16 De spina alba (= *Silybum maritimum* L.): bedeguar, bedeguarim (acc.)

17 De asparago (= *Asparagus officinalis* L.): asparagus

19 De aniso (= *Pimpinella anisum* L.): anisum

21 De ammi (= *Ammi maius* L.): amoeos (corrupte)

22 De aro (= *Arum maculatum* L.): larum
24 De arctio (= *Arctium lappa* L.): lappa maior
28 De adiantho (= *Adiantum capillus-Veneris* L.): gallitrichum (errate)
31 De aristolochia (= *Aristolochia clematis* L.): aristologia (*Corydalis bulbosa* (L.) DC): aristolochia rotunda (in Germania)
32 De ampelo leuce (= *Brionia cretica* L.): brionia
34 De amarantho (= *Helichrysum arenarium* (L.) Moench): Stichas citrina vel Stichados citrinum (= … argentea L.): flos amoris
39 De andrachne (= *Portulaca oleracea* subsp. *sativa* Haw.): portulaca
40 De asphodelo (= *Lilium martagon* L.): affodillus
41 De atraphaxi (= *Atriplex hortensis* L.): atriplex
51 De buglosso (= *Borago officinalis* L.): borrago
53 De bellide (= *Bellis perennis* L.): consolida minor (= *Leucanthemum vulgare* Lam.): consolida media
54 De brathy (= *Juniperus sabina* L.): sabina
56 De ballote (= *Ballota nigra* L.): marrubium nigrum, marrubiastrum, prasium fetidum
61 De bicio (= *Vicia sativa* L.): orobus (errate)
62 De blito (= *Chenopodion polyspermum* L.): beta (errate)
71 De galeopsi (= *Scrophularia auriculata* L.?): ficaria, milemorbia, castrangula
73 De glechone (= *Mentha pulegium* L): pulegium
75 De glechysyda (= *Paeonia officinalis* L.): paeonia
76 De gingidio (= *Anthriscus cerefolium* (L.) Hoffm.): cherefolium
79 De genista (= *Cytisus scoparius* L.?): genesta, genestra
82 De dipsaco (= *Dipsacus sativus* (L.) Honck): virga pastoris, cardus fullonum
83 De daphnoide (= *Daphne mezereum* L.): laureola
87 De daphne alexandrina (= *Ruscus hypoglossum* L.): uvularia, bonifacia, pagana lingua
88 De ephemero non letali (= *Convallaria majalis* L.): lilium convallium
89 De helenio (= *Inula helenium* L.): enula
90 De eupatorio (= *Agrimonia eupatoria* L.): agrimonia, bruchwurtz
91 De elilisphaco (= *Salvia officinalis* L.): salvia
92 De serpyllo (= *Thymus serpyllum* L.): serpyllum
99 De eruca (= *Diplotaxis tenuifolia* (L.) DC): eruca
101 De erebintho (= *Cicer arietinum* L.): cicer
105 De helleboro albo (= *Veratrum album* L.?): elleborus albus
106 De helxine (= *Parietaria officinalis* L.): parietaria, paritaria
107 De erythrodano (= *Rubia tinctorum* L.): rubea tinctorum
110 De hediosmo (= *Mentha* sp.): menta
111 De hemionitide (= *Phyllitis scolopendrium* (L.) Newman): scolopendrium, lingua cervina
112 De eryngio (= *Eryngium campestre* L.): iringus
113 De thridace (= *Lactuca sativa* L.): lactuca
117 De io (= *Viola odorata* L.): viola
123 De hipposelino (= *Smyrnium olusatrum* L.?): olusatrum
124 De ixo (= *Viscum album* L.): viscum
127 De capno (= *Fumaria officinalis* L.): fumus terrae
128 De cici (= *Ricinus communis* L.): cataputia maior, pentadactylon, palma Christi

129 De cirsio (= *Anchusa officinalis* L.): buglosum, lingua bovis
130 De coriano (= *Coriandrum sativum* L.): coriandrum
131 De cassutha (= *Cuscuta* sp.): cuscuta
133 De betonica altera (= *Dianthus cartusianorum* L.): tunica
134 De colchico (= *Colchicum autumnale* L.): hermodactylus
135 De clematide daphnoide (= *Vinca maior* L.): pervinca
136 De cardamo (= *Lepidium sativum* L.): nasturtium sativum
137 De crino (= *Lilium candidum* L.): lilium album
138 De cucurbita sativa (= *Lagenaria siceraria* subsp. *asiatica* (Kobiakova) Heiser): cucurbita
139 De colocynthide (= *Citrullus colocynthis* L.): coloquintida
140 De cotonea malo (= *Pyrus cydonia* L.): citonia
142 De caryo basilico (= *Juglans regia* L.): nux
143 De caltha (= *Calendula officinalis* L.): calendula
145 De centaurio minore (= *Centaurium erythrea* L.): centauria minor
146 De cyamo (= *Vicia faba* L.): faba
150 De caro (= *Carum carvi* L.): carvi
153 De coccimelea (= *Prunus domestica* L.): prunus
154 De conio (= *Conium maculatum* L.): cicuta
155 De cynoglosso (= *Echium vulgare* L.): cynoglossum
156 De cnico (= *Carthamus tinctorius* L.?): cartamus
157 De cenchro (= *Panicum miliaceum* L.?): milium
158 De crambe sativa (= *Brassica oleracea* L.): caulis
163 De crommuo (= *Allium cepa* L.): cepa
165 De calamintha (= *Nepeta cataria* L.): calamentum
166 De crithe (= *Hordeum vulgare* L.): hordeum
167 De croco (= *Crocus sativus* L.): crocus
171 De cyclamino (= *Cyclamen purpurascens* Miller?): cyclamen, panis porcinus, panis terrae, arthanita
172 De cypero (= *Cyperus longus* L.): cyperus, cypirus
173 De lathyro (= *Euphorbia lathyris* L.): cataputia minor
174 De leucoio (= *Hesperis matronalis* L.): chieri, keirim
178 De lino (= *Linum usitatissimum* L.): linum
179 De lichene (= *Marcantia polymorpha* L.?): hepatica
180 De linozoste (= *Mercurialis annua* L.): mercurialis
181 De libanotide coronaria (= *Rosmarinus officinalis* L.): rosmarinus
186 De lithospermo (= *Lithospermum officinale* L.) = milium solis
189 De larice (= *Larix decidua* Miller): largo
190 De melissophyllo (= *Mellitis melissophyllum* L.): melissa
191 De marathro (= *Foeniculum vulgare* Miller): foeniculum
192 De melanthio (= *Nigella sativa* L.): nigella
193 De malache (= *Alcea rosea* L.?): malva
194 De myrice (= *Myricaria germanica* (L.) Desv.): tamariscus
195 De mecone rhoeade (= *Papaver rhoeas* L.): papaver rubeum
197 De mecone ceratide (= *Glaucium flavum* Crantz): papaver cornutum
198 De moro (= *Morus nigra* L.): morus celsi
203 De nymphaea (= *Nymphaea alba* L.): nenuphar
204 De napy (= *Sinapis alba* L.?): sinapis, sinapi
205 De nerio (= *Nerium oleander* L.): oleander

206 De oxyacantha (= *Berberis vulgaris* L.): berberis
207 De osyride (= *Linaria vulgaris* L.): linaria, herba urinaria
208 De ocymo (= *Ocymum basilicum* L.): basilicum
209 De origano (= *Origanum vulgare* L.): origanum, origanus
212 De oenanthe (= *Filpendula vulgaris* L.?): filipendula, saxifraga rubea
213 De oxy (= *Oxalis acetosella* L.): trifolium acetosum, alleluya
215 De ormino (= *Salvia sclarea* L.): gallitricum
216 De orobo (= *Lathyrus sativus* L.): orobos
217 De oreoselino (= *Petroselinum crispum* (Mill.) A.W. Hill): petroselinum
222 De parthenio (= *Anthemis cotula* L.?): cotula foetida
223 De polygonato (= *Polygonatum multiflorum* L.): sigillum Salomonis
224 De polypodio (= *Polypodium vulgare* L.): polypodium
225 De prasio (= *Marrubium vulgare* L.): prasium
227 De pteride (= *Dryopteris filix-mas* (L.) Schott.): filix
228 De peucedano (= *Peucedanum officinale* L.): foeniculus porcinus
229 De persica arbore (= *Prunus persica* Batsch.): persica malus
230 De peplo (= *Euphorbia peplus* L?): esula rotunda
235 De polygono mare (= *Polygonum aviculare* L.): corrigiola, centumnodia
236 De pegano (= *Ruta graveolens* L.): ruta
243 De perfoliata (= *Bupleurum rotundifolium* L.): persicaria
244 De porro (= *Allium porrum* L.): porrum
246 De pyrethro (= *Anacyclus pyrethrum* (L.) Link.): pyrethrum
251 De pyro (= *Triticum aestivum* subsp. *vulgare* (Vill.) Mac Key): triticum
256 De raphano (= *Raphanus sativus* L.): raphanus minor
263 De seride (= *Cichorium endivia* L.): endivia
265 De strycno (= *Solanum nigrum* L.): solatrum
266 De symphyto (= *Symphytum officinale* L.): consolida maior
267 De sicy sativa (= *Cucumis sativus* L.): cucumis, cucumer, cucumis citrinus
268 De sicy sylvestri (= *Ecballion elaterium* L.): cucumer asininus
271 De satyrio basilico (= *Gymnadenia conopsea* (L.) R. Br in W.T. Aiton): palma Christi
275 De smilace levi (= *Calistegia sepium* (L.) R. Br.): volubilis maior
277 De sisymbrio (= *Rorippa nasturtium-aquaticum*): nasturtium aquaticum
278 De stratiote millefolia (= *Achillea millefolium* L.): millefolium
279 De sarracenica solidagine (= *Senecio Fuchsii* Gmelin?): cartafilago
280 De saxifraga (= *Asplenium ruta-muraria* L.): capillus Veneris (perperam)
284 De selino hortensi (= *Apium graveolens* L.): apium hortense, apium sativum
285 De sisero (= *Pastinaca sativa* subsp *sylvestris* (Mill.) Rouy et Camus): baucias
289 De syce (= *Ficus carica* L.): ficus sativa
291 De smyrnio (= *Levisticum officinale* L.): levisticum
292 De silphio (= *Peucedanum ostruthium* (L.) W. D. J. Koch): osteritium, ostritium
297 De serpentariis officinarum (= *Polygonum bistorta* L.): serpentaria

299 De stichade (= *Lavandula stoechas* L.): stichas arabica
300 De struthio (= *Saponaria officinalis* L.): condisi
301 De scilla (= *Drimia maritima* (L.) Stearn): squilla
303 De seseli massiliensi (= *Ammi visnaga* (L.) Lam.): silermontanum, seseli aethiopicum, seseli peloponnesiacum
307 De trichomane (= *Aspleniumviride* Huds.): polytrichon (errate)
308 De fenugreco (= *Trigonella foenum-graecum* L.): foenumgrecum
309 De telephio (= *Sedum telephium* L.): crassula maior
314 De tithimalis (= *Euphorbia helioscopia* L.): esula
316 De triphyllo (= *Trigonella coerulea* (L.) *Ser* in DC): trifolium
317 De trifolio (= *Trifolium pratense* L.): trifolium
320 De thymo (= *Thymus vulgaris* L.): thymus
322 De hyperico (= *Hypericum perforatum* L.): hypericum
323 De hyoscyamo (= *Hyoscyamus niger* L.): iusquiamo
325 De hyssopo (= *Hyssopus officinalis* L. 3220): hyssopus, hyssopum
327 De phlomo (= *Verbascum thapsus* L.): tapsus barbatus
329 De phu germanico (= *Valeriana phu* L.?): valeriana
330 De phaco (= *Lens culinaris* Medicus): lens, lenticula
332 De chelidonio maiore (= *Chelidonium majus* L.): chelidonia
333 De chelidonio minore (= *Ranunculus ficaria* L.): scrophularia minor
334 De chamaedry (= *Teucrium chamaedrys* L.?): chamaedris
336 De chamaecisso (= *Glechoma hederacea* L.): hedera terrestris
338 De chamaeleone (= *Carlina acaulis* L.): cardopatium
340 De chamaepity (= *Ajuga* sp.): iva
341 De psyllio (= *Plantago afra* L.): psillium
342 De pseudonardo (= *Lavandula latifolia* Medicus?): spica

Termes du *vulgus*:

12 De Alismate (= *Alisma plantago-aquatica* L.): Barba sylvana
13 De artemisia (= *Tanacetum parthenium* L.?): matricaria
16 De spina alba (= *Silybum maritimum* L.): carduus Mariae
18 De ononide (= *Ononis spinosa* L.): resta bovis
30 De aconito (= *Aconitum vulparia* Rchb): luparia
32 De ampelo leuce (= *Brionia cretica* L.): viticella
53 De bellide (= *Bellis perennis* L.?): primula veris
58 De bryo (*Humulus lupulus* L.): humulus; bruscandula (Italia, ruri)
76 De geranio (= *Erodium cicutarium* L.): rostrum ciconiae (vulgus et herbarii)
(= *Geranium rotundifolium* L.): pes columbinus
77 De gongyle (= *Campanula rapunculus* L.): rapunculus
88 De ephemero non letali (= *Convallaria majalis* L.): lilium convallium
89 De helenio (= *Inula helenium* L.): campana (rustici), enula campana
98 De heptaphyllo (= Potentilla erecta L.): tormentilla
110 De hediosmo (= *Mentha longifolia* (L.) Huds.): menta equina
130 De coriandro (= *Coriandrum sativum* L.): coriandrum
135 De clematide daphnoide (= *Vinca maior* L.): vincapervinca
136 De cardamo (= *Lepidium sativum* L.): cressio hortensis (barbari herbarii)
146 De cannabe (= *Cannabis sativa* L.): canapus (vulgo, barbari)
153 De coccimelea (= *Prunus domestica* L.): pruneola, prunula

158 De crambe sativa (= *Brassica oleracea* L.): caulis
175 De lapatho (= *Chenopodium bonus-henricus* L.): Bonus Heinricus (vulgato nomine)
176 De limonio (= *Pyrola rotundifolia* L.): pyrola
213 De oxy (= *Oxalis acetosella* L.): panis cuculi
221 De xylo (= *Gossypium herbaceum* L.): cotonum, cotonium
250 De periclymeno (= *Lonicera periclymenon*): caprifolium, matersilva
263 De seride (= *Taraxacum officinale* Weber): dens leonis, rostrum porcinum
265 De strycno (= *Solanum nigrum* L.): cuculus, uva lupina, uva vulpis, morella
267 De sicy (= *Cucumis sativus* L.): anguria (privatim) (= *Cucurbita pepo* L.): zucco marin
270 De satyro trifolio (= *Platanthera bifolia* (L.) Rich.): testiculus vulpis, testiculus sacerdotis
272 De succisa (= *Succisa pratensis* L.): morsus diaboli
276 De sisymbrio (= *Mentha aquatica* L.): mentha aquatica
285 De sanctae Barbarae herba (= *Barbarea vulgaris* R. Br.): Barbarae herba (a superstitione)
288 De sisero (= *Pastinaca sativa* L.?): pastinaca
300 De struthio (= *Saponaria officinalis* L.): saponaria
310 De Trinitatis herba (= *Viola tricolor* L.): herba Trinitatis
335 De chamaecyparisso (= *Santolina chamaecyparisso* L.): cypressus

Termes des *herbarii*:
12 De Alismate (= *Alisma plantago-aquatica* L.): Barba sylvana
14 De aparine (= *Galium aparine* L.): aspergula
27 De alcea (= *Malva alcea* L.): Simeonis herba (officinis ignota, vulgares herbarii et empirici)
30 De aconito (= *Paris quadrifolia* L.): uva versa, uva vulpina (vulgus herbariorum)
45 De asclepiade (= *Asclepias vincetoxicum* L.): hirundinaria
50 De bechio (= *Tussilago farfara* L.): ungula caballina
52 De buphthalmo (= *Matricaria perforata* Mérat): oculus bovis, oculus vaccae, cotula non foetida
53 De bellide (= *Bellis perennis* L.?): primula veris
59 De veronica (= *Veronica officinalis* L.): veronica (vulgus herbariorum)
75 De glechysyda (= *Paeonia officinalis*): germ. *Ninivenwurtz* (vulgares herbarii pro mare, Peonienblum pro foemina)
76 De geranio (= *Erodium cicutarium* L.): rostrum ciconiae (vulgus et herbarii), acus pastoris
83 De daphnoide (= *Daphne mezereum* L.): laureola
96 De erysimo (= *Sinapis arvensis* L.): sinapis silvestris (vulgares herbarii)
102 De echio (= *Anchus arvensis* (L.) M.Bieb.): buglossum sylvestre (vulgares herbarii)
110 De hediosmo (= *Mentha spicata* L.): menta romana, menta saracenica
115 De thlaspi (= *Lepidium campestre* (L.) R.Br. in W. T. Aiton): nasturtium tectorum, sinapi rusticum
116 De thermis (= *Lupinus albus* L.): lupinus sativus
121 De hippuri (= *Equisetum palustre* L.): cauda equina (recentiores herbarii)
131 De cassutha (= *Cuscuta* sp.): cuscuta
168 De cyclamino (= *Cyclamen purpurascens* Miller?): arthanita
181 De lepidio (= *Lepidium latifolium* L.): piperitis (vulgus herbariorum)
195 De mecone rhoeade (= *Papaver rhoeas* L.): papaver rubeum

199 De myrrhide (= *Anthriscus sylvestris* (L.) Hoffm.): cicutaria (vulgares herbarii)
222 De parthenio (= *Anthemis cotula* L.?): cotula foetida
226 De peristerone (= *Sisymbrium officinalis* (L.) Scop.): verbena
237 De potentilla (= *Potentilla anserina* L.): potentilla
238 De prunella (= *Prunella vulgaris* L.): prunella (vulgus medicorum et herbariorum)
239 De pentaphyllo (= *Potentilla alba* L.): quinquefolium
256 De raphano (= *Armoracia rusticana* Gaertner Meryer et Scherb): radix
257 De sanicula (= *Astrantia major* L.): osteritium (errate)
263 De seride (= *Cichorium intybus* L.): cichorea
275 De smilace levi (= *Calistegia sepium* (L.) R. Br.): volubilis maior
283 De sancti Iacobi flore (= *Senecio jacobaea* L.): sancti Iacobi flos
297 De serpentariis officinarum (= *Polygonum bistorta* L.): bistorta, colubrina
320 De thymo (= *Thymus vulgaris* L.): serpyllum romanum
322 De hyperico (= *Hypericum perforatum* L.): perforata
327 De phlomo (= *Primula veris* subsp. *suaveolens* Guterm. et Ehrend): herba paralysis, herba arthetica
329 De phu germanico (= *Valeriana phu* L.?): valeriana maior (= *Valeriana vulgaris* L.?): valeriana vulgaris

Termes du "*vulgus medicorum*", "*medici*":

57 De batrachio (= *Ranunculus auricomus* L.): pes corvi
238 De prunella (= *Prunella vulgaris* L.): prunella (vulgus medicorum et herbariorum)
311 De teutlo (= *Beta vulgaris* L.): sicla

Termes des *iurisconsulti*:

221 De xylo (= *Gossypium herbaceum* L.): lana ligni (pour le coton; le type lexical est celui qui a été choisi par l'allemand; les autres formes sont dans la liste *vulgus*)

On observera que Linné, dans sa dénomination des plantes, a souvent préféré la forme qui était peu correcte.

Laurence MOULINIER-BROGI

L'UROSCOPIE EN VULGAIRE DANS L'OCCIDENT MÉDIÉVAL: UN TOUR D'HORIZON

Introduction

L'image du médecin portant un flacon d'urine à hauteur de ses yeux pour examiner son contenu est emblématique du médecin médiéval[1]: dans tel manuscrit du *Trésor* de Brunetto Latini, n'est-ce pas un urinal qui symbolise la 'fisique', située en dessous du droit canon et au-dessus du droit civil[2]? Pourtant la science des urines, d'une part n'a pas toujours été un des piliers de la sémiologie médicale et, d'autre part, n'a pas toujours eu la préséance que suggère le stéréotype du médecin à l'urinal. Dans l'Antiquité, l'inspection des urines n'était qu'un des éléments de l'examen clinique, sans faire l'objet de traités particuliers. Ni Hippocrate ni Galien n'avaient légué de système à propos des urines, et la nouveauté vint de Byzance, avec la consécration de l'analyse des urines comme une méthode de diagnostic décisive qui devait permettre de prévoir 'les choses les plus cachées et les plus abstruses': mise au point au VIe ou au VIIe siècle dans l'entourage du Byzantin Théophile Protospathaire, qui souhaitait pallier la négligence de Galien, la science des urines gagna durablement l'Occident latin à partir du XIe siècle, époque où, d'une part, le *De urinis* de Théophile fut traduit du grec et où, d'autre part, le *Liber urinarum* d'Isaac Israeli (m. 955) fut mis d'arabe en latin par Constantin l'Africain.

L'uroscopie fut un des intérêts majeurs des maîtres dans la Salerne du XIIe siècle: outre Maurus (v. 1130-1214), qui composa un commentaire au traité de Théophile et des *Regulae urinarum*, d'autres maîtres salernitains ont laissé des écrits qui sont autant de contributions importantes à

1. Voir par exemple le ms. Paris, BnF, lat. 14732, dont l'initiale peinte au fol. 71ra montre un médecin vêtu comme un clerc en train d'examiner les urines, ou encore la peinture sur laquelle s'ouvre le ms. lat. 11860. Je tiens à remercier Nicolas Weill-Parot pour sa relecture attentive du présent article.

2. Cf. Murray Jones P., *Medieval medical miniatures* (London, 1984), fig. 21, p. 57.

l'élaboration d'une doctrine sémiologique. Puis, assez rapidement, science des urines fit son entrée dans les textes au programme des facultés de médecine: le *De urinis* d'Isaac et celui de Théophile figurent parmi les lectures au programme de la licence à Paris à la fin du XIIIe siècle (1270-1274)[3], et elle fut reconnue comme un enseignement à part entière dans d'autres lieux l'études, ce qui suscita à son tour une abondante production: à Montpellier par exemple, où le *De urinis* de Théophile fait partie de la liste des cours parmi lesquels les maîtres devaient choisir, d'après les nouveaux statuts dont se dota l'université en 1340[4], tout se passe comme si, pour devenir maître en médecine, il fallait avoir laissé un traité d'uroscopie ou un commentaire à une autorité en la matière, principalement Théophile mais aussi Gilles de Corbeil († v. 1223).

On sait donc quelle place importante la science des urines occupa dans la pensée et la pratique médicales, à partir du XIIe siècle et bien au-delà du moyen âge, au point que beaucoup finirent par réduire la sémiologie du corps malade à l'uroscopie. Or, certains aspects de l'histoire de l'uroscopie médiévale méritent encore d'être creusés, telle la question de la vulgarisation de ce savoir, un terme par lequel on peut entendre à la fois *volgarizzamento*, mise en langues vulgaires, et diffusion dans les milieux non-spécialistes.

Ces vulgarisations sont parfois complexes et l'on rencontre également, aux derniers siècles du Moyen Âge, des compositions faites directement en langue vernaculaire, ce qui ne manque pas de soulever de nombreuses questions: quelle est la différence entre rédiger en latin ou dans une autre langue? Le passage en vulgaire implique-t-il banalisation, simplification du contenu? L'attitude des auteurs change–t-elle en fonction des destinataires? Et y a-t-il divers types de publics distincts pour ces textes, savants d'un côté, 'profanes' de l'autre? C'est ce que nous tâcherons de voir à travers un panorama de ces questions générales, avant de nous consacrer à l'examen d'un exemple particulier.

1. Variété de l'uroscopie latine

Le *De urinis* de Théophile avait été traduit à la même époque et dans le même milieu, sans doute salernitain, que le traité *Du pouls* attribué à

3. Jacquart D. & Micheau F., *La médecine arabe et l'Occident médiéval* (Paris, 1990, rééd. 1997), p. 172. Voir à ce sujet Wallis F., 'Inventing Diagnosis: Theophilus' *De urinis* in the Classroom', in: R. French & C. O'Boyle (éds), *Medical Teaching and Classroom Practice in the Medieval Universities*, *Dynamis* 20 (2000), p. 31-73.

4. Jacquart & Micheau, *La médecine arabe*..., p. 192.

un certain Philaret et issu en réalité d'un opuscule pseudo-galénique, et ces deux œuvres eurent par la suite une destinée commune: tous deux devinrent en effet partie intégrante de l'*Articella*, cette anthologie de textes de médecine hippocratico-galénique constituée à Salerne, qui se diffusa très rapidement dans le reste de l'Europe et devint un canon en vigueur jusqu'à la Renaissance. Il est donc très fréquent de trouver dans les manuscrits ces deux textes copiés l'un à la suite de l'autre, mais nous laissons délibérément de côté les commentaires à l'*Articella*, qui constituent un domaine en soi.

Au sein de la production latine relative à l'urine, il faut surtout relever l'existence, en grand nombre, de manuscrits constituant de véritables collections de traités uroscopiques, mêlant auteurs connus et noms moins célèbres, voire textes anonymes, comme le ms. Vatican, B.A.V., Pal. Lat. 1117, contenant divers courts textes de petits maîtres en la matière, dont un certain 'Guillelmus', ou le ms. Pal. Lat. 1228, qui renferme un *De urinis dignoscendis* anonyme, fol. 1r-26r, suivi d'un *De urinarum judiciis*, fol. 27v-39v, attribuable à Bartholomeus de Montagnana ou à Zacharias de Feltris[5], auquel succède le *De cautelis medicorum* de Bernard attribué à Gordon, fol. 40r-41r.

Citons aussi, à titre d'exemple, le ms. Munich, Bayerische Staatsbibliothek, Clm 363, du XVe siècle, où figurent: Matteo da Verona, *De urinis*, fol. 86r[6]; Antonio da Janua, *De urinis*, fol. 93r; Pseudo-Arnaud de Villeneuve, *De urinis*, fol. 95r; des pronostics anonymes tirés des urines, fol. 97r; et enfin un *De hypostasti urine*, également anonyme, fol. 100v. Ou encore le ms. Munich, BSB, Clm 267, du XIVe siècle, qui contient: Guillelmus Anglicus, *De urina non visa*, fol. 46-48; Guillelmus Anglicus, *De signis egritudinum*, fol. 49r; Gautier Agilon, *Iudicia urinarum*, fol. 68r; et un *De urinis* anonyme, fol. 75v.

Le ms. Oxford, Bodleian Library, Digby 29 (XVe siècle), contient pour sa part, outre un *De urinis* anonyme (fol. 75v), et celui de Gilles de Corbeil (fol. 76v), un traité sur les couleurs des urines, latin et anglais (fol. 73r; 114v), ainsi qu'un *De urinis*, anglais et latin, fol. 125, 130v. Un autre *De urinis* anglais et latin se trouve par exemple dans le ms. Bodleian, Digby 95, du XVe siècle, aux fol. 81r-111v: de fait, le bilinguisme n'avait rien de rare dans ce domaine, que ce soit à l'échelle du texte[7], ou

5. Un texte que l'on trouve également dans le ms. Vatican, B.A.V., Pal. Lat. 1303, fol. 114r-118r.

6. Voir aussi le ms. Vatican, B.A.V., Pal. Lat. 1251 (1463-64), aux fol. 141v-143r.

7. On attend ainsi l'édition critique, par Massimo Oldoni et Daniela Patti, d'un *De urinis* latin-italien d'inspiration salernitaine conservé à la Bibliothèque de Palerme. Voir Oldoni M., 'Monastic Nephrology in the School of Salerno and in an unpublished treatise

à celle du recueil, comme dans le ms. Paris, BnF, lat. 11135, où se succèdent un *De urinis* latin anonyme, fol. 27-28v, et un autre en allemand ('Hystoria bona de iudicio urine'), fol. 29-36v.

Soulignons pour finir que des chapitres sur les urines faisant originellement partie d'une œuvre médicale plus complète circulaient de manière indépendante, et citons entre autres le *Liber aureus* composé par Johannes Afflacius, disciple de Constantin l'Africain. Ce traité comprenait quatre parties dans sa version complète, à savoir une pathologie générale *de capite ad calcem*, des chapitres sur la fièvre, des chapitres sur l'urine et un antidotaire. Significativement, sur dix manuscrits ayant transmis cette œuvre, deux seulement présentent les chapitres sur l'urine. Ceux-ci circulaient pourtant, mais de manière autonome, comme un traité à part; c'est ce que montre bien l'édition du *Liber aureus* parue à Bâle en 1536, où les chapitres d'uroscopie sont séparés du texte et présentés sous le titre propre de *De urinis liber compendiosus sed multa bona complectens*[8].

À propos d'une œuvre postérieure, le *De conservatione vitae humanae* de Bernard de Gordon, Luke Demaître remarque à son tour que le petit nombre de manuscrits complets de l'œuvre en latin suggère une circulation réduite de la version d'ensemble[9], phénomène sur lequel l'histoire des traductions apporte un éclairage complémentaire: seuls les deux premiers livres de cet ouvrage, consacrés à la phlébotomie et aux urines, furent traduits en hébreu, alors que les parties traitant du pouls et du régime proprement dit ne demeuraient accessibles qu'en latin[10]. On extrayait donc volontiers de certaines œuvres un *De urinis*, voué à une diffusion autonome et éventuellement à une translation. Ajoutons que si la question des traductions de ces traités est inégalement creusée selon les auteurs, les langues et les pays, et si l'on ne dispose pas à l'heure actuelle

in Middlelatin and Italian volgare of the manuscript 2Qq C63 in the Public Library of Palermo', *Journal of Nephrology* 17 (2004), p. 334-336.

8. Voir à ce sujet Veit R., 'Le *Liber aureus* de Iohannes Afflacius et ses rapports avec d'autres textes salernitains', dans: D. Jacquart & A. Paravicini Bagliani (éds), *La Scuola Medica Salernitana. Gli autori e i testi* (Firenze, 2007), p. 447-464.

9. Voir Demaître L., *Doctor Bernard de Gordon, Professor and Practitioner* (Toronto, 1980); sur le Livre des urines de Bernard de Gordon, voir aussi Demaître L., 'Bernard de Gordon et son influence sur la pensée médicale aux XIVe et XVe siècles', in: D. Le Blévec (éd.), *L'université de médecine de Montpellier et son rayonnement (XIIIe-XVe siècles), Actes du colloque de Montpellier (Université Paul-Valéry-Montpellier III), 17-19 mai 2001* (Turnhout, 2004), p. 103-132, ici p. 112-113.

10. Nicoud M., 'Les traductions vernaculaires d'ouvrages diététiques au Moyen Age: recherches sur les versions italiennes du *Libellus de conservatione sanitatis* de Benedetto Reguardati', in: J. Hamesse (éd.), *Les traducteurs au travail, leurs manuscrits et leurs méthodes* (Turnhout, 2001), p. 471-493, ici p. 477.

d'un recensement des traités d'uroscopie en langue vulgaire, le phénomène concerna l'ensemble des langues de l'Europe médiévale: le français, le catalan[11], l'italien, l'allemand, l'anglais, et même l'hébreu[12].

2. L'uroscopie en vulgaire

2.1. Les productions originales

On retrouve apparemment dans ces traités en vulgaire, toutes aires linguistiques confondues, des traits caractéristiques de la production du même genre en latin, notamment le grand nombre d'anonymes, la tendance à l'autonomisation de chapitres consacrés aux urines, et le tropisme amenant à la constitution de recueils. Encore faut-il distinguer entre traductions, dont les auteurs restent le plus souvent inconnus, et productions originales dans telle ou telle langue.

Dans l'aire germanique, par exemple, l'*Arzneibuch* composé à la fin du XIIIe siècle par Ortolf von Bayerland, est désormais assez bien connu, quoiqu'on sache peu de choses sur le personnage: 'von Beierlant geborn', et 'ein arzet in Wirzeburc', dit-il dans l'introduction de son *Arzneibuch*. Certains, comme Karl Sudhoff, y ont vu un personnage imaginaire, mais deux documents de 1339 semblent attester qu'il habitait alors près de la cathédrale de Würzburg dans une des *Domherrenhöfe*. Il était chirurgien, puisqu'il est qualifié de *cirulogus*, et peut-être

11. Citons par exemple les manuscrits suivants recensés par Guy Beaujouan: Tolède, Biblioteca Capitular, 97-23, f. 6-12v (*Orinas, pols*, en catalan d'après Galien); Salamanque, BU, 2262, f. 236-63 (*Tractado de las orinas, de los polsos et de otras senales);* Barcelone, BU, 239 (Urines, sangsues, saignées, en catalan); Séville, Biblioteca capitular y colombina, 7-4-27, f. 11-16 (traité en catalan commençant par 'les horines son conegudes per los metges…'). Cf. Beaujouan G., *Les sciences médiévales d'Espagne et d'alentour*. V. *Manuscrits médicaux du Moyen Age conservés en Espagne* (Ashgate, 1992).

12. Sur ce dernier domaine linguistique, voir par exemple les stimulantes réflexions de Rotschild J.-P., 'Motivations et méthodes des traductions en hébreu du milieu du XIIe à la fin du XVe siècle', in: G. Contamine (éd.), *Traduction et traducteurs au Moyen Age, Actes du colloque international du CNRS organisé à Paris, IRHT, 26-28 mai 1986* (Paris, 1989), p. 279-302. Voir aussi les volumes suivants de la collection *Grundriss der romanischen Literatur des Mittelalters* publiée à Heidelberg, mentionnant des traductions en vulgaire d'ouvrages médicaux: VI/2, *La littérature didactique, allégorique et satirique*, dir. H.R. Jauss (1970), p. 198-199; VIII/2, *La littérature française aux XIVe et XVe siècles*, dir. D. Poirion, p. 310-320; IX/2, *La littérature de la péninsule ibérique aux XIVe et XVe siècles*, dir. W. Mettmann (1983), p. 93-98; X/2, *Die italiänische Literatur im Zeitalter Dantes und am Übergang vom Mittelalter zur Renaissance*, dir. A. Buck (1989), p. 151.

remplissait-il des fonctions officielles, municipales, comme l'appellation *chirurgicus Herbipolensis* le suggère[13].

On lui doit en tout cas une œuvre articulée, construite, hiérarchisée. Ortolf connaissait le latin ('elliu arzetbuoch, diu ich in latîne ie vernam') et avait vraisemblablement étudié à l'université. Il traduit, mais fait surtout une paraphrase des auteurs qu'il utilise; et, à son tour, il s'intéresse à l'uroscopie, en deux chapitres de son troisième traité (tr. 3, kap. 46-47), où il s'appuie principalement sur Isaac et Gilles de Corbeil.

Selon un phénomène évoqué plus haut à propos de la production latine, les chapitres sur l'urine d'Ortolf circulèrent séparément du reste de son *Arzneibuch*, constituant un *De urinis* autonome, ainsi dans le ms. Vatican, B.A.V., Pal. Lat. 1321, XIVe-XVe s., fol. 33r-67v, ou dans le ms. Paris, BnF, lat. 6952, fol. 354r: 'Ein hübscher Tractat von dem Urteyl des Harns durch Maister Ortolff im Bayerland'.

Dans ce dernier manuscrit, le texte d'Ortolf est entouré d'autres écrits relatifs à l'urine, selon une tendance au regroupement sensible dans d'autres aires linguistiques, par exemple dans le domaine italien; le ms. Florence, Biblioteca Riccardiana, 2170 [N.IV.5], renferme ainsi un *Dell'orina* attribué à Maurus de Salerne, fol. 1r-14r, et un *De l'orina* anonyme, fol. 155r-159r[14], et le ms. Florence, BN, XV 27 Magliabechiano, contient pour sa part trois traités d'uroscopie mis en vulgaire: non seulement une traduction des *Regulae urinarum* de Maurus mais aussi celle d'un *De urinis* attribué à Jean de Parme ('Libro delle orine', fol. 36-37[15]), ainsi que celle d'un *De urinis* attribué à Michel Scot ('Sulle orine', fol. 37v-45[16]).

Ortolf était donc instruit, *litteratus*, possédant le latin; le public visé par sa *Fachschrift* était en revanche constitué de *wuntarzet*, non formés à l'université et à qui il donnait les moyens d'accéder au savoir qui y était dispensé, mais l'étude de la transmission manuscrite de cette œuvre révèle que, jusqu'à la fin du Moyen Âge, son lectorat ne cessa de s'élargir, des

13. Cf. Keil G., 'Ortolf von Baierland', in: *Die deutsche Literatur des Mittelalters: Verfasserlexikon* (Berlin – New York, 1981), Bd 3, col. 67-82, voir aussi Riha, O., Brunner H., Dickerhof H. & Huschenbett D., *Ortolf von Baierland und seine lateinischen Quellen: Hochschulmedizin in der Volkssprache*, Wissensliteratur im Mittelalter 10 (Wiesbaden, 1992).

14. Voir Elsheikh M.S., *Medicina e farmacologia nei manoscritti della biblioteca Riccardiana di Firenze* (Firenze, 1990), p. 51-52.

15. Inc. 'L'orina di colui la quale a febre e ella sarà rossa si sposa congrassezza, significa corrompimento di sangue...' Expl. 'Explicit de urinis m. Jovani de Parma'.

16. Inc.: 'Della notizia e pronosticatione dell'orine sicondo Michele Scoto cosi de sano come delli infermi'. Sur le traité des urines de Michel Scot et ses traductions en italien, voir Ackermann S., *Michael Scotus, ein Universalgelehrter des 13. Jahrhunderts*, Diss. (Frankfurt, 1987), p. 53-57.

baigneurs et barbiers jusqu'aux médecins académiques, en passant par les laïcs cultivés et intéressés à la médecine, des destinataires bourgeois voire nobles, comme le prince électeur Ludwig V[17].

Antoni Ricart, qui fut médecin des rois d'Aragon entre 1395 et 1422, année de sa mort, est un cas différent, à la fois par sa formation et ses attributions[18], et parce qu'il composa à la fois en latin et en langue vulgaire. On trouve, en effet, parmi les écrits de ce médecin du XVe siècle, outre des œuvres en latin (notamment un *Libellus de quantitatibus et proportionibus humorum* et un *Opusculum de arte graduandi*), des écrits catalans qui ont attiré l'attention avant les œuvres latines[19] et qui sont conservés dans le ms. Vatican, B.A.V, lat. 4797. Copié en 1476, ce codex contient différents écrits, dont un *De urinis* attribué à 'Galièn de Cremona', traduit par un anonyme en *De la coneixença de les orines,* et occupant les fol. 52-55[20]. Puis vient un *De pulsibus* en catalan, sans doute dû à Antoni Ricart et intitulé 'Conexençà dels polsos' (fol. 55-56v), auquel fait suite, fol. 57r-65r, un *De urinis* en catalan, intitulé *De la conexença de les urines* et nommément attribué à A. Ricart dans la rubrique suivante, fol. 57: 'En nom de deu e dela verga meria. Comença lo tractat de horinas lo qual ha ordonat Mestra entonj Ricart mestra en medicina'. Lui succède, fort logiquement, un *De pulsibus* en vernaculaire (*Conexença dels polsos*) associé lui aussi au nom d'Antoni Ricart, fol. 65r-70v; l'incipit de ce *Tractat de conoscença de polsos segons Antoni ricart* est pour sa part: 'En nom de deu e dela gloriosa mara sua. Comensa lo tractat de conaxença de polsos segons mestra antoni Ricart e comensa la definicjio'. Jeanne-Marie Dureau-Lapeyssonie tenait ces deux derniers traités pour des œuvres de jeunesse qu'Antoni Ricart aurait reniées par la suite[21], et plus récemment, Lluìs Cifuentes a repris en partie cette hypothèse, estimant qu'il pourrait s'agir d'apocryphes ou d'œuvres de jeunesse[22].

17. Keil, 'Ortolf von Baierland ..., Bd. 3, col. 67-84, col. 75.

18. Il semble avoir ajouté l'activité de professeur à ses fonctions de praticien, puisque son nom est mentionné lors de la réorganisation du *studium* de Barcelone, où il enseigna dès 1401; cf. Dureau-Lapeyssonnie J.-M., 'L'œuvre d'Antoine Ricart, médecin catalan du XVe siècle. Contribution à l'étude des tentatives médiévales pour appliquer les mathématiques à la médecine', in: G. Beaujouan, Y. Poulle-Drieux & J.-M. Dureau-Lapeyssonie, *Médecine humaine et vétérinaire à la fin du Moyen Age* (Paris, 1966), p. 171-364, ici p. 180.

19. Dureau-Lapeyssonnie, 'L'œuvre d'Antoine Ricart...', p. 197.

20. Un texte copié d'après le Vat. Lat. 4797 dans le ms. Barcelone, Bibl. Univ., 239, fol. 14.

21. Dureau-Lapeyssonnie, 'L'œuvre d'Antoine Ricart...', p. 198.

22. Voir Cifuentes L., *La ciència en català a l'Edat Mitjana i el Renaixement* (Barcelona – Palma de Majorca, 2001), p. 90. Sur les traductions d'œuvres médicales en catalan, voir aussi Cifuentes L., 'Université et vernacularisation au bas Moyen Age: Montpellier

Le reste du codex est occupé entre autres par un *Tractat de flebotomia* anonyme (fol. 71-89v); une *Suma de simples medicines e compostes*, traduction de la *Summa de simplicibus et compositis medicamentis* de Bernard de Gordon, (fol. 90-113v); un *Inventari o collectori de cirurgia*, fol. 113v-114v (une traduction anonyme de Gui de Chauliac, mais dont on connaît deux réviseurs, Bernat de Casaldòvol et Jeroni Masnovell); un *Tractatus aliquorum medicinarum aliquorum morborum et remedia illorum* mêlant catalan et latin et peut-être tiré d'une œuvre de Bernard de Gordon, fol. 116-187v; un *Receptari* anonyme, fol. 196-198v, et, pour finir, un *Régiment pour temps d'epidemie*, fol. 287r-293r (*Petit tractat sobre el regiment que es deu tenir en cas d'epidèmia*), qui a pour incipit: 'Comensa un petit tractat per lo Reverent mestra Arnau de Vila Nova sobra lo regiment qujs deu tenir en temps de hepidèmja'.

On a donc affaire ici à un recueil homogène quant à la langue et en partie quant au type de textes, mais le public visé par ces compositions en catalan n'apparaît pas clairement. En revanche, la piste d'œuvres de jeunesse composées par un médecin qui devait par la suite laisser des écrits en latin semble à retenir.

On sait par ailleurs, notamment grâce aux travaux de Marilyn Nicoud sur la diététique médiévale, qu'un même praticien put composer un texte en deux langues, ainsi au XIIIe siècle Taddeo Alderotti et son *Libello per chonservare la sanità del corpo* dont les deux versions concomittantes, latine et italienne, seraient dues à l'auteur lui-même. Au XVe siècle un autre médecin, Michele Savonarola, fit pour sa part alterner les deux langues dans toute son œuvre, usant du latin pour les ouvrages plus théoriques destinés à un public savant, et de l'italien pour des traités pratiques visant un lectorat plus large. Aurait-on là sinon un modèle du moins un schéma applicable aux traités d'uroscopie en vulgaire? Faut-il se représenter la production uroscopique comme le lieu d'un même partage des eaux, avec des traités en latin pour les savants, et d'autres en vulgaire pour un public élargi, allant des 'dilettanti' aux praticiens insuffisamment formés, tels les barbiers, matrones ou apothicaires[23]?

2.2. La délicate question du lectorat

On peut se demander si les apothicaires n'ont pas partagé, en se mêlant d'uroscopie, certaines fonctions ou prérogatives des médecins comme le

et les traductions catalanes médiévales des traités de médecine', in: D. Le Blévec (éd.), *L'université de médecine de Montpellier et son rayonnement (XIIIe-XVe siècles), Actes du colloque de Montpellier (Université Paul-Valéry-Montpellier III), 17-19 mai 2001*) (Turnhout, 2004), p. 273-290.

23. Nicoud, 'Les traductions vernaculaires…', p. 471-493.

suggèrent des indices ténus. Le premier de ces indices est textuel: dans une lettre à un vieil ami, probablement l'archevêque Rainald de Capoue, un membre de la chancellerie pontificale raconte le séjour qu'il fait à Subiaco, avec la curie d'Innocent III pendant l'été 1202. Et, lorsqu'il en vient à décrire leur lieu de séjour et les nuisances dont il souffre, il mentionne un 'apothicaire mirant les urines' et troublant la villégiature de la curie par le bruit désagréable de son mortier:

> Ab altera parte, que nascentem respicit phebum, minister quidam ypotecarius commoratur, ut tanto mane urinas in firmamentum possit subtilius intueri, quanto ipse sol perspicaciori quadam inmissione ad eundem suos radios per foramen ostendit. Ipse vero a mortario non recedens indelectabili sonitu nos incessanter offendit et multos decipere nititur per pulveres corrodendos, quos nulla racione frustraret, nisi eorum, que anxius terit, substanciam inmutasset[24].

Il s'agit là d'un texte à manipuler avec précaution, à la fois par sa qualité littéraire, ironique, et donc volontiers caricaturale, et aussi par la particularité de son vocabulaire: à la fin du texte, en effet, ce vacancier mécontent de son sort semble employer l'adjectif 'ypotecarius' comme synonyme de 'medicus' lorsqu'il évoque en ces termes la maladie de son destinataire et les ordres que lui a donnés le pape Innocent III à ce sujet:

> quod si forte, quod non credimus, ipsius mandatis negligens extiteris te auctoritate ypotecaria qua fungitur, a cibis et potibus necessarius ulterius noveris esse suspensum[25].

Le passage cité en premier lieu, avec son évocation des poudres broyées dans le mortier pour différentes préparations, ne laisse néanmoins guère de doute sur la nécessité de comprendre 'minister ypotecarius' comme 'apothicaire'.

Tournons-nous à présent vers les manuscrits: dans le ms. Florence, Biblioteca Nazionale Centrale, XV 27 Magliabechiano, un manuscrit du XVe siècle qui contient aux fol. 1-35v un *volgarizzamento* du *De urinis* de Maurus, une note marginale, fol. 45v, précise que le recueil fut compilé par 'Agostino di Niccolò di Filippo speziale l'8 aprile alle ore 17 del 1481'[26]. Le *speziale* était un vendeur d'herbes médicinales, et ce terme est souvent synonyme d'apothicaire. Or ce métier eut plus d'un point de

24. Hampe K., 'Eine Schilderung des Sommeraufenthaltes der römischen Kurie unter Innozenz III.', *Historische Vierteljahrsschrift* 8 (1905), p. 509-535, ici p. 531. Je remercie vivement le Professeur Paravicini-Bagliani de m'avoir communiqué cette référence.

25. Hampe, 'Eine Schilderung ...', p. 535.

26. Pasca M. (éd.), *La scuola medica salernitana. Storia, immagini, manoscritti dall'XI al XIII secolo* (Napoli, 1988), p. 96.

contact avec l'activité exercée par les médecins, avec qui ils partagèrent certains risques: ceux qui étaient au service des princes, notamment, étaient directement impliqués dans la préservation de leur santé, et pouvaient devenir les premiers suspects en cas d'empoisonnement, tel l'apothicaire d'Amédée VII de Savoie (†1393), emprisonné, torturé et exécuté (à tort)[27].

D'autres personnages, sans exercer cette activité à un niveau aussi prestigieux, avaient parfois une véritable responsabilité dans la santé des patients et le cas de la petite ville de Manresa étudié par Michael McVaugh suggère que les malades préféraient parfois aller les trouver pour obtenir des conseils, pour leur propre médication, plutôt que de chercher l'aide d'un médecin[28]. En Italie en particulier, des associations entre médecins et apothicaires n'étaient pas rares, dès la fin du XIVe siècle à Lucques[29], et sans doute plus tôt à Florence: dans cette dernière ville, l'*Arte dei medici e speziali* était l'un des sept arts majeurs dans l'organisation corporative de la Florence communale, ainsi organisés peut-être depuis le milieu du XIIIe siècle[30].

Mais le problème est plus complexe qu'il n'y paraît, ne serait-ce que parce qu'exercer la fonction d'apothicaire ne signifiait pas ignorer le latin. Comme le rappelle Danielle Jacquart à propos du Paris des XIVe et XVe siècles, 'le latin restait la langue des techniciens, des médecins comme des apothicaires', ces derniers devant posséder assez de latin pour utiliser l'*Antidotaire de Nicolas* et une version autorisée de *Qui pro quo*[31].

D'un autre côté, on a trace de médecins universitaires ouverts à une culture plus laïque, plus 'vulgaire'. Si l'on se fonde cette fois sur un indice encore plus modeste, à savoir l'ex-libris du codex Rome, Accademia dei Lincei, 43 B 28, 'liber iste est magistri Gentilis', l'hypothèse de traités en vulgaire ayant intéressé même les médecins savants n'est pas à exclure. Il ne fait en effet guère de doute que le 'Gentilis' ici évoqué est le médecin Gentile da Foligno (†1348), dont l'intérêt pour Maurus, notamment

27. Alexandre-Bidon D., *La mort au Moyen Age. XIIIe-XVIe s.* (Paris, 1998), p. 190.

28. McVaugh M., 'Le coût de la pratique et l'accès aux soins au XIVe siècle: l'exemple de la ville catalane de Manresa', *Médiévales* 46 (2004), p. 45-54.

29. Boisseuil D., *Le thermalisme en Toscane à la fin du Moyen Age* (Roma, 2002), p. 179 n. 75.

30. Voir *Statuti dell'arte dei medici e speziali*, éd. R. Ciasca (Firenze, 1922), p. 320-321.

31. Jacquart D., *La médecine médiévale dans le cadre parisien* (Paris, 1998), p. 298 et p. 307.

pour son uroscopie, est connu: il le cite en particulier dans son commentaire 'super prima fen quarti Canonis Avicenne'[32].

La question s'avère donc assez complexe: des apothicaires ont pu exercer la même activité que les médecins en se mêlant d'analyser les urines, et des recueils de traités d'uroscopie en vulgaire ont pu répondre aux besoins de ces praticiens, mais pas nécessairement: les apothicaires pouvaient en effet se doubler de latinistes, et inversement, des médecins savants faire une place dans leurs lectures à des traités en vulgaire.

De ces premiers indices, il appert qu'on peut tabler que les traités sur les urines furent volontiers traduits à cause de leur caractère systématique, ces traités se présentant avant tout comme un catalogue de couleurs ayant pour correspondants un certain nombre de symptômes – et l'on connaît de fait quelques traités spécifiques sur les couleurs des urines. La part assez faible qui y était faite à la théorie médicale, en d'autres termes leur nature didactique, leur facilité d'accès pour des lecteurs relativement étrangers au champ médical peut également expliquer l'existence de nombreuses traductions. Mais plutôt que d'envisager toujours l'ignorance du lectorat comme motivation principale des traducteurs, ne faudrait-il pas renverser la problématique et imputer la mise en langue vulgaire de ces écrits à une soif de culture accrue ou nouvelle? Le cas des apothicaires et celui, certes encore isolé, de Gentile da Foligno, présentés plus haut comme des utilisateurs de telles translations, montre que l'ignorance du latin ne saurait à elle seule expliquer que l'on ait recouru à des traductions; aujourd'hui encore, n'est-il pas naturel, même dans les milieux les plus cultivés, voire chez ceux dont la lecture est l'activité principale, de lire un texte dans sa langue maternelle si une traduction en existe? Plutôt que de restreindre le lectorat à un groupe handicapé par l'ignorance du latin, ne peut-on donc voir dans cet essor des 'vulgarisations' de textes aussi spécifiques un reflet de la curiosité, de la multiplication des centres d'intérêt de la société des derniers siècles du moyen âge[33], voire de nouveaux besoins liés à l'évolution de la pratique médicale et à

32. Cité par De Renzi S., *Storia documentata della scuola medica di Salerno* (Napoli, 1857²), p. 334.

33. Voir ainsi l'étude de Sante Polica sur le personnage de Michele di Giovanni Guinigi, un Lucquois qui remplit l'office d'une véritable bibliothèque de prêt de livres en tous genres pour toute la ville dans la première moitié du XVᵉ siècle; Polica S., 'Le commerce et le prêt de livres à Lucques dans la première moitié du XVᵉ siècle', *Médiévales* 14 (1988), p. 33-46.

la 'médicalisation de la société' sensible depuis le XIIIe siècle dans les pays méditerranéens[34]?

On ignore quelle est au juste la plus ancienne vulgarisation de traités sur l'urine. Quoi qu'il en soit, au XIVe siècle assurément, l'uroscopie se signale par son plurilinguisme, comme si, parallèlement au latin et parfois à ses côtés dans certains manuscrits, des langues vernaculaires, romanes et germaniques, avaient pu alors accéder au statut de langue de science. Certes, le latin resta jusqu'à la fin du Moyen Âge le grand favori de ce genre de traité, mais les traductions voire les compositions en vulgaire attestent une extension du lectorat: l'uroscopie concernait désormais un public plus large que celui des médecins formés au commentaire de Théophile à l'université, ce dont le traité d'astrologie médicale de Guillaume l'Anglais fournit une illustration extrême.

3. Un cas particulier: *De urina non visa*

Le *De urina non visa* composé par Guillaume l'Anglais à Marseille en 1219 constitue certes un cas un peu limite, puisqu'il portait jusqu'à un point ultime le lien entre uroscopie et astrologie en proposant de déterminer les caractéristiques de l'urine (principalement sa couleur et sa substance) sans la voir, en se fondant sur l'état du ciel et des planètes. Ses sources sont, entre autres, Théophile, Ptolémée (*Quadripartitum*), Messahallah, Albumasar, et Alcabitius[35], mais malgré sa technicité, ce traité jouit d'une grande popularité au moyen âge; c'est ce qu'attestent à la fois la reprise qu'en donnèrent différents auteurs, le grand nombre de manuscrits conservés, sans doute lié au fait qu'à Bologne, par exemple, d'après les statuts de 1405, il représentait une lecture obligatoire à la faculté des arts, mais aussi les traductions que l'on en connait à ce jour, intégrales ou partielles. Le ms. Vatican, B.A.V., Pal. Lat. 1416, fin XVe s., fol. 73v-74v, contient ainsi des extraits latin-allemands du *De urina non visa (Weet dattie uryn woert ghesoden in die levere...)*, suivis d'une *Figura celi*. Copié dans les anciens Pays-Bas, ce recueil d'astrologie comporte fol. 210v

34. Sur cette notion, voir notamment McVaugh M., *Medicine before the Plague. Practitioners and their Patients in the Crown of Aragon, 1285-1345* (Cambridge, 1993), et Nicoud M., 'Introduction', in: L. Moulinier-Brogi & M. Nicoud (éds), *Ethique et pratiques médicales aux derniers siècles du Moyen Age, Médiévales* 46 (2004), p. 5-10.

35. Voir dernièrement sur le sujet Jacquart D., 'L'opuscule sur le jugement des urines attribué à Hermès', in: P. Lucentini, I. Parri & V. Perrone Compagni (éds), *Hermetism from Late Antiquity to Humanism. La tradizione ermetica dal mondo tardo antico all'umanesimo*, Atti del convegno nazionale di studi, Napoli, 20-24 novembre 2001 (Turnhout, 2003), p. 461-476.

un explicit indiquant la main de 'Godefridus de Molendino, Treveris in medicina practicans' (1442)[36] – un médecin, donc, auteur d'une *Practica pestilentiae*, transmise par le ms. Pal. Lat. 1177, fol. 39r-51v.

On sait aussi qu'il exista une version française au XVe siècle du *De urina non visa* dans le ms. Turin, Biblioteca Nazionale, cod. M. IV. 11, fol. 125a-131v[37]. Bien que ce manuscrit ait disparu dans l'incendie de la Bibliothèque de Turin en 1904, on en a conservé une description relativement précise, et l'on sait notamment qui copia ce texte: 'Explicit le livre de l'orine, escript trop en hate a Namur le jour saint Thumas, devant Noel, finet au soir, empres Malgarnie demorant a ce temps l'an de grace 1463... par moy Iohannes de Borlees dis de Moust[ier]'. La transcription de Jules Camus a permis de sauver son prologue et son explicit, qui s'avèrent des traductions fidèles des mêmes passages dans la version latine, d'après les différents manuscrits que j'ai pu en consulter[38].

Dans le manuscrit de Turin, qui ne contenait que quelques courts textes en latin (une liste de drogues, fol. 151, un petit traité de nigromancie, fol. 198-206, un fragment de livre d'astronomie, fol. 260 et quelques recettes et invocations éparses), les trois premiers traités en français ont été copiés par le même scribe: outre le 'Livre de l'orine', il s'agit des 'jugemens qui appartiennent à médechines' (fol. 118-124v), et du 'Régime de santé d'Alebrant' (fol. 10-117v), c'est-à-dire du *Livre de Physique* d'Aldebrandin de Sienne, premier exemple de composition en langue vulgaire dans le domaine diététique composé en 1256, qui finissait ainsi, fol. 117v: 'Explicit a Namur, emprez les beghines, present Gertrud Jherome et Susanne la petite, le merquedi devant Pasque florie, alle chand[eille] trop en haste finet, droit a 8 heure, l'an de grace 1463. Johs de Moust[ier?]'.

36. Sur ce codex, voir Schuba L., *Die Quadriviums-Handschriften der Codices Palatini Latini in der Vatikanischen Bibliothek* (Wiesbaden, 1992), p. 201 ss.

37. Cf. Giacosa P., *Magistri Salernitani nondum editi, Catalogo ragionato della esposizione di Storia della medicina aperta in Torino nel 1898* (Turin, 1901), p. 506-507.

38. Camus J., 'Un manuscrit namurois du XVe siècle', *Revue des langues romanes*, 4e s., t. VIII (1895), p. 27-43, p. 31-32: 'Afin que je ne soie redargue pour cause d'ignorance, ou pour cause d'envie, mon germain cousin, qui aulcunes fois as estudiiet avec moy a Marseille, je baille a cheulx qui apres moy vendront, pour la grace de toy et des aultres estudians appetens ceste science, ung trattié, lequel tu avoies tant de fois désiré, ainsi come en perpetuele memoire de moy, lequel ne fut edit ne trouvé especialment d'aulcun anchiens; ou quel, selonc la force des estoilles et des signes, le jugement de l'orine non veue, avec la nature de l'enfermeté et l'issue de la maladie, est contenu. Pour la quele cause je, Guilleame de la nacion d'Angleterre, medechin par profession, pour raison de la science de medechine astronomien, maintenant cytoien de la dicte cité de Marseille, ai jugié à imposer mon nom ad ce present traictiet, et affin que l'ordonnance des choses viegne a l'encontre, j'ai presigné les capitles de ce present livret: le premier capitle, de la quadruple voie de la speculation d'astronomie. Le seconc, de l'ascension des effectz des souveraines influences en cascune des choses basses...

Le copiste était apparemment lié au milieu béguinal, mais nous n'en savons pas plus pour l'instant.

L'astrologie médicale fut surtout en faveur auprès de 'demi-savants': ce n'est pas parce qu'un médecin s'intéressait à l'astrologie qu'il l'intégrait à sa science médicale[39], et les maîtres de la faculté de médecine fixaient en général, dans leurs ouvrages proprement médicaux, des limites strictes à leur double compétence[40]. Il n'en reste pas moins remarquable qu'un ouvrage comme le *De urina non visa*, qui supposait une certaine culture astrologique, ait trouvé sa place à côté d'un texte aussi diffusé et emblématique de la littérature médicale en vulgaire que le 'Régime du corps' (ou *Livre de Physique*) d'Aldebrandin de Sienne.

4. La fortune de Maurus en vulgaire

En tout état de cause, nombre de versions vernaculaires de traités d'uroscopie appellent toujours toute l'attention, de telle version italienne du *De urinis* de Maurus de Salerne à la version anglaise d'écrits de Gautier Agilon, un des commentateurs de Gilles de Corbeil[41], et c'est sur la fortune en vulgaire, notamment italienne, de l'uroscopie de Maurus que nous allons nous arrêter ici.

Gundolf Keil a montré que dès le XIIe siècle avait vu le jour un abrégé du *De urinis* de Maurus, un 'kurzer Harntraktat' qui fut traduit en moyen haut allemand avant même le début du XIIIe siècle[42]. Ce petit traité se répandit rapidement dans tout l'espace germanophone et, fait remarquable, il fit même l'objet de réversions en latin révélant l'intérêt qu'il continuait de revêtir pour la médecine savante[43]. Il connut en cela, malgré

… Je trueve derechief Venus entrer sur les raies, et estre brullee du soleil, doncques dis je que il moroit dedens 2 mois et 8 jourz. Et moy, aveucque les aultres, heubz ycellui en cure, et il estoit etropicque et tisique.', Des. 'heubz ycellui en cure et il estoit etropicque et tisique.'

39. Voir Thorndike L., 'Medical Astrologers and Medieval Astrology', *Viator* 6 (1975), p. 295-308.

40. Cf. Weill-Parot N., *Les 'images astrologiques' au Moyen Âge et à la Renaissance. Spéculations intellectuelles et pratiques magiques (XIIe-XVe siècle)* (Paris, 2002), p. 447.

41. Par exemple dans les mss Cambridge, University Library, 346, f. 83-89, et Londres, British Library, 1635.2.

42. Voir Keil G., *Der Kurze Harntraktat des Breslauer Codex Salernitanus*, Diss. (Bonn, 1969) et Keil G., *Die urognostische Praxis in vor-und frühsalernitanischer Zeit* (Fribourg e. Br., 1970).

43. Lenhardt F., 'Maurus', in: *Die deutsche Literatur des Mittelalters*: *Verfasserlexikon* (Berlin – New York, 1981), Bd 3, col. 201-203, ici col. 202. Sur ce phénomène, voir par exemple Vernet A., 'Les traductions latines d'œuvres vernaculaires au Moyen Age', in: G. Contamine (éd.), *Traduction et traducteurs au Moyen Age, Actes du colloque international du CNRS organisé à Paris, IRHT, 26-28 mai 1986* (Paris, 1989), p. 225-241.

d'importantes différences entre ces œuvres, un sort comparable à celui du *Livre de physique* d'Aldebrandin qui, rapidement traduit en flamand, catalan et surtout en italien (langue dans laquelle on en connaît quatre versions, pour un total de 50 manuscrits), fut mis en latin au XVe siècle. Les œuvres initialement composées en vulgaire étaient destinées à un public relevant d'une aire géographique restreinte, et leur traduction en latin permettait l'augmentation du nombre de lecteurs potentiels, le latin se présentant alors comme une langue de *vulgarisation* scientifique[44]. Il semble donc important de souligner que l'uroscopie de Maurus connut ce sort dans le monde germanique.

Sa fortune italienne est en revanche encore très mal connue, alors que l'on dénombre à ce jour au moins trois versions italiennes du *De urinis*, toutes du XVe siècle à moins que la dernière ne date du XIVe, nous y reviendrons, dans les mss Florence, Biblioteca Riccardiana, 2170 [N.IV.5]; Florence, Biblioteca Nazionale Centrale, XV 27 Magliabechiano; et Rome, Accademia nazionale dei Lincei, 43 B 28.

Faute d'avoir pu, à cette date, voir les manuscrits de Florence ni même disposé de leurs reproductions malgré des demandes répétées, nous ne sommes pas en mesure de dire s'il s'agit d'une unique traduction ou de plusieurs textes différents. Nous nous concentrerons sur le seul témoin que nous ayons pu observer, afin de donner un aperçu de l'intérêt de l'étude d'un tel *volgarizzamento* pour deux problèmes spécifiques: la traduction de termes techniques et la fidélité à l'original latin.

Le ms. Rome, Accademia dei Lincei, 43 B28

Ce manuscrit, composé de 14 + CCXV folios de papier, mesure 218 sur 140 mm. Copié par différentes mains sur deux colonnes, il se signale par sa rubrication d'un bout à l'autre et des initiales calligraphiées en rouge et bleu, tandis que la présence de réclames et la numérotation des cahiers évoquent un travail de professionnel. Quant à l'ex-libris évoqué plus haut ('liber iste est magistri Gentilis'), il invite, avec l'emploi d'un verbe au présent, à revoir la question de la datation: si A. Petrucci, dans son *Catalogo inventario dei mss. Corsini II 1299-2650* manuscrit et dactylographié date le ms. du XVe siècle, la mention du premier folio pousse à lui ajouter un siècle d'âge, dans la mesure où Gentile da Foligno, son possesseur putatif, mourut en 1348.

44. Cf. Bisson S., 'Le témoin gênant. Une version latine du *Régime du corps* d'Aldebrandin de Sienne', *Le latin dans le texte*, *Médiévales* 42 (2002), p. 117-130, ici p. 119.

Les trois traités rassemblés ici sont successivement:

1° Pietro Ispano, *Somma della medicina... nominata Tesoro dei Poveri*, aux fol. Ir-CVIv, Inc.: 'Qui comincia la Somma delle medicine del maestro Pietro di Spagna nominata Tesoro dei Poveri'. Expl.: 'puote ogni cosa per ineffabile potentia. Amen. Deo gratias'[45].

2° Mauro Salernitano, *Regole dell'orine*, fol. CVIIIr-CLIXv, d'après le catalogue de Petrucci. En réalité, le traité de Maurus (Inc.: 'Qui incominciano le regole dell'orine del maestro Mauro da potere conoscere delle specie dele infermità procedenti de ciascuno dei quatro humori') s'arrête au fol. CXXXIXva, où l'on lit: 'Explicit cure et regule urinarum et aliarum curarum maximarum magistri Mauri, et Deo gratias'. S'ouvre alors, copiée par la même main, une succession de remèdes, tous signalés par une rubrique et une lettrine: le premier, annoncé par la rubrique 'suffummicamento', s'intitule 'alla sordagine', et le second est annoncé par la rubrique 'Unguento de maestro Johanni Damasceno'.

Les folios CLX-CLXIIv sont blancs.

3° Giovanni di Damasco, *Delle infermità*, fol. CLXIIIr-CCVv. Les folios CLXIIIr-CLXIVv sont occupés par la table des chapitres ('Qui comincia la tavola duna parte de libro di maestro johanni damascieno...'), le fol. CLXV est blanc, et le traité attribué à Jean Damascène ne commence qu'au fol. CLXVI. Inc.: 'Qui si tratta...' Expl.: 'explicit liber magistri Johannis de Amassein. Qui scrixit scribat, cum domino semper vivat. Gratias agamus, Deum adoremus'.

Le folio CCVI est couvert pour sa part de recettes de collyres en latin. Les folios CCVII-CCXV sont blancs, et ont été renumérotés à l'encre fuchsia par une main moderne.

Quant aux 14 premiers folios numérotés en chiffres arabes, ils consistent principalement en une table des matières du traité de Petrus Hispanus en italien, jusqu'au fol. 4r, suivie de celle de l'œuvre de Maurus, fol. 4v-7v.

La question des couleurs des urines donnera un aperçu de l'intérêt de ce texte. La coloration de ce fluide n'avait d'autre raison d'être que les proportions des humeurs, d'où l'importance de distinguos parfois fort subtils: ainsi la couleur noire, précédée de la couleur verte, signifiait la mort par adustion ou excès de chaleur, tandis que la couleur noire précédée de la nuance livide, dénotait en revanche la mortification par le

45. Sur les traductions du *Thesaurus pauperum* de Pierre d'Espagne en italien, dont celle qui fut longtemps attribuée au notaire Florentin Zucchero Bencivenni, voir Rapisarda S., 'I volgarizzamenti italiani del *Thesaurus pauperum*', in: A. Engelbert, M. Pierrard, L. Rosier & D. Van Raemdonck (éds), *Actes du XXIIe Congrès International de Linguistique et Philologie Romanes*, Bruxelles, 23-29 juillet 1998 (Tübingen, 2000), p. 107-121.

froid[46]. La liste complète de ces couleurs, du blanc au noir en passant par glauque, gris cendré, pâle, citrine, rouge, etc.[47], oscillait donc entre 19 et 20 selon les auteurs à cause de la couleur noire qui pouvait être comptée deux fois, selon qu'elle était précédée du vert ou du livide[48] – et ces nuances furent même parfois subdivisées à leur tour, ainsi au XVe siècle par le médecin Pierleone da Spoleto qui en distinguait 42[49].

En matière de noms de couleurs, l'uroscopie a enrichi la langue latine du néologisme 'subcitrinus', 'citrine pâle'[50], et y a introduit, via Théophile, des mots grecs qu'elle ne possédait pas, comme *charopos*, 'gris cendré'[51], et *inopos*, 'rouge vineux', ou qui présentaient un autre type d'emploi, comme *kyanos*, 'rouge tirant sur le bleu, pourpre'[52]. Mais ces trois derniers termes ont posé problème aux vulgarisateurs et il est intéressant de constater que, dans une traduction en anglais comme dans une version en italien, ils ont été laissés tels quels. Dans le ms. Londres, British Library, Sloane 7, par exemple, manuscrit anglais du XVe siècle décrivant 20 matules, deux sont représentées fol. 59v: on y voit différentes bandes de couleurs et des particules en suspension à différents niveaux; à gauche est écrit 'kyanos', et à droite 'inopos', sans autre forme de traduction[53]. Dans le codex de l'Accademia dei Lincei qui nous intéresse, la liste de ces couleurs se trouve aux fol. CXr-CXIr, où chaque désignation de couleur est précédée d'un pied de mouche alternativement rouge ou bleu. Et comme dans le manuscrit anglais, on constate que les désignations de couleur héritées du grec ont été reproduites telles quelles, excepté une modification graphique pour *kyanos* devenu *Quianos*.

46. Voir Vosswinkel P., *"Der schwarze Urin". Klassisches Leitsymptom der Medizingeschichte im Spannungsfeld von sinnlicher Wahrnehmung und wissenschaftlicher Erkenntnis*, Med. Habil. (Aachen, 1990).

47. Voici la liste de ces nuances: *Albus, Lacteus, Glaucus, Charopos, Subpallidus, Pallidus, Subcitrinus, Citrinus, Subrufus, Rufus, Subrubeus, Rubeus, Subrubicundus, Rubicundus, Inopos, Kyanos, Viridis, Lividus* et *Niger*, certains auteurs distinguant entre *Niger adustus* et *Niger mortiferus*. Pour une représentation figurée de ces nuances dans certains mss, voir par exemple le ms. Vatican, Pal. Lat. 1229, fol. 5.

48. Vieillard C., *L'urologie et les médecins urologues dans la médecine ancienne, Gilles de Corbeil, sa vie, ses œuvres, son Poème sur les urines* (Paris, 1903), p. 50-51.

49. Voir Rotzoll M., *Pierleone da Spoleto: Vita e opere di un medico del Rinascimento*, trad. it. (Firenze, 2001).

50. Voir Roques M., *Additions et corrections d'Antoine Thomas au Glossaire de Du Cange* (Bruxelles, 1952), p. 149: les *Regulae urinarum* de Maurus sont le seul exemple cité pour le vocable *subcitrinus*.

51. Le seul exemple fourni par Mario Roques pour *karopos* est tiré d'un *Liber de signis pronosticis* manuscrit, et donc directement emprunté au domaine de la sémiologie médicale; voir Roques, *Additions et corrections ...*, p. 128.

52. Le terme *cyanos* désignait le jaspe bleu chez Pline; cf. *Le Grand Gaffiot, Dictionnaire latin-français*, nouvelle éd. sous la dir. de P. Flobert (Paris, 2000), p. 465.

53. Plate IV, in: Murray Jones, *Medieval medical miniatures...*

Voici comment se présente la liste au fol. CXrab:

I colori delle urine sono venti cioè

[Alba	
[Glauco	[Lacteo
[Karopos	[Subpalido
[Palido	[Subcitrino
[Citrino	[Subruffo
[(fol. CXva) [Ruffo	[Subrubeo
[Rubeo	[Subrubicundo
[Inopos	[Quianos
[Verde	[Livido
[Nero	[Purpureo

Impossibilité de traduire ou désir de garder intactes des notions-clé, avec pour ainsi dire leur marque d'origine? On penche plutôt pour la seconde hypothèse, d'une part parce que la langue italienne dispose du mot 'ciano' à partir du XVe siècle au moins (si l'on en croit le *Dizionario Italiano* de F. Sabatini et V. Coletti), pour désigner un type de bleu, et d'autre part parce que le traducteur anonyme n'est pas en peine d'italianiser ailleurs des termes certes latins, mais directement hérités du grec comme les noms des fièvres: ainsi au fol. CXXVva, qui correspond par son contenu à une partie du *De urinis* de Maurus intitulée *De febribus continuis* dans l'édition De Renzi[54], 'sinochus' devient 'sinoco', 'epigmasticus' devient 'et pimasticus' (mais 'epimastico' dans la rubrique), 'augmasticus' est rendu par 'aumastico' (mais 'augmastico' dans la rubrique), et 'homotenus' par 'homoteno' ('homotano' dans la rubrique). 'Causon' en revanche (un mot désignant une fièvre ardente) n'a pas d'équivalent italien, comme on le voit au fol. CXXVIIra: 'et faxi causon de collera putrefatta nelle sottilixime vene della bocca, dello stomaco, e del mone (sic, pour 'polmone') e del fegato'[55].

Le *De urinis* latin de Maurus faisait suivre ces mentions de couleur de termes de comparaison, du type 'albus est sicut aqua clara, lacteus est sicut serum lactis, glaucus est sicut cornu lucidum album', etc. Dans l'ensemble, le traducteur anonyme les a fidèlement reprises et une petite différence qu'il introduit par rapport au latin suggère qu'il est plus proche de l'état primitif de ce texte que son unique édition disponible. À propos de la phrase 'Karopos e come colore di pelo di camelo e questo significa frigidita mediana', censée traduire 'karopos est sicut color pilorum

54. Cf. Maurus, *Regulae urinarum*..., p. 27.

55. Comparer avec Maurus, *Regulae urinarum*..., p. 29: 'Fiat autem causon de colera putrefacta in subtilissimis venis oris, stomachi, pulmonis, cordis et epatis.'

camelorum, et iste color significat frigiditatem intensam'[56], il paraît en effet certain que c'est l'édition De Renzi qui est dans l'erreur, comme on pourra s'en rendre compte grâce au petit tableau suivant:

> Texte édité[57]:
> Albus est sicut aqua clara, lacteurs est sicut serum lactis, glaucus est sicut cornu lucidum album; et isti colores significant frigiditatem intensam. Karopos est sicut color pilorum camelorum, et iste color significat frigiditatem intensam
>
> Codex, fol. Cxva:
> Albo e come aqua kiara
> Latteo e come siero de latte
> Glauco e come corno lucido
> E questi colori significano frigidita intensa o grande frigidita
> Karopos e come colore di pelo di camelo e questo significa frigidita mediana

À suivre le texte édité, on croirait que les urines couleur 'albus', 'lacteus', 'glaucus' et 'karopos' traduisent toutes un froid intense dans le corps, en d'autres termes qu'elles sont interchangeables quant à leur valeur de signe, et l'on comprend mal pourquoi ces différentes teintes font l'objet de deux assertions disjointes et dissonantes. Il est donc très vraisemblable que le texte latin d'origine portait 'frigiditatem medianam' au lieu de la seconde occurrence de 'intensam', et que le traducteur italien l'a suivi fidèlement. Notons aussi que l'italien est un peu plus développé que le latin dans quelques cas précis; ainsi la phrase 'urina in colore pallida vel subpallida, karopos, lactea, glauca, vel alba, in substantia tenuis imo tenuissima, et virgulata, vitium splenis ex oppilatione significat'[58] est-elle somme toute un peu plus complète dans sa version vulgaire, fol. CXXXVIIra: 'l'orina in colore palida o sottopalida, karopos, glaucula, lattea a alba in substantia sottilixima limpida o quasi vergolata e per lungo tempore procedente, cosale appare cosi questa come la sopradetta, significa vitio e opilatione de milza.'

Sans vouloir ici multiplier les exemples, on dira que maints passages montrent soit que le traducteur a aménagé le texte, soit qu'il a eu affaire à un texte différent de celui que l'on peut lire imprimé; en d'autres termes, par rapport à son modèle supposé, la version italienne se distingue de deux façons: abréviation de certains développements théoriques et disposition différente, ce que montrera pour finir un aperçu sur la fin du codex.

56. Maurus, *Regulae urinarum...*, p. 6.
57. Cf. Maurus, *Regulae urinarum...*, p. 6.
58. Maurus, *Regulae urinarum...*, bas de la page 49.

On rencontre en effet, fol. CXXXVIIra, deux derniers paragraphes annoncés par des rubriques dont l'une ne correspond à rien dans le texte latin ('de le differentie de l'orine di femine da uomini'), tandis que l'autre marque la fin du texte dans l'édition De Renzi – plus précisément dans un des manuscrits qu'il avait étudiés pour ce faire, le ms. Paris, BnF, lat. 6963 qui s'achève sur les 'Pillule magistri Mauri', fol. 13vb. On peut donc dire que la fin du traité en vulgaire correspond à celle d'un codex et donc d'un état du texte latin au moins, mais il faut remarquer que la dite recette est agencée dans un ordre inverse dans notre témoin italien: alors que le manuscrit parisien fournit la recette, puis présente ses indications, notre texte fait le contraire et commence ainsi: 'le pillole del maestro Mauro, le quali valgono a paraletici e maniaci e malinconi e a tutti i ceffalitici e a cancrosi, a scabio (fol. CXXXVIIv)-si e a flama salso e a lebbrosi e ali artetici et a podragi(sic) e a quartanarii et alla memoria et servano il vedere e aguzzano la sordagine cacciano, e lo stomaco confortano, et valgono alla milça e cacciano tutti i freddi humori crudi e viscosi. E fanosi in questo modo: Recipe turbinti, hermodactilo, esula, etc'[59].

L'impression dominante est toutefois que ce traité propose une version quelque peu allégée du *De urinis* tel qu'on le connaît par son unique édition. Prenons pour ultime exemple les fol. CXXXVIv-CXXXVIIra qui correspondent en gros, mais en gros seulement, aux pages 49-51 de l'édition. Là où Maurus propose différents remèdes et recettes pour soigner l'hydropisie, le manuscrit italien ne traduit les dites recettes que jusqu'à ce qui correspond à la ligne 10 de la page 48 chez De Renzi, c'est-à-dire une boisson à base de 'serum caprinum', et passe ensuite à la rubrique 'della milza per l'orina vedra il diffetto', c'est-à-dire au paragraphe intitulé 'de splene' p. 49 de l'édition latine.

De fait on se rend compte, à y regarder de plus près, et notamment en étudiant la table des matières de l'œuvre de Maurus, située entre les fol. 4v[60] et 7v[61] du manuscrit, que des recettes éditées l'une à la suite de

59. Cf. Maurus, *Regulae urinarum...*, p. 51: 'Valent iste pilule paralyticis, epilepticis, melanconicis omnibus, cefalargicis, cancrosis, scabiosis, et de flegmate salso leprosis, artheticis, podagricis, quartanariis; conservant memoriam, surditatem expellunt, stomachum confortant et splen, omnes frigidos humores crudos, viscosos, venenosos.'

60. 'Qui comincia la tavola del seguente libro di questo volume, cioe le regole da conoscere lorine e di che specie dimostrano le infermita procedenti ke ne seguitano de quatro humori. E delle specie delle febri onde sorgono e de rimedi contra exe del maestro Mauro.'

61. 'Explicit tabula cure et regule urinarum et aliarum curarum magistri Mauri. Laus tibi deus.' 'Incipit tabula secunda de multis rebus a quantitatibus eorum.'

l'autre par Salvatore De Renzi dans une rubrique 'cura' ménagée au sein de la plupart des chapitres, sont ici séparées du traité et rassemblées en une collection distincte après les *Regulae urinarum* à proprement parler à partir du folio CXXXIXvb. Ainsi s'explique entre autres que la traduction italienne donne le sentiment d'être une version abrégée des *Regulae urinarum*.

Cet exemple précis de *volgarizzamento* ne nous a rien appris sur des questions évoquées plus haut comme celle de l'identité du traducteur, celle de ses motivations et du public visé, ou de la date à laquelle fut réalisée la traduction. Il confirme en revanche les incertitudes qui pèsent à mes yeux sur l'édition du traité de Maurus depuis que j'ai entrepris d'étudier les manuscrits latins qui nous l'ont transmis, mais attend à son tour d'être confronté avec les deux autres témoins italiens. On peut tout de même transformer ce constat en l'élargissant, et souligner la valeur que peuvent revêtir les traductions vernaculaires pour qui s'attache à reconstituer un texte latin d'origine, plus diffusé, et ayant donc pu susciter davantage d'interventions ou remaniements – un constat qui excède évidemment les limites du monde latin, et qu'a pu faire par exemple William F. Ryan, à la suite de Mario Grignaschi, à propos de la traduction du *Secretum secretorum* en vieux russe[62].

Les quelques remarques ici rassemblées ne sauraient donc être que les prémisses d'une recherche plus complète, qui devrait aboutir à terme à la fois sur l'histoire de la fortune en vulgaire d'un traité d'uroscopie remarquable et sur un réexamen complet de sa tradition textuelle latine.

Université Paris 8 – Vincennes – Saint-Denis

62. Ryan W.F., 'Questions de traduction scientifique en vieux russe: le *Secretum secretorum*', in: G. Contamine (éd.), *Traduction et traducteurs au Moyen Age, Actes du colloque international du CNRS organisé à Paris, IRHT, 26-28 mai 1986* (Paris, 1989), p. 85-95, ici p. 92: 'Il est certain que le texte russe, malgré son orientation médicale très prononcée, ne correspond pas au texte de Gaster, tel que le caractérise Spitzer.'

Silvia TONIATO

LE LEXIQUE MATHÉMATIQUE AU MOYEN ÂGE ENTRE LATIN ET LANGUES VERNACULAIRES: QUELQUES PROBLÈMES POSÉS PAR LES TRADUCTIONS

0. Introduction

Les premiers textes mathématiques en français et en anglais connus à présent[1] font l'objet de cet article: il s'agit de l'*Algorisme*[2] conservé dans les manuscrits fr. 2021 de la Bibliothèque Nationale de France à Paris, (XIIIe siècle, ff. 154r-155v) et 2200 de la Bibliothèque Sainte-Geneviève (XIIIe siècle, ff. 150r-151r), et des traités *The Crafte of Nombrynge*[3] et *The Art of Nombrynge*[4], respectivement conservés dans les manuscrits Egerton 2622 (XVe siècle, ff. 136r-156r) de la British Library et Ashmole 396 (XVe siècle, ff. 48r-56r) de la Bodleian Library. Les trois textes traduisent des algorismes latins et demeurent anonymes.

Après une brève présentation de leurs sources et de leurs caractéristiques, nous analyserons ces textes au niveau lexical et au niveau du contenu par rapport aux sources: les écarts feront l'objet de quelques considérations, parce qu'ils constituent à notre avis de petites fenêtres par lesquelles nous pouvons observer quelques éléments de la culture mathématique propre aux traducteurs.

1. Les sources et l'organisation des traductions

Les textes dont nous nous occupons sont des 'algorismes'. L'algorisme est soit la science, soit un texte qui comprend l'ensemble des informations

1. Vitale Brovarone A., 'Considérations sur la production de textes mathématiques en France et sur leur diffusion manuscrite et imprimée', *Le Moyen Français* 24-25 (1989), p. 23-53.

2. 'Le plus ancien traité français d'algorisme', éd. V. Mortet, *Bibliotheca Mathematica* 9 (1908-1909), p. 55-64.

3. 'The Crafte of Nombrynge', éd. R. Steele, in: *The Earliest Arithmetics in English* (London, 1922), p. 1-32.

4. 'The Art of Nombryng', éd. R. Steele, in: *The Earliest Arithmetics in English* (London, 1922), p. 33-51.

nécessaires pour opérer avec les chiffres indo-arabes dans un système décimal, à savoir leur signification et la façon de les lire et de les utiliser pour écrire les nombres et exécuter des calculs.

Apparemment, l'ensemble de ces notions atteint l'Europe entre la moitié du XIIe et le début du XIIIe siècle, à travers le *Liber Abbaci* de Léonard de Pise[5] et la version latine de deux traités d'al-Khuwarizmi (825): un algorisme dont la traduction est attribuée soit à Adelard de Bath, soit à Robert de Chester[6], et une algèbre traduite par Robert de Chester[7]. Le cadre de la diffusion de la pratique algoristique[8] pendant le moyen âge est toutefois beaucoup plus complexe, puisqu'elle date au moins de la fin du Xe siècle[9].

Loin de vouloir tracer ici une histoire de cette diffusion, nous nous limitons à souligner que les algorismes considérés comme les plus importants pour le rôle qu'ils ont joué dans ce procès – à savoir ceux d'al-Khuwarizmi et de Léonard de Pise aussi bien que le *Carmen de Algorismo* d'Alexander Villadei (XIIIe siècle)[10] et le *Tractatus de Arte Numerandi* de Johannes Sacrobosco (XIIe siècle)[11] – diffèrent assez entre eux du point de vue de l'organisation des matières et de la pratique algoristique elle-même. Bref, si la façon de lire les nombres et de les écrire est toujours la même et les quatre opérations fondamentales sont toujours présentes, l'ordre de leur exposition ainsi que les procédés de résolution peuvent changer beaucoup d'un traité à l'autre; la multiplication et la division peuvent être distinguées du redoublement et de la division par deux; parfois on a aussi des règles pour exécuter des opérations sans avoir besoin de les écrire; les calculs de la somme des progressions et

5. *Il Liber Abbaci di Leonardo Pisano pubblicato secondo la lezione del codice Magliabechiano C. I. 2616*, éd. B. Boncompagni, Scritti di Leonardo Pisano matematico del secolo decimo terzo 1 (Roma, 1857-1861).

6. *Algoritmi de numero indorum*, éd. B. Boncompagni, in: *Trattati d'aritmetica* (Roma, 1857), I, p. 1-23.

7. *Robert Chester's Translation of al-Kwarizmi's al-jabar*, éds B. Hughes & F. Steiner (Stuttgart, 1989).

8. Avec 'pratique algoristique' je réfère aux systèmes de calcul avec les chiffres indo-arabes exécutés sur n'importe quel type de support (tablettes avec poudre ou cire, parchemin, etc.).

9. Le sujet est étudié de façon approfondie dans Smith D.E. & Karpinski L.C., *The Hindu-Arabic Numerals* (Boston-London, 1911).

10. 'Carmen de Algorismo', voir Annexe.

11. Iohannes Sacrobosco, 'Ioannis de Sacro-Bosco Tractatus de Arte Numerandi', éd. J.O. Halliwell, in: *Rara Mathematica* (New York, 1977), p. 1-26. Nous signalons aussi une édition plus récente, que nous ne connaissions pas encore au moment de la rédaction de cet article: *Algorismus Iohannis de Sacro Bosco*, éd. F.S. Pedersen, Corpus Philosophorum Danicorum Medi Aevi, X.1 (Havniae 1983), p. 165-261.

des racines ne sont pas toujours présents, ni les calculs avec les fractions; une combinaison de chiffres et de lettres est parfois utilisée au lieu des fractions pour exprimer la moitié d'un nombre; la théorie à la base des règles énoncées, lorsqu'elle est mentionnée, n'est pas toujours abordée de la même façon[12].

L'hétérogénéité des textes mathématiques (et, entre eux, celle des algorismes) ne doit pas surprendre: les mathématiques ont elles aussi une histoire qui, loin d'être une séquence de découvertes qui se posent sur une même ligne de développement, est faite d'intuitions et d'inventions qui essaient de satisfaire les besoins et les curiosités des êtres humains – de la connaissance du ciel et de la terre au commerce, aux jeux etc. – par la quête des relations constantes entre les objets qu'elles mettent en place et qui ouvrent, à partir de différents points de vue, des espaces de connaissance dans un univers exprimé par des relations formelles. Le choix du moyen le plus économique pour satisfaire une même exigence est bien sûr conditionné par des circonstances qui ne sont pas toujours les mêmes, et la possibilité de différenciation des procédés augmente quand on ne dispose pas d'un langage formel bien développé et universellement employé – ce qui nous reconduit à nos propos, et à l'aube des pratiques arithmétiques vernaculaires.

Les premiers algorismes rédigés en français et en anglais traduisent le *Carmen de Algorismo* et le *Tractatus de Arte Numerandi*[13].

1.1. Le *Carmen de Algorismo*

Le *Carmen de Algorismo* d'Alexander Villadei (1220) est la source du *The Crafte of Nombrynge* et, avec une quasi-certitude, de l'*Algorisme* français.

Le *Carmen* donne des instructions en hexamètres qui concernent la façon d'utiliser les chiffres indo-arabes pour lire et représenter les nombres, et pour résoudre les opérations d'addition, de soustraction, de redoublement, de division par deux, de multiplication et de division, et pour le calcul des racines carrées et cubiques. Le choix d'une exposition en vers, qui est fonctionnelle à l'apprentissage par cœur du texte, et l'absence de définitions et de considérations théoriques, dénoncent le caractère strictement pratique de cette œuvre.

12. Pour les pratiques mathématiques vernaculaires des deux siècles successifs nous assistons au même phénomène; cette diversification, à ma connaissance, n'a pas encore été étudiée de façon systématique.

13. Du point de vue de la pratique algoristique les deux textes sont entre eux tout à fait semblables.

Le *Crafte of Nombrynge* récrit les vers latins en les groupant par noyaux thématiques; chaque groupe est suivi de sa traduction, enrichie d'un commentaire qui comprend des exemplifications pratiques du déroulement des calculs, dont le *Carmen* énonce seulement la règle générale, et des adjonctions par rapport au contenu là où la formulation du texte latin est aperçue comme insuffisante au propos. Le *Craft* s'arrête à la multiplication, même si on peut imaginer qu'à l'origine au moins la division devait être présente: après l'énonciation de la règle qui dit qu'il faut vérifier une multiplication par une division, on lit: 'But 3et þou hast not þe craft of dyvision, but þow schalt have it afterwarde.'

L'*Algorisme* français est au contraire très synthétique et incomplet par rapport à la source: alors qu'il énonce que '6 parties sont d'augorisme: assambler et abatre, doubler, dimidier, multeplier, [deviser]'[14], il ne compte effectivement que l'addition, la division et le calcul des racines cubiques. Il est plutôt difficile de l'analyser à cause de sa brièveté, mais, comme on le verra dans la suite, il donne l'impression d'être une sélection raisonnée de la matière du *Carmen*.

1.2. Le *Tractatus de Arte Numerandi*

L'algorisme de Iohannes Sacrobosco est, par rapport au *Carmen*, plus riche du point de vue des contenus, puisqu'il comprend la somme des progressions comme l'une de ses parties et qu'il aborde la matière mathématique aussi d'un point de vue pour ainsi dire théorique: le système décimal est décrit dans son mode d'emploi et soigneusement expliqué dans ses principes[15]; les instructions pour le déroulement des calculs sont toujours précédées d'une définition de l'opération dont il est question, et souvent on explique les raisons de la façon d'opérer. Ce texte est, d'après ce que nous avons pu constater jusqu'à présent, un point de repère constant pour les textes mathématiques plus tardifs, et même si, surtout en Italie, il est aperçu comme insuffisant du point de vue de la pratique des calculs, quand il s'agit de définir une opération, c'est celui-ci qu'on cite presque toujours à la lettre.

14. Mortet, 'Le plus ancien traité…', p. 61.
15. Les fractions de l'unité ne sont néanmoins pas considérées comme des parties décimales, mais comme parties sexagésimales.

L'*Art of Nombryng* traduit le *Tractatus* de façon très fidèle; il se détache de la version latine seulement pour ajouter des exemples là où une illustration du texte est jugée nécessaire, alors que les exemples de la source sont donnés uniquement de façon généralisée (à savoir: 'Si igitur velis numerum addere numero, scribe numerum ...; ex tali ergo additione aut excrescat digitus, aut articulus, aut numerus compositus ...'[16] etc. et jamais: 'si tu veux ajouter 54 à 312 dispose ces nombres comme ci dessous ...' etc.)[17]; les exemples ajoutés dans la version anglaise n'arrivent toutefois pas à illustrer chaque passage consacré au déroulement des calculs.

2. Les traductions

Avant de procéder à l'analyse des aspects qui sont, à notre avis, les plus pertinents de ces arithmétiques, un petit préambule s'impose: la quantité de traductions vernaculaires de textes mathématiques (et peut-être aussi la présence de textes vernaculaires tout court) dans une langue particulière, ou bien leur absence, ne vaut pas comme seule mesure du niveau, voire de la diffusion des compétences mathématiques dans cette communauté linguistique.

Parmi les textes mathématiques du moyen âge connus de nos jours, ceux en langue italienne sont les plus nombreux. Les traductions en constituent toutefois une toute petite portion (15 sur 300 environ, dont la plupart sont des parties du *Liber Abbaci*, le plus souvent les chapitres 14 et 15)[18], sans doute parce que, en Italie, les mathématiques étaient bien développées. Par contre, on n'a pas de textes vernaculaires

16. Éd. Halliwell, p. 5-6.

17. Il y a deux exceptions: l'une pour expliquer la multiplication entre deux *digiti* (éd. Halliwell, p. 12-13): 'Verbi gratia, si vis scire quot sunt quater in octo, vide quot sunt unitates intra octo et decem, denario simul computato, et patet quod sunt duo; subtrahatur ergo quaternarius a quadraginta bis et remanent 32, et haec est summa totius multiplicationis'; l'autre qui montre l'emploi des quatre règles assignées pour la somme des *progressiones*, terme qui dans ce texte désigne les séries des nombres naturels, des nombres pairs et des nombres impairs (éd. Halliwell, p. 19): 'verbi gratia 1, 2, 3, 4, multiplica quinarium per binarium et exibunt decem, summa totius progressionis [...] 1, 2, 3, 4, 5, multiplica quinarium per ipsum ternarium et resultabunt quidenarium, summa totius progressionis [...] 2, 4, 6, multiplica quaternarium per ternarium et resultabunt duodecim, summa totius progressionis [...] 1, 3, 5, multiplica ternarium per se et exibit novenarius, summa totius progressionis'.

18. Van Egmond W., *Practical Mathematics in the Italian Renaissance. A Catalog of Italian Manuscripts and Printed Books to 1600* (Firenze, 1980).

pour l'Espagne avant 1482[19], alors que c'est précisément en Espagne qu'on trouve la plus ancienne attestation connue à présent des chiffres indo-arabes parmi les textes en alphabet latin [20]. Les premiers textes mathématiques en langue provençale datent du XVe siècle[21], même si, au cours de l'introduction à son *Liber Abbaci*, Léonard de Pise mentionne un 'modus Provincialis' parmi les systèmes de calcul qu'il a connus et qu'il juge moins efficace par rapport au 'modus Indorum' qu'il exposera dans son œuvre[22].

En France et en Angleterre, la production de textes mathématiques vernaculaires manuscrits est aussi assez pauvre[23]. Pour la France, avant le XVe siècle, on ne compte que trois textes: l'*Algorisme* qui fait l'objet de cet article, un algorisme en vers (Oxford, Bodleian Library, Selden Supra 26, XIIIe siècle)[24], et un texte anglo-normand, encore en

19. Francesch Sanct Climent, *Summa de la Art de Arismetica* (Barcelona, 1482); Francesc Santcliment, *Summa de l'art d'aritmetica,* éd. V. Malet (Vic, 1998).

20. Il s'agit du *Codex Vigilianus*, retrouvé dans le cloître d'Albeda et qui date de 976.

21. Deux arithmétiques occitanes sont connues: l'une est conservée dans le ms. nouv. acq. fr. 4140 de la Bibliothèque Nationale de France, et est étudiée de façon approfondie dans Sésiano J., 'Une arithmétique médiévale en langue provençale', *Centaurus* 27 (1984), p. 26-75; l'autre est le *Compendion de abaco* de Frances Pellos, imprimé à Turin en 1492 et édité par Lafont R. & Tornerie G., 'Francs Pellos, *Compendion de l'abaco*, texte établi d'après l'édition de 1492', *Éditions de la Revue des Langues Romanes* (Montpellier, 1967), avec un commentaire philologique et mathématique. Il faut mettre en évidence que le premier de ces deux textes nous offre l'exemple le plus ancien connu à présent où une solution négative d'une équation est acceptée tout court en tant que donnée mathématique, et non pas discutée sur la base de la possibilité de son existence dans l'expérience concrète (Cf. Vitale Brovarone, 'Considérations sur la production de textes…', p. 33-36; Sésiano J., 'The Appearance of Negative Solutions in Mediaeval Mathematics', *Archive for History of Exact Sciences* 32 (1985), p. 105-150).

22. *Il Liber Abbaci di Leonardo Pisano…*, éd. Boncompagni, p. 1.

23. On n'a même pas de textes imprimés avant le XVIe siècle – à ce propos nous renvoyons à Vitale Brovarone, 'Considérations sur la production de textes…', qui est notre référence principale même pour les textes manuscrits. Nous signalons aussi deux bases de données: l'une est *Jordanus*, realisée par l'*Institute for the History of Science* de Munich et le *Max Planck Institute for the History of Science* de Berlin, qui recueille les manuscrits mathématiques jusqu'au XVe siècle d'après les catalogues d'un grand nombre de bibliothèques du monde (et, sans souci d'être exhaustif, les manuscrits scientifiques tout court jusqu'au XVe siècle), qui est disponible sur le site http://jordanus.ign.uni-muenchen.de/cgi-bin/iccmsm?seite=home&sprache=en (même s'il peut arriver parfois d'avoir quelques difficultés à achever les recherches parmi ses données); l'autre est *RISTORANTI*, réalisée sous la direction d'A. Vitale Brovarone, Université de Turin, qui donne les références des manuscrits scientifiques latins et vernaculaires, et aussi des manuscrits vernaculaires de n'importe quel sujet, au moins jusqu'au XVe siècle, signalés par les catalogues: http://filemaker01.unito.it/fmi/iwp/res/iwp_home.html.

24. Éd. par E.G.R. Waters, 'A Thirteenth Century Algorism in French Verse', *ISIS* 11 (1928), p. 45-84.

vers, du XIVe siècle[25]. Au XVe siècle, le nombre de textes augmente un peu: nous en avons une quinzaine à peu près, parmi lesquels signalons le célèbre *Triparty en la science des nombres* de Nicolas Chuquet[26], deux manuscrits qui sont très proches au niveau textuel, l'un conservé à Cesena[27], l'autre à Turin[28], et l'*Arismetique* de Jean Adam[29]. Pour l'Angleterre, le nombre d'ouvrages d'arithmétique vernaculaires est plus petit: pas de manuscrits avant le XVe siècle (sauf un fragment du XIVe: London, British Library, Sloane 213, f. 120-123), où l'on ne compte que nos deux traductions, un texte sur le calcul avec les jetons[30], et trois textes cités par R. Steele[31], dont deux sont fragmentaires[32].

Nous verrons toutefois que les premières traductions en français et en anglais ne peuvent pas être considérées comme des œuvres qui naissent dans un cadre de mauvaise connaissance de la pratique algoristique.

2.1. *The Crafte of Nombrynge* (XVe siècle, anonyme)

Du point de vue lexical, le texte est très conservateur par rapport au latin, d'où sont dérivés la plupart des mots. Dans le cas où il y a plus d'un mot ou plus d'une expression pour rendre le sens du latin, et une différence de dérivation entre ces synonymes, on remarque que les emprunts sont utilisés de préférence pour la traduction littérale des vers (par rapport au schéma ci-dessous, *adden*, *signifien*, *algorym* / *augrym*), alors que dans les exemples pratiques les mots qui ne dérivent pas du latin sont plus fréquents[33].

25. Edité par L.Ch. Karpinski & Ch. Staubach, 'An Anglo-Norman Algorism of the Fourteenth Century', *ISIS* 23 (1935), p. 121-152.

26. Marre A., 'Notices sur Nicolas Chuquet et son triparty en la science des nombres', *Bullettino di Bibliografia e di Storia delle Scienze Matematiche e Fisiche* 13 (1880), p. 555-660 et 693-814; 14 (1881), p. 413-460.

27. Cesena, Biblioteca Malatestiana, ms. S-XXVI-6. Le deuxième texte contenu dans ce manuscrit a été récemment édité par Spiesser M., *Une arithmétique commerciale du XVe siècle: le Compendy de la practique des nombres de Barthélemy de Romans* (Turnhout, 2003).

28. Torino, Archivio di Stato, sezione I, Archivio Biscaretti, mazzo 29, n. 3.

29. Paris, Bibliothèque Sainte-Geneviève, 3143. Cf. Thorndike L., 'The Arithmetic of Jean Adam 1475', *American Mathematical Monthly* 33 (1926), p. 24-48.

30. 'Accomptynge by Counters', éd. R. Steele, in: *The Earliest Arithmetics in English* (London, 1922), p. 52-65.

31. Steele, *The Earliest Arithmetics*..., p. v.

32. Cambridge, University Library, LI. IV. 14 (III.), f. 212-142; Oxford, Bodleian Library, 790 G. VII. (2653), f. 146-154; London, British Library, Egerton 2852, f. 5-13.

33. Nous avons quelques difficultés à établir si les expressions telles que *casten, drauen together, drauen out*, etc., sont des calques des expressions latines correspondantes, car nous ne savons pas à partir de quel moment elles sont employées en arithmétique (si

Voici quelques exemples des mots traduits de plusieurs façons[34]:

1) AJOUTER

Addo	adden
	casten / casten together
	drauen / drauen together

2) SOUSTRAIRE

Demo	drauen out
	taken out
	keepen

3) AVOIR LA VALEUR DE

Significo /	signyfien
signo	betoknen / toknen

4) RÉSULTER APRÈS UNE ADDITION OU UNE MULTIPLICATION

Cresco /	arisen
excresco	comen of /
	comen there-of

5) EMPRUNTER

Demo	borowen / borowen out

6) ALGORISME

Algorismus	Algorym / Augrym
	Craft of nombryng

Dans le cas de *demo*, la spécialisation du lexique de la source est inférieure par rapport à celui de la traduction, qui distingue deux opérations reconnues comme différentes par deux mots différents[35].

Du point de vue textuel, les adjonctions sont, comme nous l'avons dit, consacrées aux corrections et aux exemplifications du déroulement des calculs. Voici quelques exemples[36]:

[SOMME: 10084 + 17743]

> In þis ensampul ben alle þe foure cases[37]. Cast 3 to foure þat wol be seven, do away 4 & write þere seven. þan cast 4 to þe figure of 8 þat

cela s'est fait d'après l'influence des pratiques diffusées en langue latine ou pas). *Craft of nombryng* est en revanche un calque de *algorismus*, à partir d'une fausse étymologie (Algorismus *comes of* 'algos', grece, quid est 'ars' latine, *'craft' on englis, and* 'rides', quid est 'numerus' latine, *'a nombur' on englys;* inde dicitur 'algorismus' per addicionem huius sillabe '-mus' et subtraccionem '-d-' et '-e-', quasi 'ars numerandi' – 'The Crafte...', éd. Steele, p. 1).

34. À gauche les mots de la source, à droite leurs traductions.

35. Cette réflexion ne concerne que les deux textes et met en évidence la compétence du traducteur; il ne faut pas en déduire directement que le lexique mathématique latin soit moins spécialisé par rapport à l'anglais: le *Tractatus de Arte Numerandi* utilise le verbe *mutuare* pour 'emprunter'.

36. Pour les exemples nous choisirons ici et ailleurs des portions de texte relatives aux chapitres d'addition, de soustraction et de multiplication, de façon qu'ils soient bien compréhensibles malgré la différence des procédés par rapport à notre façon de calculer aujourd'hui. Les exemples sont choisis en tout cas de façon que leurs caractéristiques soient représentatives des phénomènes qu'on trouve n'importe où dans le texte.

Pour ceux qui ne connaissent pas du tout le vocabulaire arithmétique utilisé au moyen âge, il faut dire que *digitus* indique un nombre compris entre 1 et 9, que *articulus* indique un multiple de 10 et que *compositus* est un nombre plus grand que 10 et non multiple de 10.

37. À savoir les 'quatre cas' qui peuvent se produire pendant la résolution d'une addition: (1) le résultat de la somme (ou bien de la somme partielle) est un nombre plus petit

wel be 12, do away 8 & sette þere 2 þat is a digit, and sette þe articul of þe composit, þat is ten, upon þe cifers hed & reken it for him selfe þat is on. þan cast one to a cifer & hit wulle be but on, for no3t & on makes but one. þhan cast 7 þat stondes under þat on to hym & þat wel be 8, do away þe cifer and þat 1 & sette þere 8, þan go forthermore. Cast þe other 7 to þe cifer þat stondes over hym, þat wul be bot seven for þe cifer betokens no3t. Do away þe cifer & sette þere seven; & þen go forthermore & cast 1 to 1, & þat wel be 2, do away þe hier 1 & sette þere 2 þan hast þou do. And yf þou have wel ydo þis nomber þat is sett here-after wel be þe nomber þat schalle aryse of alle þe addicion as here: 27827.[38]

Pour que l'exemple soit clair, il faut dire que les algorithmes de résolution du *Carmen* (et du *De Arte Numerandi* aussi, comme on le verra après) prévoient un support d'écriture sur lequel on peut effacer et récrire aisément les nombres plusieurs fois, que les chiffres du résultat sont écrits à chaque passage au lieu des chiffres du terme de l'opération qui est écrit en haut, et que les chiffres retenus sont écrits et non pas mémorisés. Les différentes étapes décrites dans l'exemple peuvent être représentées comme ci-dessous:

				1				
10084	→	1008**7**	→	100**27**	→	10**827**	→	
17743		17743		17743		17743		

17827	→	**27827**
17743		17743

Pour ce qui est du langage, on peut voir très clairement la présence d'expressions figées qui accompagnent le déroulement des calculs (par exemple *cast* x *to* y *that wol be* z, *do away* p *and write there* q – où nos *x, y, z, p, q* prennent évidemment la places des chiffres – qui se répète sans variation significative chaque fois qu'il faut ajouter deux chiffres, effacer celui d'en haut et écrire le résultat à sa place).

Comme on le voit aussi de nos jours[39], les mots qui accompagnent les calculs se fixent en refrains qui, dans un même milieu linguistique, ne

que 10; (2) le résultat est plus grand que 10, mais pas un de ses multiples; (3) le résultat est un multiple de 10; (4) parmi les chiffres du terme en haut il y a un 0.

38. 'The Crafte…', éd. Steele, p. 10. L'un des exemples de la disposition en colonnes dans l'*Art of Nombring* (éd. Steele, p. 36) reprend les mêmes nombres, mais ne résout pas l'addition.

39. 'Quatre fois deux' est 4 × 2, 'quatre multiplié deux fois' est apparemment équivalent, mais on ne l'utiliserait pas en faisant des calculs.

changent perceptiblement que dans la longue durée, ou selon les différentes pratiques de compte. Là où le langage pratique est codifié, il y a une pratique algoristique consolidée. L'un des deux premiers textes mathématiques en langue anglaise naît donc dans un milieu qui possédait depuis quelque temps une pratique algoristique qui permettait non seulement de traduire correctement cette source latine, mais aussi d'y reconnaître des 'défauts' et de les corriger, soit du point de vue lexical, soit par rapport au contenu.

Le deuxième exemple d'adjonction nous confirme dans notre hypothèse: il s'agit d'une règle pour calculer le produit de deux multiples de dix 'per normas subtiles absque figuris'[40]:

> Articulum si per reliquum vis multiplicare,
> in proprium digitum debe[bi]t uterque resolvi
> articulus, digitis post per se multiplicatis
> ex digitis quociens retineris multiplicatum
> articuli faciunt tot centum multiplicati.
>
> Here he teches þe furst rewle, þe quych is þis: yf þou wel multiplie an articul be another, **so þat both þe articuls bene with inne an hundreth**, þus þou schalt do. Take þe digit of bothe þe articuls, for every articul hase a digit, þen multiplye þat on digit by þat other and loke how mony unytes ben in þe nounbre þat comes of þe multiplicacion of þe 2 digittes & so mony hundrythes ben in þe nounbre þat schal come of þe multiplicacion of þe ylke 2 articuls. [...]
>
> Take hede I schall telle þe **a generalle rewle** whan þou hast 2 articuls and þou wold wete qwat comes of þe multiplicacion of þem 2. [...][41]

Le traducteur voit la nécessité de préciser que la règle de la source est valable pour un nombre limité de cas (seulement si les deux multiples de dix sont plus petits que cent); après avoir traduit ce procédé il propose une 'règle générale' pour multiplier entre eux n'importe quels 'articles'.

La présence récurrente dans les adjonctions d'expressions telles que 'take hede', 'take gode hede', 'it is lyght as dyche water', mais surtout la matière des adjonctions (définitions, corrections, exemples pratiques) et le choix même du *Carmen* comme source, avec ses vers mnémotechniques, suggèrent que le but de cette traduction est didactique. L'évidence que la traduction ne dérive pas ses connaissances arithmétiques de sa

40. Ces calculs pouvaient être à priori résolus soit par le compte digital, soit mental; nous n'avons pas d'éléments suffisants pour établir avec certitude le type de calcul sans plume (littéralement 'sans besoin d'écrire les chiffres') auquel le texte fait référence.

41. 'The Crafte...', éd. Steele, p. 28-29; les vers latins, comme nous l'avons dit, sont intégrés dans le texte. Pour les citations du texte latin, ici et ailleurs, voir aussi l'Annexe.

source, et la possibilité que la réalisation d'un traité pour enseigner la pratique de la source ait été jugée nécessaire, nous emmènent à nous interroger soit sur les raisons de l'apparition si tardive des premiers textes mathématiques vernaculaires en anglais, soit sur la circulation de pratiques arithmétiques différentes, comme celle du compte avec les jetons[42] ou d'autres encore que nous ne connaissons pas.

Nous n'avons en ce moment-ci aucune explication satisfaisante pour ce retard au delà des considérations d'A. Vitale Brovarone:

> Comme la péninsule ibérique, l'Angleterre nous présente une situation très curieuse. Si dans les milieux universitaires les mathématiques ont eu une période de splendeur, surtout au XIIIe siècle, [...] la production de textes en langue vulgaire est très petite, voire nulle.
>
> [...] Les régions à développement lent se sont, dans quelques cas, rattachées à une tradition vieille – et vieillie, comme celle d'Alexandre de Villedieu ou de Ioannes Sacrobosco – [...]. Les régions qui étaient douées d'une activité commerciale vivace, et qui n'ont pas produit des textes de mathématiques en langue vulgaire, se sont servies des vieux traités en latin transmis par l'enseignement scolaire: on peut bien comprendre et le long silence, et le recours, dans les premiers textes en langue vulgaire, aux vieux manuels, si familiers et si bien connus[43].

Par rapport à la circulation d'autres pratiques, une suggestion pourrait nous parvenir par le biais de ce dernier exemple:

> Subtrahis aut addis a dextris vel mediabis.
>
> þou schal be-ginne in þe ryght side of þe boke or of a tabul. Loke were þou wul be-gynne **to write latyn or englys** in a boke & þat schalle be called þe lyft side of þe boke, þat þou writest toward þat side schal be called þe ryght side of þe boke[44].

Si pour indiquer la gauche il ne suffit pas ici de dire que c'est la partie du livre où l'on commence à écrire, mais on ajoute que c'est la partie où l'on commence quand on écrit en latin ou en anglais, nous pouvons peut-être supposer que les destinataires de ce texte, ou au moins une partie d'entre eux, ou bien le traducteur, savaient écrire aussi en d'autres langues pour lesquelles on commence du côté droit de la page, comme l'hébraïque ou l'arabe. Quant à la question de savoir si une 'autre langue' peut impliquer en ce cas une 'autre pratique arithmétique', nous ne pouvons pas le dire à partir de ces seules données; l'état de nos recherches

42. Ce type de compte est décrit par exemple dans le texte *Accomptynge by Counters* (éd. Steele), et dans l'arithmétique de Jean Adam.
43. Vitale Brovarone, 'Considérations sur la production de textes...', p. 42-45.
44. 'The Crafte...', éd. Steele, p. 7.

ne permet pas de formuler des hypothèses sur le contexte et les modalités de réalisation de cette traduction.

2.2. L'*Algorisme* français (XIIIe siècle, anonyme)

La traduction française pose quelques problèmes d'interprétation, surtout à cause de sa brièveté, puisqu'elle ne compte que trois feuillets.

Au niveau lexical, le langage est très conservateur par rapport au latin. Cependant, on traduit *multiplicare* par *mener* à côté de *multiplier*, *subtrahere* par *abatre* à côté de *sustraire*.

En ce qui concerne le contenu, le texte paraît d'abord incomplet, puisqu'il ne traduit que les chapitres de numération, d'addition, de division et d'extraction des racines cubiques. En outre, il est parfois incorrect: ainsi, le calcul des racines est introduit de la façon suivante: 'Se tu multiplies aucun nombre par soi meisme, ce nombre ki en istera sera cubes'[45], alors que le nombre est un cube s'il est multiplié deux fois par lui-même; lorsqu'il est multiplié une seule fois par lui-même, il est carré; ou s'agirait-il seulement d'une faute d'écriture[46]? Toutefois, ces erreurs sont les seules que nous ayons trouvées dans le texte.

Bien que très synthétique, l'*Algorisme* nous donne plutôt l'impression que son traducteur et ses destinataires connaissaient déjà les chiffres indo-arabes et leur mode d'emploi.

La signification des chiffres est ainsi énoncée[47]:

> [0. 9. 8. 7. 6. 5. 4. 3. 2. 1
> Prima significat **unum**, **duo** vero secunda,
> tercia significat **tria**, sic procede sinistre
> donec ad extremam venias que **cifra** vocatur]
>
> 9. 8. 7. 6. 5. 4. 3. 2. 1. Li premiere fait **1**, la seconde fait **2**, la tierce fait **3** et les autres ausi jusc'à la darrainne, qui est apelée **cyfre**: 0.

L'interprétation des chiffres est expliquée par les chiffres eux-mêmes: il n'y avait donc pas du tout besoin d'explication, ce qui à cette époque est assez surprenant. Voici ce qu'on dit à ce même propos dans un algorisme contemporain à notre texte[48]:

45. Mortet, 'Le plus ancien traité...', p. 61.
46. La faute est la même dans les deux manuscrits, il ne s'agit donc pas, au moins à ce niveau, d'une faute de copie.
47. 'Carmen...', éd. Steele, p. 72; Mortet, 'Le plus ancien traité...', p. 60.
48. Waters, 'A Thirteenth Century Algorism...', p. 45-84, v. 96-116.

[…]
9, 8, 7, 6, 5, 4, 3, 2, 1.
Ches letres sunt de tel sentence
Qu'il veulent que l'en les conmenche
A escrire ve[r]s la par[t] destre,
En escrivant vers la senestre.
Un es[t] itel conme .i. 'i',
Et **.ii.** resenblent totedi
Une faucille; **trois**, enmi
Tornés, cil qui est betornés.
Quatre est un 'o' qui a .ii. piés.
.v. par 'h' envers escrivés.
Por **.vi.** un 'G' tornés metés,
Et por **.vii.**.i. 'v' adentés.
Metés .i. **wit** en une croche
…………………………..
Quant or ses chascune figure
Conoistre selon sa nature,
Et [que] chascune senefie,
Lors sachiés, si ne doutés mie,
Que chascune est premier en nonbre
Tant come vaut sans autre nonbre.
[…]

Au passage suivant de notre texte on lit[49]:

[Donec ad extremam venias, que cifra vocatur,
Que nil significat, sed dat signare sequenti.
Quelibet illarum, si primo limite ponas,
Simpliciter se significat; si vero secundo
Se decies, sursum procedas multiplicando,
Namque figura sequens quevis signat decies plus
ipsa locata loco quam signat pretereunte.]

Se tu mes cascune de ces figures u premier lieu ele fera soi simplement, se tu la més u secont lieu ele fera 10 fois soi, et en tant lieu que tu le meteras ele fera 10 tans plus qu él lieu devant; mais cyfre ne fait riens, mais **ele fait les autres figures multeplier**.

La traduction n'est pas faite mot à mot sans souci pour le sens du texte: dire que zéro 'fait multiplier' les nombres qu'il suit (ou qui le suivent, si on considère d'abord la première position à droit, selon l'habitude des arithmétiques médiévales), au lieu de dire tout simplement, comme le fait le vers latin, qu'il en établit la valeur, dénonce du côté du traducteur une lecture bien consciente de la matière de sa source.

49. 'The Crafte…', éd. Steele, p. 4-5 (voir aussi 'Carmen…', éd. Steele, p. 72); Mortet, 'Le plus ancien traité…', p. 60-61.

Les chapitres consacrés aux opérations décrivent les procédés de calcul dans leurs lignes essentielles de façon correcte et précise, avec peu de mots et, d'autant plus pour cette raison, avec une maîtrise du langage remarquable – ce qui nous étonne surtout dans le cas des racines cubiques, où l'algorithme est assez complexe. L'auteur de cette traduction devait être au moins un calculateur très capable, et nous croyons pouvoir exclure un but didactique pour ce texte, qui, pour un néophyte, serait presque incompréhensible.

À notre avis on peut lire cet *Algorisme* comme une sélection raisonnée de la matière du *Carmen*, une série de notes prises par un auteur compétent à propos des algorithmes auxquels lui, ou ses commanditaires, étaient intéressés, pour une raison particulière[50].

2.3. *The Art of Nombryng* (XVe siècle, anonyme)

Comme dans les textes que nous avons déjà analysés, cette traduction anglaise du *De Arte Numerandi* est strictement dépendante du latin du point de vue lexical; quand il s'agit de traduire les définitions, l'ordre des mots aussi reproduit le modèle latin, comme en témoignent les exemples suivants:

> [Additio est numeri vel numerorum aggregatio ut videatur summa **excrescens**.]
> Addicioun is of nombre, other of nombres, unto nombre or to nombres aggregacioun, that me may see that that is come therof as **excressent**[51].
>
> [Subtractio est, propositis duobus numeris, maioris ad minorem excessus inventio; vel sic: subtractio est numeri a numero ablatio ut videatur summa **excrescens**.]
> Subtraccioun is of 2 proposede nombres the fyndyng of the excesse of the more to the lasse; other: subtraccioun is ablacioun of o nombre fro a-nother, that me may see a some **left**[52].

Toutefois, nous observons une fois de plus une spécialisation du langage par rapport à la source; l'anglais traduit le latin *excrescens* par deux mots différents: *excrescent* a pour le traducteur le sens de 'quelque chose qui résulte en un accroissement', et non pas de 'ce qui résulte' tout court. Par ailleurs, le résultat d'une soustraction est plutôt le nombre 'qui reste' après une diminution (*left*).

50. À priori ce texte peut être aussi une sélection faite à partir d'une traduction française complète du *Carmen*.
51. 'Tractatus de Arte...', éd. Halliwell, p. 5; 'The Art...', éd. Steele, p. 35.
52. 'Tractatus de Arte...', éd. Halliwell, p. 6-7; 'The Art...', éd. Steele, p. 36.

Dans le cadre d'une stricte fidélité à la source, du point de vue textuel, l'*Art* n'échappe pas à quelques 'fautes' de traduction, qui tombent dans l'une des catégories suivantes:

a) fautes de traduction dans le contexte de quelques définitions à propos des nombres carrés, des nombres cubiques et des racines;
b) fautes d'interprétation des procédés indiqués par la source pour le déroulement des opérations, de façon très évidente pour les additions, les soustractions, les divisions par deux, les redoublements et les multiplications.

Si les fautes du type a) peuvent aisément s'expliquer par la faible connaissance des puissances et des racines de la part du traducteur, les fautes du type b) ne peuvent en revanche se comprendre par l'hypothèse d'une mauvaise compétence arithmétique. En effet, la source décrit les procédés de calcul de façon générale et sans aucun exemple avec des nombres précis; la version anglaise traduit presque littéralement ces instructions et ajoute souvent un exemple où les nombres sont disposés en colonnes, pour montrer comment il fallait procéder dans la pratique. Néanmoins, les termes ne sont pas toujours bien choisis par rapport au sens du texte source, ni les exemples non plus.

En guise d'illustration, nous proposons un cas tiré du passage consacré à la soustraction: nous présentons d'abord la source, à laquelle nous avons ajouté une représentation à l'aide des nombres, et ensuite la version anglaise elle-même[53].

> Subtrahe ergo primam figuram inferiori ordinis a figura sibi supraposita, et illa aut erit par, aut maior, aut minor. Si par, **ea deleta loco eius ponatur cifra**, et hoc propter figura sequens ne minus significet.
>
> 2 **0**
> 2 2
>
> Si maior supponatur, **tunc deleantur tot unitates quot contineat inferior figura et residuum loco eius ponatur**[54].

53. 'Tractatus de Arte...', éd. Halliwell, p. 7, 'The Art...', éd. Steele, p. 37.

54. On voit clairement qu'il faut effacer les chiffres du nombre en haut à la fin du chapitre, où est expliquée la manière de vérifier une soustraction: 'Si autem probare volueris utrum bene feceris necne, **figuras quas prius subtraxisti adde superioribus**, et concurrent eaedem figurae – si recte feceris – quas prius habuisti' ('Tractatus de Arte...', éd. Halliwelle, p. 8). Revenant à l'exemple auquel cette note renvoie, on peut dire, par conséquent, que soit *inferior* a fautivement remplacé *superior*, soit *deleo* signifie dans ce contexte ou 'effacer' (comme on le voit ailleurs dans l'exemple), ou 'soustraire', selon les cas. *Deleo* est aussi utilisé plus spécifiquement dans le *Tractatus* pour parler d'une soustraction qui donne zéro comme résultat (voir par exemple la p. 23). L'auteur du *Art of Nombryng* a interprété ce verbe uniquement dans le sens de 'soustraire'.

8 **2**
6 6

Si minor, quam maior numerus de minori subtrahi non potest, mutuetur unitas a figura proxima sequente quae valet decem respectu precedentis figurae; ab illo ergo denario et a figura a qua debuit fieri subtractio simul iunctis subtrahatur figura inferior **et residuum ponatur in loco figurae deletae**

24 **16**
6 6

Si vero figura a qua mutuanda est unitas sit unitas, **ea deleta** loco eius scribatur cifra ne figurae sequentes minus significent, deinde operare ut prius

312 3**09**
3 3

White-draw therfor the first of the lowere ordre fro the first of the ordre above his hede, and that wolle be other more or lasse, oþer egalle. Yf it be egalle or even the figure sette beside, put **in his place**[55] a cifre.

20
22
2

And yf it be more put away þerfro als many of unitees the lower figure conteynethe, and writ the residue as thus:

22
28
6

And yf it be lesse, by-cause the more may not be with-draw ther-fro, borow an unyte of the next figure that is worthe 10. Of that 10 and of the figure that ye wolde have with-draw fro be-fore to-gedre ioynede, wiþ-draw þe figure be-nethe, and put the residue **in the place of the figure put a-side, as þus**:

18
24
06

And yf the figure wherof me shal borow the unyte be one, **put it a-side** and write a cifre in the place þerof, lest the figures folowing faile of thaire nombre, and þan worche as it shewith in this figure here:

55. *Loco eius* est interprété par le traducteur comme ‘dans la position correspondante’, ce qui de façon indépendante du contexte pourrait être correct, alors que dans ces cas cela signifie proprement ‘à sa place’.

309
312
3

Manifestement, le traducteur ne comprenait pas qu'il fallait effacer les chiffres du nombre le plus grand et écrire ceux de la différence à leur place, et il ne comprenait pas non plus qu'il fallait écrire les chiffres retenus au lieu de les mémoriser; néanmoins il prouve son habilité dans la construction de ses exemples, qui ne sont jamais incorrects du point de vue strictement arithmétique. Ce phénomène est présent, de façon plus ou moins évidente, dans tout le traité.

Nous croyons que la possibilité que le traducteur avait une pratique de calcul différente par rapport à celle exposée dans la source permet d'expliquer les fautes de traduction présentes dans l'*Art of Nombrynge*. Nous pouvons penser qu'il écrivait le résultat au-dessus, sans remplacer les chiffres de l'un des termes de l'opération, mais en les effaçant avec un trait de plume, ou en-dessous sans rien effacer, et qu'il mémorisait, ou peut-être s'aidait-il de ses doigts pour les chiffres retenus[56], ce qui s'adapte mieux à l'exécution des calculs sur parchemin ou sur papier.

La question sur le but de cette traduction reste ouverte, bien qu'une fonction didactique soit évidemment assez improbable.

3. Conclusions

Nous avons vu que la petite quantité de traités mathématiques en langue vernaculaire, ou le retard de leur apparition, en France et en Angleterre, n'empêche pas que les premiers textes – parmi ceux qui nous sont parvenus – font preuve d'une bonne compétence de la part des auteurs, et que la 'pratique arithmétique' circulait donc de façon indépendante de la diffusion des manuscrits[57].

Nous avons constaté aussi que ces textes, qui sont des traductions, se détachent d'une certaine façon de leurs sources – soit pour en corriger les erreurs, soit pour opérer une sélection des contenus, soit parce que la différence du support d'écriture peut modifier le procédé de déroulement

56. Ce qui correspond à l'un des systèmes en vigueur pour écrire les opérations au XVe siècle.

57. Si les procédés de calcul pouvaient s'apprendre par les textes latins, on ne peut pas dire la même chose à propos des 'formules' linguistiques qui les accompagnent.

des calculs. Les traducteurs lisent donc leurs sources tout en étant conditionnés par un bagage de connaissances arithmétiques et 'algoristiques' qui est déjà formé, qui peut se détacher de façon évidente de celui des sources, et qui a influencé de façon perceptible le travail de traduction[58].

Università di Torino

ANNEXE

Bien que le *Carmen de Algorismo* soit l'un des textes les plus importants pour l'arithmétique médiévale, ses trois éditions modernes (voir ci-dessous) ne sont pas satisfaisantes: à côté de fautes de langue, elles présentent des non-sens mathématiques qui ne sont pas à notre avis attribuables au texte rédigé par Villadei.
L'impossibilité de préparer tout de suite une véritable édition (les témoins sont très nombreux), la volonté d'inclure ce texte si important dans le corpus de notre thèse de doctorat, ainsi que la nécessité d'avoir un texte fiable, nous ont amenée à essayer à mettre au propre le *Carmen* par le biais d'une comparaison des trois éditions et de la connaissance de ses algorithmes de calcul (ils sont identiques à ceux de l'arithmétique de Sacrobosco, dont l'édition de Pedersen est bien soignée).
Renvoyant à la thèse pour le texte complet et l'explication des critères de rédaction adoptés (*Lexicon Algorismi. Il vocabolario scientifico delle prime aritmetiche medievali. Latino, francese, italiano, inglese – sec. VII-XV*, Università di Torino 2006-2007), nous proposons ici les extraits cités dans l'article avec leur apparat critique.

Éditions et leurs manuscrits de référence:

H – J. O. Halliwell, *Rara Mathematica*, Hildesheim-New York, 1977, pp. 72-83 – Mss: London, British Library, Royal 8 C. iv, Royal 12 E. i., Cot. Vitell. A. i. (primo capitolo); Cambridge, Trinity College, O. v. 4, ii. 45, i. 31, St. John's College, F. 18; University Library, Mm iii 11 (2310), Ii I 13 (1690), Ii I 15 (1692); Oxford, Bodleian Library, Bodley 57, Fairfax 27, Digby 15, 22, 81, 97, 98, 104, 190, Bodley 4to. D. 21. Jur. Fragments: London, British Library, Sloane 513, 1620, 2397.

58. Merci à M. Alessandro Vitale, M. Jonathan Hunt, M. Dario Bressanini, M. Riccardo Pizzinato, Mlle Maria Mamone et à mon père.

On ne sait pas quel manuscrit Halliwell a choisi pour son édition, ni le degré de fiabilité qu'il attribuait aux autres. Après avoir demontré la paternité de Villadei dans l'introduction, Halliwell présente ainsi sa liste de manuscrits: 'MSS. of it are very numerous: I will mention a few for guidance' (p. vi).

S – R. Steele, *The Earliest Arithmetics in English*, London, 1922, p. 72-80 – Mss: London, British Library, 8 C. iv, 12 E. i.; British Library, Egerton 2622.

C – R. Steele, *The Earliest Arithmetics in English*, London, 1922, p. 3-32 (il s'agit des vers du *Carmen* présents dans *The Crafte of Nombrynge*; ici l'éditeur écrit en italiques les abréviations résolues) – Ms: London, British Library, Egerton 2622.

I

Hec algorismus ars presens dicitur, in qua
Talibus Indorum fruimur bis quinque figuris:
0, 9, 8, 7, 6, 5, 4, 3, 2, 1.
Prima significat unum, duo vero secunda,
Tercia significat tria, sic procede sinistre
Donec ad extremam venias, que cifra vocatur,
Que nil significat, sed dat signare sequenti.
Quelibet illarum, si primo limite ponas,
Simpliciter se significat; si vero secundo
Se decies, sursum procedas multiplicando,
Namque figura sequens quevis signat decies plus
Ipsa locata loco quam signat pretereunte;
Quam precedentes plus ultima significabit.

3 – absent dans C
5 – tercia: H *tertia* – sinistre: H *sinistra*
6 – venias: S *venies*
7 – SH *Que nil significat; dat significare sequenti.* Absent en C, où 12 est suivi de *Nil cifra significat sed dat signare sequenti*
11-13 – absents dans H

II

Septem sunt partes, non plures, istius artis:
Addere, subtrahere, duplare, dimidiare,
Sextaque dividere est, sed quinta est multiplicare,
Radicem extrahere pars septima dicitur esse.

Subtrahis aut addis a dextris vel mediabis,
A leva dupla, divide multiplicaque,
Extrahe radicem semper sub parte sinistra.

31 – dividere est sed quinta est: C *dividere sed quinta* – sed: S *set*
34 – multiplicaque: C *multiplica*
35 – absent dans C

III

Articulum si per reliquum vis multiplicare,
In proprium digitum debebit uterque resolvi
Articulus; digitis post per se multiplicatis,
Ex digitis quociens retineris multiplicatum
Articuli faciunt tot centum multiplicati.

143 – reliquum: C *reliquum reliquum*
144 – debebit: C *debet*
145 – digitis post per se multiplicatis: C *digitos post se multiplicantes*; S *digitos post per se multiplicantes*
146 – digitis: C *digitus* – retineris multiplicatum: C *retenerit multiplicari*; S *teneret multiplicatum*

Hiltrud GERNER

LA TRADUCTION FRANÇAISE DE QUELQUES TERMES D'ASTRONOMIE DU *COMPENDIUM THEOLOGICAE VERITATIS* (ENVIRON 1265) DANS *LE SOMME ABREGIET DE THEOLOGIE* (1481)

0. Introduction

Un projet d'inventaire du lexique de l'astronomie/astrologie avant 1500 m'a incitée à comparer le *Compendium theologicae veritatis*[1] et la *Somme abregiet de theologie*. En effet, après la parution en 1996 de l'ensemble des sept livres de la *Somme*, par Christine Michler, j'ai pu relever dans le livre II, intitulé *De la creation des choses*, et plus particulièrement dans les chapitres 3 à 9, le vocabulaire de la cosmologie et intégrer les exemples repérés dans mon inventaire en cours dont j'extrais les matériaux de cet exposé.

1. La transmission des deux textes

Avant d'aborder l'étude des mots, je voudrais rapidement situer les deux œuvres: le *Compendium theologicae veritatis*, rédigé par Hugo Ripelin de Strasbourg aux environs de 1265, a connu une exceptionnelle diffusion à travers toute l'Europe; à ce jour, on a pu identifier 620 manuscrits latins du XIIIe au XVe siècle[2]. Pour la période de 1300 à 1500, sept versions allemandes, transmises dans 32 manuscrits, ont été recensées[3]. C'est en Allemagne que la diffusion en langue vernaculaire a été la plus large. En France, face à une centaine de manuscrits en latin[4], on ne trouve que deux manuscrits français, émanant probablement d'une dictée parallèle

1. *Compendium theologicae veritatis*, liber secundus, in: *Bonaventura opera omnia*, éd. A.C. Peltier (Paris, 1866), Liber secundus, p. 88-94.
2. Steer G., *Hugo Ripelin von Strassburg. Zur Rezeptions- und Wirkungsgeschichte des 'Compendium theologicae veritatis' im deutschen Spätmittelalter* (Tübingen, 1981), p. 14.
3. *Le Somme abregiet de theologie*: *die altfranzösische Übersetzung des 'Compendium theologicae veritatis' Hugo Ripelins von Strassburg*, éd. Chr. Michler (Wiesbaden, 1996), p. X.
4. *Le Somme abregiet de theologie*..., p. X n. 30.

à partir d'un document-source qui est perdu. On suppose que la traduction ainsi que les deux copies qui nous sont parvenues ont été réalisées à la demande d'Anthoine de Chourses, lieutenant ou gouverneur du roi Louis XI à Béthune[5]. L'inventaire de la bibliothèque du commanditaire montre une prédilection pour la littérature contemporaine en langue française, dont l'éventail comprend la théologie, la jurisprudence, l'histoire, les sciences et les arts. La *Somme* trouve donc tout à fait sa place parmi ces manuscrits. Il semble néanmoins que la connaissance et la diffusion de ce texte n'aient guère dépassé le cercle béthunois de son propriétaire.

2. Le traducteur

À aucun endroit, l'auteur de la version française ne se prononce sur sa conception de la traduction, mais, appliquant la règle d'Oresme de ne pas s'écarter du texte original, il suit fidèlement son modèle latin, souvent mot à mot. Toutefois, habitué à l'exégèse des textes sacrés, comme le montre sa traduction des autres livres de la *Somme*, il exerce ses capacités pédagogiques dans des passages plutôt techniques, mettant à profit ses connaissances en astronomie en paraphrasant la formulation du texte d'origine ou en l'expliquant par des synonymes savants ou populaires.

3. Les dénominations des signes du Zodiaque

Lorsque le *Compendium* clôt le chapitre IV par l'énumération des douze signes du zodiaque en citant les vers: 'Sunt Aries, Taurus, Gemini, Cancer, Leo, Virgo / Libraque, Scorpius, Arcitenens, Caper, Amphora, Pisces', notre traducteur reprend cette même liste en latin, puis il continue ce chapitre sur deux pages avec un développement sur le déplacement du Soleil dans le cercle zodiacal, suivi de la réfutation des croyances païennes. Il s'agit là d'une des amplifications les plus importantes de la *Somme*.

Il est intéressant de constater que dans cette partie ajoutée, le traducteur garde le nom latin du signe du zodiaque, en indiquant parfois l'équivalent en français (fait rare: pendant longtemps, même dans les textes français, cette terminologie reste dans la langue d'origine), mais l'appellation

5. *Le Somme abregiet de theologie. Kritische Edition der französischen Übersetzung von Hugo Ripelins von Strassburg 'Compendium theologicae veritatis'*, éd. Chr. Michler (München, 1982), p. 21.

latine n'est pas toujours la même que celle qu'il venait de citer d'après le *Compendium*. Voici cette nouvelle liste des signes suivis, le cas échéant, de la traduction française (en gras dans la colonne de gauche les termes latins qui diffèrent de la liste citée d'après l'original latin, en italique, dans la colonne du milieu, les mots qui peuvent être considérés comme équivalents français):

Latin	**Définition française**	**Commentaire**
Aries	(il a aucunes proprietés de cellui qui s'appelle ung *rain*)	Il faut probablement lire *ram*, le nom moyen néerlandais de 'bélier'
Ø	*tor*	
Gemini	(qui vault a dire double ou iterer et conjoindre)	On trouve dans les textes contemporains: *deux freres, gemins, jumeaux, gemeaux.*
Cancer	*escrevice*	On trouve dans les textes contemporains: *cancer*
Ø	*lyon*	
Virgo	(ainsi apellé par similitude. Car comme une pucelle vierge est sterile, brehaigne ...)	On trouve dans les textes contemporains: *vierge*. *IntrAstr.*[6] c. 1270, XXI.6: *Virge.*
Libra	*balance*	
Scorpius	*scorpion*	
Sagittarius (au lieu de *arcitenens*)	*arcier*	On trouve dans les textes contemporains: *archer*, et dès le XIIIe siècle *sagittaire* (*IntrAstr.* c. 1270, Evrart de Conty[7], Simon de Phares[8]).
Capricornus (au lieu de *caper*)	(... est pareil en figure d'un *bouc*)	On trouve dans les textes contemporains: *capricorne* (*IntrAstr.* c. 1270, Pèlerin de Prusse[9], Evrart de Conty,

6. Dörr S., *Der älteste Astronomietraktat in französischer Sprache:* L'Introductoire d'astronomie. *Edition und lexikalische Analyse* (Tübingen, 1998).
7. Evrart de Conty, *Le Livre des Eschez amoureux moralisés*, éds F. Guichard-Tesson & B. Roy (Montréal, 1993).
8. Simon de Phares, *Le Recueil des plus celebres astrologues*. T. 1: édition critique, éd. J.-P. Boudet (Paris, 1997).
9. *Pèlerin de Prusse on the Astrolabe. Text and Translation of his* Practique de astralabe, éds, trad. E. Laird & R. Fischer (Binghamton-New York, 1995).

		Fusoris[10], Simon de Phares) ou *chevre cornue* (Foulechat[11]).
Aquarius (au lieu de *amphora*)	Ø	On trouve dans les textes contemporains: *aquaire*.
Pisces	*poissons*	

Compte tenu de cette incohérence dans la reprise des noms latins, il est fort possible que ce passage vienne d'une autre source que je n'ai pas réussi à identifier.

4. Comparaison: *Compendium – Somme abregiet* / textes contemporains

Nous avons choisi une douzaine de lexèmes dans le texte latin pour lesquels nous présentons en vis-à-vis leur traduction dans la *Somme abregiet.*

Chaque passage est suivi d'une ou de plusieurs attestations du même concept dans des traités antérieurs ou contemporains, afin de tester si le traducteur s'est servi d'un vocabulaire déjà disponible ou s'il a tenté de créer sa propre terminologie. Dans ce qui suit, nous nous interrogerons donc sur la question de savoir si les termes techniques de la *Somme* appartiennent en propre à la langue et s'ils dépassent la création individuelle, car l'entreprise du traducteur 'ne serait qu'un idiolecte dont la réalisation n'a rien à voir avec la langue si les termes qu'il emploie ne sont pas présents dans d'autres œuvres avec le même sens[12].'

4.1. Machina mundi *'Assemblage et construction de l'univers'*

Compendium	**Somme**
corporalis ***mundi machina*** tota consistit in duobus… (89).	le corps du monde et la ***fabrique du monde*** corporel est constitué et consiste en deux choses… (78).

10. Jean Fusoris, *Traité de cosmographie* (ms. B.N. fr. 9558), éd. L.O. Grundt (Bergen, 1973).

11. Denis Foulechat, *Le Policratique de Jean de Salisbury (1372). Livres I-III. Edition critique avec introduction, notes et glossaire*, éd. Ch. Brucker, Publications romanes et françaises 209 (Genève, 1994), p. 164.

12. Ducos J., *La Météorologie en français au Moyen Age (XIIIe-XIVe siècles)* (Paris, 1998), p. 20.

Les attestations de *fabrique* dans ce sens sont rares en français; on trouve *mondain fabrique* dans un contexte religieux: '... qui (...) regit la rote orbiculaire, la terre et ce *mondain fabricque* où sont rois, ducz, marquis et comtes' (MOLINET, *Chroniques*, 1474-1506)[13].

Ce syntagme est emprunté à la locution latine *mundana fabrica* (aussi *fabrica mundi, mundialis fabrica*) des époques patristique et médiévale. Par ailleurs, le terme *machine* dans des syntagmes tels que *machine du monde*, traduction littérale de *machina mundi*, ou *machine du ciel, machine céleste*, est disponible et attesté dans les textes français même avant Oresme:

- ... por ce que nule chose ne defausist en si grant ovrage cum est la ***machine***, ce est la ***façons del monde***, por la quele raison les fortunes des choses mortels sunt diversefiees par l'aministrement del ciel (*Introductoire d'astronomie* c. 1270, I.8).
- toute le ***machine dou ciel*** (*Compilacions de le science des estoilles*, a. 1324)[14].
- Et donques je met premierement que toute la ***machine*** corporelle ou toute la masse de tous les corps du monde est devisee en .II. parties (ORESME, *Le Livre du ciel et du monde*, c. 1377)[15].
- toute la ***machine celeste*** depent du Primier, qui est espirituel, qui a creé le ciel et la terre (*Songe verg.*, 1378)[16].
- ilz virent tourner si très velocieusement les cieux en ceste grande ***machine celeste*** et si unifformement (SIMON DE PHARES, *Le Recueil des plus celebres astrologues*, c. 1494-1498)[17].
- ou que toute la ***machine du monde*** se disolvist (SIMON DE PHARES, *Le Recueil des plus celebres astrologues*, c. 1494-1498)[18].

FEW VI-1, 10a: s.v. *machina*: Mfr. *machine* 'assemblage et construction de l'univers' (Oresme c. 1375).

L'exemple de *Introductoire d'astronomie* de circa 1270 antédate celui du FEW.

13. Jean Molinet, *Chroniques*, éds G. Doutrepont & O. Jodogne (Bruxelles, 1935), t. 1, p. 224. Je remercie Béatrice Stumpf de m'avoir communiqué cette attestation.

14. Léopold d'Autriche, *Li Compilacions de le science des estoilles, Books I-III, edited from MS French 613 of the Bibliothèque Nationale, with Notes and Glossary*, éd. F.J. Carmody (Berkeley-Los Angeles, 1947), p. 59. [Traduction de la *Compilatio de astrorum scientia* de Léopold d'Autriche]

15. Nicole Oresme, *Le Livre du ciel et du monde*, éds. A.D. Menut & A.J. Denomy (Traduction avec introduction par A.D. Menut) (Madison-London, 1968), p. 520.

16. *Le Songe du Vergier, édité d'après le manuscrit Royal 19 CIV de la British Library*, éd. M. Schnerb-Lièvre (Paris, 1982), t. 1, p. 67.

17. Simon de Phares, *Le Recueil des plus celebres astrologues...*, f. 11v.

18. Simon de Phares, *Le Recueil des plus celebres astrologues...*, f. 79r.

4.2. Primum mobile *'Sphère extérieure qui englobe toutes les autres et les entraîne'*

Compendium	**Somme**[19]
Nomine autem coeli crystallini, sive aquei, intelligitur illa pars materiae primae, quae secundum Philosophos formata in duos orbes, quorum superior est ***primum mobile*** (90a).	Par le nom du ciel cristallin et aquatique *comme eaue* se entent ceste partie de la matere premiere laquelle selon les philosophes est formee en deux cercles desquelz le plus hault est apellé ***primum mobile***, *c'est a dire* ***le premier corps mobile*** *qui se* ***moeut*** (78).
Istae proprietates tam coelo conveniunt empyreo, quam ***primo mobili*** (90a).	Ces proprietés competent au ciel empire *ouquel sont les angeles et au ciel qu'on apelle* le ***premier meuble*** (79).
Cursu, quem habent [planetae] ab influxu ***primi mobilis***, moventur directe; sed proprio cursu moventur oblique ... (91a).	Selon le cours qu'ilz ont et font par le ***premier moeuble*** *qui est le ciel par dessus le ciel des estoilles* et par la *vigueur et* influence *ou par infusion* d'icellui ilz se meuvent directement, mais selon leurs cours propre, ilz se meuvent obliquement *en toupiant* (83).
Motus ***primi mobilis***, scilicet nonae sphaerae, ab oriente in occidentem ... (91a).	Le mouvement ou le cours du ***premier moeuble*** qui est la .IX.e spere d'orient par occident ... (83).

D'autres attestations avant 1500:

- En aprés, au dessus du firmament est le .IXe. ciel, que on appelle ***le premier ciel mobille***, ouquel cielz n'y a nulles estoilles, mais il se muet regulierement d'orient en occident sur les poles du monde par chascun jour naturel un tour. (FUSORIS, *Traité de cosmographie*, 1432)[20].
- ... estoient contenues et enveloppées IX autres grandes et merveilleuses speres et nomma, après son pere, icelui Enos ceste dixieme plus haulte *celum cristalinum* ou ***premier mobille*** (SIMON DE PHARES, *Le Recueil des plus celebres astrologues*, c. 1494-1498)[21].
- ... par la violance et rapt du ***premier mobil*** qui elle et toutes les autres speres tourne et circumvolue d'Oriant en Occident et iterum en Oriant en ung jour et une nuit naturel (SIMON DE PHARES, *Le Recueil des plus celebres astrologues*, c. 1494-1498)[22].

19. Les extensions et commentaires du traducteur sont mis en italique.
20. Jean Fusoris, *Traité de cosmographie*..., p. 31.
21. Simon de Phares, *Le Recueil des plus celebres astrologues*..., f. 12r.
22. Simon de Phares, *Le Recueil des plus celebres astrologues*..., f. 133v.

FEW VI-3, 3a: s.v. *mobilis*: Fr. *premier mobile* m. 'ciel enveloppant et faisant mouvoir tous les autres cieux' (1579 – Rich 1680).

À la première apparition du syntagme *primum mobile*, le traducteur l'intègre tel quel, suivi de son équivalent français *premier corps mobile*, et, comme cela ne lui semble pas encore suffire, il explique l'adjectif *mobile* dans une relative: *qui se moeut*. Par la suite, à trois reprises, il se sert de la traduction *premier meuble* non sans omettre de préciser la position de cette sphère dans la hiérarchie des cieux emboîtés.

Je n'ai pas trouvé d'autres attestations de l'expression *premier meuble*; c'est *premier mobile* qui s'est imposé, ce dont on trouve déjà trace chez Fusoris en 1432 et, plus tard, chez Simon de Phares. La première date donnée par le FEW est 1579.

4.3. Motus rationalis *'Mouvement de la plus haute sphère'*

Compendium	Somme
Motus primi mobilis, scilicet nonae sphaerae, ab oriente in occidentem, rediens iterum in orientem, dicitur ***rationalis***, ad similitudinem motus rationis in microcosmo, quando fit consideratio a Creatore per creaturas, rediens in Creatorem, et ibi sistendo (91a-b).	Le mouvement *ou le cours* du premier moeuble qui est la .IX.e spere d'orient par occident retournant de rechief en orient, est apellé ***mouvement raisonnable*** a la samblance du mouvement de raison qui est au monde mineur *qui est apellé microcosme selon le grec* quant la consideration procede du Createur par les creatures, retournans au Createur et illec se arrestent (83).

Autre attestation en français:

Il [Dieu] li dona [au ciel] ***movement raisonable***, il li dona vertu de lumieres et li commist les natures et le muement des choses de ça desouz, autresi comme a un governeour de sa volenté, por ce que nule chose ne defausist en si grant ovrage cum est la machine, ce est la façons del monde, por la quele raison les fortunes des choses mortels sunt diversefiees par l'aministrement del ciel (*Introductoire d'astronomie* c. 1270, I.8).

Attestations en latin:

Sacrobosco, *Tractatus de Sphaera* (1220): Unde sciendum quod primus motus dicitur motus primi mobilis: hoc est nonae sphaerae sive caeli ultimi: qui est ab oriente per occidentem rediens iterum in orientem qui etiam dicitur ***motus rationalis***: ad similitudinem motus rationis qui est in microcosmo. Id est in homine scilicet quando fit consideratio a creatore per creaturas in creatorem ibi sistendo[23].

Juan Gil de Zamora, *Historia Naturalis* (a. 1320): Unde sciendum quod primus motus dicitur motus primi mobilis, hoc est, sphere none. Qui motus est ab Oriente in Occidentem, rediens iterum in Orientem. Qui eciam dicitur ***motus racionalis*** ad similitudinem racionis, qui est in microcosmo, scilicet, in homine, id est, in minori mundo, quando fit consideracio ad creatorem per creaturas, in creatore insistendo[24].

FEW VI-3, 167b: s.v. *movere*: *mouvement* 't. d'astronomie, révolution, marche réelle ou apparente des corps célestes' (seit Est 1538 s.v. meatus).

Le FEW atteste *mouvement* dans son acception générale depuis circ. 1190 mais, dans le domaine de l'astronomie, sa première référence date du XVIe siècle et renvoie à une œuvre lexicographique, ce qui nous prouve que cet usage est déjà bien implanté, mais que les relevés de la terminologie scientifique et technique sont lacunaires, et ce d'autant plus lorsqu'il s'agit de lexèmes appartenant aussi à la langue courante.

4.4. Galaxia *'La Voie lactée'*

Compendium	**Somme**
	Galaxia, voie lactie, voie paree
quanto conjunctiores, tanto in communi lucidiores; sed singulae minus apparentes, ut patet in **galaxia**; claritate et quantitate differunt, ac virtute. ***Galaxia*** est multitudo parvarum stellarum, quasi contiguarum illi loco orbis, ubi diffunditur lumen Solis. Fit autem eclipsis … (91a)	D'autant qu'elles [les étoiles] sont plus voisines, d'autant en commun et ensemble sont plus luisantes, mais en particulier moins apparentes, comme il appert en celle partie qui se apelle ***galaxia*** ou sont plusieurs estoilles petites comme contigues et prochaines a cellui lieu du cercle ou se fine et termine la lumiere du soleil. ***Galaxia*** *proprement est un cercle blanc comme laict et est une voie qu'on voit en le spere reonde que le soleil avironne comme blanche, et par le transpassement d'icellui soleil, par sa clareté le fait reluire et par ainsi est ung lieu plus cler blanc comme laict, apellé*

23. *Johannis de Sacrobosco*, *Tractatus de Sphaera* [Venise c. 1478], éd. R. de Andrade Martins (Campinas, 2002), f. 7v.

24. *Johannis Aegidii Zamorensis*, *Historia Naturalis*, éd. A.D. Garcia & L.G. Ballester (Salamanca, 1994), t. II, p. 1272.

> *la* ***voie lactie****, blance comme laict, et aussi* ***galaxia*** *ou selon la nomination du peuple la* ***voie paree****.* L'eclipse du soleil se fait … (83).

Cf. pour l'extension: Isidore III, 46: *Lacteus circulus via est quae in sphaera videtur, a candore dicta, quia alba est*[25].

D'autres équivalents français avant 1500:

- … nous diron de la ***voie du lait***, que les vielles [gens] appellent la ***voie saint Martin*** (MAHIEU LE VILAIN, c. 1280 [ms. 3e quart XIVe siècle])[26].
- … la ***voie du lait***, que l'en appelle la ***voie saint Martin*** (MAHIEU LE VILAIN, c. 1280 [ms. 3e quart XIVe siècle])[27].
- Sa nature est d'enluminer Et pource apert cest lieu plus cler. Il est ***cercle de lait*** nommé, Quar de clarté est renommé (*Propriétés des choses selon le Rosarius*, c. 1330)[28].
- ***Galaxe*** est le plus bel et le plus blanc cercle qui soit ou ciel qui va par le millieu du ciel et commence en orient jusques vers septentrion en passant par le signe du cancre et du capricorne et retourne arriere en sont point. Ce cercle est appellé le ***cercle de lait*** par sa blancheur et sa noblesse (…) selon l'opinion du peuple commun, ***galaxe*** est la trace du soleil qu'il laisse ou ciel apres luy quant Il passe mais ce est faulx par le dict d'Aristote … (CORBECHON, *Des propriétés des choses*, 1372 [éd. 1485-1486])[29].
- *Galaxia* (…). circulus lateus: c'est la ***voie de saint Jaques*** ou ciel (Aalma, c. 1380)[30].
- … et dit cely Ovide qu''ilz vindrent au palaiz dessusdit par une voye haute qui manifestement se moustre ou ciel, de nuit, quant le temps est seris et l'air pur et net, laquelle ***voye est blanche come lait***, et pour ce est elle appellee ***Lactee***, c'est a dire *voye* semblable a lait' (EVRART DE CONTY, *Le Livre des Eschez amoureux moralisés*, c. 1400)[31].

25. Isidore de Séville, *Etymologiae sive Origines*, livre III, 46.
26. Mahieu Le Vilain, *Les Metheores d'Aristote, traduction du XIIIe siècle*, éd. R. Edgren (Uppsala, 1945), p. 8.
27. Mahieu Le Vilain, *Les Metheores…*, p. 22.
28. *Les Propriétés des choses selon le Rosarius (B.N. f. fr. 12483)*, éd. A. Zetterberg. Édition revue et complétée par S. Sandqvist (Lund, 1994), p. 71.
29. Jean Corbechon, *Des propriétés des choses,* Reproduction de l'édition de Lyon: Guillaume Le Roy [1485-1486] BNF, Gallica. Livre VIII: *Du Ciel et du monde*, p. 8.
30. *Aalma*, éd. Mario Roques, in: *Recueil général des lexiques français du moyen âge (XIIe-XVe siècle)*. I. *Lexiques alphabétiques*, Bibliothèque de l'école des Hautes Études 269 (Paris, 1938), t. 2, p. 161.
31. Evrart de Conty, *Le Livre des Eschez…*, p. 521.

- … fut dit ainsy pour une ***voye blanche*** qui par samblant de nuit se moustre ou ciel souvent, qui est aussi que un cercle qui d'orient en occident est ainsy estendus, et est des philosophes ce cercle appellé ***Galacia***, pour ce qu'il est ainsy come lait blanc, et les poetes aussi en dient moult de choses; et c'est aussi la *voie* que le peuple appelle moult souvent le ***Chemin de saint Jaques*** (EVRART DE CONTY, *Le Livre des Eschez amoureux moralisés*, c. 1400)[32].
- ***Galace*** est cellui appellez, Qui moult est grans et biaux et lez, ***Cercle de lait*** maint l'appellerent Pour sa blancheur; de terre apperent Ses traces, quant il fait sanz nue Temps seri et nuit est venue (CHRISTINE DE PIZAN, *Le Livre du chemin de long estude*, 1402-1403)[33].
- Les douze signes ne teut pas Ne le char ne la pouchiniere Ne Feton qui en son trespas Fist ou ciel la ***blanche charriere*** (MARTIN LE FRANC, *Champion des dames*, 1440-1442)[34].

FEW IV, 23b: *s.v. galaxias*: Mfr. nfr. *galaxie* f. 'voie lactée' (t. d'astron., seit 1557, RF 32, 72).

FEW V, 114a: *s.v. lac*: Mfr. *voye laitine* 'voie lactée' Baïf.

FEW V, 124a: *s.v. lacteus*: Mfr. *voye laitée* 'voie lactée' (Baïf; Rons) (…); mfr. *ceinture laitée* Baïf; *cercle laicté* Marot. – Nfr. *voie de lait* (D'Aubigné – Ac 1798); *chemin de lait* (Fur 1690 – Trév 1771)… Nfr. *voie lactée* 'bande blanchâtre au ciel' (seit Oud 1640).

FEW V, 10b *s.v. Jacobus*: Mfr. nfr. *chemin Saint-Jacques* 'voie lactée' (Rab 1532; EstL 1583, 14b)… nfr. *chemin de saint Jacques* (seit Cotgr 1611).

Rien à l'article *via* qui renvoie à *lacteus*.

Rien à l'article *parare*.

La diversité des appellations populaires, d'ailleurs souvent des hapax, témoigne de l'intérêt avec lequel les Anciens observaient ce phénomène céleste; elle reflète également la multitude des mythes et légendes qui s'y attachent.

La désignation donnée comme populaire par le traducteur '… [lieu] appelé (…) selon la nomination du peuple la **voie paree**', semble être le seul exemple connu, comparable à l'attestation isolée chez Mahieu Le Vilain: '… que les vielles [gens] appellent la **voie saint Martin**' ou à

32. Evrart de Conty, *Le Livre des Eschez…*, p. 522.

33. Christine de Pizan, *Le Livre du chemin de long estude. Publié pour la première fois d'après sept manuscrits de Paris, de Bruxelles et de Berlin*, éd. R. Püschel (Berlin-Paris, s.d. [1881]), p. 83.

34. Martin Le Franc, *Le Champion des dames*, éd. R. Deschaux (Paris, 1999), t. 3, p. 118.

la '**blanche charriere**' de Martin Le Franc, désignation qui fait allusion à Phaéton, qui avait demandé à son père Hélios la permission de conduire le char du Soleil lequel, en se renversant, embrasa le ciel.

La glose de *galaxia* (ci-dessus en italique), longue addition du traducteur de la *Somme*, essaie de donner toutes les informations disponibles sur ce phénomène: *c'est un cercle* (...) *c'est une voie*, combinant ici deux concepts: *cercle* (hérité des Grecs) qui sous-entend la sphéricité de la terre – comme par ailleurs le cercle équinoxial et les cercles tropiques – et les différentes traductions de latin *via* (l'expression *via lactea* se trouve pour la première fois dans les *Métamorphoses* d'Ovide I, 168) comme *voie/chemin,* ... appellations qui comportent l'idée de largeur[35]. C'est la notion de 'voie' qui s'est imposée dans la plupart des langues européennes (all. *Milchstrasse,* angl. *milky way,* fr. *voie lactée,* ital. *via lattea,* roum. *calea lactee*).

On remarque également dans la *Somme* une surabondance d'allusions à la blancheur de la galaxie: *blanc comme laict – voie* (...) *comme blanche – lieu plus cler, blanc comme laict*, tandis que le texte-source ne donne aucune information sur son aspect.

Avant 1500, les lexèmes les plus répandus sont v*oie* ou *chemin de saint Jacques*; je n'ai trouvé aucune attestation de *galaxie* antérieure à celle du FEW (1557); les graphies relevées chez Corbechon (*galaxe* 1372) et chez Christine de Pizan (*galace* 1402-1403), ainsi que toutes les autres expressions comme *voie du lait* (après 1270), *voie de saint Jacques* (c. 1380), *chemin de saint Jacques* et *voie lactee* (c. 1400) antédatent celles du FEW ou les complètent (*voie saint Martin, voie paree*).

4.5. Ascensus, descensus

Compendium	**Somme**
	monture, descension
Operantur quoque corpora superiora distinctivam significationem dierum secundum lumen Solis, et distinctionem mensium secundum lumen Lunae, ac distinctionem annorum secundum motus Solis in **aliquo circulo**, atque distinctionem temporum secundum varium	Item les corps celestiens de dessus euvrent et font la congnoissance des distinctions des jours selon la lumiere du soleil et les distinctions des mois selon le mouvement de la lune, font aussi la distinction et la diversité des annees selon le mouvement du soleil ou

35. Le Boeuffle A., *Astronomie Astrologie: Lexique latin* (Paris, 1987), p. 162.

cursum planetarum, distantiam, et cursum, ***ascensum***, et ***descensum***, retrogradationem, et statum (90b).

cercle de travers et aussi la diversité des temps selon le divers cours des planettes, selon leur distance de nous et selon qu'ilz s'approcent ensemble, selon leur ***monture*** et ***descencion***, selon leur station *ou arrestement* et selon leur reculement (80).

D'autres équivalents français avant 1500:

- Et par l'***aprochement*** et l'***eslognement*** des planetes, la diversetez del tems et la qualitez des elemenz et les natures des choses se variassent par certaine loi et par certaine raison pardurablement (*Introductoire d'astronomie* c. 1270)[36].
- ... et les livres de Jehan Mansur, *Des retrogradacions et* ***humiliacions***, et ung livre nommé en arabicq *Asselcoc* et ung autre livre, contenant les utilités de astrologie en vers (SIMON DE PHARES, *Le Recueil des plus celebres astrologues*, c. 1494-1498)[37].

Cf. aussi l'extrait suivant nº 6.

Monture: sens non attesté dans l'article **montare* du FEW VI-3.

Descension: forme et sens non attestés dans l'article *descendere* du FEW III.

Il me semble que *monture* et *descension* désignent ici – toujours dans la vision géocentrique de l'époque – le mouvement d'une planète vers son apogée ou son périgée, mouvement qui dans *L'introductoire d'astronomie* est traduit par les termes *eslognement* et *aprochement*. Une autre tournure intéressante, mais jusqu'ici isolée, pour exprimer la notion de périgée est *humiliation* chez Simon de Phares. Elle s'insère néanmoins dans la tradition latine comme le montre l'étude du lexique latin d'André Le Bœuffle: '*Humilitas* et *humiliari* expriment aussi la notion de périgée planétaire; l'approche de la terre était conçue comme une descente'[38]. L'antonyme latin est *altitudo* ou un lexème de la famille de *ascendere*[39].

Monture peut donc être considéré comme un calque sémantique et *descension* comme un emprunt au latin. Ces multiples désignations prouvent bien que la terminologie reste floue (cf. aussi dans l'exemple ci-dessous *acces* ou *aprocement* versus *depart* ou *eslongement*).

36. Dörr, *Der älteste Astronomietraktat...*, I.19.
37. Simon de Phares, *Le Recueil des plus celebres astrologues...*, f. 79v.
38. Le Boeuffle, *Astronomie Astrologie...*, p. 151 s.v. *humilitas.*
39. Le Boeuffle, *Astronomie Astrologie...*, p. 38-39 s.v. *altitudo.*

4.6. Circulus obliquus

Compendium	Somme
	Cercle oblique, crom et traversal
ab Aristotele vero, in secundo *de Generatione,* dicitur ***circulus obliquus*** [iste circulus signifer], ubi dicit quod secundum accessum et recessum Solis in ***obliquo circulo*** fiunt generationes et corruptiones in rebus (90).	Aristotiles ou livre 'De generation et corruption' apelle le ***zodiac*** le ***cercle oblique***, *c'est a dire qui va de travers*, et dist ainsi que selon l'acces *ou aprocement* et le depart *ou eslongement* du soleil ou ***cercle oblique, crom et traversal*** se font les generations et corruptions ez choses (80).

D'autres équivalents français avant 1500:

- Et quant tu oïs dire les assencions des signes, enten les ascensions du cercle de le equateur, premierement u cercle direct (et droit) en le premiere maniere, u en le seconde maniere u ***cercle oblique*** (et travesain), les quelles elle [la sphère] fait illueques avoeques l'elevacion des signes. (*Compilacions de le science des estoilles*, a. 1324)[40].
- … donques s'ensuit il que le ciel du soleil qui est aussi comme un membre de tout le ciel, ait en soy .II. ames ou intelligences au moins, c'est a savoir une qui le meut o tout le firmament par mouvement journal, et l'autre qui le meut par son propre mouvement selon les signes du ***zodiaque***. (ORESME, *Le Livre du ciel et du monde*, c. 1377)[41] (+ 4 autres occurrences).
- … et se nous adjoustons au cercle dessusdit [i.e. l'écliptique] deux autres cercles equidistans a ly et eslongiés chascun de .VJ. degrés, l'un d'une part et l'autre d'autre, ce sera lors le cercle que on appelle le ***zodiaque*** ou le ***cercle oblique*** ou le ***cercle des signes*** dont la latitude est de .xij. degrés, dessoubz lequel se meuvent toutes les sept planetes sanz yssir hors des mectes de sa latitude. (EVRART DE CONTY, *Le Livre des Eschez amoureux moralisés*, c. 1400)[42].
- … Et pour ce dit Aristote que par le mouvement du soloil et des autres planetes ou ***cercle oblique***, c'est a dire ou ***zodiaque*** dessusdit, se font les generacions et les corrupcions en ce bas monde. (EVRART DE CONTY, *Le Livre des Eschez amoureux moralisés*, c. 1400)[43].
- … la quelle ***ecliptique*** est le chemin et voie du soleil car iamais le soleil ne part de desoulz celle ligne et ainsi est tousiours ou milieu

40. *Li Compilacions…*, p. 60.
41. Nicole Oresme, *Le Livre du ciel…*, p. 316.
42. Evrart de Conty, *Le Livre des Eschez…*, p. 47.
43. Evrart de Conty, *Le Livre des Eschez…*, p. 50.

> du ***zodiaque*** mais les autres planetes tousiours sont d'un coste ou d'autre de ceste ligne (*Le Compost et kalendrier des bergiers*, 1493)[44].

FEW XIV, 666a: s.v. *zodiacus*: Afr. *dyodake* m. 'zone du ciel parallèle à l'écliptique et comprenant les 12 constellations principales que le soleil semble parcourir dans l'année' (ca. 1240), fr. *zodiaque* (seit RoseM) ...

FEW VII, 270a: s.v. *obliquus*: *cercle oblique* 'écliptique' (1632, Rotrou); le sens 'zodiaque' n'y est pas attesté.

L'appellation *cercle oblique* manque souvent de précision; dans le cas ci-dessus, l'ambiguïté existe déjà dans le texte latin. Seuls le premier exemple d'Evrart de Conty et celui du *Compost et kalendrier des bergers* distinguent nettement la ligne de l'écliptique – c'est-à-dire le chemin apparent du Soleil – de la bande zodiacale dans laquelle évoluent les planètes et qui s'étend sur une largeur d'environ 6° de part et d'autre de l'écliptique. Tous les exemples de *cercle oblique* cités ici sont des antédates par rapport au FEW.

4.7. Retrogradus, progressivus, stationarius, (epicyclus)

Compendium	**Somme**
	Reculant, progressif, stationnaire
Planetae dicuntur quandoque ***retrogradi***, quandoque ***progressivi***, quandoque ***stationarii***, prout in ***epicyclis*** suis diversimode moventur (91a).	Les planettes sont aucune fois dittes ***reculantes*** *alant en reculant ou de travers*, aucune fois et le plus sont ***progressives*** *alant directement devant elles*, aucune fois ***stacionaires*** *estans sans proceder*, aucunement selon qu'en leurs **eclipses**[45] se meuvent de diverses manieres (83).

D'autres équivalents français avant 1500:

– Le treble regard sorent des hauz planetes quant il virent que entre le Soloil et .j. d'els ot .cxx. degrez. Quar donques estoit faiz ***retrogrades***, se il estoit devanz ***direcz***, c'est ***adroiz***, ou il estoit faiz adroiz et aloit son droit cours, se il estoit devant ***retrogrades*** (*Introductoire d'astronomie* c. 1270)[46].

44. *Le Compost et kalendrier des bergiers*. Reproduction en fac-similé de l'édition de G. Marchant (Paris, 1493), éd. P. Champion (Paris, 1926), fol. 61r.
45. On constate au passage une erreur de traduction (ou erreur de copiste?): *epicycle – eclipse*.
46. Dörr, *Der älteste Astronomietraktat*..., LXIX.24-25.

- Et se la hauteur du planete devers orient est mendre et devers occident greigneur que la hauteur de l'estoile, doncques le planete est ***direct*** et ***hastif***; et se il avient au contraire, doncques le planete est ***retrograde***. Et se les hautesses par pluseurs journees sont equales, doncques le planete est ***stacionere*** (PÈLERIN DE PRUSSE, Astrolabe, 1362)[47].
- Et donques est son mouvement plus isnel une fois que autre, et aussi des autres planetes des unes plus, des autres moins, selon ce que elles sont meues de plus ou de moins de mouvemens et environ centres ou sus poles differens et distans plus ou moins. Et selon ce, aucunes sont ***stacionaires*** et plus ***directes*** et ***retrogrades*** et en telles disposicions qu'i ne peuent estre sanz irregularité de mouvement (ORESME, *Le Livre du ciel et du monde*, c. 1377)[48].
- Item come les planetes se moustrent ***retrogrades*** aucunesfoiz et autresfoiz ***directes*** au contraire. Et telle fois est aussy qu'elles se monstrent estre aussy comme ***en estat,*** pour laquelle chose les astronomiens furent meuz de mettre ou ciel les cercles ecentricles et les petiz ***epicicles*** aussy comme Ptholomee les ymagine et met. (EVRART DE CONTY, *Le Livre des Eschez amoureux moralisés*, c. 1400)[49].
- Savoir devez que les Planètes fors le Soleil ont troiz mouvemens, c'est assavoir ***rétrograde*** qui vault à dire ***retournant***; montant en hault que l'en appelle ***stationaire***, et ***directif*** quant la Planète se meut dessuz son ***épicicle***. (LA HAYE, *Poëme sur la grande peste*, 1426)[50].
- ... je lui demanderoi volentiers quel fruit ou utillité peut venir à l'omme de sçavoir seullement le cours du ciel et des corps celestes, quant ne à quelle heure et minute et par quelz signes ilz passent, (...) s'ilz sont (...) ***retrogrades*** ou ***stacionnaires***... (SIMON DE PHARES, *Le Recueil des plus celebres astrologues*, c. 1494-1498)[51].

FEW X, 347a: *s.v. retrogradus*: mfr. *rétrograde* 'qui va ou paraît aller contre l'ordre des signes (de corps célestes)' (seit 14. jh.).

FEW XII, 245b: *s.v. stationarius*: 'qui reste au même point, sans avancer ni rétrograder (d'une planète)' (ca. 1350-DG, auch Oresme).

FEW III, 90a: *s.v. directus* n'enregistre pas le sens astronomique.

En analysant ces exemples, on constate que dans la plupart des traités français depuis le XIIIe siècle, les trois mouvements des planètes sont

47. *Pèlerin de Prusse...*, p. 52.
48. Nicole Oresme, *Le Livre du ciel...*, p. 412.
49. Evrart de Conty, *Le Livre des Eschez...*, p. 110.
50. Olivier de La Haye, *Poëme sur la grande Peste de 1348*, publié d'après le ms. de la Bibliothèque du Palais de Saint-Pierre par G. Guigue (Lyon, 1888), p. 224.
51. Simon de Phares, *Le Recueil des plus celebres astrologues...*, f. 4v.

désignés par les termes *retrograde, stationnaire* et *direct*, ou, éventuellement pour les deux derniers, par un lexème de la même famille morphologique (comme *directif, en estat*). En revanche, le traducteur de la *Somme* introduit *reculant* suivi d'une paraphrase et, en s'inspirant du terme latin, il emploie *progressif* à la place de *direct* tout en l'explicitant par 'en allant directement'; seul *stationnaire* correspond au vocabulaire technique déjà largement utilisé dans les écrits astronomiques.

Les occurrences relevées dans *l'Introductoire d'astronomie* antédatent les attestations du FEW; par ailleurs, ni Tobler-Lommatzsch, ni Godefroy, ni le FEW n'attestent le sens technique de *direct* en astronomie.

4.8. Aequinoctialis

Compendium	**Somme**
	Cercle/zone equinoxiale
Circulus quidam est in sphaera intersecans ***aequinoctiale***, qui intersecatur ab eodem in duas partes aequales, et una medietas declinat versus Septentrionem, alia versus Austrum: et dicitur ille circulus Zodiacus ... (90b).	... et cellui mesmes cercle [l'écliptique] (...) est entrecoppé du **cercle** *ou de la* ***zone equinoxiale*** *qui fait les jours et nuys equaulz* et le coppe et fend en deux parties equales et pareilles dont une partie decline et va en descendant envers Septentrion *qui est du senestre costé du firmament*, et l'aultre moitié vers midi *dont vient un vent qui s'appelle auster duquel se nomme ladicte partie de midi et dist on la partie australe qui est du dextre costé du firmament ou du ciel*, et est apellé le susdit cercle zodiacus, *zodiac et est selon l'interpretation ainsi appellé a zoe grec qui en latin vault a dire vie* (80).

D'autres équivalents français avant 1500:

- ... li quels cercle devise la chaude zone en .ij. moitiez, et l'apelerent ***equinoctial*** por la raison de l'iveleté des jorz et des nuiz (*Introductoire d'astronomie* c. 1270)[52].
- Et ce apert clerement, car il descript mendre cercle en 1 jour naturel quant il est es solstices que quant il est en l'***equinoctial***, c'est a savoir ou plus lonc jour ou ou plus court, que quant la nuit et

52. Dörr, *Der älteste Astronomietraktat...*, VII.6.

le jour sont equalz (ORESME, *Le Livre du ciel et du monde*, c. 1377)[53].

- … lors descript le solail par son mouvement sur terre un cercle qui est appellés ***equinoctial*** pour le equalité (…) et en cestui tamps aussi est il aussi comme equalité de esté et d'yver (EVRART DE CONTY, *Problèmes d'Aristote*, c. 1380)[54].
- Item il est assavoir que l'***equinocial*** trenche en deux moitiez le zodiaque au commencement d'Aries et de Libra, et la moytié decline vers septentrion et est appellée septentrionnelle (…) et l'autre moitié est appellée meridionnelle (FUSORIS, *Astrolabe*, c. 1407-1412)[55].
- Aries commence ou l'***equinoctial*** croise le zodiaque et quant le soleil y est commence decliner cestadire approcher de septentrion et vers nous et se extend vers orient (*Le Compost et kalendrier des bergiers*, C., 1493)[56].

FEW XXIV, 219a *s.v. aequinoctium*: *équinoctial* adj. 'relatif à l'équinoxe' (1379); *cercle équinoxial* (Rich 1680-Ac 1932).

Pas d'attestation de *zone equinoxiale*.

À la différence des autres exemples cités, le traducteur de la *Somme* emploie le terme *equinoctial* adjectivement, suivi, comme toujours, d'une amplification du texte. Par la suite, il dévie sur une explication de l'écliptique. Il ne la nomme pas mais il l'introduit par 'Il y a ung cercle en la spere, c'est a dire en la reondeur de la demeure ou le soleil est situé et tient sa mansion et son estage…', citant au passage un vent qui vient du sud, pour revenir à l'écliptique qui se confond maintenant avec le zodiaque.

4.9. Interstitium

Compendium	**Somme**
Sphaera ignis tria habet ***interstitia***, scilicet supremum, quod vocatur igneum, et medium, ac infimum, quod vocatur olympium. Similiter aer tria habet interstitia, scilicet supremum, quod vocatur aethereum … (90a).	Le spere ou le cercle du feu a trois ***interstices***, *c'est a dire trois* ***stations*** *ou* ***entervalles*** *ou* ***divisions***: le souverain qui est apellé ardant tout en feu, le moyen et le bas qui s'apelle olimpicum, *c'est a dire ciel cler et reluisant*. L'air aussi a trois ***interstices***: le souverain qui s'appelle ethereum, *c'est a dire la souveraine partie de l'air fervent et chault* (78).

53. Nicole Oresme, *Le Livre du ciel…*, p. 412.
54. XXVI, 25, ms. Paris, BNF, fr. 564, f. 218v.
55. Jean Fusoris, *La Composicion de l'astralabe*, in: É. Poulle, *Un Constructeur d'instruments astronomiques au XVe siècle: Jean Fusoris* (Paris, 1963), p. 95-124, ici p. 113.
56. *Le Compost…*, fol. 60v.

FEW IV, 763a: *interstitium*: *intertisse de temps* (1495) 'intervalle de temps'; *interstice* 'région (de l'air)' (1528 Oronce, *Theorique des ciels*).

Remarque: Aucun autre exemple d'*interstice* n'a été trouvé dans la base textuelle Frantext (218 textes de 1330-1500), ni dans la base des 13 lexiques du moyen français[57].

La traduction française d'*interstitium* suscite plusieurs remarques: l'auteur transpose le terme latin en français mais, craignant ne pas être compris de son public, il le fait suivre d'une accumulation d'éléments synonymiques exprimant l'idée de 'partie d'un ensemble': 'c'est-à-dire trois *stations* ou *entervalles* ou *divisions'*. *Interstisse* désigne bien ici, comme dans le texte latin et comme dans l'attestation française citée par le FEW, une région du ciel, en entendant par cette expression un espace d'une certaine largeur et non pas la séparation entre deux espaces. L'énumération des trois interstices confirme cette interprétation.

L'apparition du mot dans la *Somme* antédate l'attestation du FEW (1528), dont l'emploi se situe exactement dans le même champ sémantique. L'encyclopédie *De proprietatibus rerum* (1230-1242) de Bartholomaeus Anglicus, œuvre largement diffusée dans toute l'Europe, emploie également *interstitium* au livre VIII, tout au début du chap. V:

Bartholomaeus:	Corbechon:
Aether graece, splendor dicitur latine. Aether secundum Isidorum, est superior regio supremi *interstitii* aeris & ignis, ubi est perpetuus splendor luminis siue lucis[58].	Ether en grec est a dire splendeur ou clarté en latin. Et est selon Ysidore la plus haulte region de l'air ou est la splendeur et la clarté du feu perpetuellement sans jamais faillir (éd. 1485-1486).

Corbechon a omis *interstitium* dans sa traduction, mais d'autres auteurs ont pu reprendre ce terme. Par la suite, avec le changement de la vision de l'univers, cette acception du mot disparaît. On peut supposer que la notion d'*interstice* présente dans les traités latins de cosmologie a pu aussi faire son chemin dans les textes français; le fait que nous ne disposions que de ces deux exemples est probablement dû au manque d'éditions modernes de traités techniques du moyen âge.

57. CNRS ATILF/Université Nancy 2: Base de Lexiques du Moyen Français (BLMF/DMF1): http://www.atilf.fr/blmf/.

58. Bartholomaeus Anglicus, *De rerum proprietatibus* (Frankfurt, 1601; réimpr. Frankfurt, 1964), p. 380.

4.10. Repercussio

Compendium	**Somme**
(... et in supremo [interstitio] est calor et lux propter solis propinquitatem.) Similiter est et in infimo, scilicet propter radiorum ***repercussionem*** a terra.	La plus basse partie de l'air est clere et chaulde et ce est par la ***reverberation*** ou ***reflexion*** des rays du soleil *lesquelz, quant ilz touchent la terre et ne le pevent percier ou penetrer pour sa grosseur, ils se retournent et en reployant remontent. Et ainsi la chaleur se double pour la* ***reflexion*** *et* ***doubleté*** *du ray.*
In medio autem interstitio, ad quod non potest pertingere ***repercussio*** radiorum, est frigiditas et obscuritas, et ibi dicuntur habitare daemones ... (90a).	Mais en la stacion moyenne de l'air est froidure et obscureté, pour ce que la ***reflexion*** et ***reploiement*** des dits rays [du soleil] ne puet parvenir jusques a icelle partie et lieu ouquel on dist que les diables habitent ... (78).

D'autres attestations avant 1500:

- Et selon ce que recite Averroïz, un appellé Avennartha fist un especial tractié a monstrer que la lumiere que la lune a du soleil n'est pas par fraction ou par ***reflexion*** (ORESME, *Le Livre du ciel et du monde* c. 1377)[59].
- Six attestations de *reflexion* dans ce sens trouvées chez Oresme: reflexion ou refraction (de la lumiere).
- La substance de l'arc est engendree de la refulgence et de la ***reflexion*** des rays du souleil en une nue plaine de pluye et de rousee (CORBECHON, *Des propriétés des choses*, 1372)[60].
- ... aucunes radiacions et sintilles des estoilles bailloient ***reverberacion*** à sa face, et, pour ceste merveille, lui fut, après sa mort, eslevé sepulcre merveilleux et lui donnerent ce nom Zoroastes (SIMON DE PHARES, *Le Recueil des plus celebres astrologues*, c. 1494-1498)[61].

repercussion: FEW X, 265b: – 1314 Henri de Mondeville 'Action des humeurs qui refluent au-dedans du corps; action des médicaments répercussifs' – 1348 réflexion, renvoi (de la chaleur, des sons, etc.) – dep. Bersuire. Dans le DMF[62]: – c. 1380 Evrart de Conty, *Problèmes d'Aristote* 'Choc

59. Nicole Oresme, *Le Livre du ciel...*, p. 456.
60. Jean Corbechon, *Des propriétés...*, XI, 5, 186v.
61. Simon de Phares, *Le Recueil des plus celebres astrologues...*, f. 15v.
62. http://www.atilf.fr/dmf.

en retour, réflexion (de la vue)'. – 1400 Evrart de Conty, *Eschez amoureureux moralisés* ' Réflexion, renvoi des sons'.

repercuter: FEW X, 265a: dep. c. 1370 Oresme 'réfléchir, renvoyer (la chaleur, les sons, etc.)'.

doubleté: FEW III, 186 '(taffetas) dont le tissu offre une fleur de deux couleurs'.

reflexion: FEW X, 191b Mfr.frm. f. 'phénomène par lequel un corps quelconque est renvoyé sur sa trajectoire' (1380-Trév 1771) (...) mfr. *refliction* 'phénomène par lequel la lumière est renvoyée par un obstacle' (14e s.) mfr. frm. *réflexion* (dp. 1530, Palsgr 261) (...) 'phénomène par lequel un rayon calorique est renvoyé' (dp Besch 1845).

reploiement: FEW IX, 71, *plicare* (ca. 1190).

reverberation: FEW X, 354a, '... reflexion et diffusion de la lumière et de la chaleur' (seit HMond) [l'exemple est *reverberation du venin*] – *feu de reverberation* 'feu de reverbère' (c. 1500).

Les deux emplois de ***repercussio*** du texte-source sont abandonnés en faveur d'une série de réduplications synonymiques: ***reverberation*** ou ***reflexion, reflexion*** *et* ***doubleté, reflexion*** *et* ***reploiement***, le tout encore appuyé par les verbes *retourner, reploier*, *remonter* et *se doubler*. Pourtant, depuis au moins Pierre Bersuire, puis Nicole Oresme et Evrart de Conty, des équivalents français dérivés de *repercutere* sont en usage dans ce domaine. Néanmoins, le traducteur de la *Somme* ne cède pas à la facilité de l'emprunt et choisit trois fois *reflexion,* terme qui s'impose par la suite et qui est également bien attesté dans ce sens à la fois chez Corbechon et chez Oresme, bien que ces auteurs ne soient pas mentionnés dans les exemples relevés par le FEW X, 191 s.v. *reflexio*.

4.11. Albus (?)

Compendium	Somme
	A sub montant / A sub descendant
Albus duplex est, ascendens scilicet, et descendens. Ascendens fit ex partibus vaporis inflammati in supremo areis [sic], et videtur ascendere, sicut si evolarent scintillae de fornace. ***Albus*** vero descendens est ignis retentus in nube, qui tamen expellitur obviante	Une aultre emprainture est qui se dist ***a sub***, mais il est de deux manieres. L'un qui monte et l'aultre qui descend. ***A sub montant*** est quant les vapeurs et fumees sont enflammees et alumees en la plus haulte region de l'air et montent en approcant la spere du feu et est a veoir

nube frigida, et cadit continue, propter quod videtur quod stella cadens sit longa, quia relinquit vestigia casus post se (93).

comme se estincelles volassent d'une fournaise. ***A sub descendant*** est feu enclos en la nuee lequel feu est bouté hors la nuee par l'approcement d'aucune nuee froide et adont ist par force et choit continuelement en descendant. Par ce semble une estoille cheant et atraine avec lui une longue queue en relenquant aprez lui ung long train (88).

D'autres attestations avant 1500:

- La quarte impression de l'air est de Aristote appellee ***asub*** et est double, l'une est montant qui est engendres en la plus haulte /184r/ partie de l'air de deliees fumees chaudes et seiches en la haulte partie de l'air et qui ont grant longueur et autant de large et sont enflambees et alumees par le feu et par le mouvement du ciel et va en montant pour cause de la subtilleté de sa matere et pour la force de sa chaleur. L'autre est de Aristote appellee ***assub*** le dessendant et est engendres de fumees grosses, chaudes et seiches en la souveraine partie de la moyenne region de l'air et a autant de long comme de large et est enflambee par le feu et par le mouvement du firmament, descent aval pour le froit qui est autour et pour la groisseur de sa mateire. Ces impressions sont celles que le pueple dit que ce sont estoilles qui chieent du ciel ou qui y montent (CORBECHON, *Des propriétés des choses*, XI,2)[63].
- Ce ciel aerain et les estoylles qui sont appelees ***asus***, qui ont similitude d'estoy(i)lles: et sont dictes cheoir du ciel selon la commune opinion quant ***asus*** descendant [sic]. (...) Ou les estoylles sont dictes cheoir pour ce qu'ilz mectront de hors flambez de feu ainsi comme crins (BATALLIER, *Légende dorée*, 1476)[64].

Selon Ducos *assuub* est la transcription du pluriel de *as-sihab* qui veut dire 'flamme' et est utilisé métaphoriquement dans le sens 'étoile filante'[65];

Le texte latin présente une forme erronée *albus* pour *assuub*[66]; le copiste n'a peut-être pas compris le mot arabe et l'a remplacé par une graphie proche existant dans le lexique latin de l'astronomie. En latin classique,

63. Jean Corbechon, *Des propriétés...*, XI.2. Je remercie Joëlle Ducos de m'avoir communiqué cette attestation.

64. Jean Batallier, *La Légende dorée*. Édition critique, dans la révision de 1476 par Jean Batallier, d'après la traduction de Jean de Vignay (1333-1348) de la *Legenda aurea* [de] Jacques de Voragine (c. 1261-1266), éd. B. Dunn-Lardeau (Paris, 1997), p. 99.

65. Cf. Ducos, *La météorologie...*, p. 49 n. 97 et p. 430 (Vocabulaire aristotélicien).

66. Je remercie Jean-Loup Ringenbach d'avoir vérifié la graphie de ce mot dans le dictionnaire de H. Wehr.

Albus peut désigner, soit le Soleil, soit Vénus, soit encore des étoiles brillantes[67]. Le traducteur de la *Somme* disposait probablement d'une version correcte; cf. la citation de la description de ce phénomène tirée de la *versio vetus* de Gérard de Crémone: 'si sunt partes illius vaporis inflammati continue ad invicem, prolongatur et est ex eo *assuub*'[68].

Ce mot n'a pas d'entrée dans le tome XIX du FEW.

4.12. Corona

Compendium	**Somme**
	Couronne
Corona est albus circulus, qui quandoque apparet circa Solem, vel Lunam, aut stellas alias; et causatur ex reliquiis nubium per planetarum radios attractis et illuminatis (93a).	Ung aultre impression ou empressement est en l'air de dessus qui s'apelle ***corone*** et est ung cercle lequel appert aucune fois contre le soleil ou contre la lune ou contre les aultres estoilles et est cause des reliques et residus des nuees lesquelz residus sont attrais par les rays des planettes et illuminees par iceulz (88).

Autre attestation avant 1500:

> et ces autres empraintures qui appairent en figure de cercle pour l'especeté de l'air, c'est a savoir l'arc ou ciel et la ***couroune*** du soleil et de la lune (MAHIEU LE VILAIN, c. 1280 [ms. 3e quart XIVe siècle])[69].

Le FEW II-2, 1210a *s.v. corona* atteste *couronne* en 1600 dans le sens 'halo du soleil' (chez Olivier de Serres); l'attestation suivante est de 1690, date de l'entrée du mot dans le dictionnaire de Furetière. Par conséquent, avec l'exemple relevé chez Mahieu le Vilain, nous disposons de deux attestations antérieures.

5. Conclusion

Par cette analyse de quelques lexèmes extraits du livre II de la *Somme* et leur comparaison avec les dénominations du même champ conceptuel

67. Cf. Le Boeuffle, *Astronomie Astrologie...*, p. 37.
68. Cité d'après Ducos, *La météorologie...*, p. 49 n. 97.
69. Mahieu Le Vilain, *Les Metheores...*, p. 4.

relevées dans des textes antérieurs ou contemporains, nous pouvons constater l'émergence progressive d'un lexique spécifique de la cosmologie durant les XIVe et XVe siècles. La confrontation des commentaires explicatifs permet en même temps d'éclairer le processus d'adaptation et de vulgarisation de la terminologie latine en français et de mettre en évidence le rôle de la langue vernaculaire dans la transmission du savoir. Leur examen montre une certaine conception somme toute assez cohérente de la glose en fonction de la nature du terme choisi en français. Voici quelques principes qui semblent avoir guidé notre traducteur:

a) Lorsqu'il maintient la forme du document-source dans le texte français, l'initiation à la terminologie se fait souvent en trois étapes par approches successives: le mot – le plus souvent latin – est d'abord paraphrasé par une forme savante, puis repris par un mot ou une expression de la langue courante. Exemples:
   primum mobile / c'est a dire *le premier corps mobile qui se moeut / premier meuble.*
   galaxia / voie lactie / voie paree.
   ...et est apellé le susdit cercle *zodiacus*, *zodiac* et est selon l'interpretation ainsi appellé a *zoe* grec qui en latin vault a dire *vie* (procédé analogue chez Evrart de Conty: 'le cercle que on appelle le *zodiaque* ou le *cercle oblique* ou le *cercle des signes*').
   (exception faite de *assuub* qui n'a pas d'équivalent en français et pour lequel le traducteur n'indique pas non plus l'origine comme il le fait pour *zodiac*, cf. ci-dessus).
b) Lorsqu'il choisit un terme emprunté au latin (*oblique, progressif, stationnaire, equinoctial, interstice*), il emploie la paraphrase et/ou la réduplication ou accumulation synonymique. Exemples:
   progressives alant directement devant elles.
   stacionaires estans sans proceder.
   interstices, c'est a dire trois *stations* ou *entervalles* ou *divisions.*
c) Lorsqu'il remplace le mot latin par un dérivé d'un étymon différent (par exemple *repercussio* traduit par *reflexion*), il met en œuvre les mêmes procédés que ceux cités sous b).
d) Lorsqu'il traduit le mot latin, il se dispense de commentaire.
   motus rationalis – mouvement raisonnable.
   ascensus – monture.
   descensus – descencion.

Dans ces gloses, les procédés stylistiques mis en œuvre sont les mêmes que chez la plupart des traducteurs de l'époque:

– reprise du concept par un autre mot du même radical:
le corps du monde et la fabrique du monde corporel (cf. 4.1)
– introduction par une relative:
primum mobile, c'est a dire le premier corps mobile qui se moeut (cf. 4.2)
– apposition:
reculantes alant en reculant ou de travers (cf. 4.7)
***stacionaires estans** sans proceder* (cf. 4.7)
– réduplications synonymiques ou accumulations:
*cercle **oblique**, **crom** et **traversal*** (cf. 4.6)
interstices**, c'est a dire trois **stations** ou **entervalles** ou **divisions (cf. 4.9)
etc.

Quant au choix des lexèmes, dans de nombreux cas, l'auteur de la version française écarte l'emprunt au latin sous sa forme francisée, comme par exemple, *machine, rétrograde, rétrogradation, répercussion* – mots pourtant attestés dans des textes contemporains – pour choisir un terme que l'on peut qualifier d'emploi rare ou isolé dans ce domaine, comme *fabrique*, *reculant* et *reculement* (au lieu de *retrograde* et *retrogradation*), *doubleté* et *reploiement* (qui ne sont pas des termes techniques dans la série synonymique qui traduit *repercussio*); mais, par ailleurs, on trouve *reverberation* et *reflexion* qui, eux, s'implanteront dans la langue. Les commentaires et les amplifications, particulièrement abondants dans cette partie de la *Somme*, montrent que le traducteur maîtrise parfaitement son sujet et qu'il prend plaisir à présenter cette science à un public laïc, cultivé certes, mais non spécialiste. On peut donc qualifier cette traduction de vulgarisation scientifique.

J'ai tenté de mettre en évidence l'équivalent choisi en français et les termes utilisés dans les gloses. Quel qu'en soit le degré de technicité, il me semble important d'enregistrer tous les lexèmes rencontrés aussi bien dans les traités et les encyclopédies que dans les textes littéraires afin de dresser l'inventaire des dénominations de chaque concept; c'est en confrontant un maximum d'attestations et en établissant des statistiques sur leurs occurrences que nous pourrons au mieux retracer la naissance d'une terminologie.

ATILF CNRS & Nancy-Université

Tony HUNT

THE OLD FRENCH TRANSLATION OF THE 'FOUR MASTERS GLOSS' IN WELLCOME MS. 546

Editors of scientific texts face a particular problem when the agreed principles of textual criticism yield a result which knowledge of the natural sciences forbids us from accepting. What are they to do? Declare that their author was not in all respects a competent scientist and retain his scientific errors? Or should they correct his text, introducing, as tactfully as possible, what their author ought to have said (conceivably what he in fact intended to say) into their edition? The historicism involved in valuing the study of what people actually read (sometimes nonsense) and believed, as revealed by individual manuscripts (none of which is ever flawless), is in such instances understandably subordinated to what it would have been better (conceivably right), as judged by a variety of criteria, for them to have read and believed. The deficiency may lie with the author, possessing an imperfect knowledge of his subject, or else with a scribe or scribes who traduced his knowledge, or, in the case of translations of scientific writings, with an inefficient translator, and thereafter inevitably with *his* scribe, thus totalling four stages or layers of transmission. The detection of purely scribal error will, of course, be greatly aided by palaeographical skills, that of translator's error by knowledge of the languages of both source and target texts, and of scientific error by historical knowledge of the subject. This is a lot to ask of a single scholar.

The commonest source of error in medical manuscripts relates to dosage: pharmaceutical abbreviations of weights and measures are subject to misinterpretation; then the *local* value of any identified measure (*uncia, drachma, scrupulum*) must be determined, and the roman numerals accompanying that measure carefully checked (here divergence of the manuscripts is the norm). More generally, two broader expressions of quantity are perilously confusable: *nimis* and *minus*, only a minim different. Palaeographical error lies behind a lot of easily confusable items in lists of materia medica: *salvia* / *saliva*, *menta* / *memitha*, *mirra* / *mirta*, *sine timore* / *sine tumore*. The letters *ro*. after a name may be for *rotunda* (e.g. *aristologia ro.*) or for *rosa* as a separate item (*aristologia, rosa*). Names may be connected or separated by conjunctions *et*, *vel*, *ut*, which can be easily confused by scribes. But in theory, at least, science, palaeography and linguistics can converge in recovering a reliable text, even when we are uncertain of its exact identity[1].

Now we move to the linguistic problems. A translation can only be assessed in relation to the (exact) text it purports to be translating. Ideally, therefore, what is needed is a critical edition of the (Latin) source text with an exhaustive critical apparatus so that the translation can be controlled at any given point against the readings of all the MSS of the source text and an affiliation with some of them established. Naturally, few editors produce a scientific text with this need in mind, not least because it consumes a great deal of space and there is the danger that readers may not see the wood for the trees. Instead, most editors seek to recover what the author wrote and not what haphazardly circulated under his name. The only practical remedy is usually for the would-be editor of a translation to procure microfilms of as many of the source MSS as possible. Many translations contain readings which do not reflect those in a critical edition of the source but all of which can be explained by variant readings in individual manuscripts[2]. The very idea of a critical edition of a medical treatise may be regarded as unattainable, the reconstitution of a definitive text (whatever the meaning given to the term) illusory, since medical texts were so often compilations which in their various accretions, interpolations, omissions and revisions display something of what literary scholars call 'mouvance', but which in many texts is merely an other name for adaptability. The task of establishing textual identity[3], textual boundaries, and textual accuracy can be daunting. Certainly in the case of theoretical discussions based on easily consultable authorities like Hippocrates, Galen and the Arabic physicians, the correct content is probably recoverable. But in the case of therapeutic receipts which fill the pages of almost all medieval medical treatises, who can

1. See Hunt T., 'An Anglo-Norman Medical Treatise', in: P.E. Bennett & G.A. Runnalls (eds.), *The Editor and the Text* (Edinburgh, 1990), p. 145-64 and 'Old French Translations of Medical Texts', *Forum for Modern Language Studies* 34 (1999), p. 350-57.

2. For the importance of MS London, B.L., Sloane 1124 for the Anglo-Norman translation of Platearius's *Practica brevis* see Hunt T., *Anglo-Norman Medicine* 1 (Cambridge, 1994), p. 162. In the 'Four Masters' Gloss', f. 55ra 'tenir pallement a fame qui a son tens', 'pallement' (= parlement) seems to be a euphemism for sexual intercourse. The Latin variants are *colloquio (mulieris menstruose)* and *collegio*.

3. See the excerpt I printed from Oxford, Bodl. Libr., Auct. F.5.31 in 'An Anglo-Norman Medical Treatise' (above n. 1).

claim accurately to discriminate between the various (changing) ingredients and judge their efficacy?

How does the work presented here fit into this framework of desiderata? MS 546 in the Wellcome Library for the History and Understanding of Medicine, London, is a volume of 91 leaves copied by a single scribe writing a French Gothic hand of the first half of the fourteenth century. It contains a number of vernacular medical texts, of which three exemplify different types of problem posed by translations. These three texts are: (i) the only known (but unedited) Old French translation of the 'Four Masters Gloss'; (ii) an Old French translation, accompanied by the Latin text, of approximately one third of Gilles de Corbeil's *De urinis* (ff. 22va-28vb)[4], thus raising the question of the relationship of the vernacular to the Latin text provided; (iii) an abridged Old French version of Jordanus Rufus's *De medicina equorum* (c.1250), rejected as a witness by Brigitte Prévot, editor of one of the Old French translations[5], largely on the grounds that whilst sometimes being in accord with the oldest Latin manuscripts, it reshapes the material in such a way as to constitute a *remaniement*. Here, of course, we must distinguish the characteristics of a translation at the macro-level (arrangement and coverage of material) and micro-level (accuracy of individual readings).

The Wellcome MS 546 transmits an Old French translation, of c.1300, of what is commonly known as the 'Four Masters Gloss', itself something of an enigma, the end result of an almost indeterminable continuity of copying, supplementing, glossing and commenting on the work which stands at the head of Western surgery in the Middle Ages, Roger Frugard's *Chirurgia*, a treatise of the twelfth century (c.1170) which is itself subject to much textual disagreement and doubt. Despite attempts to link him with Salerno, which in fact seems to have produced little surgical writing[6], Roger Frugard is really a North Italian associated with

4. See *Aegidii Corboliensis carmina medica*, ed. L. Choulant (Lipsiae, 1826), p. 3-18.

5. See Prévot B., *La science du cheval au moyen âge. Le Traité d'hippiatrie de Jordanus Rufus* (Paris, 1991) and cf. Prévot B. & Ribémont B., *Le cheval en France au moyen age* (Orléans, 1994). Prévot's thesis (microfiche reproduction, Lille, 1989), provides the Latin source text from BNF lat. 5503 (s. xiv). A Reims manuscript (Bibl. mun. 991 written in Avignon in 1390) contains a Latin text and the Old French translation, which is based on BNF fr. 25431 (an. Gaignières 82) of c.1300.

6. But see the so-called 'Bamberg Surgery' discussed by Corner G.W., 'On early Salernitan surgery and especially the "Bamberg Surgery"', *Bull. Hist. Med.* 5 (1937), p. 1-32 and a student notebook discovered by Henry Sigerist, 'A Salernitan Student's Surgical Notebook', *Bull. Hist. Med.* 14 (1943), p. 505-16. A later (13th century) Salernitan Surgery is recorded in Giacosa P., *Magistri Salernitani nondum editi* (Torino, 1901), p. 380-86. Nothing here stands comparison with the work of Roger.

Parma[7], where he lectured without producing, himself, a surgical treatise. The treatise by which he is known was put together by Guido of Arezzo the Younger, after he had completed his one *Liber Mitis*, acting in collaboration with a number of Roger's pupils and utilising lecture notes which were then revised and edited into a compilation. Guido prefixed to this redaction of Roger's teaching a preface, *Post mundi fabricam...* and probably wrote the prefaces to each of the four books which make up the treatise, as well as bearing responsibility for the organisation of the whole. There is no published inventory of the manuscripts, which run to over thirty, of naturally varying value, so varying that the idea of a critical edition may not be viable. What they generally term the 'Chirurgia magistri Rogerii' was edited by Salvatore de Renzi from the Juntine edition of 1546 with reference also to MS Paris, BNF, lat. 7035[8], whilst Charles Daremberg edited the versified text, much amplified, furnished by a compendium in MS Paris, BNF, lat .8161A[9]. Sudhoff preferred MSS Munich, Bayerische Staatsbibliothek, Clm. 376 (written in southern Italy) and Florence, Bibl. Naz., cod. Magliabecchianus J.10.16 (possibly from Salerno) from the end of the twelfth and the beginning of the thirteenth centuries respectively[10], and produced what seems to be a good text for comparative purposes.

Once the 'Chirurgia Rogerii' became available a Salernitan connection was rapidly made and c.1200 a continuous commentary on Roger was made which is sometimes known as 'the first Salernitan gloss'. Some time later (1230-50) a disciple of Roger's, Roland of Parma – possibly after producing a preliminary set of marginal 'additiones' – worked up his own material into a new text of the *Chirurgia* which has become known as the 'Rolandina' and sometimes appears with the title 'Cyrurgia Rogerii cum additionibus Rolandi'. What is essentially a new and revised edition of Roger was then itself subjected to a welter of glossing and annotation, to form, with the incorporation of the 'First Salernitan Gloss', what is described in the manuscripts as 'The Glosses of the Four Masters on the surgeries of Roger and Roland', a work which has been edited[11], but with the same deficiencies as afflicted

7. See Pazzini A., *Ruggero di Giovanni Frugardo maestro di chirurgia a Parma e l'opera sua* (Roma, 1966).

8. Cf. de Renzi S., *Collectio Salernitana* 2 (Napoli, 1853), p. 425-93. The Paris text lacks most of the additions in the Juntine edition which are printed by de Renzi.

9. In de Renzi S., *Collectio Salernitana* 4 (Napoli, 1856), p. 39-125.

10. In *Beiträge zur Geschichte der Chirurgie im Mittelalter* 2 (Leipzig, 1918), p.156-236.

11. By Charles Daremberg in *Collectio Salernitana* 2 (Napoli, 1853), p. 497-724 and republished as a separatum a year later. References are to the pagination of the latter. See also Lodispoto A., *Brevi glosse dei quattro maestri sulla chirurgia di Ruggero e Rolando* (Roma, 1961).

the editing of Roger – that is, precarious editorial deductions from inadequate knowledge of the manuscripts and their relative status. The Four Masters in question are probably a fiction designed to confer authority and are thought to allude to Archimattheus, Petroncellus, Platearius and Ferrarius. It seems probable that the 'Four Masters Gloss' was produced in France, since a receipt speaks of 'extremitates arboris que gallice dicitur *pisac*' (Daremberg, p.141; omitted from the Old French). There is also an allusion to the touch of the King of France curing scrophula (see Daremberg, p. 101 & 108). We hear of a form of 'anthrax' 'qui [All Souls ms. communiter] dicitur a vulgo bonum malum sive *le bon* [S mal]'(Daremberg, p. 88)[12]. Individual MSS yield confirming evidence: 'Cambridge ms., musculus... in tybiis stricte sumptum in vulgari dicitur 'soriz' (Daremberg p. 53, MS Oxford, Bodl. Libr., Ashmole 1398 p.119; omitted in the Old French)[13].

It is against this confusingly fluid corpus of derivative material, representing the 'Four Masters Gloss', imperfectly known because of the small number of the surviving manuscripts which have been studied, that we are obliged to consider the process of vernacularisation. There is no doubt that the five Old French translations of Roger's *Chirurgia*, of which two have been edited, are based closely on the available Latin texts. What is the situation with the 'Four Master Gloss'? First of all, we have from c. 30 surviving manuscripts, an edition of only one of them, and that one quite unrepresentative. Charles Daremberg 150 years ago discovered a manuscript in the Bibliothèque de la Mazarine which he took to his heart and printed in Savatore de Renzi's *Collectio Salernitana*[14]. Instead of the conventional topical disposition of ailments *a capite ad calcem*, found in the other MSS, the Mazarine copyist-compiler re-jigs the contents of the Gloss on a pathological basis e.g. wounds (all parts of the body), apostumes etc. This makes it extremely inconvenient as a text with which to control the Old French translation. Resort must be had to other MSS. I have therefore used four manuscripts as the basis of comparison with the translation (Oxford, Bodleian Library, Ashmole 1398, e museo 19; All Souls College, 76, and Cambridge, Gonville and Caius College, 105/57). Three of these were known to Daremberg.

The Old French translation itself is part of a confusing network of relations. At a certain point (Daremberg p.155) there is reference to a simple

12. MS Wellcome 546 f. 24ra: 'Puis qu'il nos convient dire d'antrace nos en diron qui est apellee bon mal ou bubete'.

13. Daremberg, p. 53 prints 'in vulgari dicitur sorr3' with a note where he guesses that the word may be 'sorce/sura/ fr. sorice.' Note elsewhere (f. 50vb) Wellcome has 'en sorriz' where Latin has 'in musculis'.

14. See n. 11 above.

and commonplace embrocation introduced as 'secundum Rogerium embroca facta de...'[15] Now in the Old French translation we read, f. 67rb, '[Per]nez ici l'embroque mestre Rogier *Poutrel* por fere porreture'. Rogier Poutrel was the author of, or authority for, a compilation, mainly of medical receipts, given the name *cyrurgia*, which survives in a Picard translation (in MS Vatican, B.A.V., Reg. lat. 1211), possibly by a certain Jean de Prouville (we do not know whether he edited it or translated it or both) c.1300[16]. After an opening which is firmly based on Salernitan teaching, the rest of Rogier's compilation is largely based on Book 2 of Roger Frugard's *Chirurgia* (the 'Rogerina') and the 'Four Masters Gloss'. Around 1300, therefore, we have in the same area the production of a compilation, ostensibly indebted to Rogier Poutrel 'cirurgien et abbé et maistre en phisique et en cirurgie' (ed. Södergård p. 16, f. 5v l.8), and a translation of the Four Masters Gloss which names him.

In presenting the Old French translation I have considered three fundamental questions.

(i) First, coverage and disposition of the text. Does the translation cover the whole of the Latin text? The answer is yes, qualified simply by the observation that there is a tendency, familiar to all medievalists, to abbreviate progressively, so that the work is less thorough towards the end. Not surprisingly the translator is prone to excise theoretical discussions (e.g. of apostumes) and, like so many writers and scribes in the Middle Ages, an opening thoroughness and exactitude gradually give way to briefer treatment and even omission. Nevertheless the translation follows exactly the sequence of chapters in the Gonville and Caius and the All Souls MSS. The Rolandina is divided into four books but chapter divisions in the MSS are notoriously unstable. If we take the most orderly of the Latin MSS of the 'Four Masters Gloss' that I have used, the Cambridge MS, it divides the text into c. 75 sections by a combination of four features: rubrics, incipits (drawn from the Rolandina), the formula 'Consequenter / hic determinat actor'[17], and the use of coloured initials. Any

15. Ashmole p. 92 secundum Rogerinum; E museo f. 20vb secundum Ro.; omitted from All Souls.

16. See *La Chirurgie de l'abbé Poutrel: texte picard de 1300 environ*, ed. O. Södergård (Lund, 1980) and de Tovar Cl., 'A propos de la *Chirurgie* de l'abbé Poutrel', *Romania* 103 (1982), p. 345-362. Note MS London, Lambeth Palace, 456 (s. xv) f. 216v which has 'Cy commence ung extrait de la surgie de l'abbé Poutrel'.

17. Ca. 30 times; variants include 'Hic d.', 'In hac parte', 'In hoc capitulo', 'in diversis capitulis', 'agit actor', 'intendit determinare', 'determinandum est', 'Hic tractat de...', 'Hic sequitur de rubore oculorum', 'Hic breviter de expedit actor'.

one of these may function alone, but the vast majority of divisions are marked by all four[18]. Smaller internal entries are marked by parafs and *Item*.

The Old French translation offers a shorter text but marks slightly more divisions (85) through rubrics, some of which are lost in the text itself, and decorated initials (on two occasions the latter have not been inserted). There are no Latin incipits (the text is not designed for use in conjunction with the original). The Latin original's formula 'Determinat actor de' (cf. f. 62rb 'de ce ne determine pas l'autor') referring to Roger / Roland is always converted (c. 45 cases) to a personalized variant i.e. commentator or translator: '(Or) nos convient dire / Nos diron aprés'[19], 'Ici diron, no deliverron'. References to the source retain the expression 'actor' ('l'autor', 'si comme l'autor le devise'), but impersonal constructions like 'sicut dictum est' are personalized ('sicomme j'ei dit devant', 'sicomme nos avon dit'). Neither the Cambridge Latin MS nor the Old French translation contains indexes or marginal headwords, so that finding one's way around the text must have been extraordinarily difficult. Nevertheless, the translation's coverage is impressive: frequent small omissions are often due to eyeskip ('saut du même au même'), larger ones arise from a desire to simplify detailed classification and theoretical discussion, and a handful of long omissions may well have resulted from missing pages in the exemplar. An interesting detail concerning the Wellcome manuscript is the scribe's absolute 'horror vacui', an apparent horror of the smallest gaps or spaces. Writing in two columns, he often uses a fat minim, sometimes with a cross stroke, to fill the line end even when the space is only a couple of millimetres.

(ii) A second variable feature of translations is the supplementary provision of aids to understanding, for example glosses:

> galbanum in aqua tepida, c'est en eue detrempé; use un letuere qui a non diasené; picis, c'est a dire poiz .i. libre; fabaria c'est orpin; repercussir, c'est chacier arriere; popaver rous – c'est nomee que le papaver est trové entre le forment; particuler, c'est par parties; la circumstance de l'uil, c'est a dire l'avironement de l'uil; pulpam pomi cocti sub cineribus vel sub carbonibus, ge croi que ce soit le cheure de la pome; un bufon, c'est un grapaut; nasturcii, c'est

18. Eight incipits function alone, seven 'determinat actor' formulas, and four cases of rubric only.

19. See also 'Ici nos convient dire', 'En ceste partie nos diron', 'Briement nos deliverron de' (Lat. 'Hic breviter se expedit actor de').

creson; lie de tonele, c'est la gravele; eresypila, c'est feu d'enfer; herpes cingulus, c'est a dire herpes cengle ou herpes araigne; veine du chief qui est organique apelee; le lieu soit lavé de lessive fete de sarment de vigne – tu ardras le sarment et de ceule cendre tu feras ta lessive.

Unique is the addition in the opening section:

il apert que cole porrie nuist non pas en tant come la cole, mes en tant comme il fait fievre, *c'est a savoir de ce que est porrie entré.*

(iii) Then there is lexis. How comprehensively does the translator find vernacular equivalents? The answer is strikingly often. All vernacular *receptaria*, whether embedded in larger compilations or not, to some extent illustrate the phenomenon of code-switching[20]. Even the opening more discursive sections of the French Gloss contain some untranslated Latin words[21].

se puis i vient, si est apelee *ulcus*. *Ulcus* si est feite de perdicion de sustance
se selonc le lonc, si est dite t[r]encheure, se selonc le lé, si est dite *incisio*
si i vient *herpestiomenus*, fistole et crenque si comme les plaies *musculorum*
les plaies fetes en *ysophago* et *trachea* [et] arteres
et note que se sanc vient, *emorosagia*, soient mis pateaus... dom il avient sovent – *emorosagia* – que les chiés des granz arteres sont trenchiés

As most of the code-switching occurs in lists of ingredients in the recipes (Latin 'recipe' is normally used, very rarely 'receif') it may be that many of the Latin terms taken over by the translator were never vernacularized: naturalisation is complex and encysts common, that is words naturalised as to use, but not as to form, inflexion or pronunciation: certainly the switch sites in the French are never flagged (e.g. latine, or gallice in the Latin gloss). What is the difference between an unflagged switch and a loanword it may be asked:

jus de corili (cf. jus de coudre), popaver rous, oile ladaminum, catheractera d'umors reumatiques, jus de filicis, ovec ole lentico, entre cristallinum et le blanc, pren farine ciceris, pren ciceris rubei, ciceris roge, de l'un et de l'autre elleborum, semence de psilii, de solatri,

20. See Hunt T., 'Code-Switching in Medical Texts', in: D.A. Trotter (ed.), *Multilingualism in Later Medieval Britain* (Cambridge, 2000), p. 131-47.
21. Unlike the *Cirurgie Poutrel*, which employs many regionalisms, the 'Four Masters Gloss' uses very few.

emplastre de galbanum arsenic, pren lumbricos terrestres, pren ciminum, entre cristallinum et vitreum, un grain de gallitricon

Latinisms: amistion, branchis, cantarides, cartillage, cateracion, cefalica, corrosion, diaframe, embroquation, encatismum, encision / incision, erugination, esternutation, estorsions, excoriation, formication, linition, mandibule, minution, mundification, muscilagines, oncion, perdicion, repercussir

rare words: ale(i)nement, bubete, bufon, canon, chacie, conchie, defaillable, enbarrement, espatule / espatuel, fabarie, lermier, medecinement, melangie, moistiseté, moistisor, mordicacion, oreilun, pincescarioles, quartier (flap of skin'), roisne, spa(s)me, vergelure, visc de quesne

encysts: ameos, baccas lauri, boli, epithimi, sandali albi et rubei, storax calamite, tapsi barbasti, tinnitum, viride es (cf. vert de griz)

Latin forms on their own: calcucemenon, lapidis caucis, lignum coruli, melonis, mirobolani conditi, uve passe ab arillis mundate

Errors are few. In the summary of the Gloss's contents Guido of Arezzo appears as 'Gui de Reins' (f. 51rb).

The Old French translation in Wellcome 546 is essentially a rationally written compendium of therapeutic medicine, with some elementary prognostic indications, quite without the popular use of charms and incantations which characterize many medical treatises[22].

University of Oxford St Peter's College

22. Cf. Joret Ch., 'Les incantations botaniques', *Romania* 17 (1888), p. 337-54. As for the reference to the 'King's touch' and taking the patient suffering from scrofula to a fountain on St John's Eve, these are side-by-side in the Latin and simply reproduced.

Sara MARRUNCHEDDU

LA TRADUCTION FRANÇAISE DU *MOAMIN* DANS SES RAPPORTS AVEC LA VERSION LATINE DE THÉODORE D'ANTIOCHE

Le traité de fauconnerie connu comme *Moamin* nous offre la possibilité de faire quelques réflexions à propos de l'importance de la figure du traducteur et des buts d'une traduction, qui nous donnent des textes différents entre eux, bien qu'ils partent de la même source.

La traduction du *Moamin* en latin pour Frédéric II est achevée par Théodore d'Antioche en 1241 et l'empereur la corrige personnellement pendant le siège de Faenza. On peut donc observer qu'il s'agit d'une traduction qui naît dans le milieu de la cour, commandée par un souverain, et que le commanditaire connaissait si bien la matière traitée qu'il est à même de réviser l'œuvre du traducteur[1].

La situation se répète dans la phase suivante, le passage du latin au français, comme nous en témoigne le prologue de la version française, dont nous connaissons deux manuscrits: l'un qui se trouve à Venise, et qui a été édité par Hakan Tjerneld[2], l'autre conservé à Bruxelles, inédit,

1. Sur le rôle de Théodore d'Antioche, voir Burnett C., 'Master Theodore, Frederick's II Philosopher', in: *Federico II e le nuove culture. Atti del 31° Convegno storico internazionale: Todi, 9-12 ottobre 1994* (Spoleto, 1995), p. 225-285; Burnett C., 'Antioch as a link between Arabic and Latin culture in the Twelfth and Thirteenth Century', in: I. Draelants, A. Tihon & B. Van den Abeele (éds), *Occident et Proche-Orient: Contacts scientifiques au temps des croisades. Actes du colloque de Louvain-la-Neuve, 24 et 25 mars 1997* (Turnhout, 2000), p. 1-19; Kedar B.Z. & Kohlberg E., 'The intercultural career of Theodore of Antioch', *The Mediterranean Historical Review* 10 (1995), p. 164-76.

Sur Frédéric II et son traité de fauconnerie, voir l'introduction à Frédéric II, *De arte venandi cum avibus*. Texte établi, traduit et commenté, éd. A.L. Trombetti Budriesi (Roma - Bari, 2000); Van den Abeele B., 'Il *De arte venandi cum avibus* e i trattati latini di falconeria', in: A. Paravicini Bagliani & P. Toubert (éds), *Federico II e le scienze* (Palermo, 1994), p. 395-409; Cardini F., 'Federico II e il *De arte venandi cum avibus*', in: *Politica e cultura nell'Italia di Federico II* (Pisa, 1986), p. 213-232; Haskins Ch.H., 'The *De arte venandi cum avibus* of the Emperor Frederick II', *English Historical Review* 36 (1921), p. 334-355.

2. Tjerneld H., *Moamin et Ghatrif. Traités de fauconnerie et des chiens de chasse* (Stockholm - Paris, 1945). Le manuscrit à la base de l'édition porte le code *a* (Venise, Bibl. di San Marco, Str. App. 14 (= 279).

dont nous nous occupons[3]. On y lit le nom du commanditaire royal, Henri de Sardaigne[4], et on peut observer que, même dans ce cas, celui-ci est à même de réviser le texte et il est aussi sollicité par le traducteur Daniel de Lau[5] (ou Daniel de Cremone) afin qu'il corrige l'œuvre:

> E se je mespreign en aucune chose, je li pri humblemant com a mon lige segnor q'il le me perdoing par sa franchise, car ce nen avendra par foi ne pas por estude qe je n'i mete enterinemant, ainz avendra por la povre sciençe de moi, e q'il li mende e corrie en toç les leuz ou il en avra mestiers, com cil qi bien est tant sages e tant porvoianz en totes reisons et œuvres qi a prince, estreit de tant noble sanc com il est, apertiegnent, q'il le savra bien fere e raisnablement sanz nulle defaute.

Il s'agit donc d'un usage de la cour qui est à l'origine de ces traductions, même si on traduit pour quelqu'un qui connaît déjà la matière et qui peut contribuer à la version définitive de la traduction. Mais, contrairement à la version latine et aux italiennes[6], dans le *Moamin* français le traducteur parle de lui-même et de son œuvre, avec la conscience de l'importance de son travail. À ce propos, la comparaison avec les autres prologues connus est significative, à savoir le prologue latin et celui des versions italiennes, grâce auxquels émergent les attitudes des traducteurs.

Le prologue latin[7] est un éloge de la chasse, vue comme l'activité la plus indiquée pour un souverain pendant son *otium*. Théodore d'Antioche ne parle ni de lui-même ni du commanditaire, si ce n'est pour la référence à l'excellence spirituelle du roi qui se dédie à l'art de la chasse, tel Frédéric II. Il parle de lui-même seulement dans ce passage:

3. Bruxelles, Bibliothèque Royale, IV 1208. Le manuscrit sera indiqué par la suite comme *k*. La division du texte fait référence à la thèse doctorale inédite de Marruncheddu, S. *La versione franco-italiana di un trattato di falconeria : il* Moamin *di Daniele di Lodi nel confronto con la tradizione latina e i volgarizzamenti italiani*, Università di Torino, 2008.

4. Sur le roi Henri de Sardaigne et son goût pour la fauconnerie, voir Boscolo A., *La figura di re Enzo* (Sassari, 1950); Frati L., *La prigionia del re Enzo a Bologna, con appendice di documenti* (Bologna, 1902); Frati C., 'Re Enzo e un'antica versione francese di due trattati di falconeria', in: *Miscellanea tassoniana di studi storici e letterari* (Bologna - Modena, 1908), p. 61-81.

5. On a peu d'informations sur Daniel de Lau, mais, désormais, personne ne doute de son existence historique. Les traits linguistiques du texte, en outre, nous ramènent avec certitude au lombard-vénitien. Sur Daniel de Lau, voir Tjerneld, *Moamin et Ghatrif. Traités*..., p. 28-30; Roncaglia A., 'La letteratura franco-veneta', in: *Storia della letteratura italiana, Il Trecento*, II (Milano, 1965), p. 743-782.

6. Le *Moamin* a connu deux traductions italiennes: la version toscane (XIVe siècle) et la napolitaine de Iammarco Cinico (XVe siècle). Les manuscrits *c* (Venise, Bibl. di San Marco, It. III, 22 (= 5049), témoin de la première version, et *b* (Florence, Bibl. Laurenziana, Ashb. 1249), témoin de la deuxième, ont été édités par Glessgen M.D., *Die Falkenheilkunde des Moamin im Spiegel ihrer Volgarizzamenti: Studien zur Romania Arabica* (Tübingen, 1996).

7. Glessgen, *Die Falkenheilkunde des Moamin*..., p. 335-336. Nous utilisons pour la comparaison le manuscrit *I* (Vaticano, B.A.V., Reg. Lat. 1617) selon cette édition.

> Hec considerans Moamyn falconarius composuit hunc librum in arabicum de venatione et divisit in quattuor tractatus quem magister Theodorus philosophus mandato Cesaris transtulit in latinum.

La version napolitaine[8] présente la *dedicatio*, dans laquelle le traducteur loue son seigneur et formule le vœu d'obtenir sa bienveillance, mais il parle en termes génériques, selon la tradition. Il est intéressant de remarquer, ici, pourquoi l'œuvre a été traduite:

> [...] ho voluto imponere fine a la tradutione del dignissimo Moamyn arabico falconario, per monstrare che, avenga non mi dilecti di cacçia, cosa certamente regia, pure me ho sforçato,
>
> traducendo de latino in vulgare dicto autore, prestare utilitate a vostri sacri et rapaci aucelli, quale legendo potrà Vostra Maestà con doctrina et cura del nobile Moamyn cognoscere le generationi de tucti ucelli de rapina, et signali de loro bontade et defecti et governarli et preservarli da morbi et e quelli curarli et farli apti et valenti al gratissimo et curioso volato, per lo quale se fanno segnori de l'altissimo et inmenso aere, dal quale precipitano ad guisa de fulgure ogni altro ucello in terra[9].

C'est une motivation que Théodore n'aurait jamais pu avoir, parce que Frédéric II était déjà expert en l'art de la fauconnerie; il faut donc l'attribuer au traducteur.

La version toscane, enfin, n'a pas de véritable prologue, mais un bref incipit qui ne contient ni le nom du commanditaire ni celui du traducteur[10]:

> Al nome di Dio amen. Comincia il libro degli uccelli di preda e degli cani ydonei a chacciare e delle loro proprietà e costumi e delle loro malattie colle cure che convengono, il quale fu di Coramomellino, re di Cartagine, translatato d'arabico in latino per maestro Theodoro, medico dello 'mperadore Federigo[11].

8. Glessgen, *Die Falkenheilkunde des Moamin*..., p. 87.

9. '(...) j'ai voulu terminer la traduction du très digne Moamyn fauconnier arabe, pour montrer que, même si je ne pratique pas la chasse, chose qui appartient aux rois, je me suis efforcé, en traduisant cet auteur du latin en langue vulgaire, d'être utile à vos oiseaux sacres et rapaces; Votre Majesté, en le lisant avec attention et doctrine, pourra connaître du noble Moamyn les générations de tous les oiseaux de rapine, les marques de leur bonté ou de leurs défauts et pourra les entretenir, les garder des maladies, les soigner, les rendre doués et habiles au vol, pour lequel ils deviennent les seigneurs du très haut et immense air, d'où ils font tomber tous les autres oiseaux en terre comme la foudre.'

10. Glessgen, *Die Falkenheilkunde des Moamin*..., p. 179.

11. 'Au nom de Dieu, amen. Ici commence le livre des oiseaux de proie et des chiens aptes à la chasse, de leurs propriétés et habitudes, de leurs maladies avec les soins convenables, lequel [livre] appartenait à Coramomellino, roi de Carthage, traduit de l'arabe en latin par maître Théodore, médecin de l'empereur Frédéric.'

Dans cette perspective on peut remarquer la diversité du texte français, qui contient:

a) une brève préface qui présente le commanditaire et le traducteur;

b) le véritable prologue, dans lequel le traducteur décrit le stade précédent (de l'arabe au latin), sollicite la révision d'Henri de Sardaigne et s'excuse pour les erreurs qu'il peut faire en invoquant deux arguments. Premièrement, il écrit 'tot soie je de povre letreure et de povre sciençe garnis', qui est en contraste avec 'metre Theodres, phylosophes granz': c'est le topos de l'inadéquation du rédacteur, qui se répète lorsqu'il écrit 'avendra por la povre science de moi'. Les mêmes expressions seront reprises au début du livre IV:

> (...) selonc ce qe, o la povre letreure qi en moi est, je poi avoir plus veraie expositions de tot ce qe mestres Theodres mist en suen livre, si qe je n'en ai leisié rien q'il meist a metre el mien, selonc ma conscience.

Le traducteur ajoute une deuxième justification, qui est bien plus intéressante parce qu'elle concerne le travail du traducteur: 'tot soit greveuse chose a ma lange profferre le droit françois por ce qe lombard sui'. La difficulté, donc, ne consiste pas seulement dans la transposition d'une langue à l'autre, mais d'une langue à une autre qui n'est pas la langue maternelle du traducteur. Daniel de Lau est lombard, et en tant qu'homme du Nord de l'Italie il sait que son français pourrait être influencé par la langue qu'il parle et que ce fait pourrait compromettre le bon résultat de sa traduction.

c) une introduction, où Daniel de Lau expose le contenu de l'œuvre et fait l'éloge du roi Henri et de la noblesse de sa famille.

Pour mieux comprendre le travail du traducteur, il faut voir le texte et les stratégies de traduction qui ont été employées. La version de Daniel de Lau présente, par rapport au texte latin et aux traductions italiennes, une série de différences au niveau du contenu qu'on peut attribuer au traducteur.

Au début de chaque chapitre, par exemple, une phrase introductive résume le thème dont on a parlé dans le chapitre précédent et/ou avance le sujet du suivant; ces phrases ne se trouvent ni dans la version latine, ni dans les textes italiens. C'est comme si le traducteur essayait d'aider le lecteur, en explicitant la structure de l'œuvre, en résumant, en mettant en évidence les arguments, en créant en quelque sorte un petit apparat didactique.

Cette exigence du traducteur est bien plus évidente dans le manuscrit de Venise, qui présente ce que Daniel de Lau appelle 'rubrices', de véritables petits chapitres dans lesquels on anticipe les sujets des chapitres suivants et qui, probablement, ont été ajoutés pour une plus grande efficacité du point de vue didactique, avec une solution tout à fait inédite dans la tradition du texte.

Pour voir quelques exemples de ces additions, on peut se référer au livre I, dans la version du manuscrit de Bruxelles:

a) La petite introduction qui suit se trouve après la table au début du livre I et avant le chapitre sur les 'générations' des oiseaux de proie. Il s'agit d'une addition exclusive de ce manuscrit qui sert à bien souligner la structure de l'œuvre, dans un but didactique:

> Aprés ce que ie vus ait fet a entendre en qantz traitéz est cist livres parti et qantz chapitres a li premiers traitiéz en soi, vos devisserai et demostrerai distintemant par ordre chescun traitiéz par soi et les chapitres qui a lui apertient, et commencierai au premier traitiéz dont ie vus ai los chapitrez contéz (I Tab,16).

b) Dans le cas suivant, la modification de la partie initiale du chapitre est fonction de l'intention didactique du traducteur français, mais n'ajoute pas d'informations essentielles pour la compréhension; les versions italiennes, comme nous pouvons le voir, traduisent le latin d'une façon plus littérale:

l I 13,2	De modo tenendi et portandi avem, et si teneatur et portetur super manum dexteram [...].
k I 13,3	Apréz ce que vos avez oi dez enfermitéz des oissiax queles elez sunt et commant l'en les doite conoistre, vos ferai a savoir breumant commant l'en les doit les oissieax porter et tenir.
	(51.3) Mestrez Theodres dist qe se li oissel se portast sor la main destre (...).
c I 13,2	Della maniera di portar l'uccello e di tenerlo dicemo; se huom lo porta in sul pugno diricto, cioè dextro [...].

c) Le traducteur pose dans l'exemple suivant une limite à la règle donnée sur le comportement des aigles. Le texte latin, ainsi que les versions italiennes, se limitent à dire que les aigles n'attaquent pas les autres oiseaux quand ils sont sauvages. La traducteur français y ajoute une observation: l'aigle, s'il est intéressé à une proie, attaque même dans son environnement naturel.

l I 8,3	et nescimus aquilarum rationem cur volant in ipsas nisi hanc, quoniam dum sunt in deserto non faciunt hoc.
k 8,8	Li autre reison ne li sevent li sage assigier qe ceste, qar il dient que li austor et li autre oissel raptor quant il sunt

sauvage el desert ne sunt pas touchié de li et qu'ele ne li fet nul mal, s'il n'avenist par ce q'il aportassent rapine.

a I 43,3 (…) qant il sunt sauvage, elles ne volen onques sor eax, ne mal ne lor font, se ce ne fust autresi por occaison de proie q'il eussent prise.

c I 8,3 e veruna altra cagione sapemo assegnare perché l'aquile corrano suso a questi uccelli di preda, se non questa, imperò che quando sono in nel diserto e son salvatichi non usan di ciò fare.

d) Le dernier exemple est intéressant parce que l'exigence didactique amène le traducteur français à rendre son texte plus actuel: il insère une comparaison avec la course de chevaux dans un contexte familier au lecteur, avec une allusion au tournoi qui est encore plus évidente dans le manuscrit de Venise:

l I 2,6 (…) oportet ut aves rapide regantur ut regitur equus, curere debens in stadio de pinguedine in macritate et e contrario.

k 6,15 (…) li oissel raptor se dovient tenir et governer a la guisse et a la mainere que li preudom tient et governet son cheval quant il le volt fere corre el cors por gaaingnier le paille, zo es de graisseze en megrece et de megrece en grassice.

a I 24,2 (…) li oissel se doit governer e tenir (…) a la mainere et a la guise qe li preudom governe et tient son ceval qant il le velt fere metre au cors qe l'en establist en maintes citez de Lombardie por gaagnier le paile, ce est torner le de graisceçe en magreçe (…).

Dans cette perspective il faut relever aussi une stratégie du traducteur qu'on ne rencontre ni dans le texte latin, ni dans les autres versions vernaculaires: le recours à l'autorité, qui est constant dans l'œuvre. La modalité qui revient le plus souvent est celle de brèves formules, telles que 'Mestres Theodres dit que', 'selonc ce qe Mestres Theodres devisse' et d'autres semblables, qui, grâce à leur répétion, rappellent toujours au lecteur le point de repère.

Comme on peut remarquer, l'autorité à laquelle Daniel de Lau a recours n'est pas le 'Moamin fauconier' auquel on fait allusion au début du texte, comme on pourrait s'y attendre de façon légitime, mais Théodore d'Antioche, le traducteur qui le précède. Et le recours à Théodore comme autorité arrive à substituer le traducteur à Moamin, donc à l'auteur. On lit:

> L'entention de mestre Theodres en cest traitié segonde est de narrer les enfermetéz rescousez dels oissieax qui vivent de rapine et les meçines que l'en en doit usser por els guerir (*k* II Intr, 1).
>
> Ou 'en cestui segont [traitié] fu proposemant de mestre Theodres de traitier des mezinez dez enfermités rescousez qui ont sovant li oissell de rapine' (*a'* II Prol, 2).

Théodore d'Antioche est vu comme un modèle parfait, qui exclut toute possibilité d'erreur, comme on le voit dans les exemples suivants:

> Més ie vos assignierai la maniera et le mou en tel guise en la quantité des oissieax et des mecinez ensi com la <mist> mestrez Theodres, que vos ne poréz faillir se vos ia voléz et savéz bien oserver (*k* II 1,14).
>
> Més ce dist bien mestre Theodres qu'il n'i laissa rien a translatier et a diviser delz traitiéz delz ancienz, aincois l'en aiosta da soi de ce q'il espreva a son tenz selonc sa puissance (*k* II Prol, 9).

La traduction est bonne parce qu'on a rapporté tout ce qui était écrit dans la version précédente, et Théodore avait aussi enrichi le texte avec les observations qui résultaient de sa propre expérience. Le bon traducteur, donc, ne se limite pas à traduire, mais il enrichit là où cela lui paraît opportun, parce qu'ainsi il peut donner une véritable contribution à la diffusion des connaissances, et surtout d'une connaissance de spécialistes telle que la fauconnerie.

Le choix de Théodore d'Antioche comme modèle n'est pas fortuit; il est dû à la nécessité du traducteur d'aider ses lecteurs, en proposant comme modèle un personnage mieux connu par le public occidental, et qui jouissait d'un grand prestige au niveau culturel, bien sûr plus reconnaissable et indiqué que le fauconnier Moamin[12].

Il y a aussi une deuxième raison pour expliquer ce choix, qui nous ramène au milieu de la cour où cette traduction est née. Cette traduction a été commandée par le roi Henri; il était opportun, donc, pour le traducteur, de choisir comme point de repère un intellectuel que le roi

12. L'identification de Moamin est encore objet de débat. D'après Viré F., 'Sur l'identité de Moamin le fauconnier', in: *Comptes rendus de l'Académie des Inscriptions et Belles Lettres* (Paris, 1967), p. 172-176, il s'agissait de Hounayn ibn Ishaq, sur la base de deux formes du nom de Moamin dans les manuscrits latins. B. Van den Abeele, dans son article 'Inspirations orientales et destinées occidentales du *De arte venandi cum avibus* de Frédéric II', in: *Federico II e le nuove culture. Atti del 31° Convegno storico internazionale: Todi, 9-12 ottobre 1994* (Spoleto, 1995), p. 363-392, a ramené les formes indiquées par Viré à des lectures erronées ou à des fautes du traducteur, et a attribué plus de crédit à l'hypothèse de Fradejas Rueda J.M., *Muhammad ibn 'Abd Allah ibn 'Umar al-Bayzar. Libro de los animales que cazan* (Madrid, 1987), qui identifie Moamin avec Muhammad ibn 'Abd Allah ibn 'Umar al-Bayzar, mort à Bagdad en 859-860, et signalé comme auteur d'un traité de fauconnerie.

connaissait très bien, puisqu'il s'agissait d'une grande personnalité de la cour de son père. De cette façon Daniel de Lau rend hommage, bien sûr apprecié par le roi Henri, au climat d'impulsion à la culture que Frédéric II avait voulu dans sa cour.

La comparaison avec la version latine nous permet d'observer le travail du traducteur du point de vue linguistique. Ce qu'on remarque d'abord en lisant le texte est que le français de Daniel de Lau est une langue assez hétérogène, où l'on trouve des latinismes, des arabismes, des italianismes et des mots tirés du provençal. L'élément italien est prépondérant, et, en particulier, la traduction est influencée par les dialectes du Nord de l'Italie. Il s'agit d'un des premiers exemples de ce qui est appelé 'littérature franco-italienne'[13].

Ces caractéristiques, dans notre cas, étaient évidentes dans l'édition Tjerneld, fondée sur le seul manuscrit connu à l'époque, mais elles sont bien plus évidentes si on prend en compte le manuscrit de Bruxelles. La découverte d'un deuxième manuscrit ouvre des perspectives très intéressantes, parce qu'au niveau du texte la traduction est plus fidèle au latin, tandis qu'au niveau linguistique la traduction semble moins contrainte et nous donne un texte contaminé par les dialectes du Nord de l'Italie. Grâce à ce deuxième manuscrit on a l'attestation de mots et de tournures dont

13. Sur la collocation du *Moamin* dans la littérature franco-italienne, voir Roncaglia, 'La letteratura franco-veneta...', p. 747.

Sur la littérature franco-italienne, voir Viscardi A., *Letteratura franco-italiana* (Modena, 1941); Viscardi A., Nardi B. & T., Vidossi G. & Arese F., *Le Origini* (Milano - Napoli, 1956), p.1055-1219; Holtus G., 'Zur franko-italienischen Sprache und Literatur: Forschungsbericht 1959-1974', *Zeitschrift für romanische Philologie* XCI (1975), p. 491-533; Rosellini A., 'Il cosiddetto franco-veneto: retrospettive e prospettive', *Filologia moderna* II (1977), p. 219-303; Zingarelli N., 'Letteratura franco-italiana nei sec. XIII e XIV', *Atti della Società italiana per il progresso delle scienze* IV (Pavia, 1933), p. 139-156; Lomazzi A., 'Letteratura franco-veneta', in: V. Branca (éd.), *Dizionario critico della letteratura italiana*, vol. II (Torino, 1974), p. 125-132.

Sur les caractéristiques linguistiques des textes franco-italiens, voir Meyer P., 'De l'expansion de la langue française en Italie pendant le Moyen Age', in: *Atti del Congresso Internazionale di scienze storiche* IV (Roma, 1904), p. 61-104; Viscardi A., *Letteratura franco-italiana* (Modena, 1941); Ruggieri R.M., 'Origine, struttura e caratteri del franco-veneto', *Orbis* X (1961), p. 20-30; Vidossi G., 'L'Italia dialettale fino a Dante', in: *Le Origini* (Milano-Napoli, 1956), p. 78-79; Wunderli P. & Holtus G., 'La renaissance des études franco-italiennes. Rétrospective et prospective', in: G. Holtus, H. Krauss & P. Wunderli (éds), *Testi, cotesti e contesti del franco-italiano: atti del 1° simposio franco-italiano (Bad Homburg, 13-16 aprile 1987); in memoriam Alberto Limentani* (Tübingen, 1989), p. 3-23; Holtus G., 'Approches méthodiques d'une description linguistique du franco-italien', *Festschrift K. Baldinger* II (Tübingen, 1979), p. 854-875; Holtus G., 'Aspects linguistiques du franco-italien', in: *Essor et fortune de la Chanson de geste dans l'Europe et l'Orient latin, Actes du IXe Congrès International de la Société Rencesvals 1982* (Modena, 1984), p. 802-806.

il n'y aurait aucune trace autrement. Les exemples que nous allons citer sont donc tirés du manuscrit de Bruxelles.

Dans cette version il arrive, par exemple, que le même objet ou concept soit identifié de façon différente, avec des mots qui trahissent différentes origines; cela signifie que, dans un texte de nature scientifique tel que celui-ci, le vocabulaire est plus riche et offre au lecteur la possibilité de définir et d'identifier les éléments par des voies différentes. Il nous offre la possibilité de comprendre quels étaient le vocabulaire et les moyens linguistiques dont le traducteur disposait.

Dans les exemples suivants, nous présentons la traduction des mots de la version latine telle qu'on la trouve dans toutes les variantes du manuscrit de Bruxelles. Nous donnons ensuite la comparaison avec les versions italiennes pour témoigner ce qui était l'usage italien:

a) *carnem* > *carne; char*
carne (*b, c*)
b) *favo mellis* > *faumiel, favomiel*; 'gaufre'
favo de melle (*b*), *brescha di mele* (*c*)
c) *lapides* > *pietres, pieres*; 'pierre' ou 'calcul'
pietre (*b, c*)
d) *virtutem* > *vertude, vertu*; 'qualité caractéristique' d'un oiseau
virtù (*b, c*)
e) *zuccarum, çuccarum* > *zucar, zucre*; 'sucre'
zuccaro (*b, c*)

Dans la plupart des cas cependant, la traduction n'hésite pas entre deux possibilités, mais utilise seulement le mot d'origine italienne:

a) *bistardos* > *bistartz*, *Otis tarda*, 'outarde'
bistarde (*c*), *biscardi* (*b*)
L'ancien français connaît uniquement le mot 'ostarde'[14], que Daniel de Lau n'emploie pas.
b) *milvos* > *milf*, *Milvus milvus*, 'milan'
e piglia milvi, cioè milvani (*b*), *nibii* (*c*)
Le terme *milan* n'était pas employé en ancien français; l'expression *e piglia milvi, cioè milvani*, qui se trouve dans *b,* confirme l'hypothèse que *milf* est un italianisme.

Dans certains cas le traducteur utilise le terme français dans une nouvelle acception, qui dérive de l'italien[15]:

14. FEW I, 188b.
15. Battaglia S., *Grande dizionario della lingua italiana* (Torino, 1961-2001); Pfister M., *Lessico etimologico italiano* (Wiesbaden, 1979-2004).

a) *accipere > brancher*
piglia et abranca (b), abrancha (c)
En français *brancher* signifie uniquement 's'appuyer', 'se percher'; le traducteur l'emploie donc dans le sens de l'ancien italien *brancare*, 'saisir' ou 'agripper'[16].

b) *tendens* + accusatif de la couleur (ex. *ad rubedinem*) > *traiant a rosece*
declinante al russo (b), chi trae al rosso (c)
L'emploi de *trere* avec la valeur de 'tirer (sur une certaine nuance de couleur)' est typique de l'ancien italien *trarre*[17].
Il n'est pas incorrect en français; nous le trouvons, par exemple, dans les traductions françaises du *De falconibus*[18]:

Color pedum optimus croceus multum vergens ad albedinem et
DF I: la coulors des piés jaune tendans mout a la blanchour et
DF II: les piez bien jeaunes tendans a blancheur
DF III: celle de ses piez est jauine tirant ung peu a blancheur

Toutefois, comme c'est une tournure qui a une grande diffusion dans le territoire italien, on peut supposer qu'ici il s'agisse d'un italianisme.

Souvent la contamination avec les dialectes italiens s'arrête au niveau phonétique. Il suffit de penser à des mots tels que *chavre* pour *chevre* (< *capra*), *chaval* pour *cheval* (< *cavallo*), *galine/a* pour *geline* (< *gallina*), où c'est le passage à *-e* qui manque, ou aux résultats de [k] et [g] latins, typiques du côté italien et provençal, tels que *capitre*, *carne*, *caudez*, *calor*, *camelz*, *cavre*, *cardine*, *carde*, *cose*, *calce*, *canule*. On peut mentionner, en outre, les nombreuses terminaisons en *–a* au féminin: *quarta, pennas*, *vacha*, *terra*, *assa fetida*, *fissuras*, *podraga*, *cera*, *stella*, *assara* et d'autres semblables[19].

On peut citer encore, parmi les autres: *mascle* pour *masle* (it. *maschio*), *sapor* pour *savor* (it. *sapore*), *furment* pour *frument* (it. *formento*).

L'influence des dialectes italiens sur la langue du traducteur se manifeste aussi au niveau morphologique et syntactique. Il est intéressant de remarquer, par exemple, l'hésitation sur le genre des substantifs abstraits

16. Battaglia, *Grande dizionario…*, II, 354c.
17. Battaglia, *Grande dizionario…*, XXI, 198b.
18. *De falconibus* II.1. Nous remercions An Smets pour cet exemple tiré de sa thèse de doctorat: *Les traductions françaises du* De falconibus *d'Albert le Grand. Analyse lexicale d'un dossier inédit* (Leuven, 2003).
19. Rohlfs G., *Grammatica storica della lingua italiana e dei suoi dialetti* (Torino, 1966).

en *–or* (ex. *clamor, vapor, blancor, froidor, puor* et *valor*), due au fait que l'usage du masculin est caractéristique de l'italien, surtout des régions centrales; en français et dans les attestations médiévales des dialectes de l'Italie du Nord, par contre, le genre est féminin. On peut penser aussi à l'oscillation d'un mot tel que *gazzella,* qui devient soit *gazel* soit *gazelle*: l'ancien italien, en fait, connaissait *gaçello* au masculin[20].

Le français de son côté n'est pas homogène, ce qui s'explique par l'histoire même de la langue[21]. Au moyen âge le *francien* ou *françois* remplace les autres parlers d'oïl et devient progressivement la langue de la France, avec l'affermissement de la monarchie dans le pays, même si les autres dialectes continuent à subsister.

Daniel de Lau, donc, dans sa traduction, accepte des mots et des tournures qui dérivent de différents dialectes de la France, mais qui d'après lui peuvent être utilisés, parce qu'ils sont, tout compte fait, français et vivants à l'époque. Dans les exemples qui suivent on peut remarquer les deux possibilités offertes par le même texte:

a) *anser* > *ouche* (< **auca*) résultat français
auche, résultat provençal (*auca*), 'oie'
b) *lacus* > *lac*, résultat français
lauç, résultat que l'on trouve dans plusieurs vallées des Alpes, entre la France et le Piémont[22], 'lac'
c) *pernix* > *perdiz*, résultat français
perdrix, résultat majoritaire dans les dialectes du Nord et de l'Est de la France, 'perdrix'

Le texte présente aussi des graphies françaises qui appartiennent à des phases chronologiques différentes de l'évolution de la langue: même dans ces cas, le traducteur les utilise parce que d'après lui elles sont également françaises. Cela arrive surtout pour les formes verbales, telles que le futur simple des verbes *estre* (*ert*, *iert*, *sera*) et *doner* (*donra*, *dorra*), ou l'impératif de *voir* (*veitéz*, *voiéz*, *veéz*, *véz*).

La version franco-italienne, enfin, est riche de latinismes et d'arabismes. Les latinismes nous sont transmis presque uniquement par le manuscrit de Bruxelles, comme on peut le voir en le comparant avec le manuscrit de Venise (*a*) et les versions italiennes (*b* et *c*). Ils sont relatifs aux trois domaines dont nous donnons quelques exemples:

20. Battaglia, *Grande dizionario…*, VI, 623a.
21. Pope M. K., *From Latin to modern French with especial consideration of Anglo-Norman* (Manchester, 1952).
22. *Laus* dans FEW V, 126a.

a) anatomie et pratiques médicales
- *meat* (*de son anelit*) < *meatu anelitus*, ‘passage’ du souffle
  pertus par ont le rent s’alaine (*a*); *quella via de onde fiata* (*b*), *via onde passa l’alito* (*c*)
- *cauteriçatio* < *cauteriçatio*, ‘brûlure causée par une cautérisation’
  cuiture (*a*); *cauteriçatione* (*b*)
- *ruptura* < *ruptura*, ‘fracture’
  rompimant (*a*); *ruptura* (*b*), *rompitura* o *roctura* (*c*)
- *desiuntura* < *disiunctura*, ‘luxation’
  desjongemant (*a*); *disiunctura* (*b*), *dislocatura* (*c*)

b) mots relatifs au temps
- *vesper* < *vesper*, ‘soir’
  vespre (*a*)
- *hora* < *hora*, ‘heure’
  ore (*a*)

c) plantes ou substances avec des propriétés thérapeutiques
- *amigdale* < *amigdala*, ‘amande’
  amigdale (*a*); *amendola* (*b*), *mandorla* (*c*)
- *capilz veneris* < *capillis veneris*, ‘cheveux de Vénus’
  capils veneris (*a*); *capilli veneri* (*b*), *capelvenere* (*c*)
- *oile petroleo* < *oleo petroleo*, ‘pétrole’
  oile dou petroleo (*a*); *oleo petroleo* (*b*), *olio petroleo* (*c*)

Le même discours vaut pour les grécismes, plus rares, que l’on retrouvait déjà en latin, adaptés graphiquement et circonscrits aux noms d’espèces végétales ou animales. Quelques exemples:

a) *acaron*, gr. ἄκορον, lat. *acharon*, qui identifie l’*Acorus calamus*.
   accharon (*a*); *acharon* (*b*); *acoro* (*c*)
b) *silladeph*, gr. μηλαδέλφι, lat. *philadelfus*; on suppose le *Hieraaëtus fasciatus*, ‘aigle de Bonelli’[23]
   phyladep (*a*); *phyladelpho* (*b*), *filiadelfun* (*c*)

Le traducteur choisit de transmettre de façon inaltérée les arabismes qui se trouvaient dans la version latine, avec des variantes exclusivement de

23. L’identification est proposée par plusieurs auteurs, tels que Glessgen, *Die Falkenheilkunde des Moamin…*, p. 493-494; Viré F., ‘Essai de détermination des oiseaux de vol, mentionnés dans les principaux manuscrits arabes médiévaux sur la fauconnerie’, *Arabica* 24 (1977), p. 138-149; Tjerneld, *Moamin et Ghatrif. Traités…*, p. 392.

nature graphique. On utilise l'arabe dans des contextes techniques, fortement spécialisés, pour indiquer:

a) des oiseaux
   - *aynolu* < *anūq/anīqī, Falco biarmicus*, 'lanier', en latin *aynoki* (AEIMR) ou *aynoli* (DH).
     aynclen (*a*), *smiriglio* (*b*), *anoleym* (*c*)
   - *bayda* < *baydaq*, rapace non identifié[24], en latin *baydac* (AF) ou *baylac* (T).
     bayda (*a*), *balach* (*b*), *luyda* (*c*)

On peut mentionner, parmi les autres: *cotayn, nymin, palem, sabech, xemi* et *zummeg* qui indiquent, respectivement, le faucon hobereau, l'épervier à pieds courts, le faucon émerillion, une espèce d'épervier non identifiée, l'épervier shikra et l'aigle de Bonelli.

b) des maladies
   - exunag < *šannağ*, 'trouble digestif avec spasme'. Le terme est utilisé dans toute la tradition manuscrite.
   - *sudhaam*, 'contusion'. La version française rapporte l'arabe *suda*, qui signifie 'douleur à la tête'. Dans les versions latines, le terme remplace le moins connu *sadma*, présent dans le Ghatrif, qui signifie 'contusion', et la contamination entre les deux termes se répète dans les versions vernaculaires. Il est *saddaam* en *a* et *saadari* en *c*.

c) des plantes ou substances avec des propriétés thérapeutiques
   - *baclurug* < *bādarūğ*
     Le texte latin présente *baclurug idest basilicon* (T) et *badurus idest basilicon* (I). Dans ce cas particulier, le traducteur veut donner l'équivalent français et rendre le terme plus compréhensible pour le lecteur: 'baclurug ou soit basiliches'. Le texte contient cependant même le mot *basilicon*, qui est latin, mais proche de l'italien *basilico* et qui pouvait être lui aussi familier au lecteur.
     On retrouve la même attitude chez le traducteur italien de *c*: *bacluruch, cioè basilicon* (*b*) et *baclurug, cioè basilicho* (*c*).
   - *altit* < *hiltīt*, 'asa foetida', gomme-résine de l'ombrellifère *Ferula assa-foetida*.

24. Il s'agit, peut-être, d'une espèce semblable à l'épervier ou de l'épervier levantin.

Dans ce cas, le traducteur se limite à citer le texte latin et donne le terme arabe comme équivalent de *assa fetida*: 'asse fetide, la quel s'apelle en hebreu altit'[25].

– *secbyneg* < *sakbīnag*, 'sagapénum', une gomme-résine produite par une férule qui n'est pas identifiée. Même ici le traducteur, en suivant le latin qui avait *sethynec* (ou *sechineg*) *idest serapinum*, donne l'équivalent français, que le lecteur connaît bien: 'secbyneg, ce est serapin'.
Cette attitude est adoptée aussi en *b*, qui a *sethinec idest serapino*, tandis que *c* arrive à remplacer complètement l'arabisme par le mot italien *serapino*.

Pour conclure: les latinismes et, surtout, les arabismes, ne sont pas occasionnels, mais circonscrits à des milieux techniques. Le traducteur se préoccupe aussi de fournir au lecteur l'équivalent le mieux compréhensible quand il s'aperçoit que le terme technique est, peut-être, peu connu. Dans les autres cas, il traduit en français, dans les conditions dont nous avons parlé.

Cela signifie que la traduction n'est pas perçue comme une transposition passive, mais comme un travail important, digne de soin, et le texte représente un matériel sur lequel le traducteur, pour en favoriser la compréhension et la diffusion, n'a pas seulement le pouvoir, mais aussi le devoir d'intervenir.

Università di Torino

25. Cette substitution de l'arabe est caractéristique de la version franco-italienne.

An SMETS[1] – Magali TOULAN

LES ACCESSOIRES DES FAUCONS ET DES FAUCONNIERS DANS LES TRADUCTIONS FRANÇAISES DU *DE ARTE VENANDI CUM AVIBUS* DE FRÉDÉRIC II ET DU *DE FALCONIBUS* D'ALBERT LE GRAND

0. Introduction

La tradition latine des traités de fauconnerie culmine au XIIIe siècle avec deux ouvrages majeurs: le *De arte venandi cum avibus* de Frédéric II et le *De falconibus* d'Albert le Grand. Il s'agit de véritables manuels où le fauconnier trouve à la fois des données ornithologiques, thérapeutiques et cynégétiques. Les deux premiers éléments occupent une place importante chez Albert le Grand, tandis que Frédéric II s'intéresse essentiellement à la cynégétique. Ainsi, le *De arte venandi* est le traité qui consacre le plus de place aux accessoires dont le fauconnier a besoin pour dresser les oiseaux[2]. Si Albert le Grand ne leur accorde qu'un intérêt limité, il mentionne tout de même les principaux accessoires de l'équipement de l'oiseau (le chaperon, les jets et le touret), et aussi ceux du fauconnier (la filière, le leurre et le perchoir)[3]. Les dénominations de ces accessoires dans les traités latins et dans leurs traductions françaises font l'objet de cette contribution.

1. Cette contribution s'inscrit dans le projet de recherche 'La traduction des traités biologiques médiévaux', subventionné par le fonds de recherche (Onderzoeksfonds) de la K.U.Leuven (OT/01/06) et effectué au sein du groupe de recherche 'Traduction et changement linguistique'.

2. Cette attention particulière chez Frédéric II s'exprime entre autres dans la présence des définitions, que nous reprenons dans la présente contribution, alors qu'Albert le Grand définit rarement les accessoires.

3. Pour la distinction entre les accessoires qui appartiennent à l'équipement de l'oiseau ou à celui du fauconnier, nous suivons la répartition proposée par Van den Abeele B., *La fauconnerie au Moyen Âge, Connaissance, affaitage et médecine des oiseaux de chasse d'après les traités latins*, Sapience 10 (Paris, 1994), p. 105-121. À cet endroit se trouve également la terminologie employée dans les autres traités latins.

1. Présentation du corpus et de l'objet de recherche

1.1. Le *De arte venandi cum avibus*

Le *De arte venandi cum avibus* de Frédéric II de Hohenstaufen nous est parvenu en deux versions, dont la première, en six livres, est très probablement copiée d'après l'œuvre originale de l'empereur, malheureusement perdue[4]. La seconde version ne contient que les deux premiers livres de la première, complétés par des ajouts du roi Manfred, fils de l'empereur. Cette version dite manfrédienne est intéressante à plusieurs titres: elle est conservée dans le manuscrit le plus ancien (Vatican, BAV, Pal. Lat. 1071, seconde moitié du XIIIe siècle)[5] et offre ainsi le témoin le plus proche de la version originale pour les deux premiers livres. En outre, les illustrations du manuscrit sont très riches. Enfin, c'est cette version qui a été traduite en français à deux reprises.

En fait, le *De arte venandi cum avibus* n'a été traduit qu'en français[6]. Les deux traductions existantes sont indépendantes et toutes deux faites sur la base du manuscrit du Vatican. La première traduction, conservée dans le ms. Paris, BNF, fr. 12400 (= *Chace Oisiaux I*), est complète et très fidèle à l'original latin dont elle reproduit même les illustrations[7]. Elle date d'avant 1310 et a été copiée trois fois[8]. La seconde traduction, conservée dans le ms. Paris, BNF, fr. 1296 (= *Chace Oisiaux II*), ne contient que le second livre et n'est pas illustrée. Le manuscrit date de ca. 1470, mais il s'agit très probablement d'une copie d'une traduction plus ancienne, datée provisoirement du deuxième tiers du XIVe siècle[9].

4. Cette version est conservée dans cinq manuscrits complets, dont le plus ancien (Bologna, Biblioteca Universitaria, Lat. 717) date de la fin du XIIIe siècle – début du XIVe siècle.

5. Il n'existe qu'un seul autre manuscrit de la version manfrédienne, une copie du manuscrit du Vatican de la fin du XVIe siècle. Celui-ci se trouve à Vienne, dans l'Österreichische Nationalbibliothek, sous la cote 10948.

6. J.M. Fradejas Rueda a cependant démontré que l'auteur espagnol Juan Manuel s'est basé (entre autres) sur Frédéric II pour composer son traité cynégétique. Voir Fradejas Rueda J.M., 'La influencia del *De arte venandi cum avibus* de Federico II en el *Libro de la caza* de Juan Manuel', in: J.M. Fradejas Rueda (éd.), *Los libros de caza,* Estudios y ediciones 6 (Tordesillas, 2005), p. 41-54.

7. Federico II, *De arte venandi cum avibus. L'art de la chace des oisiaus. Facsimile ed edizione critica del manoscritto fr. 12400 della Bibliothèque Nationale de France* (Napoli, 1995).

8. Voir les mss. Cambridge (Mass.), Houghton Library of Harvard University, f. Typ. 129; Genève, Bibliothèque publique et universitaire, fr. 170; et Stuttgart, Württembergische Landesbibliothek, HB.XI 34a.

9. Cf. Toulan M., *Edition critique et analyse lexicale de la seconde traduction française du* De arte venandi cum avibus *(Bn fr. 1296)* (thèse de doctorat non publiée Strasbourg, 2005).

Quant au contenu, le premier livre offre une ornithologie générale, tandis que le second, commun aux deux traductions, traite des faucons et offre quelques données de base sur la fauconnerie[10]. L'originalité du traité de Frédéric II réside dans les thèmes abordés: il ne contient pas de chapitres thérapeutiques, alors que les traités de fauconnerie médiévaux étaient essentiellement des réceptaires[11]. Signalons également que, malgré sa renommée, le traité de l'empereur est très peu diffusé[12].

1.2. Le *De falconibus*

Presque au même moment[13], mais sans qu'il y ait de relation directe entre les deux ouvrages[14], Albert le Grand rédigea le *De falconibus*, qui sera ensuite inséré dans son ouvrage encyclopédique plus large, le *De animalibus*[15]. La partie la plus importante du traité cynégétique d'Albert le Grand est constituée par sa 'hiérarchie des faucons' (comprenant dix espèces de faucons nobles, trois espèces de faucons non nobles et quatre espèces mixtes). Cette partie est précédée d'observations sur la différence entre les oiseaux rapaces et les autres oiseaux (la forme, la couleur, la façon de voler et la voix), et suivie de remarques sur l'affaitage ou l'entraînement à la chasse, ainsi que de quelques chapitres thérapeutiques. Bien que le *De falconibus* soit beaucoup plus court que le *De arte venandi*, ce traité présente un atout important par rapport au travail de l'empereur: sa diffusion plus large. En effet, nous connaissons quatre manuscrits indépendants contenant ce traité[16], de même que 53 copies du *De animalibus* le comprenant également[17]. En outre, des traductions en

10. Le livre III de la version complète traite de l'affaitage des faucons, alors que les trois derniers livres sont tous consacrés à un mode de chasse particulier avec des faucons gerfauts, sacres et pèlerins.

11. Van den Abeele, *La fauconnerie au Moyen Âge*..., p. 22.

12. Pour des renseignements supplémentaires sur le texte latin et les traductions françaises, voir Willemsen C.A., *Über die Kunst mit Vögeln zu jagen – Kommentar zur lateinischen und deutschen Ausgabe* (Frankfurt-am-Main, 1970).

13. La rédaction du *De falconibus* se situerait dans les années 1240, cf. Smets A., "*Des faucons*". *Les quatre traductions en moyen français du* De falconibus *d'Albert le Grand. Analyse lexicale d'un dossier inédit* (Thèse de doctorat non publiée) (Leuven, 2003), p. 21.

14. Albert le Grand renvoie quelques fois à Frédéric II, mais ces références sont incorrectes, cf. Smets, "*Des faucons*"..., p. 28.

15. *De animalibus libri XXVI. Nach der Kölner Urschrift*, éd. H. Stadler, Beiträge zur Geschichte der Philosophie des Mittelalters 15-16 (1916-1920) (*De falconibus*: p. 1453-1492).

16. Il s'agit des codex Milano, Biblioteca Ambrosiana, B 91 sup., fol. 110r-120r; Oxford, Bodleian Library, Rawlinson D 483, fol. 1r-47v; Oxford, Oriel College, 28, fol. 36v-50v; et Wien, Österreichische Nationalbibliothek, 5315, fol. 257r-272v.

17. Smets, "*Des faucons*"..., I, p. 32-35. Parmi ces manuscrits figure l'autographe du *De animalibus*, qui est conservé dans le Stadtarchiv de Cologne, sous la cote W258a.

furent réalisées en français, en allemand, en italien, en anglais et en catalan, déjà à l'époque médiévale[18]. En ce qui concerne les quatre traductions françaises, nous les avons nommées *Des faucons I*, *Des faucons II*, *Des medecines des faucons* (traduction partielle) et *Des faucons III*. Elles sont toutes conservées à la Bibliothèque nationale de France à Paris[19] et datent respectivement de la fin du XIVe, du milieu du XVe, de la seconde moitié du XVe et du début du XVIe siècle[20].

Pour effectuer une étude comparée des techniques de traduction dans les traités de fauconnerie d'Albert le Grand et de Frédéric II, nous avons choisi le champ sémantique des accessoires. Albert le Grand mentionne les accessoires suivants: le chaperon, les jets et le touret, la filière, le leurre et le perchoir, que l'on retrouve tous dans le traité de l'empereur[21]. Notre objectif est de comparer les noms latins utilisés dans les deux textes et la manière dont ils ont été traduits en français. Le fait que

18. Pour les traductions allemandes voir Lindner K., *Von Falken, Hunden und Pferden: Deutsche Albertus-Magnus-Übersetzungen aus der ersten Hälfte des 15. Jahrhunderts*, Quellen und Studien zur Geschichte der Jagd 7-8 (Berlin, 1962) et Müller J.-D., 'Naturkunde für den Hof. Die Albertus-Magnus-Übersetzungen des Werner Ernesti und Heinrich Münsinger', in: J.-D. Müller (éd.), *Wissen für den Hof. Der spätmittelalterliche Verschriftungsprozess am Beispiel Heidelberg im 15. Jahrhundert* (München, 1994), p. 121-168. Quant aux trois versions italiennes, une seule a été éditée: voir Holmér G., 'Traduction en ancien italien de quelques chapitres du *Liber de animalibus* d'Albert le Grand. Le manuscrit V.E.506', *Studia neophilologica* 38 (1966), p. 211-256. Une brève présentation des traductions françaises, allemandes et italiennes se trouve dans Smets A., 'La réception en langue vulgaire du *De falconibus* d'Albert le Grand', in: G. Donavin, C. Poster & R. Utz (éds), *Medieval Forms of Argument: Disputation and Debate*, Disputatio 5 (Eugene, Oregon, 2002), p. 189-199. La traduction catalane a été présentée par Garcia Sempere M., 'El *Llibre dels ocells de caça*: una recopilación de saber cetrero', in: J.M. Fradejas Rueda (éd.) *Los libros de caza...*, p. 55-68. Les deux traductions anglaises, qui sont conservées dans le *Kerdeston Hunting Book* et le *Kerdeston Hawking Book*, ont été acquises par la British Library en 2006. Les manuscrits les contenant ont reçu les cotes Add. 82948 et Add. 82949 (données confirmées par la conservatrice K. Doyle).

19. Voici les cotes: naf. 18800 pour *Des faucons I*, fr. 2003 pour *Des faucons II*, fr. 25342 pour *Des medecines des faucons* et fr. 1304 pour *Des faucons III*.

20. Là où le modèle latin des traductions françaises du *De arte venandi* est connu, nous ne connaissons pas les manuscrits latins sur lesquels les traducteurs du *De falconibus* ont travaillé. Signalons toutefois que dans le champ sémantique des accessoires, cette observation n'entraîne pas beaucoup de conséquences, étant donné que les modèles employés par les traducteurs semblent suivre ici presque toujours une version proche de celle de l'autographe latin.

21. Le *De arte venandi* mentionne en outre les accessoires suivants: la longe (*longa*, II.36), la clochette (*canpanella* / *nola*, II.38) et la chemise (*maleolus*, II.54) pour l'équipement du faucon; le gant (*ciroteca*, II.79), la gibecière (*carneria*, II.80), le tiroir (*tiratorium*, II.140), la brochette (*virga* / *virgula*, II.202), le chapeau (*pileus* / *galerus*, IV.59 / III.36) et le tambour (*timpanum*, VI.115) pour l'équipement du fauconnier. Voir Federico II di Svevia, *De arte venandi cum avibus*, éd. A.L. Trombetti Budriesi (Roma-Bari, 2000), p. 323, 326, 340, 354, 420, 478, 728, 592 et 1060.

les textes latins et les traductions françaises s'échelonnent entre le XIIIe et le XVIe siècle permettra d'apprécier l'évolution de l'usage des différents mots[22].

2. Les accessoires du faucon

2.1. Le chaperon[23]

Le chaperon est une 'petite coiffe de cuir dont on couvre la tête du faucon afin qu'il reste tranquille'[24]. Nous constatons tout de suite une différence entre les textes latins à ce sujet. En effet, Frédéric II se limite au seul mot *capellus*, qu'il définit ainsi:

> Est autem capellus quoddam formatum de corio secundum formam capitis falconum ad continendum caput ipsorum usque ad collum, rostro et naribus exclusis ab eo. De corio vero convenientius fit quod neque nimis sit durum neque nimis molle[25].
>
> 'Le chaperon est un objet fait en cuir ayant la forme de la tête des faucons et devant la contenir jusqu'au cou, le bec et les narines restant à l'extérieur. On le confectionne au mieux avec un cuir qui ne soit ni trop dur ni trop souple.'[26]

22. Nous signalons que nous ne prenons en compte que trois des quatre traductions du *De falconibus*, étant donné que *Des medecines des faucons* se limite à deux chapitres thérapeutiques dans lesquels les accessoires ne sont pas mentionnés. De même, pour Frédéric II, nous limitons notre analyse au second livre du *De arte venandi*, le seul qui soit commun aux deux traductions françaises. Les renvois au texte latin du *De arte venandi* sont faits d'après l'édition d'A.L. Trombetti Budriesi; pour le *De falconibus*, nous renvoyons aux pages de l'édition de Stadler et à la numérotation de Smets. L'édition de Trombetti Budriesi prend le manuscrit de Bologne comme manuscrit de base, donne les variantes du manuscrit du Vatican pour les livres I et II et reproduit en annexe les ajouts du roi Manfred. La numérotation des paragraphes dans le *De falconibus* est basée sur les pieds de mouche dans *Des faucons I* et a ensuite été transférée vers le texte latin et les autres traductions vernaculaires afin de faciliter la comparaison entre les différentes versions. La datation des textes dans lesquels apparaît tel ou tel mot est faite d'après le Complément bibliographique du DEAF. Cf. Möhren F., *Dictionnaire étymologique de l'ancien français. Complément bibliographique* (Tübingen, 1993).

23. Pour plus d'informations sur le chaperon, cf. Van den Abeele B., 'Aux origines du chaperon. Les instruments du fauconnier d'après les traités médiévaux', in: R. Durand (éd.), *L'homme, l'animal domestique et l'environnement, du Moyen Âge au XVIIIe siècle* (Nantes, 1993), p. 279-290. Une reproduction d'un chaperon se trouve dans Willemsen C.A., *Das Falkenbuch Kaiser Friedrichs II.*, Bibliophilen Taschenbücher 152 (Dortmund, 1987), fol. 104v, 105, …

24. Van den Abeele, *La fauconnerie au Moyen Âge…*, p. 112.

25. Federico II, *De arte venandi…*, éd. Trombetti Budriesi, II. 256, p. 524.

26. Frédéric II de Hohenstaufen, *L'art de chasser avec les oiseaux. Traduction intégrale en français du traité de fauconnerie* De arte venandi cum avibus, trad. A. Paulus & B. Van den Abeele, Bibliotheca cynegetica 1 (Nogent-le-Roi, 2000), p. 285.

De son côté, Albert le Grand ne donne pas de définition, mais utilise quatre mots différents: *capellus*, *capitegium*, *mitra* et *pileus*[27]. L'emploi de quatre substantifs témoigne de l'incertitude du dominicain au sujet de la désignation de ce nouvel accessoire. Il s'agit en effet d'un nouvel accessoire, parce que, bien que le chaperon fût connu de longue date dans le monde arabe, il avait fait son entrée en Europe seulement au XIIIe siècle, quand Frédéric II en importa l'usage après la croisade de 1228-1229. Albert le Grand est le second auteur à le mentionner[28].

Dans les traductions de l'ouvrage frédéricien, *Chace Oisiaux I* rend toujours *capellus* par *chapel*[29]. C'est également la traduction la plus courante dans *Chace Oisiaux II*, où le traducteur emploie d'abord *chapel* seul, 339 fois au total, puis cinq fois le couple *chapel ou chaperon* et une seule fois *chaperon* seul à la toute fin du manuscrit. Quant aux traductions du *De falconibus,* nous rencontrons quatre fois *chapel* dans *Des faucons I,* alors que *Des faucons II* et *III* emploient toujours le substantif *chaperon*, respectivement deux et cinq fois. Ces occurrences sont en outre complétées par les verbes *chaperonner* et *deschapperonner* dans *Des faucons III* (cf. infra).

Le substantif *chapel* apparaît en français au XIIe siècle avec le sens de 'chapeau'[30]. Selon le FEW[31], la première attestation du sens cynégétique de ce substantif se trouve dans *Chace Oisiaux I.* Il faut cependant corriger cette datation, parce qu'il existe une attestation plus ancienne, plus précisément de 1298, et cela chez Marco Polo[32].

En ce qui concerne le mot *chaperon*, il est aussi attesté en français au XIIe siècle, désignant un type de coiffure[33]. Dans le sens cynégétique, on le trouve essentiellement dans des traités de fauconnerie à partir des *Livres du roy Modus et de la royne Ratio*, œuvre datée du troisième quart

27. Le substantif *mitra* figure aux paragraphes III.3-4 et *capitegium* au paragraphe IV.2 (*De animalibus...*, éd. Stadler, p. 1456), alors que *capellus* et *pileus* se présentent au paragraphe XVII.1 (*De animalibus...*, éd. Stadler, p. 1471-1472).

28. Van den Abeele, *La fauconnerie au Moyen Âge...*, p. 112-113.

29. Étant donné que nous ne disposons pas d'une édition électronique de la première traduction, nous ne connaissons pas le chiffre exact pour les occurrences de *chapel* dans cette version: le glossaire énumère dix occurrences, suivies de la mention *etc.* Cf. Federico II, *De arte venandi...*, éd. Trombetti Budriesi, p. 585.

30. *Trésor de la langue française. Dictionnaire de la langue du XIXe et du XXe siècle (1789-1960)* (Paris, 1971-1994), dorénavent TLF; ici 5, p. 520.

31. Wartburg W. von, *Französisches Etymologisches Wörterbuch. Eine darstellung des galloromanischen sprachschatzes* (Bonn - Basel, 1922-), dorénavant FEW; ici 2_1, p. 288.

32. Benedetto L.F., *Marco Polo: Il Milione*, Comitato geografico nazionale italiano 3 (Firenze, 1928), p. 87.

33. TLF 5, p. 527.

du XIVᵉ siècle[34]. Il semble donc que l'emploi de *chapel* et *chaperon* dans les deux textes se fasse de façon chronologique et permette ainsi de suivre le développement du sens d'un mot courant, néanmoins remplacé par un autre dont l'emploi devient strictement technique.

Finalement, Albert le Grand emploie également une seule fois le verbe *pilleare* (XVII.1)[35], dont le rapport avec le substantif *pileus* est évident. Dans *Des faucons I,* ce verbe est rendu par la paraphrase *remetre son chapiau*, dans *Des faucons II* ce passage est omis et dans *Des faucons III,* le traducteur a opté pour le verbe *chapperonner*. Ce traducteur a d'ailleurs également employé l'antonyme de ce verbe, à savoir *deschapperonner,* comme traduction de l'ablatif absolu *mitra deposita* (III.4). Le verbe *chaperonner* date déjà du XIIᵉ siècle, bien que son emploi en ancien français soit rare. Son utilisation dans le contexte cynégétique doit se situer avant 1614[36]. Pour *de(s)chaperonner*, les dictionnaires avancent la datation d'environ 1465, d'après un exemple tiré du *Livre de faulconnerie* de Jean de Francières et cité par F. Godefroy[37], une occurrence qui précède celle de *Des faucons III* de quelques décennies.

2.2. Les jets et le touret

Passant de la tête aux pieds du faucon, nous trouvons tout d'abord les jets. Ceux-ci sont de 'petites courroies de cuir fixées aux tarses de l'oiseau, et qui sont tenues en main ou fixées à la longe'[38]. Frédéric II emploie le substantif *iactus* pour désigner cet objet qu'il définit ainsi:

> Iacti sunt laquei de corio facti, imponendi pedibus falconum, ut cum eis retineantur et iactentur ad predandum, qui ob hoc iacti dicuntur, quia cum eis iaciuntur falcones et emittuntur ad predandum[39].
>
> 'Les jets sont des lacets de cuir à attacher aux pieds des faucons pour les retenir et les jeter vers une proie. On les appelle "jets" parce

34. *Les livres du roy Modus et de la royne Ratio. Publiés avec introduction, notes et glossaire*, éd. G. Tilander, Société des anciens textes français 74 (Paris, 1932), e.a. chapitre 91.
35. *De animalibus…*, éd. Stadler p. 1471.
36. TLF 5, p. 528.
37. Godefroy F., *Dictionnaire de l'ancienne langue française et de tous ses dialectes du IXᵉ au XVᵉ siècle composé d'après le dépouillement de tous les plus importants documents manuscrits ou imprimés qui se trouvent dans les grandes bibliothèques de la France et de l'Europe et dans les principales archives départementales, municipales, hospitalières ou privées* (Paris, 1969), 10 vol., dorénavent Gdf; ici 9, p. 320.
38. Van den Abeele, *La fauconnerie au Moyen Âge…*, p. 106. Pour une reproduction des jets, voir Willemsen, *Das Falkenbuch…*, fol. 62v et 63v.
39. Federico II, *De arte venandi…*, éd. Trombetti Budriesi, II. 31, p. 316.

qu'avec eux, les faucons sont "jetés" et envoyés pour prendre une proie.'[40].

Le mot *iactus* est absent chez Albert le Grand. Cette soi-disant absence des jets peut cependant s'expliquer de deux façons, dont la première est l'emploi du terme général *ligatura* ('entravement') avec le sens spécifique de 'jets'[41]. Une seconde explication possible se trouve dans le passage 'gyros liga in cruribus eius prope pedes interiecto spatio unius palmae (...)'[42], dans lequel *gyros* a le sens de 'anneau'[43]. Albert le Grand a copié cette phrase du *Liber de natura rerum* de Thomas de Cantimpré ('liga giros in cruribus eius prope pedes interiecto spatio unius palme')[44], mais la source de ce texte, à savoir le paragraphe 38.7 de l'*Epistola Aquilae, Symmachi et Theodotionis ad Ptolomeum regem*, contient la proposition suivante: 'liget ei gictos prope pedes quantum tenet palma'[45]. Le mot *gictos* signifie évidemment 'jets'. On pourrait donc conclure que Thomas de Cantimpré et ensuite Albert le Grand se sont probablement trompés, surtout si on sait que Frédéric II recommande également la longueur d'un *palmus* pour les jets[46].

Quant aux versions vernaculaires, les deux traductions françaises du *De arte venandi* rendent *iactus* par *jet* qui est bien attesté avec le sens cynégétique depuis ca. 1185[47]. Cette forme se rencontre pas moins de 95 fois dans la seconde traduction[48]. Cette traduction se rencontre également une seule fois dans *Des faucons II* et *III* pour *ligatura*, alors que le traducteur de *Des faucons I* a opté pour un terme général, à savoir *loyures* ('lanières'). Dans *Des faucons II*, le rapport avec le substantif *ligatura*

40. Frédéric II de Hohenstaufen, *L'art de chasser...*, trad. Paulus & Van den Abeele, p. 209.

41. Il s'agit des faucons qui *sine ligaturis* entrent et sortent de la maison (IV.4; *De animalibus ...*, éd. Stadler, p. 1457).

42. 'Attache les anneaux à ses jambes, près des pieds, et laisse l'espace d'une paume entre les anneaux' (XXIII.39; *De animalibus...*, éd. Stadler, p. 1492).

43. Latham R.E., *Dictionary of Medieval Latin from British Sources* (London, 1975-), p. 1120.

44. Thomas Cantimpratensis, *Liber de Natura Rerum*, éd. H. Boese (Berlin, 1973), 5.50, p. 201.

45. Van den Abeele B., *Les traités de fauconnerie latins au Moyen Age* (Thèse de doctorat non publiée) (Louvain-la-Neuve, 1991), annexes p. 54.

46. Van den Abeele, *La fauconnerie au Moyen Âge...*, p. 106.

47. Kluckow F., *Hue de Rotelande, Protheselaus* (Göttingen - Halle, 1924), v. 7491. Le sens non cynégétique ('action de jeter') est attesté quelques décennies plus tôt (1155; TLF 10, 689).

48. Faute d'une édition électronique de la première traduction (cf. supra), nous pouvons seulement signaler que le glossaire de la première version énumère seize occurrences de *jet*, suivies de la mention *etc*. Cf. Federico II, *De arte venandi...*, éd. Trombetti Budriesi, p. 592.

reste quand même clair, parce que *get* y figure dans le couple *getz ne aultre lieure*.

La traduction de *gyros* s'est révélée plus compliquée, à cause de la confusion entre *gictos* et *gyros* qui est mentionnée ci-dessus. D'ailleurs, contrairement au traducteur anglais[49], aucun traducteur français n'a essayé de corriger la source sur ce point. Le traducteur de *Des faucons I* a rendu *gyros* par *tourés*. De cette façon, il a introduit un autre accessoire, à savoir le touret. Celui-ci est un 'anneau double qui empêche les jets d'un faucon ou toute autre courroie de s'embrouiller'[50]. Le substantif *gyros* est donc rendu correctement par 'anneaux', puisque le touret se compose d'un anneau double, mais on ne peut pas identifier les jets avec les anneaux du touret, étant donné que les jets s'attachent aux jambes du faucon, alors que le premier anneau du touret est attaché à l'extrémité des jets et l'autre à la longe. Toutefois, le recours à un terme typiquement cynégétique comme *tourés* montre que le traducteur connaissait bien le langage des fauconniers et était peut-être lui-même un praticien de la chasse au vol. Le mot *touret* date du XIIe siècle, où il avait le sens de 'poulie'[51], et est employé dans le domaine de la fauconnerie dès 1228[52].

Le traducteur de *Des faucons II* n'a pas repris le paragraphe qui contient ce mot et n'a donc pas dû chercher une solution pour rendre *gyros*, alors que le traducteur de *Des faucons III* a interprété différemment le passage et n'a pas vraiment rendu *gyros*[53].

Frédéric II mentionne également le touret qu'il décrit avec précision:

49. Le compilateur anonyme du *Prince Edward's Book*, qui a repris ce passage en se basant sur Albert le Grand, a traduit *gyros* par *jesses*. Cf. Swaen A.E.H., 'The booke of Hawkynge after Prince Edwarde Kyng of Englande and its relation to the Book of St Albans', *Studia neophilologica* 16 (1943), p. 17.

50. Gdf 7, 753. Pour plus d'informations sur le touret cf. Van den Abeele, *La fauconnerie au Moyen Âge...*, p. 109-110. Pour une reproduction, voir Willemsen, *Das Falkenbuch...*, fol. 64v, même si l'illustration ne correspond pas tout à fait à l'objet décrit. Une photo moderne se trouve dans de Chamerlat C.A., *La fauconnerie et l'art* (Courbevoie, Paris, 1986), p. 53, qui constitue une reproduction de l'ouvrage de Schlegel et Verster de Wulverhorst.

51. TLF 16, 391.

52. *Le roman de la Violette ou de Gerart de Nevers par Gerbert de Montreuil*, éd. D.L. Buffum, Société des anciens textes français 72 (Paris, 1928), p. 101.

53. Ce traducteur mentionne qu'il faut attacher un lièvre aux jambes d'un jeune autour afin que celui-ci apprenne à chasser le lièvre. Cette façon d'affaiter l'oiseau rapace ressemble aux pratiques décrites par Albert le Grand aux paragraphes XV.2 ('dimittantur aves de pedibus eorum evadere', *De animalibus...*, éd. Stadler, p. 1469), XVII.2 ('saepe evadentes desub unguibus eius', *De animalibus...*, éd. Stadler, p. 1472) et XXII.3 ('fac eam (= columbam) saepius desub pedibus eius effugere et ut iterum capiat eam', *De animalibus...*, éd. Stadler, p. 1488), c'est-à-dire à l'entraînement à l'aide d'un gibier d'escap (cf. Van den Abeele, *La fauconnerie au Moyen Âge...*, p. 133), mais nous ne croyons pas que ce soit ici le sens voulu par le dominicain.

> Est autem tornettum quoddam de duobus anulis conpositum girantibus in se invicem, et hoc modo est factum: sunt duo anuli ferrei aut enei aut argentei aut de alio metallo facti, magni ad quantitatem anulorum, qui sunt in iactis[54].
>
> 'Le touret est un objet composé de deux anneaux tournant l'un sur l'autre et il est fait de la manière suivante. Ce sont deux anneaux de fer, de bronze, d'argent ou d'un autre métal, de même grandeur que ceux qui garnissent les jets'[55].

Le latin *tornettum* a été traduit par deux mots dans les deux traductions de ce traité: *tournet* et *touret*. *Chace Oisiaux I* utilise plus souvent *touret* (neuf fois *touret* et une fois *tournet*), comme dans *Des faucons I*, et *Chace Oisiaux II* plus souvent *tournet* (neuf fois *tournet* et une fois *touret*). Étant donné que *tornettum* n'est attesté en latin que dans le *De arte venandi* et que le français *toret* existe avec le sens cynégétique depuis 1228, il est probable que Frédéric II ait forgé le mot latin à partir du français *tournet*.

3. Les accessoires du fauconnier

3.1. La filière

Lors des séances d'apprentissage de rappel au leurre (cf. infra), on attache la longe du faucon à une longue ficelle qui permet de le retenir et de le faire revenir au fauconnier[56]. Pour désigner cet accessoire, Frédéric II utilise *fileria* en alternance avec *credentia*[57]. Ces deux noms sont synonymes et définis au livre III de la version en six livres, consacré à l'affaitage au leurre et à l'escape:

> Funiculus autem iste dicitur fileria, que a quibusdam credentia dicitur, pro eo quod in primis recessibus avis de manu portantis et in primis emissionibus eius ad loyrum, ex quibus avis docetur non fugere ab

54. Federico II, *De arte venandi...*, éd. Trombetti-Budriesi, II. 37, p. 324.
55. Frédéric II de Hohenstaufen, *L'art de chasser...*, trad. Paulus & Van den Abeele, p. 212.
56. Cf. Van den Abeele, *La fauconnerie au Moyen Âge...*, p. 119. Pour une illustration, voir *Les livres du roy Modus...*, éd. Tilander, contenant la reproduction du folio 59 du ms. Paris, BNF, fr. 12399.
57. On trouve plus souvent *fileria* que *credentia* au livre III, et au livre II une fois le couple *fileria seu credentia* (P II.7) et une fois *fileria* (II.36) seul (cf. Federico II, *De arte venandi...*, éd. Trombetti-Budriesi, p. 284 et 324). Nous n'insistons pas sur le mot *funiculus*, employé onze fois aux livres III et V (y compris la citation ci-dessus), étant donné qu'il ne figure pas dans les livres traduits en français. Signalons simplement qu'il signifie 'corde' et est donc un hyperonyme de *fileria* et *credentia*.

homine, non creditur ipsi avi, quoniam ipsa, memorans suam agrestitatem, fugeret et libertatem potius eligeret quam super loyrum sedere, set creditur illi filerie que credentia dici potest[58].

'Cette cordelette est dite "filière" (*fileria*), et certains l'appellent "créance" (*credentia*) parce que, lors des premiers départs de la main et des premiers envols de l'oiseau vers le leurre, quand on lui apprend à ne pas fuir l'homme, celui-ci ne se fie pas (*non credit*) à l'oiseau: au souvenir de son état sauvage, il s'enfuirait et choisirait la liberté plutôt que de se poser sur le leurre. On se fie alors plus à la filière, qui peut dès lors être dite créance'[59].

Albert le Grand utilise *zona* pour cet accessoire (XXII.2)[60], mot d'origine grecque signifiant 'ceinture'[61].

Les traducteurs du *De arte venandi* n'emploient que *filiere*, un mot qui est attesté vers 1230 dans la séquence *fil de filieres*[62]. En ce qui concerne le sens cynégétique, il se rencontre quatre fois au XIVe siècle, la première fois dans *Chace Oisiaux I*, puis de manière continue depuis le XVIe siècle. Il est probable que *fileria* soit l'adaptation du français *filiere*, sachant que d'autres occurrences du mot latin sont seulement attestées à partir du XIVe siècle et pas dans le sens cynégétique.

Pour le *De falconibus,* il n'y a que deux traducteurs qui reprennent le passage correspondant: celui de *Des faucons I* a choisi le mot courant *corroie* qui n'est pas attesté ailleurs dans le sens de 'filière', tandis que celui de *Des faucons III* connaît le terme technique *filiere*.

3.2. Le leurre

Le leurre est un 'simulacre d'oiseau formé de deux ou plusieurs ailes attachées ensemble et tenues au bout d'une lanière, que l'on fait tournoyer pour faire revenir le faucon'[63]. Il arrive que les fauconniers 'acharnent' le leurre d'un morceau de chair en guise d'appât[64]. Cette façon de

58. Federico II, *De arte venandi...*, éd. Trombetti-Budriesi, III. 29, p. 588.

59. Frédéric II de Hohenstaufen, *L'art de chasser...*, trad. Paulus & Van den Abeele, p. 310-311.

60. *De animalibus...*, éd. Stadler, p. 1488.

61. Ernout A. & Meillet A., *Dictionnaire étymologique de la langue latine. Histoire des mots* (Paris, 1959⁴), p. 759.

62. Jean Renart, *Le roman de la rose ou de Guillaume de Dole*, éd. F. Lecoy, Classiques français du moyen âge 91 (Paris, 1963), v. 275.

63. Frédéric II de Hohenstaufen, *L'art de chasser...*, trad. Paulus & Van den Abeele, p. 303. Pour une illustration, voir *Les livres du roy Modus...*, éd. Tilander, contenant la reproduction du fol. 59 du ms. Paris, BNF, fr. 12399.

64. Al Gitrif ibn Qudama al-Gassani (VIIIe siècle), *Traité des oiseaux de vol (Kitab Dawari at-tayr)*, éds F. Viré & D. Möller, Ouvrage préparé pour l'édition par B. Van den Abeele, Bibliotheca cynegetica 3 (Nogent-le-Roi, 2002), p. 46.

rappeler l'oiseau se pratique essentiellement avec les faucons (à l'exception du gerfaut et de l'émerillon), qui y sont dressés et qui sont appelés dès lors des 'oiseaux de leurre'[65]. Les oiseaux de bas vol, tout comme le gerfaut et l'émerillon, sont des 'oiseaux de reclaim', c'est-à-dire qu'ils reviennent quand on leur tient un morceau de chair dans la main[66].

Pour désigner le leurre, Frédéric II est le seul auteur à utiliser *loyrum*, c'est-à-dire la latinisation du mot français *leurre*[67], alors qu'Albert le Grand emploie trois termes différents, à savoir *reclamatorium, revocatio* et *revocatorium*[68]. En outre, le dominicain donne cette fois-ci une définition de l'accessoire, mais sans en préciser le nom: 'circumducitur in corda quiddam ex quatuor alis vel pluribus colligatis ad modum avis colligatum cui caro recens superligatur'[69].

À notre connaissance, le nom *reclamatorium* se rencontre seulement chez Thomas de Cantimpré (1 occurrence[70]) et chez Albert le Grand, qui l'emploie trois fois[71]. Ce terme n'est repris que par deux dictionnaires, à

65. Al Gitrif ibn Qudama al-Gassani, *Traité des oiseaux*..., éds Viré & Möller, p. 46.

66. Cette opposition entre oiseaux de reclaim et oiseaux de leurre constitue même la base de la division des oiseaux rapaces chez le roi Modus (cf. Van den Abeele, *La fauconnerie au Moyen Âge*..., p. 119).

67. Cf. Van den Abeele, *La fauconnerie au Moyen Âge*..., p. 118.

68. Frédéric II emploie aussi *revocatio* (huit fois) et *revocatorium* (une fois), mais pas pour désigner l'accessoire. Au livre II du *De arte venandi* il n'y a qu'une occurrence de *revocatio* où il s'agit manifestement du sens de 'rappel à l'aide de cris', et cela dans le passage suivant: 'Sit homo magne vocis et fortis nam avis sua que aliquando stabit remota ab ipso, ad revocationem suam quam audiet, et socii sui quando erunt remoti, eius vocem, si magna fuerit, facilius comprehendant' ('Sa voix sera ample et forte, ainsi son oiseau, lorsqu'il sera éloigné, entendra mieux son rappel et ses compagnons de chasse, à distance, le comprendront plus facilement si sa voix est ample'; II.68, Federico II, *De arte venandi*..., éd. Trombetti-Budriesi, p. 350; Frédéric II de Hohenstaufen, *L'art de chasser*..., trad. Paulus & Van den Abeele, p. 221). La seule occurrence de *revocatorium* se trouve au livre III.16: 'Si non volabit illuc expedit quod revocet eum cum revocatorio que usus est revocare scilicet cum galina vel columbo'. ('S'il ne vole pas vers eux, il convient de le rappeler avec un mode de rappel auquel il est habitué, c'est-à-dire une poule ou un pigeon'; Federico II, *De arte venandi*..., éd. Trombetti-Budriesi, p. 578; Frédéric II de Hohenstaufen, *L'art de chasser*..., trad. Paulus & Van den Abeele, p. 306-307). Le premier traducteur rend *revocatio* par *son apel* (II.50.15) et le second par *rappel de sa voix* (50.17); le terme *revocatorium* figure dans un livre qui n'est pas traduit en français. Il pourrait s'agir du sens 'appât de rappel' (à distinguer du leurre). Seul le contexte permet d'apporter une précision. De manière générale, disons que *revocatio* et *revocatorium* ont le sens de 'rappel, mode de rappel'.

69. 'On tourne en rond un objet attaché à une corde, lequel objet se compose de quatre ailes ou plus, comme un oiseau, et auquel on attache de la viande' (IV.1; *De animalibus*..., éd. Stadler, p. 1456).

70. Thomas Cantimpratensis, *Liber de Natura Rerum*..., éd. Boese, 5.46, p. 197. Nous remercions madame A. Paulus d'avoir attiré notre attention sur cette occurrence.

71. Voir les paragraphes IV.4, V.6 et IX.2 (*De animalibus*..., éd. Stadler, p. 1457, 1458 et 1464).

savoir ceux de L. Diefenbach[72], avec le sens de 'leurre', et de R.E. Latham[73], avec le sens de 'rappel' (en fauconnerie) et avec la datation 'avant 1446' (mais sans indication de source). L'origine du substantif est claire, parce qu'il est basé sur le verbe *reclamare*, qui est un composé du verbe *clamare* ('crier')[74]. Le rapport avec le verbe 'crier' s'explique probablement par le fait que le fauconnier crie le plus souvent quand il leurre le faucon. Le mot *reclamatorium* peut cependant prêter à confusion parce qu'il ressemble à *reclaim* (cf. infra)[75]. Les deux autres substantifs, *revocatio* et *revocatorium*, comptent chacun une seule occurrence au paragraphe VII.7[76]. Selon le dominicain, il ne faut pas rappeler le faucon montagnard quand il est en colère, mais *dissimulare revocationem*[77]. En outre, quand le faucon montagnard fâché ne revient pas *ad revocatorium*, il n'existe pas de véritable danger parce qu'il reviendra de lui-même quand sa colère sera passée.

Dans les deux traductions françaises du *De arte venandi*, *loyrum* est traduit par *loirre*[78]. Ce mot, qui est attesté dans le sens cynégétique dès le début du XIIIe siècle[79], figure également dans *Des faucons I* et *III* comme traduction de *reclamatorium*. Le dernier traducteur se sert également du dérivé *leurrer*, qui est attesté au XIIIe siècle[80]. De son côté, le traducteur de *Des faucons II* a voulu rester plus proche du texte latin, c'est pourquoi il recourt d'abord à un redoublement synonymique: *reclamatore ou leurre* (V.6), pour se limiter ensuite au premier élément de ce binôme (IX.2)[81].

72. Diefenbach L., *Novum glossarium latino-germanicum mediae et infimae aetatis. Beiträge zur wissenschaftlichen Kunde der neulateinischen und der germanischen Sprachen* (Aalen, 1964, 1867[1]), p. 314.

73. Latham R.E., *Revised Medieval Latin Word-list from British and Irish Sources* (London – Oxford, 1965), p. 394.

74. Ernout & Meillet, *Dictionnaire étymologique*..., p. 124.

75. Cf. Van den Abeele, *La fauconnerie au Moyen Âge*..., p. 118 et 132.

76. *De animalibus*..., éd. Stadler, p. 1460.

77. *De animalibus*..., éd. Stadler, p. 1460.

78. *Chace Oisiaux I* emploie *loirres* (II.1.29) et *lurre* (II.77.21), *Chace Oisiaux II, loire* (P.34) et *loirre* (79.21). Peu présent dans le deuxième livre, cet accessoire est décrit en détail au livre III qui lui consacre 70 sections.

79. Van Hamel A.-G., *Li romans de carité et Miserere du renclus de Moiliens, poèmes de la fin du XIIe siècle*, Bibliothèque de l'École des hautes études, Sciences philologiques et historiques 61-62 (Paris, 1885), II, p. 135.

80. Pour le sens cynégétique, il faut attendre le XIVe siècle: on le trouve dans le *Castelain de Couci* de Jakemes, v. 479 (*Le roman du castelain de Couci et de la dame de Fayel par Jakemes*, éds J.E. Matzke & M. Delbouille, Société des anciens textes français 81 (Paris, 1936)). Signalons cependant qu'il existe une occurrence de l'adjectif *luirié* datant de ca. 1130 (*Le Bestiaire de Philippe de Thaün*, éd. E. Walberg (Lund - Paris, 1900), v. 1777). Le dépouillement de nouveaux textes permettra donc sans doute de rétrodater l'apparition du verbe et du substantif.

81. *Des faucons II* ne contient pas de traduction pour la première occurrence de *reclamatorium* (IV.4).

Ce mot est un latinisme qui ne figure pas dans d'autres ouvrages, et le second élément du binôme sert dès lors à l'expliquer.

Quant à *revocatorium* et *revocatio*, ces substantifs peuvent désigner aussi bien les cris pour rappeler le faucon que le leurre; par conséquent, on retrouve les deux interprétations dans les versions françaises. Ainsi, le traducteur de *Des faucons I* rend *revocatorium* par le verbe courant *apeler* (et ne reprend pas *revocatio*), optant donc pour le rappel à l'aide des cris. Dans *Des faucons II*, le substantif *revocatio* est rendu par le verbe *rappeler* et *revocatorium* par *reclaim*. En ce qui concerne le verbe *rappeler*, le second traducteur a donc également opté pour le rapport avec le verbe 'crier'[82]. La traduction *reclaim* est plus ambigüe, parce que ce substantif désigne à partir du XIIe siècle le cri par lequel on rappelle les oiseaux de chasse[83], mais à partir du siècle suivant également ce qu'on montre pour faire revenir l'oiseau[84], surtout les oiseaux de bas vol[85]. Dans *Des faucons III*, par contre, *revocatorium* est rendu par *leurre*, alors que le correspondant du substantif *revocatio* y est omis. Le traducteur de cette version est donc le seul à considérer les substantifs *reclamatorium* et *revocatorium* comme synonymes.

3.3. Le perchoir

Finalement, nous étudions un dernier objet qui ne fait pas partie de l'équipement du faucon ou du fauconnier à proprement parler, mais qui est indispensable à l'aménagement de la fauconnerie, à savoir l'endroit où on tient les oiseaux rapaces. Il s'agit du perchoir sur lequel est placé l'oiseau rapace qui ne chasse pas. Généralement, c'est une pièce de bois longue et mince, de section circulaire et reposant sur des pieds de hauteur variable. Cet accessoire est désigné dans le *De arte venandi* ainsi que dans le *De falconibus* par le nom *pertica*.

Si nos deux auteurs mentionnent le même accessoire, leur avis sur son utilité est cependant bien différent. En effet, Albert le Grand, s'opposant à sa source pour ce passage[86], déconseille au paragraphe XVII.9 l'usage

82. Pour faire revenir le faucon, le verbe *reclamer* est cependant plus fréquent dans cette version.

83. FEW 10, 153.

84. Tilander G., *Glanures lexicographiques*, Actae regiae societatis humaniorum litterarum Lundensis 16 (Lund, 1932), p. 218-219.

85. Van den Abeele, *La fauconnerie au Moyen Âge...*, p. 129.

86. Il s'agit du traité *Dancus rex* daté du début du XIIe siècle, cf. *Dancus rex, Guillelmus falconarius, Gerardus falconarius. Les plus anciens traités de fauconnerie de l'Occident publiés d'après tous les manuscrits connus*, éd. G. Tilander, Cynegetica 9 (Lund, 1963), p. 92.

des perchoirs en bois pour les faucons en captivité, sous prétexte qu'à l'état sauvage ceux-ci perchent plutôt sur des rochers ou par terre[87].

Il existe encore une seconde différence entre les deux textes latins, parce que chez Frédéric II le perchoir ou la perche n'est qu'un des 'sièges' possibles. En effet, l'empereur consacre vingt-deux sections aux différents types de 'sièges' sur lesquels on peut placer les faucons:

> Est igitur sedes falconis id super quod falco ligatur ad quiescendum quando de manu deponitur. Sedium autem una est que pertica dicitur, alia est que approbato vocabulo sedile seu seditorium nuncupatur[88].
>
> 'Le siège est ce sur quoi l'oiseau est toujours attaché pour se reposer lorsqu'on le fait descendre de la main. Parmi les sièges, l'un se dénomme perche et l'autre s'appelle de façon appropriée bloc (*sedile seu seditorium*)'[89].

Le premier type de perchoir est alors la perche qui peut être haute ou basse mais toujours de bois:

> - Alta igitur pertica debet esse lignea ampla desuper ad mensuram unius pedis[90].
>
> 'La perche haute doit être en bois et sa surface de la largeur d'un pied'[91].
>
> - Pertica vero ima duobus modis esse potest: lignea rotunda aut lignea quadrata[92].
>
> 'La perche basse peut être faite selon deux modes: de bois et ronde, ou de bois et carrée'[93].

L'autre type de perchoir est le bloc, pied massif en bois ou en pierre auquel on attache les faucons[94]. Pour le terme général, l'empereur emploie le substantif *sedes*, pour la perche *pertica* et pour le bloc surtout le nom *sedile*, mais parfois aussi *seditorium* ou *sessorium*.

87. *De animalibus...*, éd. Stadler, p. 1474.
88. Federico II, *De arte venandi...*, éd. Trombetti Budriesi, II. 106, p. 378.
89. Frédéric II de Hohenstaufen, *L'art de chasser...*, trad. Paulus & Van den Abeele, p. 231.
90. Federico II, *De arte venandi...*, éd. Trombetti Budriesi, II. 107, p. 378. Pour une illustration, voir Willemsen, *Das Falkenbuch...*, fol. 74v (et al.).
91. Frédéric II de Hohenstaufen, *L'art de chasser...*, trad. Paulus & Van den Abeele, p. 231-232.
92. Federico II, *De arte venandi...*, éd. Trombetti Budriesi, II. 110, p. 382.
93. Frédéric II de Hohenstaufen, *L'art de chasser...*, trad. Paulus & Van den Abeele, p. 233.
94. Federico II, *De arte venandi...*, éd. Trombetti Budriesi, II. 111, p. 382. Pour une illustration, voir Willemsen, *Das Falkenbuch...*, fol. 75.

Le substantif *pertica* est toujours traduit par le substantif *perche*[95] et une fois, dans *Des faucons II*, par le verbe *percher* employé de façon transitive. Le nom *perche* apparaît au XIIe siècle en français et est attesté dans le sens cynégétique depuis le XIIIe siècle[96]. Quant au verbe *percher*, l'emploi transitif se rencontre pour la première fois dans le *Ménagier de Paris*, rédigé vers 1393[97].

Chace Oisiaux I traduit toujours *sedes* par *siege* et *sedile* par *sedile*, alors que *Chace Oisiaux II* utilise *siege* à la fois pour *sedes* et pour *sedile*, ce qui l'oblige parfois à préciser de quoi il parle[98]. Le cas de *seditorium* est différent, car les deux traducteurs ont rendu le doublet latin *sedile seu seditorium* par un seul mot, *sedile* dans *Chace Oisiaux I* et *siege* dans *Chace Oisiaux II*. Enfin, *sessorium* est traduit par *siege* dans les deux traductions. Signalons encore que par la différence dans la traduction (*siege* et *sedile* versus *siege* seul), cet accessoire est le seul à refléter la tendance propre à chaque traducteur du *De arte venandi*, puisque la première traduction est généralement plus fidèle et plus proche du modèle latin que la seconde.

Dans son sens principal de 'meuble ou tout autre objet fait pour s'asseoir', *siege* est bien attesté depuis le XIIe siècle[99]. Le sens cynégétique apparaît trois fois, mais uniquement dans des traités de fauconnerie: la première fois dans *Chace Oisiaux I*, la deuxième dans *Chace Oisiaux II* et la troisième et dernière à la fin du XIVe siècle[100].

4. Synthèse

Le *De arte venandi cum avibus* de l'empereur Frédéric II est sans conteste le traité de fauconnerie le plus important de l'époque médiévale.

95. Quant au nombre d'occurrences, le glossaire de *Chace Oisiaux I* signale 5 occurrences suivies de la mention *etc* (Federico II, *De arte venandi*..., éd. Trombetti Budriesi, p. 597), *Chace Oisiaux II* compte 195 occurrences, *Des faucons I* 5, *Des faucons II* 3 et *Des faucons III* 6.

96. FEW 8, 278.

97. *Le mesnagier de Paris*, éd. G.E. Brereton & J.M. Ferrier (Oxford, 1981), p. 149, 11.

98. Cela se produit deux fois au chapitre 54: 54.36 'sur le quel siege que ly fauconniers vieut des .iij. sieges dis, c'est assavoir ou sur la basse perche ou sur la haulte ou sur le siege', et 54.37 'au quel convient des .iij., soit sieges, soit perche'. Dans les deux cas, *siege* traduit le latin *sedes* (Federico II, *De arte venandi*..., éd. Trombetti Budriesi, II. 120, p. 392).

99. FEW 11, 409.

100. Adam des Aigles, *Traité de fauconnerie d'Adam des Aigles*, éd. Å. Blomqvist, Studia romanica holmiensia 5 (Karlshamn, 1966), 19, 3.

De son côté, le *De falconibus* d'Albert le Grand mérite également notre attention bien qu'il soit beaucoup plus court. En outre, déjà au moyen âge, ces deux ouvrages furent traduits plusieurs fois en français.

Dans cette contribution, nous avons commenté les dénominations des accessoires du faucon et du fauconnier communs à ces deux traités du XIIIe siècle. Ces accessoires, pour lesquels les fauconniers médiévaux utilisaient souvent une terminologie spécifique, faisaient partie de leur vie quotidienne. Dans la plupart des cas discutés ici, le lexique cynégétique du moyen âge est encore employé par les fauconniers actuels[101].

Notre attention s'est essentiellement portée sur les deux traductions françaises du *De arte venandi* et trois des quatre traductions françaises du *De falconibus*. Mais comme un traducteur dépend toujours de son modèle, nous voulons tout d'abord signaler quelques différences entre les deux traités latins. En premier lieu, Frédéric II attache beaucoup d'importance aux différents accessoires et en donne souvent une définition détaillée, alors que les informations sommaires d'Albert le Grand se limitent généralement à quelques observations plutôt anecdotiques dispersées dans tout le traité. Cette différence de traitement a évidemment des conséquences terminologiques: ainsi, il arrive qu'Albert le Grand emploie différents termes pour désigner le même objet là où l'empereur se limite à un seul mot. Le cas le plus remarquable est sans aucun doute le nom latin du chaperon: *capellus* chez Frédéric II, *capellus, capitegium, mitra* et *pileus* / *pilleare* chez Albert le Grand. De même, le dominicain emploie le substantif *reclamatorium* pour désigner le leurre, mais *revocatio* et *revocatorium* peuvent être traduits de la même façon (à côté de la traduction plus courante de '(r)appel'), alors que Frédéric II se sert toujours de *loyrum*. À l'inverse, Frédéric II emploie aussi bien *fileria* que *credentia* (et *funiculus*) pour la filière, contre seulement *zona* chez Albert le Grand, et l'empereur emploie également différents noms pour le bloc (*sedile, seditorium* ou *sessorium*). Toutefois, contrairement à Albert le Grand, Frédéric II insiste souvent sur l'équivalence des termes, ce qui facilite la tâche du lecteur: *filera seu credentia* (II P.7) et *Funiculus autem iste dicitur fileria, que a quibusdam credentia dicitur* (III.29) pour les différents noms de la filière; *sedile seu seditorium* (II.106) pour le bloc. Une autre complication du *De falconibus* est qu'Albert le Grand emploie souvent des mots qui ne se rencontrent pas ailleurs, au moins pas dans le contexte cynégétique. Par exemple, il est le seul à employer *capitegium, mitra* et

101. Boyer A. & Planiol M., *Traité de fauconnerie et d'autourserie* (Paris, 1948), p. 65-77.

pileus pour le chaperon, *reclamatorium* pour le leurre ou encore *zona* pour la filière.

Cela met en évidence qu'au XIIIe siècle, la désignation latine des accessoires de l'équipement du faucon et du fauconnier n'est pas unifiée, ce qui ne facilite pas le travail des traducteurs.

Le nom latin le plus simple de tous ceux que nous avons discutés est probablement *pertica*, que les cinq traducteurs rendent par *perche*[102]. De même, et cela malgré une terminologie latine plus compliquée, les traducteurs ont toujours reconnu sous les différentes dénominations le chaperon, la différence *chapel – chaperon* étant seulement d'ordre chronologique. Pour les autres accessoires, les deux traducteurs du *De arte venandi* présentent toujours la même traduction correcte: *jet* pour *iactus*, *tour(n)et* pour *tornettum*, *filiere* pour *fileria* / *credentia* et *loire* pour *loyrum*. Seul le cas du siège et du bloc est plus compliqué. En respectant la distinction entre *sedes* et *sedile*, traduits respectivement par *siege* et *sedile*, le traducteur de *Chace Oisiaux I* montre qu'il reste plus proche du texte latin que celui de *Chace Oisiaux II*, qui utilise *siege* dans les deux cas.

Quant aux traductions du *De falconibus*, le traducteur de *Des faucons III* emploie une terminologie similaire à celle des traductions frédériciennes: *jet* pour *ligatura*, *filiere* pour *zona* et *leurre* pour *reclamatorium* et *revocatorium*. De son côté, le premier traducteur se sert généralement de mots plus courants: *loyure* pour *ligatura*, *corroie* pour *zona* et le verbe *appeler* pour *revocatorium*. Les seuls termes plus spécifiques sont *leurre* pour *reclamatorium* et *touret*, qui rend le substantif problématique *gyros*. La préférence pour des termes plus généraux pourrait être due aussi bien à une connaissance insuffisante du traducteur du vocabulaire technique (ce qui est cependant contredit par l'emploi des substantifs plus spécifiques comme *touret*), qu'à l'imprécision des mots employés par Albert le Grand (mais *Des faucons II* et *III* ont employé la même source) ou à une rédaction précoce, au moment où le vocabulaire cynégétique n'est pas encore fixé (mais *Chace Oisiaux I* et probablement aussi *Chace Oisiaux II* ont été rédigés plus tôt et emploient une terminologie plus spécifique). Comme chaque argument pris individuellement peut être démenti, il s'agit probablement d'une combinaison de ces facteurs. Finalement, le traducteur de *Des faucons II* montre surtout sa volonté de rester proche du texte latin, même s'il connaît la variante française plus commune. Ainsi, il rend *ligatura* par *getz ne autre lieure* et la première occurrence de *reclamatorium* est rendue dans cette version par *reclamatore ou leurre*, la seconde

102. Dans *Des faucons II*, nous rencontrons également une occurrence du verbe dérivé *percher*.

simplement par *reclamatore*. Mais au lieu de faciliter la compréhension, les binômes rendent le texte plus compliqué: l'emploi du terme simple, *getz* ou *leurre*, aurait en effet été préférable[103].

5. Conclusion: le développement du vocabulaire technique

Nos textes témoignent d'une nette évolution. Au XIIIe siècle, la terminologie propre au domaine de la fauconnerie n'est pas encore fixée. Frédéric II reconnaît qu'il ne disposait pas toujours du mot juste en latin et qu'il a utilisé celui qui lui semblait le plus approprié:

> Nam cum ars habeat sua vocabula propria quemadmodum et cetere artium et nos non inveniremus in gramatica latinorum verba convenientia in omnibus, apposuimus illa que magis videbantur esse propinqua per que intelligi possit intentio nostra[104].

> 'En effet, comme les autres arts, celui-ci comporte des termes particuliers, et si bien que nous n'avons pas trouvé dans la grammaire latine des mots convenant à tous les cas, nous avons recouru à ceux qui paraissaient les plus proches et par lesquels notre propos pouvait être compris'[105].

Albert le Grand et Frédéric II jouissaient donc encore d'une grande liberté d'expression, comme il ressort de l'observation que, pour désigner les accessoires du faucon et du fauconnier, les mots communs aux deux traités latins se limitent à *capellus* et *pertica*. Dans les traductions françaises, on trouve plus de similitudes: *chapel / chaperon, touré, jet, perche, loirre* et *filiere*. Cela montre une bonne compréhension du texte de la part des traducteurs et nous informe en même temps sur une certaine fixation du vocabulaire propre à la fauconnerie à l'époque de la rédaction de ces traductions. Si on regarde les premières attestations des sens cynégétiques des noms français, nous constatons que plusieurs d'entre elles se situent au XIIIe siècle: *chapel, toret, loirre, reclaim* et

103. Si nous comparons ces binômes avec ceux qui se présentent dans la seule traduction italienne complète (ms. Dresden, Sächsische LB, Ob 21, fol. 107v-153v, toujours inédit), nous constatons que l'utilisation des binômes s'y fait beaucoup plus dans un souci de compréhension: *mitra overo capelleto* (III.3) avant d'utiliser *capelleto* seul, *giri o circuli* (XXIII.39), *(re)clamatorio overo ludrio* (IV.4, VII.7, IX.2). À propos du leurre, signalons d'ailleurs que ce traducteur italien est le seul à ajouter une glose à la définition donnée par Albert le Grand: 'un certo composito de quatro ale overo più, a la similitudine de uno ludrio, ligato a una corda sopra lo quale se li pone la carne frescha' (IV.1).

104. Federico II, *De arte venandi…*, éd. Trombetti Budriesi, P.I.1, p. 4.

105. Frédéric II de Hohenstaufen, *L'art de chasser…*, trad. Paulus & Van den Abeele, p. 58.

perche. Le XIIIe siècle semble donc occuper une position centrale dans la formation du vocabulaire cynégétique, aussi bien en latin – l'époque de la rédaction des deux traités latins discutés ici – qu'en français[106].

Au lieu d'avoir d'abord un lexique latin qui fut traduit ensuite en français, il semble qu'il s'agisse ici d'un processus contemporain, avec des influences mutuelles. Un exemple concret est le nom *loyrum*, employé par Frédéric II, qui trahit clairement son origine française, et ce même raisonnement vaut probablement aussi pour *tornettum* et *fileria*. À ce sujet, n'oublions pas que la Sicile fut longuement en mains normandes et que Frédéric II fut l'héritier des traditions politiques normandes en Sicile, où le français était donc la langue de la cour. Par leur nature même de biais entre deux langues, les traductions ont ensuite joué un rôle important pour la création, la fixation ultérieure et la diffusion du vocabulaire technique. Comme les traductions présentées ici, en particulier *Chace Oisiaux I*, mais aussi *Chace Oisiaux II* et dans une moindre mesure *Des faucons I*, figurent parmi les premiers traités de fauconnerie en langue vernaculaire dans lesquels il est question d'accessoires, ces traductions témoignent à nouveau de l'importance du travail des traducteurs dans le développement lexical du moyen français[107].

Leuven – Strasbourg

106. En outre, la présence de ce vocabulaire dans des textes littéraires qui offrent souvent la première attestation (*jet*, *loirre* et *toret*), indique que la volerie était familière au public du XIIIe siècle. Cf. Van den Abeele B., *La fauconnerie dans les lettres françaises du XIIe au XIVe siècle*, Mediaevalia Lovaniensia 18 (Leuven, 1990).

107. Goyens M. & Van Hoecke W., 'La traduction comme source pour l'étude d'anciens états de langue', *Le Moyen français* 44-45 (2000), p. 243-264.

Géraldine VEYSSEYRE

LE *LIVRE DES PROPRIETÉS DES CHOSES* DE JEAN CORBECHON (LIVRE VI), OU LA VULGARISATION D'UNE ENCYCLOPÉDIE LATINE

0. Introduction

C'est à la demande du roi Charles V qu'en 1372, un moine augustin du nom de Jean Corbechon entreprit de traduire le *Liber de proprietatibus rerum* de Barthélemy l'Anglais[1]. Cette commande prend place au cours d'une période particulièrement active de ce règne dont les enjeux intellectuels sont bien connus[2]: elle est effectuée la même année que celle du *Polycraticus* de Jean de Salisbury par Denis Foulechat[3] et au moment où Nicole Oresme et Raoul de Presles sont en train de traduire respectivement les œuvres d'Aristote[4] et la *Cité de Dieu* de saint Augustin[5].

1. Cette date est mentionnée dans la rubrique qui ouvre le manuscrit BNF fr. 16993 ('Ci commance le livre des proprietéz des choses, translaté de latin en françois l'an mil CCC LXXII par le conmandement du roy Charles…', fol. 1r) et est confirmée par divers documents d'archives. Voir Delisle L., *Recherches sur la librairie de Charles V* (Paris, 1907), t. I, p. 92.

2. Le rôle de Charles V en tant que fondateur de la librairie royale a été mis en évidence de longue date – voir notamment Avril F. & Lafaurie J. (éds), *La librairie de Charles V* (Paris, 1968) –, de même que l'importante impulsion donnée par le roi au mouvement des traductions en français. Pour un panorama complet des commandes royales de *translacions* sous Charles V, voir notamment Dembowski P.F., 'Learned Latin Treatises in French: Inspiration, Plagiarism and Translation', *Viator* 17 (1986), p. 255-269, ici p. 262-263.

3. Brucker Ch., 'Quelques aspects du style de Denis Fouchelat, traducteur de Charles V', *Zeitschrift für französische Sprache und Literatur* 80 (1970), p. 97-101, ici p. 97.

4. C'est au cours des années 1370-1372 que Nicole Oresme achève la traduction du *corpus* moral aristotélicien, et en 1377 qu'il termine sa *translacion* de *Du ciel et du monde* (Mathieu R. & Lefèvre S., 'Nicole Oresme', in: R. Bossuat, L. Pichard & G. Raynaud de Lage (éds), *Dictionnaire des lettres françaises: le Moyen Âge*, édition revue par G. Hasenohr et M. Zink (Paris, 1992), p. 1072-1075, ici p. 1072[b]).

5. Cette *translacion* fut rédigée entre 1371 et 1375 (Beer J.M.A., 'Patronage and Translator: Raoul de Presles's *La Cité de Dieu* et Calvin's *Institution de la religion chrestienne* and *Institutio religionis christianae*', in: J.M.A. Beer & K. Lloyd-Jones (éds), *Translation and Transmission of Culture between 1300 and 1600*, Studies in Medieval Culture 35 (Kalamazoo, 1995), p. 91-142, ici p. 91 et Willard Ch.C., 'Raoul de Presles's Translation of Saint Augustine's *De civitate Dei*', in: J.M.A. Beer (éd.), *Medieval Translators and their Craft* (Kalamazoo, 1989), p. 329-346, ici p. 331).

Pourtant, le *Livre des proprietés des choses* a donné lieu à moins d'investigations que d'autres traductions du XIVe siècle, et le travail de Jean Corbechon reste moins bien connu que celui de Nicole Oresme ou d'Evrart de Conty[6]. Il est vrai que son œuvre, dont l'abondante tradition manuscrite a longtemps découragé les éditeurs[7], n'est facilement accessible que par le biais d'éditions anciennes[8]. En conséquence, hormis une thèse de l'École nationale des chartes, largement biographique et désormais ancienne[9], le *Livre des proprietés des choses* n'a donné lieu qu'à des études ponctuelles[10]. Le portrait qui se dégage de l'ensemble de ces travaux est celui d'un traducteur consciencieux, dont la seule ambition serait de transposer le message élaboré par son modèle d'un système linguistique à un autre sans en infléchir le contenu: l'ensemble de la critique s'accorde à reconnaître la

6. Sur celui-ci, voir notamment Guichard-Tesson Fr., 'Le métier de traducteur et de commentateur au XIVe siècle d'après Evrart de Conty', *Le Moyen Français* 24-25 (1989), p. 131-167; Ducos J., 'Evrart de Conty, médecin et vulgarisateur à la cour de Charles V', in: J.-L. Cabanès (éd.), *Littérature et médecine*, t. II, Bordeaux, 2000, p. 63-74 et Goyens M., 'Comprendre Aristote au Moyen Âge: le procédé de reformulation dans la traduction des *Problèmes* par Evrart de Conty', in: P. Nobel (éd.), *Textes et cultures: réception, modèles, interférences*, t. I: *Réception de l'Antiquité*, Littéraire (Besançon, 2004), p. 145-163.

7. On ne compte pas moins de 36 manuscrits de cette traduction (Seymour M.C., 'Some Medieval and French Readers of the *De proprietatibus rerum*', *Scriptorium* 28-1 (1974), p. 100-103, ici p. 100[a]). Son édition bilingue (latin/ancien français) actuellement en cours sous la direction de Baudouin Van den Abeele et Joëlle Ducos permettra de rendre accessible une version fiable du texte. Sur les prémices de ce projet d'édition, voir Van den Abeele B., Meyer H. & Ribémont B., 'En guise d'avant-propos: éditer l'encyclopédie de Barthélemy l'Anglais; vers une édition bilingue du *De proprietatibus rerum*', *Cahiers de recherches médiévales (XIIIe-XVe siècle)* 6 (1999): *Vulgariser la science: les encyclopédies médiévales*, p. 7-18.

8. On ne compte pas moins de 9 éditions incunables de l'œuvre de Corbechon (Seymour, 'Some Medieval and French Readers…', p. 100[a]), dont celle de 1492 'per Petrum Ungarum' est disponible en ligne sur le site de la BNF (www.bnf.fr/gallica). La traduction proposée par Bernard Ribémont, partielle, est quant à elle destinée plutôt au grand public qu'aux chercheurs: *Le livre des propriétés des choses, une encyclopédie au XIVe siècle*, éd., trad. Ribémont B. (Paris, 1999).

9. Herfray-Rey Cl., 'Jean Corbechon, traducteur de Barthélemy l'Anglais (1372)', *Positions des thèses soutenues par les élèves de la promotion de 1944 pour obtenir le diplôme d'archiviste paléographe* (1944), p. 59-67.

10. Voir les études de Michel Salvat, notamment 'Barthélemy l'Anglais, traités du soleil et de la lune, traduits par Jean Corbechon (1372), édition et commentaire', *Senefiance* 13 (1983): *Le soleil, la lune et les étoiles au Moyen Âge*, p. 339-357; 'Quelques échos des rivalités franco-anglaises dans les traductions du *De proprietatibus rerum* (XIIIe-XVe siècle)', *Bien dire et bien aprandre* 5 (1987), p. 101-109; 'Le ciel des vulgarisateurs: note sur les traductions du *De proprietatibus rerum*', in: B. Ribémont (éd.), *Observer, lire, écrire le ciel au Moyen Âge, actes du colloque d'Orléans (22-23 avril 1989)* (Paris, 1991), p. 301-313. Voir aussi Byrne D., '*Rex imago Dei*: Charles V of France and the *Livre des proprietés des choses*', *Journal of Medieval History* 7 (1981), p. 97-113 et Vedrenne I., 'Érudits médiévaux et théorie des climats', *Energeia: recherches doctorales* 1 (mai 1995), p. 62-80.

compétence et le scrupule de Corbechon[11], allant même jusqu'à trouver excessive sa fidélité à l'original latin[12]. Il est vrai que Corbechon, après avoir fréquenté la faculté de théologie de Paris[13], dispose à la cour royale de l'une des bibliothèques les mieux fournies de son temps. Il n'est donc pas sans savoir que le *Liber de proprietatibus rerum*, qui relevait déjà au XIIIe siècle de l'effort de synthèse[14], n'est nullement à la pointe des connaissances scientifiques de son temps. Mais s'il conserve telle quelle la teneur de l'encyclopédie latine, c'est que son objectif n'est pas de renouveler, ni même de préciser, le savoir du XIIIe siècle: Corbechon ne prétend pas découvrir de nouvelles connaissances, mais faire œuvre de vulgarisateur. Son ambition n'est donc pas seulement de faire passer le contenu de l'encyclopédie de Barthélemy du latin au français, mais aussi de rendre sa teneur accessible à un nouveau public. Aussi, s'il est vrai que les données brutes de son modèle sont peu modifiées, il n'en va pas de même du mode d'exposition de ce savoir. C'est que Jean Corbechon vise des destinataires différents: son texte n'est plus écrit, comme l'encyclopédie latine, pour des moines ou des étudiants[15], mais destiné à l'entourage politique de Charles V[16], voire à ses successeurs[17] – en tout cas à des laïcs relativement instruits,

11. En témoignent notamment les jugements de Claude Herfray-Rey – 'L'ensemble est traduit sans contresens' (Herfray-Rey, 'Jean Corbechon…', p. 64) – ou de Michel Salvat: 'Cette traduction suit le déroulement du texte de Barthélemy avec une attention qui n'est pas habituelle à son temps' (Salvat, 'Le ciel des vulgarisateurs…', p. 303).

12. 'Ce qui étonne, c'est le manque d'originalité et de curiosité de notre traducteur qui, ayant à sa portée toutes les grandes sommes de la fin du XIIIe et du XIVe siècle, n'ajoute pas un nom à la bibliographie donnée par l'Anglais' (Herfray-Rey, 'Jean Corbechon…', p. 64).

13. Herfray-Rey, 'Jean Corbechon…', p. 61.

14. Jean-Marie Fritz signale que, pour ce qui concerne l'astronomie comme la médecine, l'encyclopédie de Barthélemy l'Anglais était déjà dépassée lorsqu'elle fut achevée (Fritz J.-M., 'Bourguignons et crétins des Alpes au Moyen Âge', *Recherches et travaux* 55 (1998): *Parlers au Moyen Age, hommage à Albert Meiller*, p. 63-80, ici p. 64).

15. Pour preuve de sa diffusion en milieu universitaire à l'époque de Corbechon, le texte de Barthélemy figurait au XIVe siècle parmi les livres enchaînés de la Sorbonne (Desoby J., 'Le *Livre des propriétés des choses* de Barthélemy l'Anglais – manuscrit de la bibliothèque municipale d'Amiens', in: D. Buschinger & A. Crépin (éds), *Les quatre éléments de la culture médiévale. Actes du colloque du centre d'études médiévales de l'Université de Picardie (Amiens, 25-27 mars 1982)* (Göppingen, 1983), p. 299-308, ici p. 299). Il y fit son entrée en 1286 (Ventura S., 'Autour de la version occitane du *De proprietatibus rerum* de Barthélemy l'Anglais', in: P. Nobel (éd.), *Textes et cultures: réception, modèles, interférences*, t. II: *Interférences et modèles culturels*, Littéraires (Besançon, 2004), p. 47-62, ici p. 48-49).

16. Sur le souci exprimé par Charles V de combler l'écart culturel entre son entourage politique et les savants qu'il admirait, voir Willard, 'Raoul de Presles's translation…', p. 329.

17. Sur la volonté de Charles V de faire traduire 'les plus notables livres' pour le profit d'éventuels successeurs ne connaissant pas le latin, voir le témoignage bien connu de Christine de Pizan (*Le Livre des fais et bonnes meurs du sage roy Charles V*, éd. S. Solente (Paris, 1936), t. I, p. 42-43).

mais pas à des savants ou des érudits. Le lectorat visé est clairement évoqué dans le prologue de l'encyclopédie romane qui, s'ouvrant sur un miroir du prince, met l'accent sur le *desir de sapience* caractéristique du monarque idéal. Cette notion de *sapience*, qui allie les notions de 'compétence' et de 'sagesse'[18], fait figure de lieu commun dans les prologues des encyclopédistes[19]; il n'en reste pas moins que chez Corbechon, la volonté de répondre aux aspirations d'un public curial n'est pas un vain mot, mais semble modeler en permanence la démarche du traducteur. En effet, si la critique est unanime pour souligner sa fiabilité, c'est non seulement qu'il suit sa source de près, mais aussi qu'il multiplie les procédés visant à créer une impression de compétence: à chaque ligne Corbechon, volontiers pédagogue, cherche à faire naître la confiance de son lecteur. En outre il prévoit, au sein de son texte, un mode de lecture susceptible de convenir aux loisirs studieux des conseillers du roi[20].

Nous tenterons donc d'évaluer les interventions d'un traducteur discret, mais soucieux de satisfaire son commanditaire au point de s'éloigner parfois de la lettre et, plus souvent encore, de l'esprit de l'encyclopédie latine qui lui sert de source; c'est le livre VI, traitant *des proprietés d'omme en general et en especial*[21], qui a été retenu comme seul échantillon. De contenu assez disparate[22], il est peu technique et ne comporte guère de vocabulaire spécialisé. Moins riche en néologismes et en difficultés de tous ordres que d'autres parties de l'ouvrage, il est propice pour orienter l'analyse vers d'autres enjeux du travail de traduction: structures syntaxiques, efforts d'explicitation ou au contraire de simplification, cohérence textuelle… Ces efforts ponctuels d'adaptation devraient permettre

18. *FEW*, t. XI, p. 205^{b}, art. *sapientia*.

19. Ribémont B., 'Encyclopédisme médiéval et modernité', in: J.R. Scheidegger, S. Girardet & E. Hicks (éds), *Le Moyen Âge dans la modernité: mélanges offerts à Roger Dragonetti, professeur honoraire de l'Université de Genève* (Paris, 1996), p. 381-394, ici p. 384.

20. Peut-être ce savoir-faire, ce souci du confort du lecteur, contribuent-ils à expliquer le succès de l'encyclopédie romane de Corbechon, dont Joëlle Ducos signale qu'il rend 'perplexe', pour un ouvrage 'consacr[ant] un savoir déjà mort à l'époque de [sa] rédaction' (Ducos J., 'Le clerc et les météores: constitution et évolution d'une culture encyclopédique', *Senefiance* 37 (1995): *Le clerc au Moyen Âge*, p. 151-164, ici p. 156).

21. Manuscrit BNF fr. 16993 (désormais *F*, sigle repris à Van den Abeele, Meyer & Ribémont, 'En guise d'avant-propos…', p. 16), fol. 73vb.

22. Cette diversité est annoncée par Corbechon lui-même à l'orée de ce livre: 'puis que nous avons dit des proprietés des parties de l'omme en especial, il reste a dire de ses proprietéz en general et en especial, selon la variacion de l'aage et la distincion du sexe, c'est a dire de l'omme et de la fame, et de la diversité des choses qui sont naturelles et contre nature' (*F*, fol. 73vb). Ainsi, traitant successivement des différents âges de la vie, des oppositions entre hommes et femmes, des diverses conditions sociales, il s'achève sur une sorte de manuel dispensant des conseils de diététique et d'hygiène.

in fine de mieux cerner l'attitude du traducteur ainsi que le mode de lecture qu'il prévoit pour son ouvrage, tout au moins dans ce livre VI, qui compte parmi les moins techniques et les plus disparates.

1. Le savoir-faire d'un traducteur professionnel

Le *Livre des proprietés des choses*, par son attention permanente au texte source, correspond à la catégorie de traductions qualifiées par Peter F. Dembowski de *service translations*[23]. Ainsi, tout au long du livre VI, on ne note aucune modification majeure de contenu ou d'organisation entre le *Liber de proprietatibus rerum* et sa *translacion* romane. Mieux, les subdivisions du texte, notamment son découpage en chapitres, restent identiques[24]. Cette proximité est telle que, même si l'on ne peut guère espérer identifier l'exemplaire latin que consulta Corbechon, tant la tradition manuscrite de Barthélemy est pléthorique[25], il est manifeste que le *Livre des proprietés des choses* s'apparente davantage au manuscrit 123 de la Bibliothèque de la Sorbonne (*E*) qu'aux autres exemplaires retenus pour l'édition critique du texte latin[26]. Il permet notamment de rendre compte de quelques particularités[27] ou ajouts mineurs[28] du *Livre des proprietés* français.

23. Dans son ébauche d'histoire de la traduction médiévale, il caractérise par ce terme les traductions effectuées par un auteur '[serving] overtly and proudly the accepted authority of the Latin writers whom he rendered as faithfully as he could into the language of those who had difficulties with Latin' (Dembowski, 'Learned Latin Treatises…', p. 259).

24. L'emplacement stable des titres rubriqués entre témoins de la version latine et manuscrits de la version romane témoigne, outre le respect du traducteur lui-même pour ces repères, de l'attention que portent les copistes à ces éléments structurants.

25. L'encyclopédie latine nous reste conservée par plus de 300 exemplaires (Ventura, 'Autour de la version occitane…', p. 49), et la librairie du Louvre n'en comportait pas moins de quatre copies (Desoby, 'Le *Livre des propriétés…*', p. 299). Michael C. Seymour souligne par ailleurs le succès tout particulier qu'obtint en France l'encyclopédie de Barthélemy (Seymour, 'Some Medieval and French Readers…', p. 100[a-b]).

26. Sur ces exemplaires et leurs sigles, voir Van den Abeele, Meyer & Ribémont, 'En guise d'avant-propos…', p. 15. Nous remercions Iolanda Ventura, qui édite le livre XVII en latin et coordonne le projet d'édition de l'encyclopédie bilingue, et Joëlle Ducos, qui édite les livres IV, X et XI, de nous avoir confirmé que, pour d'autres parties du *Livre des proprietés…*, c'est aussi du manuscrit *E* que la traduction de Corbechon est la plus proche.

27. Au chapitre V, suite à l'évocation des traits distinctifs des enfants, le manuscrit BNF lat. 16099 cite le livre III du *De bestiis* d'Aristote (fol. 47rb); mais c'est manifestement une leçon identique à celle du manuscrit 123 de la Sorbonne ('libro I', fol. 56vb) que Corbechon a eue sous les yeux, puisqu'il traduit par 'ou premier livre' (*F*, fol. 76va).

28. Ainsi au chapitre III, le manuscrit BNF lat. 16099 porte la leçon 'et jam incipit se movere et pedibus calcitrare' (fol. 46va), et le manuscrit *E* la suivante: 'et jam incipit movere se et pedibus et manibus calcitrare' (fol. 55vb). La traduction de Jean Corbechon suit de près la seconde version: 'et se conmance ja a mouvoir et a hurter des piéz et des mains' (*F*, fol. 75rb).

1.1. Une traduction toujours lisible

Si Corbechon n'est pas un traducteur interventionniste, il ne peut pas non plus être qualifié de servile. En effet, quoiqu'il soit toujours aisé d'établir une comparaison terme à terme entre son original latin et sa *translacion*, une lecture suivie des deux textes révèle l'effort pédagogique du moine augustin.

1.1.1. Un effort permanent de clarté

Ce souci du traducteur se traduit d'abord par l'attention extrême qu'il porte au sens des mots, y compris pour le vocabulaire le plus courant. En effet, à chaque fois que l'emploi d'un calque pourrait induire les plus légères difficultés de compréhension, Corbechon l'évite, quitte à recourir à une périphrase si aucun terme roman ne s'impose. C'est notamment le cas pour le mot *animal*, désignant en latin tout 'corps vivant, humain ou animal'. Le mot *animal* est toujours employé dans la langue romane, mais il s'est spécialisé et n'inclut plus les êtres humains[29]. En conséquence, lorsque notre traducteur rencontre le mot *animal* sous la plume de Barthélemy, il l'explicite à plusieurs reprises, traduisant *spatium vite animalis* (*E*, chap. I, fol. 54va) par 'l'espace de la vie de la beste et de la personne' (*F*, fol. 74ra), et *animalia* (*E*, chap. XII, fol. 58ra) par '[les] hommes et les bestes' (*F*, fol. 78rb).

À plus forte raison évite-t-il les néologismes: lorsqu'un mot latin ne trouve pas d'équivalent direct en français, Corbechon prend ses distances par rapport au lexique de sa source. Et s'il n'est pas certain que sa transposition romane est suffisamment claire, il n'hésite pas à l'étayer d'une définition. Ainsi l'adjectif latin *mediocris* – 'Cibus vero mediocris est laudabilis' (*E*, chap. XX, fol. 62ra) – est-il traduit par 'moienne, entre trop et pou' – 'Derechief la viande moienne, entre trop et pou, fait a loer' (*F*, fol. 83ra). Corbechon ne cherche pas à innover en employant le calque *mediocre*, mais choisit l'équivalent *moien*; néanmoins, comme celui-ci est

29. À partir de la fin du XII[e] siècle, le mot *animal* signifie 'être animé privé de raison (par opposition à *homme*)' (*FEW*, t. XXIV, p. 590[b], art. *animal*). Jean Corbechon n'est d'ailleurs pas le seul traducteur de son époque à avoir perçu le décalage entre le mot latin *animal* et son héritage français: Nicole Oresme cite cet exemple de la difficulté de la tâche de traducteur dans le *proheme* de sa *translacion* du *Livre des Éthiques* d'Aristote (Dembowski P.F., 'Scientific Translation and Translator's Glossing in four Medieval French Translations', in: J.M.A. Beer (éd.), *Translation Theory and Practice in the Middle Ages*, Studies in Medieval Culture 38 (Kalamazoo, 1997), p. 113-134, ici p. 115). Reste à savoir si chacun des deux traducteurs a pris conscience du problème dans le secret de son atelier, ou si la pratique de Corbechon résulte d'échanges entre les savants de la cour de Charles V.

légèrement ambigu[30] – même si, dans le contexte, son sens se devine assez aisément –, il l'accompagne d'une courte glose.

Cependant, tous les ajouts du traducteur ne visent pas à parer aux éventuels décalages sémantiques entre latin et moyen français: la plupart d'entre eux ne seraient pas malvenus dans l'encyclopédie latine elle-même. En effet, pour rendre le texte plus facile à lire et à comprendre, le traducteur tend souvent à rendre explicites des éléments implicites dans sa source. Ainsi révèle-t-il le sens de tel ou tel mot par une glose plus concrète, comme lorsqu'il traduit *sexus distinctionem* (*E*, introduction, fol. 54va) par *la distinction du sexe, c'est a dire de l'omme et de la fame* (*F*, fol. 73vb). Comme c'est le cas ici, la plupart des ajouts de ce type sont signalés par des marqueurs de gloses, *c'est a dire* étant le plus courant.

Plus fréquentes encore sont les insertions visant à clarifier, dans la traduction, des mécanismes anaphoriques. En effet, tous les pronoms latins dont la référence nécessite tant soit peu de déduction tendent à être développés par un Corbechon qui, non content de se faire comprendre, semble désireux de ménager les efforts de son lecteur. Ainsi explicite-t-il volontiers les relatifs de liaison de son modèle latin[31]. De même, alors que Barthélemy multipliait les ellipses, tant pour évoquer les mécanismes semblables que les éléments antinomiques, son traducteur développe dans les deux cas ses descriptions: si l'encyclopédie latine omet un verbe déjà exprimé dans la proposition précédente, le traducteur le répète volontiers. Lorsque Barthélemy explique, à propos des aliments, que 'quod caldum est et humidum transit in naturam sanguinis et carnis, quod vero frigidum est et siccum in naturam nervi et ossis; et sic de aliis' (*E*, chap. XX, fol. 61va), sa phrase, composée de segments parallèles, est allégée pour éviter des redites. Dans la version romane, l'expression est beaucoup moins resserrée et le traducteur ne recule pas devant quelques répétitions: 'ce qui est chault et moiste se convertist en sanc et en char, et ce qui est froit et sec se convertist en ners et en os. Et ainsi les autres qualitez de la viande se convertissent en diverses parties du corps' (*F*, chap. XX,

30. Il peut renvoyer à la position dans l'espace au sens de 'qui se situe au milieu' (Gerner H., 'Lexique de la littérature didactique'; Jacquart D. & Thomasset Cl., 'Lexique de la langue scientifique', in: ATILF / Équipe 'Moyen français et français préclassique' (éd.), *Base de lexiques du moyen français*, www.atilf.fr/blmf/, site consulté le 25 janvier 2005) aussi bien qu'à la quantité, comme c'est le cas ici (cet adjectif est attesté au sens d''intermédiaire, qui relève du juste milieu' par Brucker Ch., 'Lexique de Nicole Oresme', in: ATILF / Équipe 'Moyen français et français préclassique' (éd.), *Base de lexiques du moyen français*, www.atilf.fr/blmf/, site consulté le 25 janvier 2005).

31. 'Cui succedit pueritia' (*E*, chap. I, fol. 54vb; *cui* y anaphorise le nom *infantia*) est traduit par 'Aprés enfance vient le second aage' (*F*, fol. 74ra).

fol. 82rb-82va). Dans la dernière proposition, tous les éléments grammaticaux qui étaient sous-entendus dans le texte latin figurent en toutes lettres.

L'écart entre Barthélemy l'Anglais et Jean Corbechon est tout aussi sensible dans les passages comportant des oppositions. Au lieu d'exprimer les tenants et les aboutissants de celles-ci, l'encyclopédiste du XIIIe siècle s'en tient souvent à des termes ou formules comme *econtra*, *e converso*; l'emploi de l'adjectif *contrarius* le dispense ailleurs de plus longues descriptions. Face à cette concision, la réaction de notre traducteur consiste presque toujours à développer les idées sous-jacentes aux formules resserrées du texte source:

> et [...] materia destrueretur infantis, nisi ex sanguine mulieris contrarias qualitates habentis reciperet temperamentum (*E*, chap. III, fol. 55rb) → et ainsi la matere de l'enfant seroit destruite se ce n'estoit la semence de la mere, qui est clere et froide et l'atrempe (*F*, fol. 74vb);
>
> Si fuerit virtus sanguinis in matre vincens, fetus patri assimilabitur, et e converso (*E*, chap. III, fol. 55rb-55va) → se la semence du masle est plus vertueuse que celle de la mere, l'enfant ressemblera au pere; et se la mere vaint, l'enfant lui ressemblera (*F*, fol. 74vb);
>
> vir sive masculus respectu femine calidus [est] et siccus; ipsa vero econtra (*E*, chap. XII, fol. 58ra) → il est chaut et sec, et la femelle est froide et moiste (*F*, fol. 78ra);

Toujours par souci de clarté et pour rendre la lecture plus fructueuse, Corbechon facilite autant que possible l'identification des noms propres. Ceux-ci sont beaucoup moins nombreux dans le livre VI que dans d'autres parties de l'encyclopédie[32]; mais à chaque occurrence qui pourrait soulever quelque difficulté, le traducteur ajoute de brèves mentions permettant l'identification. C'est notamment le cas pour *Helysei* (*E*, chap. XVI, fol. 59va), dont l'identité est explicitée par la mention apposée *le prophete* (*F*, fol. 80ra); de même le site du Temple, qui n'est pas précisé dans l'encyclopédie latine, est-il mentionné dans sa transposition française: 'Et secundum istum modum computat Augustinus edificationem templi' (*E*, chap. III, fol. 55vb) → 'Et par ceste maniere compte saint Augustin l'edificacion du temple de Jherusalem' (*F*, fol. 75rb). Le traducteur, en tant que lecteur du texte latin, accomplit donc un certain nombre d'élucidations et de déductions dont il fait figurer les acquis dans son encyclopédie romane: il traite sa source comme un sentier trop étroit, qu'il élargirait pour son futur public.

32. On pense notamment aux chapitres consacrés à la géographie (livre XV).

1.1.2. Un propos plus simple et plus direct

En outre, lorsque les informations de son modèle lui semblent trop abstraites, il tend à leur donner un caractère plus concret, et donc plus facile à saisir. Ainsi est-il fréquent que, par métonymie ou par raccourci, Barthélemy caractérise à l'aide de tel ou tel adjectif non pas des animés humains, mais une période de la vie ou une condition sociale. Jean Corbechon, quant à lui, tend à attribuer ces caractérisations à des mots désignant directement les hommes, quitte à ajouter des développements:

> Huic adolescentie succedit juventus, et hec inter omnes etates est media, et ideo fortissima (*E*, chap. I, fol. 54vb) → Aprés s'ensuit jennesce, qui tient le moien entre les aages, et pour tant la personne y est en sa plus grant force (*F*, fol. 74ra);

> Senecta [...] que ab Isidoro gravitas nuncupatur (*E*, chap. I, fol. 54vb) → senecte [...], et l'appelle Ysidore pesanteur pour ce que la personne en cest aage est pesante en meurs et en maniere (*F*, fol. 74rb).

C'est sans doute par souci de proposer une présentation plus concrète qu'il opère ce genre de transformations, et cette propension à stimuler l'imagination de son lecteur est plus nette encore dans d'autres développements. Ainsi, au moment où Barthélemy évoque les ravages de la vieillesse sur la santé de l'esprit, il se contente de les suggérer brièvement par cette formule: 'senes pre vetustate desipiant' (*E*, chap. I, fol. 54vb). Corbechon propose une représentation plus développée, mais aussi plus suggestive: 'les vielles gens n'ont pas si bon sens conme ilz ont eu et radobtent' (*F*, fol. 74rb). De même, alors que l'encyclopédiste latin décrit le mauvais caractère des femmes avec une discrétion qui l'honore, puisqu'il se contente d'une allusion à leur 'animus [...] amarus' (*E*, chap. VI, fol. 57ra), son traducteur les qualifie de 'plaines de barat et d'amertume' (*F*, fol. 77ra). Certes, le contenu du livre VI nécessite sans doute moins d'efforts pédagogiques que d'autres parties de l'encyclopédie dans la mesure où il traite d'éléments plus faciles d'approche; mais le travail de Corbechon n'en révèle pas moins son goût constant d'une présentation concrète, qui va parfois jusqu'à faire allusion à l'expérience quotidienne du lecteur[33]. Ce recours à des détails prosaïques n'est pas sans rappeler des livres plus techniques; ainsi Corbechon compare-t-il, dans

33. Cette tendance est particulièrement nette, au sein du livre VI, pour le chapitre XX ('de la viande'), qui fait notamment observer qu'il faut manger davantage lorsqu'on est jeune que lorsqu'on est vieux, lorsqu'on est actif que lorsqu'on est oisif, davantage aussi l'hiver que l'été (*F*, fol. 83rb-83va). De même le chapitre suivant (XXI: 'du buvrage', *F*, part. fol. 82va-85rb) fait volontiers allusion à la vie quotidienne du lecteur.

le livre VIII, la courbe du ciel au bassin d'un barbier – rapprochement absent de sa source[34].

En conséquence, il n'est guère surprenant de constater que, lorsque Corbechon rencontre chez son modèle des images à portée pédagogique, il les conserve. Mais il est intéressant de voir qu'il lui arrive de les étoffer pour en rendre la portée plus limpide, comme dans le cas suivant:

> Si vero fuerit cibus nimius [...], calorem naturalem sicut oleum lampadem extinguit et suffocat, spasmum et contractionem nervorum causat (*E*, chap. XX, fol. 61vb) → car trop de viande [...] estaint la chaleur naturelle aussi conme huille, quant il y en a trop en une lampe, estaint le feu et si fait la personne pasmer et les sens retraire (*F*, fol. 83ra).

L'ajout du traducteur (*quant il y en a trop en une lampe*) relève d'abord de l'explicitation: Barthélemy indiquait en substance que la nourriture est à l'homme ce que l'huile est à une lampe, laissant à son lecteur le soin d'appliquer le terme *nimius*, qu'il ne répétait pas, à *oleum* pour déchiffrer l'image. Corbechon développe entièrement cette analogie, au prix d'une répétition (*trop*). Mais son intervention va plus loin: elle donne à l'image une saveur et une autonomie qu'elle ne possédait pas dans l'encyclopédie latine. Cet exemple n'est pas isolé, et les ajouts du traducteur contribuent à donner du relief et des couleurs à certaines comparaisons qui, brossées à traits rapides par Barthélemy, n'avaient sous sa plume qu'une valeur pédagogique.

1.1.3. Une progression plus facile à suivre

Ce souci permanent de clarté du traducteur se manifeste également par une certaine réorganisation des informations. Certes, Corbechon respecte la subdivision du texte source en chapitres structurés de manière thématique[35]; il s'en tient aussi, en général, à l'enchaînement des informations tel qu'il figure dans son modèle latin. Mais, à échelle microstructurale, il lui arrive de rendre plus perceptible une progression, voire d'intervertir quelques développements. Ainsi, lorsque Barthélemy évoque les différentes phases de l'enfance, il entame invariablement chaque nouvelle description par les formules *huic succedit* ou *cui succedit*, qui ouvrent notamment la description du second âge de l'enfance (*pueritia*: *E*, chap. I, fol. 54vb), celle de l'adolescence (*E*, chap. I, fol. 54vb) et celle de la jeunesse ('huic adolescentia succedit juventus': *E*, chap. I,

34. Salvat, 'Le ciel des vulgarisateurs…', p. 306.
35. Voir *supra* p. 335.

fol. 54vb). Pour rendre cette énumération plus facile à suivre, le traducteur n'hésite pas à en numéroter les différentes étapes: 'Le premier aage est enfance', 'aprés enfance vient le second aage', 'aprés s'ensuit le tiers aage' (*F*, chap. I, fol. 74ra)[36]. L'ajout de ce type de mentions ne fait que systématiser certains procédés de Barthélemy: on rencontrait, en d'autres points de l'encyclopédie latine, des repères chiffrés similaires, notamment lors de l'évocation des différents stades de formation de l'embryon[37]. Mais le traducteur tend à généraliser ces marques de la progression du texte.

Il arrive même, quoique plus rarement, qu'il remanie le détail de l'ordonnancement du texte lorsqu'il n'en perçoit pas la logique. Ainsi ramène-t-il la comparaison entre les durées de formation des nourrissons à un ordre strictement croissant de durée parce que la succession n'en était guère déchiffrable chez Barthélemy:

Constantinus [...] dicit quod masculus qui in mense VIII est nasciturus XXX diebus formatur et LXX completur. Deinde subdit masculus X mensium: XLVI formatur diebus et in nonaginta completur; masculus vero IX mensium formatur in XL diebus (*E*, chap. III, fol. 55vb-56ra).	→	Constantin dit [...] que l'enfant masle qui naist en le VIIIe mois si est formé en XXX jours; et celui qui naist ou IXe mois si est formé en XL jours; et celui qui naist ou X^{e} mois est formé en XLVI jours (*F*, fol. 75va).

Il semble donc qu'une étroite comparaison entre l'encyclopédie de Barthélemy et sa traduction romane permette de nuancer l'impression de fidélité qui se dégage du travail de Corbechon: il est vrai qu'il donne à lire l'ensemble des données proposées par son prédécesseur, qu'il ne sélectionne, ne remet en cause ni ne nuance, mais il en modifie la présentation. Il propose ainsi à son lecteur un itinéraire plus facile à suivre, parce que l'avancée en est ponctuée de repères plus tangibles, que la continuité du parcours y est plus apparente. La sûreté dont fait montre notre traducteur pourrait résulter, comme le suggère Jeanette Beer, d'une concertation

36. On note néanmoins que c'est seulement pour les différentes étapes de l'enfance que Corbechon procède ainsi, puisqu'il renonce ensuite à ce système: 'Aprés s'ensuit jennesce, qui tient le moien entre les aages' (*F*, chap. I, fol. 74ra). Il est possible qu'il ait tenu à baliser avec netteté les transitions dans un passage où la transposition du vocabulaire posait des problèmes particuliers (voir *infra* p. 344).

37. 'L'enfant, tant conme il est ou ventre, est en IIII degrés: le premier est quant..., le second si est..., le tiers degré est quant... Le derrenier degré...' (*F*, chap. III, fol. 75rb) traduit simplement 'Infans ibi quattuor gradibus concreatur: primus est cum... Secundus est cum... Tertius gradus est cum... Ultimus gradus est cum...' (*E*, fol. 55vb).

quotidienne avec les autres lettrés de la cour de Charles V[38]. Quoi qu'il en soit, cette maîtrise permanente du matériau textuel et de son organisation, alliée à l'emploi d'un vocabulaire et d'images toujours limpides, met à la disposition du lecteur de l'encyclopédie vulgaire un texte non seulement confortable à lire, mais qui inspire confiance. C'est aussi qu'en cas de difficulté insurmontable ou d'hésitation, le traducteur n'hésite pas à prendre la parole pour commenter son travail, suggérant ainsi *a contrario* que les larges parties où le texte est exempt de tels commentaires sont traduites d'une manière qui le satisfait.

1.2. Une traduction qui se signale comme telle

1.2.1. Des références ponctuelles à la langue source

Lorsque, pour rendre compte avec précision des données de l'encyclopédie latine, Jean Corbechon est contraint d'innover sur le plan lexical, il n'hésite pas à introduire des *gloses* au sens où l'entend Peter F. Dembowski[39]: il prend la parole en tant que traducteur, signalant à son lecteur qu'il a sous les yeux un texte qui résulte d'une transposition linguistique, et lui faisant du même coup partager les difficultés de sa tâche. Passé le seuil que constitue le prologue, ces interventions se raréfient. Mais en cas de difficulté particulière, elles justifient l'introduction – rare – de mots savants ou la citation en latin de la source:

Huic succedit adolescentia (*E*, chap. I, fol. 54vb)	→	Aprés s'ensuit le tiers age que on appelle en latin adolescence[40] (*F*, fol. 74ra).
Ultima autem pars senectutis dicitur senium (*E*, chap. I, fol. 55ra)	→	La derrenière partie de viellesce est appellee *seni[u]m* en latin; et en françois, elle n'a point d'autre nom que viellesce (*F*, fol. 74rb).

38. 'The lexical and syntactic disparateness of French and Latin and the notion of differing stylistic levels were translators' discussion-pieces at the court of Charles V' (Beer, 'Patronage and Translator…', p. 104).

39. Dembowski, 'Scientific Translation…', p. 113.

40. Le *FEW* date la première occurrence du nom *adolescence* ('âge qui succède à l'enfance et qui commence avec la puberté') du XIII[e] siècle (t. XXIV, p. 175[b], art. *adolescentia*). Néanmoins, comme souvent, la date de première attestation du terme ne nous renseigne guère sur l'intuition des locuteurs médiévaux: Jean Corbechon, lorsqu'il emploie le mot *adolescence*, a manifestement le sentiment de recourir à un terme rare, d'où sa glose. L'expression 'en latin' qu'il y emploie témoigne d'ailleurs de son impression que ce substantif n'est pas vraiment un mot roman, mais un calque qui reste largement étranger à la langue vernaculaire.

Dans le second cas, Corbechon propose une traduction romane, *viellesce*, qui ne lui semble pas assez précise; du coup, il la fait précéder d'une citation littérale de sa source latine – qui a d'ailleurs été éborgnée au cours de la tradition manuscrite du texte roman – et d'un commentaire linguistique qui pallie la trop large extension du terme roman *viellesce*.

Outre les difficultés ponctuelles que posent certaines dénominations de ce type, le traducteur se heurte constamment au problème insoluble que constituent les notations étymologiques ponctuant toute encyclopédie. En effet, deux mots affiliés en latin – fût-ce à la mode médiévale, qui percevait et pratiquait autrement l'étymologie[41] – ne sont pas toujours transposables par deux mots romans de même famille. Dans ce cas, le traducteur n'hésite pas à laisser tout ou partie de la formule en latin de manière à proposer, dans l'encyclopédie française, deux mots dont la filiation étymologique reste vraisemblable: 'Vir a virtute virium est dictus secundum Ysidorus' (*E*, chap. XIII, fol. 58rb) est traduit par 'Homme est en latin appelé *vir* pour la vertu et la force qui est en lui, ce dit Ysidore' (*F*, fol. 78va). Le nom latin *vir* étant dépourvu de toute descendance romane, Corbechon a recours à *homme*, dont il est difficile de rapprocher un mot signifiant 'force'. Ainsi acculé, le traducteur ne transpose que partiellement la formule: le rapprochement entre *vir* et *vertu* est frappé du sceau de la vraisemblance, et l'emploi du nom *vertu*[42] lui permet de limiter au maximum l'usage du latin. Corbechon l'associe tout de même à un synonyme français qui en confirme le sens.

Ainsi, lorsque la transposition directe se ferait au détriment de la précision sémantique ou de la clarté, Corbechon n'hésite pas à rompre l'illusion que pouvait, malgré le prologue, avoir son lecteur de consulter un ouvrage directement composé en français: il revient alors au latin, dont il insère des bribes, mais en les éclairant et en justifiant systématiquement leur présence.

1.2.2. À l'impossible nul n'est tenu...

Et lorsque ces mises au point ponctuelles sont insuffisantes parce que l'inadéquation entre latin et français touche plus d'un mot, le traducteur n'ignore pas la difficulté, mais en tire profit pour glisser quelques réflexions linguistiques. Certes, on relève peu d'interventions de ce genre

41. Sur la passion médiévale pour l'étymologie et sur ses modes d'exposition, voir Guenée B., *Histoire et culture historique dans l'Occident médiéval* (Paris, 1980), p. 187-189.

42. Ce mot est attesté au sens de 'courage, ardeur, énergie' depuis le XIe siècle (*FEW*, t. XIV, p. 517^{b}, art. *virtus*).

au sein du livre VI; c'est le décalage entre la précision latine permettant de décomposer l'enfance en deux périodes distinctes – avant et après l'acquisition de la parole –, et l'indigence du français, qui ne possède qu'une seule désignation pour ces deux âges, qui donne lieu aux plus longs développements du traducteur. Le premier s'apparente aux *gloses* déjà évoquées: 'Cui succedit pueritia' (*E*, chap. I, fol. 54vb) est ainsi traduit par 'Aprés enfance vient le second aage, qui en françois n'a point de nom different du premier, mais en latin on l'appelle *puericia*' (*F*, fol. 73vb).

Le second va plus loin:

> Etas ergo in infantulis terminum habet in inchoationem etatis puerilis (*E*, chap. IV, fol. 56va) → L'aage d'enfance se fine a VII ans, et la conmance le secont aage que nous appellons enfance en françois, mais en latin on l'appelle *puericia*. Et en ce il appert qu'il y a plus grant deffaulte de langage en françois que en latin, car en latin il y a VII aages nonméz par divers noms, desquelz il n'en y a que trois en françois, c'est assavoir enfance, jennesce et viellesce. Et par ce on puet pensser quelle paine c'est de proprement translater latin en françois (*F*, fol. 76rb).

Lors de cette transition délicate entre le chapitre IV (consacré chez Barthélemy à l'*infantulus*) et le chapitre V (décrivant dans l'encyclopédie latine le *puer*), le traducteur ne jette pas l'éponge: il expose précisément le contenu sémantique du latin *pueritia*, qu'il cite en latin, et désamorce ainsi l'apparente absurdité qui pourrait se dégager, à première lecture, de la réitération du terme d'*enfance*. Néanmoins, il prend appui sur cet écueil ponctuel pour analyser la difficulté, récurrente dans les chapitres précédents, qu'implique la traduction des termes latins désignant les différentes périodes de la vie. Il prend alors du recul pour comparer le stock lexical des langues savante et vernaculaire. En outre, même s'il ne prend pas la parole à la première personne, il établit un véritable dialogue avec son lecteur, qu'il fait pénétrer dans son atelier afin de lui montrer l'insuffisance des simples outils vernaculaires face aux subtilités du latin. Cette démarche illustre les rapports établis, tout au long du texte, entre le *translateur* et ses destinataires: le contrat sans naïveté qui les lie semble stipuler que le traducteur s'engagera à fournir à son lecteur un texte clair et confortable à lire, mais que si, à bout de ressources, il ne peut désamorcer quelque chausse-trappe, il devra interrompre sa marche pour signaler la difficulté plutôt que de passer outre. C'est dire à quel point le public curial du XIVe siècle est conscient des enjeux de la traduction et accoutumé à se frotter à des textes de seconde main, qui résultent d'une transposition linguistique…

2. Traduction et vulgarisation: un texte vernaculaire plus simple et plus univoque

La constante impression de clarté et d'honnêteté qui se dégage de la méthode de Jean Corbechon est en partie un leurre, qui tend à masquer les transformations qu'il fait subir à sa source latine. Or il ne se contente pas de transposer linguistiquement le texte qui lui sert de modèle: les modifications qu'il pratique vont au-delà des explicitations, gloses et commentaires qui donnent du traducteur l'image d'un professionnel digne de foi.

2.1. Un certain gauchissement du vocabulaire

Comme on l'a vu[43], le recours au vocabulaire spécifiquement roman plutôt qu'à des calques du latin ne nuit en rien à la précision du propos[44]; en outre, son caractère systématique montre que cette pratique relève, chez Corbechon, du choix délibéré[45]. Toutefois, il n'en va pas de même d'autres transpositions qui, remplaçant un mot précis par un terme plus vague, modifient légèrement le sens tout en réduisant considérablement la diversité lexicale du texte source.

2.1.1. Multiplication des mots les plus courants

Ce trait s'observe pour plusieurs catégories grammaticales, mais il ressort de manière particulièrement frappante pour les verbes. En effet, *faire* devient le relais de bien des verbes d'action du latin, et *estre* sert à traduire divers verbes d'état:

43. Voir *supra* p. 336.

44. Ainsi les expressions *sanguis menstrualis* et *passio menstrualis* sont-elles constamment traduites par *les fleurs* (voir le chapitre III, avec plusieurs occurrences; voir aussi les chapitres IV, VII et XII). Ce terme roman relève plutôt de l'emploi courant que du vocabulaire médical, mais il n'est pas pour autant moins précis – il désigne les 'menstrues' (*FEW*, t. III, p. 643^b-644^a, art. *fluor*) – que son équivalent latin.

45. Le terme *systématique* doit néanmoins être ramené à des proportions toutes médiévales: quoique l'usage des calques, néologismes ou mots savants, soit rare, le traducteur ne se l'interdit pas totalement. On relève notamment l'emploi, accompagné d'une glose, du terme *secondine* au chapitre IV (*F*, fol. 75vb: 'une pel qui est appellee la secondine'). Ce calque du latin *secundina* (*E*, fol. 56ra) est, sous la plume de notre traducteur, un néologisme – ce mot n'est attesté au sens de 'seconde enveloppe de l'embryon' qu'à partir de 1685 (*FEW*, t. XI, p. 384^b, art. *secundus*) – ou tout au moins, si l'on doute du témoignage des dictionnaires quant aux premières attestations, un mot rare.

Verbe *faire*

membra carnosa quorum substantia ex sanguine procreatur (*E*, chap. III, fol. 55va)	→	touz les membres charnuz, desquelz la creacion se fait de sanc (*F*, fol. 75ra)
ut [...] digestio celebretur (*E*, chap. IV, fol. 56va)	→	pour faire la digestion (*F*, fol. 76ra)
Festa et convivia nuptialia celebrat (*E*, chap. XIII, fol. 58va)	→	Il fait les noces solempnelment (*F*, fol. 78vb)

Verbe *estre*

nunquam permanet in eodem statu (*E*, chap. I, fol. 54va)	→	et n'est onques en un estat (*F*, fol. 74ra)
preter ursam et leopardam quia ferociores maribus reputantur et audaciores (*E*, chap. VI, fol. 57va)	→	excepté la femelle de l'ours et du liepart, qui sont plus fieres et plus garnies que les mascles de leur espece (*F*, fol. 77ra)
que plus causa prolis quam causa libidinis in bonis matrimonii conversatur (*E*, chap. XIII, fol. 58vb)	→	qui est en mariage plus pour cause de lignie que de luxure (*F*, fol. 79ra)

Des exemples comparables pourraient être cités à propos du verbe *aler*, verbe de mouvement qui tend à succéder à divers termes latins plus précis (*ambulare, se exponere*...). Quant aux substantifs et adjectifs, ils font l'objet de transpositions analogues[46]. Au total, le vocabulaire employé dans la traduction est plus prosaïque et plus répétitif; néanmoins, cette tendance à l'appauvrissement lexical est peu perceptible à simple lecture, les mots les plus récurrents étant passe-partout.

La relative indigence du vocabulaire du *Livre des proprietés*... par rapport à son modèle latin s'explique aussi par un parti-pris de Corbechon qui le démarque de bien des traducteurs de son époque: celui de renoncer aux polynômes de synonymes. En effet, non seulement il en emploie peu, mais il simplifie ceux qu'il rencontre dans l'encyclopédie latine.

46. Ainsi les mots suivants se voient-ils substituer des termes plus vagues, ou tout au moins plus courants, lors de la traduction: *adulta* (*E*, chap. I, fol. 54vb) est transposé par *personne* (*F*, fol. 74ra), *sollicitudine* (*E*, chap. V, fol. 56va) par *penssee* (*F*, fol. 76rb), *propter nutrimenti superfluitatem* (*E*, chap. V, fol. 56vb) par *pour trop manger* (*F*, fol. 76va), etc.

2.1.2. À la recherche du mot juste: le refus de toute redondance

Pour le lecteur accoutumé aux traductions médiévales, ce trait de la méthode de Corbechon est assez surprenant. En effet, Claude Buridant a montré que, chez bien des *translateurs* médiévaux, le recours aux binômes synonymiques permettait, en faisant voisiner un terme savant et son équivalent plus courant, d'élucider des difficultés sémantiques sans interrompre le fil du discours[47]. Cet usage est d'ailleurs bien représenté chez les traducteurs de la cour de Charles V, Nicole Oresme[48] et Denis Foulechat[49] notamment. Or, Jean Corbechon prend généralement le contre-pied de cette pratique: non seulement il lui arrive très rarement de forger des doublets de synonymes[50], mais lorsqu'il rencontre la figure sous la plume de Barthélemy, il la supprime presque systématiquement. Il est vrai que l'encyclopédiste du XIIIe siècle coordonne assez fréquemment des termes de sens parfaitement superposable. Dans ce cas, Corbechon tranche impitoyablement, soit qu'il ait recours à un dérivé de l'un des deux termes:

> Ex amore et zelo ipsam corrigit (*E*, chap. XIII, fol. 58va) → Il la chastie par amour (*F*, fol. 78vb)
>
> recitant et revelant (*E*, chap. V, fol. 56vb) → revellent (*F*, fol. 76vb)
>
> unguentis et fomentis (*E*, chap. X, fol. 57va) → aucuns oingnemens (*F*, fol. 77va),

soit qu'il traduise l'ensemble par un mot totalement indépendant:

> appetunt et affectant (*E*, chap. V, fol. 56vb) → desirent (*F*, fol. 76va)
>
> In sponsa autem et uxore bona (*E*, chap. XIII, fol. 58va) → En la bonne fame (*F*, fol. 78vb)
>
> Homo autem puerum suum sive fetum diligit (*E*, chap. XIV, fol. 58vb) → Mais homme [...] aime son enfant (*F*, fol. 79ra).

Dans ce cas, le traducteur modifie le rythme de la phrase, mais conserve globalement son sens. Néanmoins, lorsque dans le modèle latin sont coordonnés deux termes d'acceptions complémentaires, même si la suppression de l'un des deux infléchit sensiblement le contenu de la phrase, le traducteur n'hésite pas non plus à simplifier, quitte à trahir

47. Buridant Cl., 'Les binômes synonymiques. Esquisse d'une histoire des couples de synonymes du Moyen Âge au XVIIe siècle', *Bulletin du centre d'analyse du discours* 4 (1980): *Synonymies*, p. 5-79, ici p. 14-15.

48. Buridant, 'Les binômes...', p. 15.

49. Brucker, 'Quelques aspects du style...', p. 101.

50. Il traduit toutefois *attenduntur* (*E*, introduction du livre VI, fol. 54va) par 'on puet bien considerer et entendre' (*F*, fol. 73vb).

quelque peu. Ainsi traduit-il le couple *venas et arterias* (*E*, chap. III, fol. 55va) par *vaines* (*F*, fol. 75ra) ce qui, au plan médical, constitue un raccourci fautif[51]. La même tendance à réduire les syntagmes binaires s'observe à deux reprises dans un autre passage: 'Plus de amissione pomi vel pupini quam de amissione patrimonii plangunt et plorant' (*E*, chap. V, fol. 56vb) → 'Ilz plaingnent plus la perte d'une pomme que de leur heritaige' (*F*, fol. 75va). En effet, quoique les deux verbes *plangere* et *plorare* renvoient à des manifestations d'affliction différentes, seul le premier, plus général, est conservé dans la version romane[52]. La réduction de *pomi vel pupini* est plus abrupte encore, le second élément désignant un 'poupon'[53]; mais les deux mots désignant des exemples interchangeables de vétilles, Corbechon n'a conservé que la première de ces deux illustrations.

Cette tendance à supprimer toute redondance transparaît ailleurs. Au plan microstructural, elle se traduit par une propension manifeste à faire table rase de toutes les figures de répétition. Ainsi tendent à disparaître les simples réitérations de termes[54], les anaphores[55], les figures de dérivations[56], etc. À plus large échelle, le traducteur tend à faire l'impasse sur

51. Ce n'est pas le décalage entre latin et langue romane qui est à l'origine de cette approximation, puisque le terme *veine* désigne depuis le XIIe siècle un 'vaisseau qui ramène le sang au cœur' (*FEW*, t. XIV, p. 266^{a-b}, art. *vena*), s'opposant donc au mot *artere*, emprunt du XIIIe siècle au latin qui renvoie dès cette date à un 'vaisseau qui distribue le sang du cœur à toutes les parties du corps' (*FEW*, t. XXV, p. 367^{b}-368^{a} et p. 369, art. *arteria*). D'ailleurs, Corbechon n'ignore pas l'écart entre les deux, puisqu'un peu plus loin dans le même chapitre, il suit Barthélemy à la lettre pour signaler que 'du cuer issent les arteres et du foye yssent les vaines' (*F*, chap. III, fol. 75ra) – qui correspond étroitement à 'de corde arterie, de epate vero vene oriuntur' (*E*, fol. 55va). Mais le raccourci signalé *supra* se rapporte à la toute première phase du développement de l'embryon, pour laquelle l'approximation du traducteur fausse peu l'encyclopédie latine. C'est donc en connaissance de cause que Corbechon a simplifié ce couple de synonymes.

52. De même, pour évoquer les astuces employées par les nourrices pour apaiser les enfants, Barthélemy emploie l'expression 'sibilis et cantilenis' (*E*, chap. IX, fol. 57va), que Corbechon réduit à 'par son chant' (*F*, fol. 77va).

53. Le mot *pupinus* doit en effet être lu comme un masculin, graphié *pup-* au lieu de *pop-*, du mot *popina*, 'poupée' (Du Cange Ch., *Glossarium mediae et infimae latinitatis* (Paris, 1900), t. VI, p. 410^{c}).

54. 'Senex [...] fatigatur quousque resolvatur cinis in cinerem et pulvis in pulverem revertatur' (*E*, chap. I, fol. 55ra) est traduit par 'Le viellart [sueffre] [...] jusques atant qu'il retourne en cendre et en la poudre dont il est pris' (*F*, fol. 74va). En dépit de l'écho à l'Évangile (*Genèse* III, 19: 'quia pulvis es et in pulverem reverteris'), le traducteur a renoncé à cette figure.

55. 'Subito rident, subito plorant' (*E*, chap. V, fol. 56vb) est traduit par 'Ils pleurent et rient soudainement' (*F*, fol. 76vb). Cet exemple n'est pas isolé.

56. 'ligat et colligat' (*E*, chap. VII, fol. 57va) est traduit par 'elle le lie' (*F*, fol. 77va); 'quando dormiunt vel dormitant' (*E*, chap. V, fol. 56vb) par 'en dormant' (*F*, fol. 76vb). Enfin 'Item maledicens et lingosus de omnibus [...] maledicit' est simplifié en 'Derechief il dit mal de chascun' (*F*, fol. 80rb).

tous les éléments qui pourraient faire double emploi. En effet, à chaque fois qu'une idée de Barthélemy débouche, dans l'encyclopédie latine, sur plusieurs illustrations, Corbechon tend à en sélectionner une seule. Ce trait est particulièrement frappant dans les chapitres consacrés au bon et au mauvais serviteur, qui donnent lieu, chez Barthélemy, à nombre de renvois au texte biblique: Corbechon élague dans cette abondante matière. Ainsi, suite à l'énumération des vices du mauvais serviteur, l'encyclopédie latine s'appuie sur deux références accompagnées de citations: 'Ecc. X: Vidi servos in equis etc., et Prov. XXIX: Per servum cum regnaverit etc.' (*E*, chap. XVI, fol. 59va). Le traducteur n'a conservé que la première de ces deux illustrations: 'Et pour ce dit Salemon: J'ay veu les servans aler a cheval et les princes aler a pié aussi comme garçons, si comme il est escript ou livre Ecclesiastique ou X^e^ chapitre' (*F*, fol. 80ra). De telles réductions sont trop récurrentes, dans ce chapitre et le suivant, pour être attribuables à des lacunes de l'exemplaire latin consulté par le traducteur[57].

En somme, toutes ces coupes suggèrent qu'à chaque fois que l'écriture de Barthélemy est trop bavarde aux yeux de Corbechon, celui-ci cherche à exprimer l'idée force qu'il décèle chez son modèle de manière synthétique. Renonçant en partie à l'expressivité et à la richesse de sa source, le traducteur semble avant tout soucieux de fournir à son lecteur un texte resserré, dans le fond comme dans la forme; son adaptation romane, manifestement destinée à un public pressé, est calibrée pour lui permettre de retirer de sa lecture un maximum de fruit en un minimum de temps[58]. À cette simplification répond également le souci de Corbechon de présenter à ses lecteurs un savoir dépourvu de toute hésitation.

2.2. Un texte autorisé: l'élimination des points douteux

L'encyclopédie latine relevant davantage de l'effort de synthèse que de l'exposition de recherches en cours, elle présente le savoir sous une forme

57. À propos des qualités du bon serviteur, Barthélemy illustre le portrait qu'il vient de dresser en citant un extrait des *Proverbes* et un passage de l'*Ecclesiaste* (*E*, chap. XVII, fol. 60rb-60va); de nouveau Jean Corbechon ne conserve que la première de ces deux références (*F*, fol. 80va).

58. Le traducteur le signalait d'ailleurs dès le prologue qu'il ajoutait, au seuil de sa *translacion*, à celui de Barthélemy: 'et par especial au temps present vous ne [...] pouéz pas veoir ne visiter [vos livres] pour cause de voz guerre et de l'administracion de vostre roiaume et de pluseurs grandes occupacions qui chascun jour sourdent et viennent a vostre magnificence, pour tant est venu en votre noble cuer un desir d'avoir le livre des proprietéz des choses...' (*F*, fol. 2ra).

ordonnée et lisse. Néanmoins, lorsqu'une information demeure douteuse ou que diverses sources sont contradictoires, Barthélemy ne répugne pas à exposer plusieurs hypothèses. Ainsi propose-t-il deux étymologies pour le nom *pueritia*: 'Est vero dicta pueritia a pubertate vel a pupilla, quia adhuc pueri sunt puri ut pupilla, ut dicit Isidorus' (*E*, chap. I, fol. 54vb). Le traducteur du XIVe siècle, effaçant la première hypothèse de son prédécesseur, se contente d'évoquer la seconde étymologie en la présentant comme certaine: 'en latin on [...] apelle [le second aage] *puericia*, et est ainsi appelé pour ce que en celui aage est aussi comme la prunelle pur selon Ysidore' (*F*, fol. 74ra). De même, à propos de l'origine du mot *morte* ou *mort*, le traducteur est beaucoup plus catégorique que l'encyclopédiste du XIIIe siècle. Il transpose en effet 'Mors dicta est eo quod sit amara, vel a Mar[t]e, qui mortem fingitur efficere' (*E*, chap. II, fol. 55ra) par 'La mort est ainsi appellee pour ce qu'elle mort amerement' (*F*, fol. 74va). Non content ici de trancher dans le vif en éliminant la piste mythologique de son modèle[59], le traducteur remanie fortement la première de ses hypothèses pour lui donner davantage de vraisemblance dans la langue d'arrivée. En effet si, dans la transposition romane, l'adjectif *amara* est bel et bien traduit par l'adverbe *amerement*, ce n'est plus ce mot-là qui est présenté comme l'étymon du substantif *mort*, mais le verbe *mordre* – dont le vocalisme est, il est vrai, plus satisfaisant. L'innovation du traducteur, pour pittoresque qu'elle soit, constitue une exception; en revanche, sa propension à présenter un texte autorisé, dont les rares alternatives ou éléments douteux sont évacués, est constante[60]. Elle constitue sans doute une nouvelle manifestation du souci du traducteur d'en user discrètement avec le temps de son lecteur: lui faisant grâce des hypothèses de travail de Barthélemy, il lui présente exclusivement la version qu'il juge la plus convaincante. C'est que Corbechon entend remanier l'œuvre de Barthélemy pour répondre à un nouvel usage de ce savoir encyclopédique.

59. Il est vrai que dans le manuscrit *E*, qui comporte en ce point la leçon fautive *Mare* (fol. 55ra) au lieu de *Marte* (BNF, fr. 16099, fol. 46ra), cette seconde hypothèse d'étymologie était difficile à comprendre; la copie latine dont disposait Corbechon posait peut-être là de semblables problèmes. Il n'en reste pas moins que de telles suppressions sont trop fréquentes pour que l'hypothèse d'incorrections liées à la tradition manuscrite puisse être généralisée.

60. En matière d'étymologie, nous n'avons relevé qu'une exception: il s'agit du mot *Virgo* ou *Vierge*, pour lequel Corbechon maintient les diverses hypothèses avancées par Barthélemy (chap. VI: *E*, fol. 57ra et *F*, fol. 76vb). Peut-être la gravité du sujet, sa dimension religieuse, l'ont-elles amené, dans ce cas, à être nuancé.

3. Le *Livre des proprietés des choses*: un nouveau mode de lecture

3.1. L'encyclopédie vulgaire comme ouvrage complet et homogène

L'ambition de l'encyclopédie latine est déjà d'atteindre l'exhaustivité et de fournir à son lecteur l'ensemble des connaissances disponibles sur la Création. Cet objectif est clairement affiché dès le prologue de Barthélemy:

> In istis XIX libellulis rerum naturalium proprietates summatim et breviter continentur; [...] in quibus de meo pauca vel quasi nulla apposui, sed omnia que dicentur de libris auctenticis sanctorum et philosophorum excipiens sub brevi compendio pariter compilavi (*E*, fol. 1rb)[61].

Mais si l'ouvrage prétend traiter de tout ce qui existe, c'est surtout pour permettre à son lecteur d'accéder de plain pied à la Bible: la description du Créateur et de la création sont d'abord conçus comme prolégomènes à la lecture du texte sacré[62]. La même intention pieuse prévaut pour la version romane, dont le prologue suit fidèlement son modèle en affichant son intention d'être exhaustive[63]. Toutefois, le rapport entre encyclopédie et Bible y est sensiblement infléchi: alors que consulter Barthélemy suppose le recours préalable ou simultané à la Vulgate, Jean Corbechon traduit et intègre à son propre ouvrage la totalité des éléments bibliques qu'il est nécessaire de connaître pour le lire avec profit.

3.1.1. Un texte clos

Lorsque Barthélemy illustre son propos en faisant appel à la Bible, il se contente, dans la plupart des cas, de citer les quelques mots du passage qui l'intéresse, les faisant suivre de la mention *etc.*[64] et d'une

61. L'effort de compilation et de synthèse de l'encyclopédiste était également suggéré dès l'ouverture du prologue, où Barthélemy affirmait avoir compilé son ouvrage pour rassembler les connaissances qui pourraient manquer à ceux 'qui naturas rerum et proprietates per sanctorum libros necnon et philosophorum dispersas non cognoverunt' (*E*, fol. 1ra).

62. Barthélemy prétend en priorité 'offrir une clé de lecture et d'interprétation aux Écritures saintes' (Van den Abeele, Meyer & Ribémont, 'En guise d'avant-propos...', p. 9). Le prologue de l'ouvrage en témoigne, puisqu'il se veut utile 'ad intelligenda enigmata scripturarum que sub symbolis et figuris proprietatum rerum naturalium et artificialium a Spiritu Sancto sunt tradite et velate' (*E*, fol. 1ra). En la matière, le *De proprietatibus rerum...* ne constitue pas une exception, bien des encyclopédies médiévales ayant été conçues comme des moyens de mieux comprendre les Écritures, voire comme des voies d'accès au Salut (Ribémont, 'Encyclopédisme médiéval...', p. 384).

63. L'encyclopédie en français y est qualifiée de 'sonme general contenant toutes materes' (*F*, fol. 2ra).

64. Voir notamment *supra* p. 349. Les copies du texte latin sont suffisamment unanimes sur cette troncature pour qu'il soit exclu d'y voir le résultat de la fantaisie de tel ou tel copiste.

référence précise à la Vulgate. Quelle que soit l'étendue de la citation biblique – certaines sont suffisamment développées pour être comprises, d'autres beaucoup trop elliptiques pour être intelligibles telles quelles –, chacune est suivie d'une marque d'incomplétude et d'un renvoi à l'autorité. C'est que l'encyclopédie latine se présente ouvertement comme un puzzle incomplet, dont les pièces manquantes doivent être cherchées ailleurs: dans la Bible le plus souvent, mais aussi dans des encyclopédies antérieures (notamment celle d'Isidore de Séville, abondamment citée), dans des auteurs de l'Antiquité classique, etc.

Au contraire, l'encyclopédie vernaculaire se veut autonome et close: lorsque le traducteur évoque d'après Barthélemy une autorité, il ne se contente pas, comme son modèle, d'une allusion accompagnée d'une référence qui suggère une lecture complémentaire. Il accomplit pour son lecteur la recherche suggérée par l'encyclopédiste du XIIIe siècle, comblant ainsi les brèches de son texte. C'est par exemple le cas, au chapitre XVI, pour la parabole des talents: illustrant la paresse parmi les travers du mauvais serviteur, elle ne figure chez Barthélemy que sous la forme d'un renvoi très elliptique, qui ne peut être compris que d'un lecteur connaissant déjà l'épisode ou se reportant en parallèle au texte biblique[65]. Corbechon le développe de manière à rendre inutile le détour par la Bible, dont il continue toutefois de fournir la référence précise: 'si comme il est contenu ou XIXe chapitre de l'Evangile saint Luc de celui servant qui prist du roy un besant d'or pour lui faire gaingner en marchandise, et il le fouy soubz terre et ne le moulteplia point' (*F*, fol. 80ra)[66]. L'ajout du traducteur est minime et peu développé, mais il permet à un lecteur qui ignorerait l'anecdote de la comprendre sans support extérieur: les motifs de la conduite du serviteur sont explicités – dans le texte latin, le fait qu'il enterre la pièce pouvait intriguer – et son échec final est révélé. De même l'allusion illustrant la nécessité de convoquer des instruments de musique pour assurer la convivialité d'un banquet ('Luce XV: Cum

65. 'Et Luce XIX dicitur: Servus piger abscondit talentum domini sui in terra et abiit…' (*E*, fol. 59va). Ensuite, le texte passe sans transition à un autre défaut du mauvais serviteur: l'effronterie ('Item protervus et cervicosus…', *E*, fol. 59va).

66. On notera que la référence à la parabole des talents est ici erronée, chez Barthélemy comme chez son traducteur: le récit qui figure dans les deux versions de l'encyclopédie correspond à l'Évangile de saint Mathieu (XXV, 14-30) plutôt qu'à celui de saint Luc (XIX, 20-22). Voir notamment l'écart entre 'Qui autem unum acceperat, abiens fodit in terram, et abscondit pecuniam domini sui. (…) Accedens autem et qui unum talentum acceperat, ait: Domine, (…) timens abii, et abscondi talentum tuum in terra' (Mt XXV, 18 et 24-25) et 'Domine, ecce mna [*sic*] tua, quam habui repositam in sudario' (Lc XIX, 20). Corbechon a conservé telle quelle cette référence erronée: sans doute, pour compléter le récit de cet épisode fort célèbre, a-t-il travaillé de mémoire, sans se reporter à la Vulgate.

audisset necesse etc.', *E*, chap. XXIII, fol. 64ra), n'est nullement reconnaissable telle quelle dans l'encyclopédie latine, sauf à avoir en mémoire l'Évangile de Luc. De nouveau le traducteur développe quelque peu la narration de l'épisode, assez pour le rendre intelligible tout en mettant l'accent sur les détails qui soulignent la pertinence de l'illustration:

> Aprés doit il avoir instrumens de musique pour resjouir la compaignie. Et pour tant lisons nous ou XVe chapitre de l'Evangile saint Luc que quant le pere qui avoit deux filz ot retrové celui qu'il avoit longuement perdu, il en fist grant feste et fist venir instrumens au manger pour fere meilleur chiere et meilleur feste (*F*, fol. 85va)[67].

Grâce à de tels développements[68], l'encyclopédie française se voit dotée d'une autonomie que le *De proprietatibus*... ne prétendait pas posséder: l'ouvrage est devenu intelligible sans nécessiter ni culture préalable, ni recours à un appui extérieur, fût-ce la Bible.

On notera enfin que, même lorsque les deux encyclopédistes fournissent à leur lecteur un récit de teneur semblable – nombre de citations étaient déjà suffisamment étoffées dans la version latine pour être intelligibles –, Barthélemy signale l'incomplétude de son texte, encourageant systématiquement son lecteur à se reporter à la source par une mention *etc.*; or le traducteur, même lorsqu'il n'étoffe nullement le récit de son modèle, élimine toutes ces marques de béance[69]. Mais il ne se contente

67. L'allusion telle qu'elle figure dans l'encyclopédie latine renvoie non pas à Lc XV et à la parabole de l'enfant prodigue, mais à Lc XIV, 15-18. Il est vrai que la citation telle qu'elle se présente dans le manuscrit *E*, juxtaposant le début du verset 15 ('[Haec] cum audisset') et un extrait du verset 18 ('Villam emi, et necesse habeo exire'), avait de quoi désarçonner même un fin connaisseur de la Bible comme Corbechon. De nouveau, celui-ci semble avoir travaillé de mémoire.

68. Une réécriture du même ordre touche l'allusion au mauvais serviteur qui bat les enfants de son maître en son absence – 'Luce XII, si dixerit: Dominus meus moram facit venire...' (*E*, chap. XVI, fol. 59va), citation libre de Lc XII, 45 –, qui est complétée dans la traduction de Corbechon: 'si comme de celui qui pensoit que son seigneur ne revendroit point et bastoit ses enfans et despendoit ses biens maisement, si comme il est escript ou XIIe chapitre de l'ewangile saint Luc' (*F*, fol. 80rb). La mention du festin donné par Assuerus – 'Hester I: cum incaluisset etc.' (*E*, chap. XXIII, fol. 63vb) –, renvoi tronqué à Esther I, 10, donne lieu à pareil développement: 'mais en especial moult de choses sont qui embellissent ou souper, lesquelles choses furent en la feste et ou souper du roy Assuere, si comme il [est] escript ou premier chapitre du livre de Hester' (*F*, fol. 85rb).

69. Voir notamment, au chapitre XI, le récit des aventures d'Agar, la servante d'Abraham et Sarah, qui fait l'objet d'un traitement aussi circonstancié chez Barthélemy (*E*, fol. 57vb) que chez Jean Corbechon (*F*, fol. 77vb), mais ne comporte dans la version romane aucune marque d'incomplétude; voir aussi, au chapitre VI, la citation prétendument tirée de Sénèque (*E*, fol. 57ra et *F*, fol. 76vb). Ces suppressions sont trop répandues dans la tradition manuscrite du texte roman pour résulter de l'intervention d'un copiste isolé.

pas de développer chacune des allusions aux autorités qui fondent l'encyclopédie pour éviter à son lecteur de se reporter à ses sources. Non seulement il referme l'œuvre sur elle-même en rendant inutile tout recours à d'autres textes, mais il s'interpose entre les auteurs qu'il cite et son lecteur.

3.1.2. Une seule voix: le refus du discours direct

L'ensemble du texte de Corbechon émane en effet d'une unique instance d'énonciation et le traducteur s'impose tout au long de son ouvrage non seulement comme le 'master of the author'[70], mais comme le porte-parole de toutes les autorités qui y sont convoquées. C'est pourquoi, lorsque Barthélemy se fait l'écho direct de saint Paul: 'ut dicit Apostolus in Cor. VI: Mulierem in ecclesia docere non permitto etc.' (*E*, chap. XII, fol. 58rb), le traducteur transpose les propos de l'Apôtre au style indirect: 'si comme dit saint Pol ou VIe chapitre de sa premiere espitre aux Chorinthiens, ou il deffent que nulle fame ne s'entremecte d'enseignier ou de preschier' (*F*, fol. 78rb). Une telle modification, fût-elle isolée, serait anecdotique. Mais sa récurrence dans nombre de passages du livre VI[71] en fait une nouvelle manifestation de l'autorité d'un traducteur qui garde la parole tout au long de son ouvrage. Elle est en outre signe d'honnêteté intellectuelle: ne citant le texte sacré qu'*in substantia*, et non dans sa lettre ni sa langue d'origine, Corbechon rappelle ainsi au lecteur qu'il n'a affaire qu'à un ersatz. Comme ses allusions aux difficultés de la tâche de traducteur[72], ces marques explicites donnent l'encyclopédie vernaculaire pour ce qu'elle est: un ouvrage de seconde main, fruit de la transposition d'un *translateur* qui ne cherche nullement à voiler son rôle d'intermédiaire. Mais si Corbechon répugne à user du discours direct, c'est peut-être aussi par souci de donner davantage d'homogénéité à son texte.

70. L'expression est due à Jeanette Beer qui explique qu'en tant qu'intermédiaire entre le texte latin et le public plus ou moins cultivé qui en lira la traduction romane, le traducteur médiéval doit manifester un certain 'degree of didactism'; son travail est donc perçu comme 'a teaching activity', et c'est à ce titre que le traducteur est qualifié de 'master' (Beer J.M.A., 'Introduction', in: J.M.A. Beer (éd.), *Medieval Translators* ..., p. 1).

71. Voir notamment la suite du même chapitre *de masculo*, où 'Scriptum enim est: Sub viri potestate eris et ipse dominabitur tui (Genesis III)' (*E*, fol. 58rb) devient 'Et si est escript ou tiers chapitre ou livre de Genesis que la fame est dessoubz la puissance de l'omme et il a la seigneurie d'elle par l'ordonnance de Dieu' (*F*, fol. 78rb).

72. Voir *supra* p. 344.

3.1.3. Un texte lisse: l'intégration des références à la syntaxe

Le traitement qu'il fait subir aux références de son modèle conforterait volontiers cette hypothèse. En effet, chez Barthélemy, les renvois à d'autres ouvrages ne sont nullement intégrés aux phrases, mais figurent en marge; l'éditeur moderne se trouve contraint de les éditer entre parenthèses, tant ils s'apparentent par leur présentation et par leur caractère marginal, à nos notes de bas de page – elles ont d'ailleurs une fonction similaire.

Dans l'encyclopédie vulgaire, elles ont désormais pour rôle principal de cautionner les informations délivrées; elles fournissent en outre au lecteur une sélection des auteurs à connaître[73]. Ne s'adressant plus au public d'étudiants rompus aux abréviations et aux références paratactiques qui était celui de Barthélemy, Jean Corbechon respecte les conventions d'écriture qu'attendent ses lecteurs, en majorité laïques. En conséquence, il insère les références de Barthélemy à la syntaxe: elles font désormais partie intégrante du texte, au même titre que les récits ou les discours. La tournure la plus fréquemment employée pour ce faire est la proposition comparative, comme après l'allusion à la servante Agar: 'Genesis XVI' chez Barthélemy (*E*, chap. XI, fol. 57vb) devient 'si comme il est escrit ou VIe chapitre du livre de Genesis'[74] (*F*, fol. 77vb)[75]. Certes, l'écart de présentation entre les manuscrits latins – à la page plus dense et aux abréviations plus drastiques, notamment en ce qui concerne les références –, et les manuscrits français, moins abrégés et plus aérés, accuse le trait. Il n'en reste pas moins que le statut des autorités au sein des deux encyclopédies, s'il est souligné par le travail des copistes, relève avant tout de la volonté de chacun des deux auteurs et que Corbechon en avait largement modifié la fonction lors de sa mise en roman de l'encyclopédie latine.

En revanche, la teneur des références change peu; tout au plus constate-t-on chez Corbechon une certaine tendance à l'exhaustivité, le traducteur rappelant souvent le titre de l'ouvrage concerné et son auteur là où Barthélemy ne fournissait que l'une de ces deux données[76] – et ce même

73. Sur la vocation de l'encyclopédie romane à être un ouvrage de seconde main, voir la note 61.

74. Le passage de XVI à VI est à mettre au compte des aléas de la tradition manuscrite.

75. Les tournures similaires sont nombreuses tout au long du livre VI: 'si comme Dieu le dit ou second chapitre du livre de Genesis' (*F*, chap. XIII, fol. 78va); 'si comme nous lisons de Laban ou livre de Genesis' (*F*, chap. XIX, fol. 82ra); etc.

76. Par deux fois au cours des chapitres II et III, les références à Aristote ('Dicit enim Aristoteles libro II' et 'et Aristoteles libro XV', *E*, fol. 55rb) sont étoffées: 'Et pour ce dit Aristote ou secont livre des bestes…' (*F*, fol. 74va) et 'Et pour ce dit Aristote ou XVe livre des bestes…' (*F*, fol. 74vb).

lorsque la référence est évidente ou récurrente[77]. Sans doute faut-il voir là chez le traducteur un instinct de pédagogue qui le pousse à l'exhaustivité. On note enfin sa perplexité face à quelques références, notamment face au renvoi de Barthélemy au *Medicus*[78] (*E*, chap. I, fol. 54vb), qui se voit gauchi en 'les plusieus' (*F*, fol. 74ra). Ce cas de figure reste cependant très marginal.

Si la teneur des références change moins que leur présentation, c'est que le traducteur, loin de remettre en cause les sources de son prédécesseur, cherche à renouveler la lecture de son texte: loin de faciliter une consultation ponctuelle et de favoriser le recours à d'autres ouvrages, comme le faisait l'encyclopédiste latin, le traducteur semble prévoir une lecture continue et linéaire de l'encyclopédie romane.

3.2. L'encyclopédie comme chemin tracé

Les encyclopédies du XIIIe siècle – et celle de Barthélemy ne fait pas exception – innovent en adoptant une présentation et en élaborant une série d'apparats qui ouvrent la voie à une lecture partielle[79]. Le *De prorietatibus rerum* est donc conçu pour permettre au lecteur de butiner; comme les dictionnaires modernes, l'ouvrage est prévu pour une lecture plutôt fragmentée que linéaire.

La version de Jean Corbechon prévoit une autre manière d'accéder aux connaissances: il s'agit non plus de consulter des fragments d'encyclopédie, mais d'en lire le texte de manière continue. L'absence, dans presque toutes les versions romanes, de tables alphabétiques thématiques[80] est un indice du mode de lecture prévu pour ces textes. En témoigne en outre chez Corbechon la forme des renvois internes, fort différents de ceux du modèle latin. En effet, alors que Barthélemy emploie le présent pour évoquer l'existence, au sein de son encyclopédie, d'éléments à rapprocher, Jean Corbechon confère à ces renvois une épaisseur temporelle

77. Lorsque Barthélemy cite Isidore de Séville (*Isidorus*), Corbechon rappelle fréquemment que, comme nombre d'encyclopédistes, c'est au 'livre des ethimologies' qu'il se réfère (voir notamment *F*, chap. II, fol. 74va; chap. V, fol. 76vb; etc.).

78. La source de Barthélemy serait en fait ici l'*Isagoge* d'Hunain (ou Hunain ibn Ishaq), d'après Seymour M.C. *et alii* (éds), *Bartholomaeus Anglicus and his encyclopaedia* (Aldershot-Brookfield, 1992), p. 78.

79. Ribémont, 'Encyclopédisme médiéval…', p. 389. Aux apparentes subdivisions en chapitres et aux tables des matières, communes à bien des manuscrits de cette époque, s'ajoutent, dans le cas de nombreux manuscrits de Barthélemy l'Anglais, des tables alphabétiques thématiques (Ventura, 'Autour de la version occitane…', p. 50).

80. Seule la traduction occitane conservée dans le manuscrit 1029 de la bibliothèque Sainte-Geneviève présente une table de ce genre (Ventura, 'Autour de la version occitane…', p. 50).

qui suggère le mode d'emploi qu'il prévoit pour son ouvrage: une lecture linéaire. Ainsi le traducteur de Charles V emploie-t-il le passé pour renvoyer à l'amont textuel: 'Filie autem conveniunt heedem [*sic*] proprietates que dicuntur de puella supra in eodem libro' (*E*, chap. VIII, fol. 57va) est transposé en 'Toutes les proprietéz de la pucelle si appartiennent a la fille, desquelles proprietéz nous avons parlé tantost ci dessus' (*F*, fol. 77rb).

Les modifications opérées sur les renvois à l'aval textuel sont plus révélatrices encore. En effet Barthélemy, qui suppose chez son lecteur une recherche d'ordre thématique, emploie fréquemment l'impératif pour l'inciter à glaner dans la suite de l'ouvrage afin d'y trouver des informations complémentaires sur le même sujet: 'Alias mortis proprietates quere infra in fine istius VI libri, in fine de infirmitate' (*E*, chap. II, fol. 55rb); 'vide infra de sulphure ubi virtutem ejus plenius poteris intueri' (*E*, chap. XXI, fol. 63ra)[81]. Dans tous les cas, l'encyclopédiste du XIIIe siècle encourage son lecteur à s'affranchir de l'organisation matérielle du texte pour se reporter ailleurs. Jean Corbechon a conservé ces renvois internes, qui confortent la cohérence de l'énoncé et peuvent aider à assimiler les informations présentées. Néanmoins, il en remanie considérablement la forme et fait de ces allusions à la suite du texte des pierres d'attente qui n'incitent plus à interrompre le fil de la lecture: soulignant le déroulement temporel de l'énonciation du texte, il emploie volontiers le futur pour donner l'illusion que la fin du texte, si elle est déjà prévue, n'est pas encore écrite:

> [...] c'est la mort a celui qui le prant et venim a celui qui en abuse, si comme nous dirons aprés quant nous parlerons d'yvresce (*F*, chap. XII, fol. 85ra);
>
> si comme il apparra ci aprés plus plainement quant on fera mencion de la nature du souffre (*F*, chap. XXI, fol. 84va);
>
> Du dormir non naturel nous ne dirons riens jusques au VIIe livre ou chapitre de la litargie, ou nous en parlerons se Dieu plaist (*F*, chap. XXV, fol. 86vb).

La dernière occurrence est celle qui pousse le plus loin l'artifice, en suggérant que l'élaboration du livre suivant reste incertaine. Bien sûr le lecteur médiéval, devant son manuscrit achevé, ne se laissait pas abuser plus que nous: la manière qu'a le traducteur de présenter comme impossible le recours à la suite du texte parce qu'elle reste à écrire n'est qu'une convention. Mais celle-ci modifie le cheminement prévu par Barthélemy

81. L'emploi de l'impératif domine pour ce genre de renvois, le verbe *quere* étant le plus fréquent. Il est parfois juxtaposé à des formules au présent indiquant la présence de données complémentaires dans la suite du texte: 'sed mors erit potius et venenum, sicut patet infra de ebrietate; quere ibi' (*E*, chap. XXI, fol. 63va).

dans son encyclopédie: alors que le texte latin autorise, voire encourage, une lecture discontinue, sa transposition romane incite à une lecture suivie, comme pour un récit. Cette découverte du texte au fil de son déroulement correspond manifestement à l'horizon d'attente de lecteurs laïques désireux de s'instruire au gré d'une découverte exhaustive.

4. Conclusion

Quoique Jean Corbechon conserve globalement les données du *De proprietatibus rerum*, qu'il n'actualise ni ne complète, ses interventions en tant que traducteur sont finalement moins inoffensives que son zèle et ses précautions pourraient le laisser croire à première lecture. En effet, il vise un public totalement différent de celui auquel s'adressait Barthélemy, et adapte son discours à un nouvel horizon d'attente: celui de l'entourage royal du XIVe siècle. Ainsi est-il beaucoup plus présent dans son texte que ne l'était Barthélemy: tel un guide, il ouvre la marche et désamorce tout chausse-trappe. La contrepartie, c'est qu'en encourageant une lecture linéaire, il n'incite plus le lecteur à prendre des chemins de traverse, le privant en partie de l'autonomie à laquelle les raccourcis aménagés par l'encyclopédiste du XIIIe siècle lui donnaient accès. Le texte, beaucoup plus ergonomique, met moins à contribution l'esprit d'initiative du lecteur ou sa culture préalable.

Aussi trouve-t-on, dans le maillage de l'encyclopédie romane, des transpositions et remaniements qui correspondent de manière sous-jacente aux caractéristiques extérieures de bien des manuscrits: la plupart des copies du texte de Jean Corbechon sont beaucoup plus luxueuses que celles du modèle latin – elles sont volontiers enluminées, recourent de manière beaucoup plus limitée aux abréviations, etc. Cet écart de présentation n'est pas une originalité du couple Barthélemy l'Anglais / Jean Corbechon: le même contraste s'observe notamment entre le *Speculum historiale* de Vincent de Beauvais et sa traduction française par Jean de Vignay[82]. Mais si la traduction de Jean Corbechon a remporté tant de succès[83], si ses commanditaires lui ont accordé suffisamment de prix pour

82. Chavannes-Mazel (Cl.A.), 'Problems in Translation, Transcription and Iconography: the *Miroir historial*, Books 1-8', in: S. Lusignan, M. Paulmier-Foucart & A. Nadeau (éds), *Vincent de Beauvais. Intentions et réceptions d'une œuvre encyclopédique au Moyen Âge. Actes du 14e colloque de l'Institut d'études médiévales, organisé conjointement par l'atelier Vincent de Beauvais et l'Institut d'études médiévales, 27-30 avril 1988*, Cahiers d'études médiévales; cahier spécial 4 (Saint-Laurent, 1990), p. 345-374, ici p. 346-347.

83. Nous n'en conservons pas moins de 36 manuscrits (voir la note 7), auxquels s'ajoutent plusieurs imprimés, ce qui représente au total 45 copies (Van den Abeele, Meyer & Ribémont, 'En guise d'avant-propos…', p. 11).

en faire exécuter des copies de luxe, c'est que le texte répondait parfaitement à leurs attentes: le travail de compilation de Barthélemy l'Anglais en faisait un parfait manuel de culture générale (culture classique, religieuse, et même matérielle pour le livre VI). Jean Corbechon a parachevé la tâche en donnant au texte l'apprêt formel propice à une lecture de loisir à la fois agréable et fructueuse.

Reste à voir si les caractéristiques qui se dégagent du livre VI peuvent être étendues à l'ensemble de l'encyclopédie, et notamment aux parties plus techniques. En effet, comme le rappelle Claude Herfray-Rey, 'Corbechon s'est lancé dans sa traduction sans système préétabli'[84], et il est vrai que si certains traits récurrents encouragent de plus amples recherches, le texte semble parfois s'évertuer par ses disparités à déconcerter toute tentative d'observation systématique[85]. Reste aussi à comparer les procédés mis en œuvre par Corbechon à ceux d'Evrart de Conty ou de Jean de Vignay, qui travaillaient pour un lectorat comparable[86]. De tels rapprochements permettraient sans doute d'affiner le portrait-robot de chaque traducteur, et aussi de mieux mesurer le rôle de leurs commanditaires dans l'élaboration des *translacions*[87].

Université Paris IV-Sorbonne
Laboratoire CEPAM Université de Nice-Sophia Antipolis

84. Herfray-Rey, 'Jean Corbechon...', p. 64.

85. Ainsi, si Corbechon supprime fréquemment les couples de synonymes de Barthélemy, il lui arrive, quoique plus rarement, de développer sous forme d'itération lexicale tel ou tel mot de son modèle. Les dents des enfants en bas âge sont par exemple qualifiées, dans l'encyclopédie latine, de 'nondum bene ordinatis' (*E*, chap. I, fol. 54vb), et de 'pas encore [...] bien ordonnees ne affermees' dans le *Livre des proprietés...* (*F*, fol. 74ra). De même, dans ce chapitre, 'decor corporis adnullatur' (*E*, fol. 55ra) est développé en 'toute la beauté du corps y est perdue et anoientie' (*F*, fol. 74va). De tels exemples, quoique minoritaires, ne sont pas rares. De même, si Corbechon s'efforce de supprimer bien des figures de répétition (voir *supra* p. 347-349), il ne manque pas d'en conserver quelques-unes, comme dans le passage suivant: 'Unde (...) ingeniata est natura homini conferre membra nobilia et diversa nobilibus operationibus et variis convenientia' (*E*, chap. II, fol. 55rb) 'Et pour ce a soutillé nature de lui donner tresnobles menbres convenables a ses nobles euvres' (*F*, fol. 74vb). Voir aussi la note 50.

86. Evrart de Conty gravitait aussi dans l'entourage de Charles V (voir les références bibliographiques de la note 6). Quant à Jean de Vignay, il travailla quelques décennies plus tôt et fut surtout patronné par Jeanne de Bourgogne; il s'adressait donc à un public du même type, et plusieurs de ses traductions trouvèrent place dans les principales bibliothèques royales et princières de la fin du Moyen Âge.

87. L'attention permanente que semble accorder Jean Corbechon aux attentes de son commanditaire telles qu'il les expose dans le prologue conforte le témoignage de plusieurs enluminures représentant Charles V en dialogue avec tel ou tel écrivain lui offrant un ouvrage (Willard, 'Raoul de Presles's translation...', p. 334). Mais si un tel rapprochement pique notre curiosité, il ne saurait suffire à élucider l'emprise exercée par le mécène sur la tâche du *translateur*, ni les circonstances dans lesquelles son influence trouvait à s'exprimer.

Francesco CAPACCIONI

INFINITI INGEGNI DA' PIU NON SAPUTI: LA PRIMA TRADUZIONE ITALIANA DEI *RURALIA COMMODA* DI PIETRO DE' CRESCENZI (LIBRO X)

'The typical Knight of the Middle Ages was far more interested in pigs than in tournaments', forse questa frase del medievalista Alexander J. Carlyle, citata da C.S. Lewis in un suo celebre saggio[1], parodia un poco la reale portata della conoscenza del mondo animale da parte dell'uomo medievale. In realtà, ogni cavaliere possedeva una solida cultura 'naturalistica' che comprendeva, tra le altre cose, la conoscenza approfondita dei rapaci, dei cavalli, dei cani e degli svariati tipi di selvaggina che era solito cacciare. Non si può quindi escludere che qualche cavaliere potesse nutrire un interesse particolare anche per i maiali, animali notevoli sotto molti aspetti. Ma è probabile che l'argomento 'suini' non ricorresse poi così spesso nei discorsi dell'aristocrazia medievale. In effetti si potrebbe mitigare il paradosso di A.J. Carlyle dicendo che i cavalieri medievali si interessavano *sia* ai tornei *sia* ai maiali.

In generale, l'Europa durante i secoli XI-XIV conobbe un forte incremento della produzione agricola, dovuta – oltre che al complesso di pratiche diffuso a partire dall'esperienza dei grandi monasteri carolingi – alla spinta verso l'ampliamento degli spazi coltivabili da parte dei grandi proprietari terrieri e dei Comuni, ma anche delle comunità dei villaggi e persino dei singoli, contadini o eremiti che fossero; grazie a queste operazioni di disboscamento e bonifica anche l'allevamento del bestiame subì un forte incremento, incremento che nel complesso interessò tutta l'economia europea.

* Desidero ringraziare per i preziosi consigli il dott. Christoph Gasser.

1. 'The percentage of the population who knew a great deal about certain animals must have been far larger in medieval than in modern England. It could not have been otherwise in a society where everyone who could be was a horseman, hunter, and hawker, and everyone else a trapper, fisher, cowman, shepherd, swineherd, goose-girl, henwife, or beekeeper. A good medievalist (A.J. Carlyle) once said in my hearing, 'The typical Knight of the Middle Ages was far more interested in pigs than in tournaments'' (Cf. Lewis C.S., *The discarded image: an introduction to medieval and renaissance literature*, Cambridge, 1964, p. 146-147).

In questo panorama, un nuovo ceto cittadino, estraneo all'aristocrazia militare e formato principalmente da mercanti, artigiani e liberi professionisti, comincia ad accumulare notevoli ricchezze e a investire nell'acquisto di proprietà terriere, iniziando così a interessarsi ai numerosi problemi che la gestione di queste proprietà presenta. In effetti questa nuova classe sociale in continua ascesa è pervasa da una curiosità profonda e vitale, una sete di conoscenza che tocca tutti i settori della vita.

L'opera di Pietro de' Crescenzi (ca. 1233 – ca. 1320) nasce proprio in questo clima, per rispondere a esigenze pratiche ormai diffuse. I *Ruralia commoda*, scritti nei primi anni del Trecento, sono composti da dodici libri introdotti da un'epistola dedicatoria a Carlo II d'Angiò. Ogni libro è dedicato a un'attività particolare: dalla cerealicoltura alla viticoltura, dall'allevamento alla coltivazione delle erbe o alla cura dei giardini, i *Ruralia commoda* non tralasciano nessun aspetto, incontrando per questo un larghissimo e duraturo successo[2]. Non dirò nulla di nuovo ricordando che il primo libro stampato a Leuven fu proprio un'edizione dei *Ruralia commoda*, la seconda per l'esattezza, realizzata nel 1474 ad opera di Giovanni di Vestfalia. Tra il quattordicesimo e il sedicesimo secolo l'opera di Pietro de' Crescenzi viene tradotta in italiano, francese, tedesco e polacco. La traduzione italiana in particolare, seguendo il medesimo destino dell'opera latina, conobbe un notevole successo. Soprattutto tra Quattrocento e Cinquecento, nell'ambito del più ampio dibattito sul tema della lingua italiana, fu molto apprezzato il toscano arcaico e solenne utilizzato dall'Anonimo. Pietro Bembo, nelle *Prose della volgar lingua*, arriva a preferire la prosa del volgarizzamento – da lui peraltro creduta opera dello stesso Pietro de' Crescenzi – a quella dello storico fiorentino Giovanni Villani[3]. Qualche anno più tardi il medesimo giudizio positivo

2. Per approfondire la conoscenza della vita e dell'opera di Pietro de' Crescenzi, qui delineate con estrema concisione, si veda: *Pier de' Crescenzi (1233-1321). Studi e documenti* (Bologna, 1933); Toubert P., 'Pietro de' Crescenzi', in: *Dizionario Biografico degli Italiani* (Roma, 1960-), s.v. Crescenzi, Pietro de', p. 649-657; Ambrosoli M., 'Lettori e chiosatori delle edizioni a stampa di Pier de' Crescenzi tra 1474 e 1561', *Rivista storica italiana* XCVI (1984), p. 360-413; Camillo E., 'Voci quotidiane, voci tecniche e toscano nei volgarizzamenti di Plinio e Pietro de' Crescenzi', *Studi di lessicografia italiana* XI (1991), p. 125-151; Santa Eugenia F.J., 'Ottave quattrocentesche sugli uccelli da caccia', *Studi di filologia italiana* LIV (1996), p. 221-260.

3. 'Furono altresì molti prosatori tra quelli tempi, de' quali tutti Giovan Villani, che al tempo di Dante fu e la istoria fiorentina scrisse, non è da sprezzare, e molto men Pietro Crescenzo bolognese, di costui più antico, a nome del quale dodici libri delle bisogne del contado, in volgare fiorentino scritti, per mano si tengono' (Cf. P. Bembo, *Prose della volgar lingua* in *Trattatisti del Cinquecento*, a cura di M. Pozzi (Milano-Napoli, 1978), I, p. 114.

è dato da Leonardo Salviati il quale, con qualche distinguo in più, ne apprezza le 'buone voci' e la 'leggiadra semplicità'. Il testo volgarizzato del de' Crescenzi entra dunque a far parte del ristretto novero dei testi di riferimento per i 'puristi' della lingua italiana. Lo stesso Sansovino, autore di un'altra versione italiana dei *Ruralia*, non partirà dall'originale latino, ma dal testo dell'Anonimo, stravolgendolo nella sintassi e nel lessico fino quasi a riscriverlo[4]. Circa quarant'anni dopo anche Bastiano de' Rossi, curando una nuova edizione dei *Ruralia* per i tipi di Cosimo Giunti (Firenze, 1605), interviene direttamente sul testo trecentesco, correggendo gli errori evidenti e smussando le asperità sintattiche e lessicali, senza apportare, però, le profonde modifiche operate dal Sansovino[5].

A fronte di un così grande successo, anzi probabilmente proprio a causa di esso, finora non è stato possibile licenziare un'edizione critica che tenga conto della grande mole dei manoscritti che tramandano sia la versione latina sia il volgarizzamento italiano. Per quanto riguarda l'opera latina l'edizione più recente si basa essenzialmente su due manoscritti (T = Vaticanus, Vat. Lat. 1529 e U = Ultraiectensis universitatis, Cod. lat. 712)[6], mentre, per ciò che concerne il volgarizzamento trecentesco, continua a essere utilizzata l'*editio princeps*, stampata a Firenze nel 1478, purtroppo vessata da numerose scorrettezze. Recentemente il procedere della recensione dei codici esistenti ha permesso di individuare almeno due versioni distinte del volgarizzamento, una, chiamata dallo studioso Francisco Javier Santa Eugenia col nome di *Utilità della Villa* e da cui deriverebbe il testo dell'incunabolo fiorentino, di più larga diffusione e più aderente al testo originale, l'altra, chiamata *Rusticano*, meno[7].

È evidente dunque quali difficoltà incontra chi si trova a confrontare due testi del genere, che per natura – in quanto opere pensate per fornire indicazioni per risolvere problemi pratici –, sono sottoposte a un lavoro continuo di rimaneggiamento e aggiornamento nel corso della loro trasmissione, una spinta centrifuga che porta alla disgregazione del concetto stesso di originale, disgregazione che appare con maggiore evidenza nella

4. *Pietro Crescentio Bolognese tradotto nuovamente per Francesco Sansovino. Nel quale si trattano gli ordini di tutte le cose che si appartengono a commodi et a gli utili della villa* (Venezia, 1561).

5. L'edizione curata da Bastiano de' Rossi (Firenze, 1605) 'è, nella storia degli studi crescenziani, un documento fondamentale, non solo per la parte che ebbe nella redazione delle prime due impressioni del Vocabolario della Crusca, ma anche perché da essa dipendono tutte le stampe successive' (Cf. Santa Eugenia, 'Ottave quattrocentesche…', p. 223).

6. Petrus de Crescentiis (Pier de' Crescenzi), *Ruralia commoda. Das Wissen des vollkommenen Landwirts um 1300*, Herausgegeben von W. Richter. Zum Druck vorbereitet von R. Richter-Bergmeier (Heidelberg, 1995-1998).

7. Santa Eugenia, 'Ottave quattrocentesche…', p. 223.

tradizione dei volgarizzamenti, opere che hanno al contempo la dignità di testi indipendenti e lo *status* di copie più o meno fedeli.

Ma procediamo dal dato testuale, l'unico che ci permette di proporre qualche argomentazione concreta. La prima e più forte impressione che scaturisce dal confronto fra il testo italiano e quello latino è che l'anonimo traduttore abbia realizzato una copia fedele dell'originale, seguendo passo passo l'andamento del dettato latino:

2,2, 23-25 Inveniuntur autem ancipitres in Alpibus quibusdam nidificare, et meliores omnibus sunt, ut fertur, qui nascuntur in Alpe de Bruca in Sclavonia[8].	II Et truovonsi gli sparuieri nell'alpi alcuni nidificare, e migliore di tutti sono, sichome si dicie, di quelli che nascono nell'alpe di Bruca in Ischiauonia[9].
15,2, 12-15 Et statim cum videt canes invenisse pasturam, aquilam consuetam et instructam dimittere debet, quae semper super canes volabit, et cum leporem viderit, subito descendet et capiet eam	XV Et incontanente quando vede e' cani havere trovata la pastura, l'aguglia usata e admaestrata dee lasciare, la quale sempre sopra e' cani volerà et, quando la lepre vedrà, subitamente disciende et prenderla.
18,3, 34-39 Et in spina infigatur faseolus vel faba mollis, solummodo unum granum, quod avis rostro capiet ac caput elevans super collum laqueum trahet plicata scalella cum virgulis, quam avis sentiens territa caput elevabit et collum et laqueo capietur.	XXVI E nella spina si ficchi il fagiuolo overo la fava molle, solamente un granello, il quale l'ucciello con bocca prenda e il capo levando il laccio sopra'l collo trarrà la piegata esca cholle verghe; il quale l'ucciello sentendo, spaventato, il capo e il collo leverà e dal laccio sarà preso.
20,4, 18-22 Item qui cum balista vel arcu sagittat, debet manum sinistram tenere firmissimam, si vult rectissime sagittare et est necesse, quod habeat balistam vel arcum optimam et sagittas rectissimas, si recte voluerit sagittare.	XXVIII Ancho quello che con saetta overo arco saetta, dee la mano mancha tenere fermissima, se dirittissimamente vole saettare [e] è di neciessità che habbia balestro o archo optimo e saette dirittissime, se dirittamente vole saettare.
26, 2, 13-15 Qui tunc evaginato cum una manu gladio locum cordis ursi perforat et sic eum occidit. P, 1, 12	XXXIV E l'uomo, coll'una mano sguainato il coltello il luogo del cuore fora e uccidelo. P

8. Il testo latino è quello edito in Petrus de Crescentiis (Pier de Crescenzi), *Ruralia commoda. Das Wissen...*, Dritter Teil: Buch VII-XII.

9. Il testo italiano è tratto da: *Libro della Agricoltura di Piero Crescentio Cittadino di Bologna ad Honore Didio et del serenissimo Re Carlo...* impressum est hoc Opus Florentie Per me Nicholaum Diocesis Vratis lavriensis Die XV Mensis Julii. Anno DMCCCCLXXVIII.

La struttura della frase italiana si appoggia quasi per intero su quella latina, anche quando ciò significa forzare la sintassi italiana:

[...] infinita addiderunt ingenia pluribus ignota.	[...] infiniti ingengni aggiunsono da' più non saputi.
1,1, 2-3 Prudentes antiqui [...] laboraverunt avium rapacium genera domesticare	I I savi antichi [...] s'affaticono de gli ucciegli rapaci le loro gienerationi dimesticare
2,1, 17 et se, quantum possunt, occultant.	II e loro, quanto possono, occultano.
5,1, 24 valde indignantis naturae	V molto d'indegnante natura
5,2, 33 Sed cum eas ceperit, quas videt [...]	V Ma quando quelle harà prese, le quali vede [...]
6,3, 36 in gulam eius	VI nella gola sua
21,1, 1-3 Lepores praecipue capiuntur canibus, ad quas inveniendas necessarii sunt canes, qui seguli vel brachi vocantur.	XXIX Le lepri spetialmente si prendono con cani alle quali trovare sono e' cani neciessarii, i quali segugi ouero bracchetti sono chiamati.
30,1, 11-13 [hamus] cum quo pisces capiuntur rapaces deglutientes hamum cum pisciculo vivo.	XXXVIII [amo] col qual i pesci si pigliono rapaci che inghiottiscono il hamo col pesciculo vivo.

Molto spesso il verbo si trova in fondo alla frase:

4,1, 29-36 sed cum optime privati et domestici facti sunt, [...]	IV Ma quando optimamente privati e dimestichi facti sono, [...]
11, 6, 9-10 Si vero carnes solum sine sanguine comedant, hoc vitium incurrere non creduntur.	XI Ma se lle charni sanza il sangue mangiano questo chotale vitio incorrere non si crede.
15,2, 10-12 Et qui aquilam venatum portat, [...]	XV E quegli che l'aguglia a ucciellar porta, [...]

Non si contano i termini che ricalcano l'originale latino fin nella grafia:

P,1, 10-11 cautelas	P cautele
P,1, 11 et suis industriis	P con loro industrie
P,1, 12 ingenia	P ingegni
P,1, 13 riducere	P riduciere
1,1, 5 scientiae	I alla scientia
2,1, 20 desistit	II si rimane
2,2, 26-27 in confinibus	II ne' confini
2,2, 27 Tridenti	II Tridenti [di Trento]
3,1, 18; 4,2, 13 in feritate	III; IV in fierezza
4,1, 32 ferventem	IV fervente
4,2, 10-11 inter moltitudinem hominum	IV intra la moltitudine degl'uomini
4,2, 22 conspectum	IV cospetto
4,3, 27-28 specialiter ad hoc facta	IV spetialmente acciò facta
4,3, 36 consideranter	IV consideratamente
5,1, 25 iverit	V yerrà
6,4, 7 in articulis alae	VI negli articoli dell'alie
16,1, 4-5 in visione	XVI nella visione
16,1, 7 brutalis	XVI bruttale
16,1, 7 rationalis	XVI rationale
16,1, 7 in rebus novis	XVI nelle cose nuove
16,1 18-19 circumvolare	XVI intorno volare

Non sempre, però, l'Anonimo riesce a cogliere il senso della frase latina; del resto, per sua stessa natura, una traduzione che proceda parola per parola rischia spesso di incorrere in errori e incomprensioni evidenti:

4,2, 21-22 et quidam in aliquo fossato absconsus ante conspectum ancipitris eam [gazam] proicit	IV Et alcuno in uno fossato nascoso innanzi al cospetto dello sparviere si gitti
5,1, 23 Semper in manu eum cibet	V sempre la mattina il cibi
12,1, 26 Er ideo de his hic tractare non expedit	XII Et imperò di quelle di cui tractere non bisogna
13,1, 7-8 Quorum aucupatio potius voluptatis quam utilitatis causa existit	XIII Chol quale, quando s'ucciella, più tosto di volontà che d'utilità permane
16,1, 3 Cum ergo deformia sint [...]	XVI conciò sia cosa che della forma siano [...]
16,2, 14-15 [Guvus et guvetta] Vivunt autem de quibuslibet carnibus et praecipue de muribus et noctuis	XVI [Gufi e coccovegie] vivono i(n) verità d'ogni carne et maximamente di topi et di coccovegie
26,5, 5 [furettus] omnes extra cuniculos pellit	XXXIV [il furetto] così tutti e' conigli fuori caccia

I due ultimi esempi, in particolare, mi sembrano utili per delineare, almeno in generale, la personalità dell'Anonimo. In entrambi i casi l'errore scaturisce da quello che potremmo definire un 'difetto' di attenzione al contesto. Nell'ultimo esempio, il traduttore confonde *cuniculus* usato col significato di 'coniglio' e *cuniculus*, 'cunicolo'. La confusione non è grave – l'errore è autorizzato dall'uso pressochè costante, da parte di Pietro de' Crescenzi, di questo termine col significato di 'coniglio' –, ma testimonia una lettura poco attenta del testo originale.

Anche nel penultimo esempio l'Anonimo traduce il termine *noctua* secondo il significato più diffuso, testimoniato dal latino classico, e cioè 'civetta', ma una maggiore attenzione al contesto avrebbe permesso di optare per un più verisimile 'pipistrello', evitando al traduttore l'errore, un po' comico, di attribuire una natura cannibalesca alla povera civetta, uccello fin troppo bistrattato, senza doverla gravare anche di questo nefando costume[10].

Spesso si verifica un'oscillazione nella resa di alcuni termini latini. Il caso più eclatante riguarda la parola *guvetta* 'civetta' (sempre lei!), tradotta con almeno tre termini diversi:

16,1, 1 guvetta	XVI coccovegia
16,1, 9 guvettae	XVI alla civetta
16,1 11 guvettam	XVI coccovegia
16,2, 18 guvetta	XVI guveta
17,13, 19 guvetta	XVII cavetta overo civetta
20,5, 25 guvetta	XX civetta
20,5, 27 guvetta	XX civetta

La stessa cosa accade anche in altri frangenti:

10. Sia la civetta sia il pipistrello appartengono all'ampia e spesso indefinita categoria degli animali notturni. Lo stesso termine *noctua* non è altro che un aggettivo: la civetta è un'*avis noctua*, così come lo è, in un certo senso, il pipistrello (cfr. Forsyth Major C.J., 'Italienische Vulgärnamen der Fledermaus', *Zeitschrift für romanische Philologie* XVII, 1893, p. 151-152). Ancora nel Cinquecento il Sansovino parlando proprio di questo animale in quella sorta di glossario da lui voluto alla fine della sua traduzione dei *Ruralia*, non sa decidersi se esso sia un uccello o un mammifero. In italiano antico (e la confusione proseguirà anche nei moderni dialetti) la parola *nottola* indica sia la civetta sia il pipistrello. *Nottola* deriva dal tardo latino *noctula* a sua volta derivato dal latino classico *noctua* 'civetta' (cfr. Battisti C. & Alessio G., *Dizionario etimologico italiano,* Firenze, 1950-1957, s.v.). La prima attestazione di *noctula* 'pipistrello' si trova in una traduzione latina di un'opera del medico efesino Sorano, vissuto nel secondo secolo dopo Cristo (cfr. *Dizionario etimologico...*, s.v.). Un altro dato di grande interesse riguardante questa confusione terminologica ci è fornito da Jacques André: 'Dans la Vulgate, *Baruch*, 6, 21, *noctua* traduit 'chauve-souris' de la Septante (*Epist. Jérémie*, 21, inséré dans *Baruch* par la Vulgate)'. (Cf. André J., *Les noms d'oiseaux en latin,* Paris, 1967, p. 110).

17,1, 2 casella	casella
17,3, 20 casellam	casetta
17,4, 28 cucullum	cocuzzolo
17,4, 31 cuculli	del cucullo
17,15, 1 domusculam	capanna
17,15, 3 domuscula	capannetta
18,3, 25 scalellae	scale
18,3, 30 scalella	scaletta
25,6, 26 scalella	scaletta
20,4, 18 balista	saetta
20,4, 20 balistam	balestro
26,5, 37 reticulum	reticella
26,5, 6 reticulos	rete

In un caso avviene anche il contrario, e un'unica parola italiana viene utilizzata per rendere più termini latini. È quanto succede con il termine *palo* che traduce *palus* 'palo', ma anche *pilus* 'lancia' e *pala* 'pala'.

In generale, come abbiamo visto, la fedeltà al testo di riferimento è estrema, quasi acritica. Proprio in ragione del tipo di traduzione adottato, però, alcuni errori presenti nel testo italiano non possono essere imputati all'Anonimo:

2,2,29-30 corvus et graculus	II cierbo e chavriuolo
11, 2, 10 macriores	XI più maturi
17,2, 10-12 Circa foveam iuxta ripas figantur pali quattuor perticarum, quae retia elevent	XVII Apresso della fossa, al lato alla ripa, ficchino pali quanto che le pertiche diríttamente lievino delle reti

17,3, 19 ad modum culminis domus paleatae	XVII Ad modo di un comignolo di casa o di paglia
30,1, 17-18 Piscis captus longe fugere vel se occultare fasciculo impeditur.	XXXVII Il pescie preso di lunge fugire overo occultarsi nel fasciculo s'intende.
20,7, 8-9 Item capiuntur passeres, praecipue iuvenes minus sagaces cum nassa seu brechoello	XXVIII E piglionsi passere spetialmente e tutti ucciegli sagaci con mano overo con vertorelo

Gli esempi mostrano con sufficiente chiarezza come la distanza tra il testo latino e quello italiano sia dovuta non alla difficoltà del traduttore di comprendere una frase (in realtà l'autore della traduzione spesso non prova nemmeno a comprendere quello che sta scrivendo, come risulta evidente da alcune frasi che sembrano prive di senso), ma piuttosto alla corruzione del testo latino originale, giunto in una versione già segnata da scorrettezze. Come mi pare ulteriormente confermare l'ultimo degli esempi, in cui si verificano la caduta di *minus* e la sostituzione di *nassa* con *mano*, 'infrazioni' che messe insieme concorrono a rendere la frase quanto meno poco chiara.

L'uso della parola *mano* per rendere il latino *nassa* non può essere catalogato come un'innovazione introdotta autonomamente dal traduttore, per il semplice fatto che l'Anonimo non innova mai, anzi, per eccesso di conservatorismo mantiene anche gli errori evidenti del testo di riferimento, come abbiamo visto. Del resto, non possiamo nemmeno considerare *nassa* un termine tecnico di difficile comprensione, tanto più che l'Anonimo, pochi capitoli più avanti (29,1, 29), traduce correttamente la frase 'item ex iuncis fiunt *nassae* rotundae ac latae' con 'anche di vinchi si fanno *nasse* ritonde e late' [corsivi miei], dimostrando di conoscere il termine in questione.

Sullo sfondo della traduzione, dunque, non possiamo non scorgere la struttura del manoscritto, il suo 'scheletro', con le sue principali caratteristiche, che possono essere brevemente sintetizzate nella tendenza ad intervenire sul testo, seppure non invasivamente, mediante operazioni di riduzione (soprattutto di parti poco importanti):

7,1, 26-27 [...] sicut in tractatu ancipitrum dictum est.	VII [...] sicome decto
12,2 10-12 Et qui aquilam venatum portat, debet esset fortis; alioquin pondus eius sustinere non posset.	XV E quegli che l'aguglia a ucciellar porta dee essere forte a quella sostenere.
17,10, 25-27 et erit esca conveniens spelta, faba, milica, lolium et semina vinacia et similia, ut frumentum motum et cicercula vel faseoli.	XVII E sarà l'esca conveniente spelda, fava, miglio e simili.
21,2, 21-22 [...] et similiter ab his, qui venantur eundem.	XXIX [...] e da' detti cacciatori.
26,1, 2-3 ideoque cum dormire volunt [scil. Elefantes], se arboribus appodiant et quiescunt.	XXXIV e imperò, quando dormire vogliono si ssi appoggiano a' grandissimi arbori
26,2, 12-13 [...] et amplexatur eum, armatusque ipsum.	XXXIV [...] e abbracciasi co·llui
29,10, 2-3 Nec de ipsa, sicut de prima, exire sciunt.	XXXVII E di niuna uscire sanno.

e modifiche del dettato:

9,1, 2 [...] scilicet duae feminae et tertius masculus	IX [...] cioè due femine e il terzuolo ch'è maschio
9,1, 31-34 et ex magnis quidam sunt nigri, quidam respective albi et quidam rubei ex coitu praedictorum invicem generati, videlicet cum trizolus unus perdita socia sua cum altera permiscetur.	IX e de' grandi alcuni sono neri et alcuni pe' rispetto bianchi e alcuni rossi per lo mischiamento insiema a luxuria intra loro gienerati, cioè, quando il terzuolo d'uno ae perduto la sua compagnia, chon un'altra si permischia.
20,10, 5-7 Item fertur, quod aves, quae gustant granum vel milicum in fece boni vini et suco cicutae maceratum et siccatum, volare non possunt et poterunt manu capi.	XXVIII Ancho si dicie che gli ucciegli che becchino grano overo miglio, se nella feccia di buon vino si metta, con sugo di cicuta macierata e seccato, il grano o miglio, se poi il darai a beccare, volare non possono, ma ebbri con mano si possono pigliare.

26,1, 3-4 Venatores has arbores resecant, sed non perficiunt	XXXIV i cacciatori questi arbori risegono, ma non affacto

Si tratta, perciò, di un manoscritto caratterizzato da alcune innovazioni, non ultime quelle che riguardano i termini tecnici:

3,1, 11 *et passim* nidarius	nidacio
3,1, 11 *et passim* ramarius	ramingo
9,1, 36 *et passim* trizoli sive tricelli	terzuoli
4,3, 35 stelionum	de' pipistrelli
20,6, 1 boscellum	frugnuolo

L'esempio di *boscellum* merita un breve approfondimento. In questo caso è interessante notare la scelta del copista, il quale non trascrive meccanicamente, segno di un'attenzione comunque desta nei confronti di ciò che ha davanti. Infatti rispetto alla sua conoscenza di uomo non del tutto ignaro di cose di caccia stride il *boscellum* crescenziano. Il 'boschetto' (*boscellum* è l'esatto corrispettivo latino) è un vero e proprio piccolo bosco, piantato appositamente in un luogo di passaggio per gli uccelli migratori e recintato da siepi. Durante il periodo della migrazione, grazie alla presenza di richiami vivi, gli uccelli di passaggio vengono attirati nel 'boschetto' dove trovano ad attenderli svariati posatoi cosparsi di vischio. Il 'boschetto', dunque, rappresenta una sorta di oasi verde lungo il cammino sempre faticoso degli uccelli che migrano da un continente all'altro, un luogo riparato e tranquillo che serve, però, a mascherare numerose terribili insidie.

Questo, in sintesi, lo scopo del 'boschetto'. La descrizione di Pietro de' Crescenzi, però, non si riferisce alla caccia col vischio, ma a una di altro genere, altrettanto diffusa, consistente nell'andare di notte con una lanterna quadrata (il *frugnolo* o *furnuolo*, dal latino *furniolus* 'piccolo forno', per la forma della lanterna) nei luoghi dove i volatili si raccolgono per riposare. La luce emessa dalla lanterna disturba gli uccelli, i quali, spaventati e disorientati, saranno facile preda dei cacciatori, che li uccideranno con delle apposite mazze. Facile gioco, dunque, per il copista, dopo aver letto la descrizione di de' Crescenzi, correggere la presunta imprecisione dell'autore, dimenticando che l'espressione utilizzata nel testo latino ('ad boscellum') non intende indicare lo strumento ma il luogo ove si pratica la caccia.

Infatti, durante l'inverno, fuori dalla stagione della migrazione, il 'boschetto', per la sua natura di luogo riparato e, almeno apparentemente, sicuro, era sfruttato proprio per la caccia col *frugnolo*, come testimonia il seicentesco trattato dell'*Uccelliera* di G.P. Olina:

> Ordinariamente si va con quest'arnese [scil. Il frugnuolo] in traccia di tordi, che però non in altro tempo s'adopera che verso il fine dell'autunno e l'inverno. S'usa andar con esso la notte per i *boschetti* [corsivo mio], nelle ragnaie e luoghi simili [...][11]

Il comportamento del copista del manoscritto latino, dunque, – mi scuso se, per mancanza di informazioni sul codice specifico usato dall'Anonimo e sulla sua tradizione, sono costretto ad appiattire tutto su un'unica figura – è piuttosto sorvegliato, le modifiche intervengono solo su punti in cui il contesto lo autorizza, o sembra farlo. Il copista, cioè, non interviene arbitrariamente ad alterare il testo, ma si preoccupa di migliorarlo quando ciò sia possibile senza precluderne l'integrità originale[12].

La sensazione generale, scorrendo la traduzione dell'Anonimo, è quella di una prosa pesantemente influenzata dal latino dell'originale, come testimoniano, per non parlare dei numerosissimi latinismi, la presenza di iperbati e il ripetersi della struttura soggetto-oggetto-predicato, tipici della lingua latina. Una così forte dipendenza può, in parte, essere spiegata con la naturale evoluzione che nel corso del quattordicesimo secolo subirono i volgarizzamenti dal latino in area italiana. La maggiore esperienza di traduzione rispetto al secolo precedente, e una maggiore consuetudine con la latinità classica, divenuta sempre più modello da seguire, fanno sì che il nuovo modo di tradurre, abbandonata la libertà di interpretazione tipica del Duecento, faccia un uso massiccio delle strutture della lingua latina,

11. Olina G.P., *Uccelliera* (Roma, 1622). Una scelta di capitoli è offerta dall'antologia di Innamorati G., *Arte della caccia. Testi di falconeria, uccellagione e altre cacce* (Milano, 1965), t. 2, p. 421-466 (458), testo da cui è stata tratta la citazione. Per un ulteriore approfondimento delle tecniche venatorie si consiglia Gasser C., *L'uccellagione nel Trentino (1850-1914)* (Trento, 1995).

12. Anche se non sempre riesce nel suo intento. In un caso particolare il copista non si avvede di una possibile correzione, avviene così che il manoscritto usato dall'Anonimo per la sua traduzione condivida con i mss. T e U un'imprecisione – presente anche nell'edizione stampata a Basilea nel 1548, antologizzata da Giuliano Innamorati nella sua raccolta di testi sulla caccia (cfr. Innamorati, *Arte della caccia...*) – riguardante il nome di un uccello:

sia nella sintassi sia nel lessico. Non è infrequente anche nella traduzione dell'Anonimo trovarsi in presenza di latinismi che non trovano giustificazione immediata nel testo originale, ma che possono essere spiegati proprio con questo gusto 'preumanistico' per la latinità. Tale gusto, però, non è sufficiente a dare conto del comportamento dell'anonimo traduttore dei *Ruralia commoda*.

Paradossalmente, nonostante la pratica di tradurre letteralmente, quasi vocabolo per vocabolo, accade che l'Anonimo dedichi scarsa attenzione proprio alla traduzione delle singole parole. Infatti è nell'oscillazione di più termini usati per tradurne uno solo o nell'uso di parole fuori contesto (seppur lessicalmente corrette) che risalta con maggior evidenza una certa trascuratezza generale. La scelta di procedere con una traduzione letterale sembra originare dalla scarsa sicurezza dell'Anonimo, forse nei confronti della lingua, forse nei confronti della materia.

Non dimentichiamo infatti il carattere e gli obiettivi dei *Ruralia commoda*, vasta opera tecnica destinata a scopi pratici. Del resto, ciò che noi

---

17,5-6
Est aliud ingenium, quo capiuntur grues, cigni, starnae et anseres, quod est tale, in ripis fluminum ab utraque parte figitur arbor altissima vel duae simul iunctae, ut longior sit arbor, habetque per totum caviculas ad ascensum et in summitate cirellam, in qua ponitur funis retis. Cuius longitudo est secundum fluminis latitudinem et arborum distantiam, et eius latitudo est, ut superius elevata usque ad medias pendeat arbores. Deinde homines usque a remotis veniant per glaream fluminis omnes volucres, quas invenerint, expellentes. Quae volantes a splendore aquae non discedent, donec in retia inciderint. Et tunc declinetur rete cum suis funibus et volucres capiuntur. Et modus iste locum non habet, nisi cum tempus fuerit nubilosum et aër obscurus, alioquin aves alevatae facile a flumine deviarent. Est et aliud ingenium, quo praecipue capiuntur anseres et starnae, quod est tale:...
L'inclusione della starna nella lista degli uccelli che possono agevolmente essere cacciati con reti poste *in ripis fluminum* in quanto, una volta spaventati, *a splendore aquae non discedent*, cioè non si allontanano dallo specchio d'acqua incappando così nelle trappole, non può essere attribuito a Pietro de' Crescenzi. Innanzitutto perché l'autore di un'opera enciclopedica che comprende un intero libro dedicato alla caccia (più un altro sull'allevamento) non può sbagliare così grossolanamente da credere che le starne, volatili assolutamente terricoli, abbiano le stesse abitudini di cigni, gru e oche. Inoltre l'errore, apparentemente solo di carattere 'ornitologico', nasconde, a mio avviso, anche un più interessante errore linguistico. A ben guardare, infatti, nella lista di uccelli acquatici stilata nel nostro testo, manca forse quello più diffuso e conosciuto: l'anatra (in latino 'anas', pl. 'anates' – forma utilizzata da Crescenzi –, ma anche 'anatres' e 'anatrae'). Senza tentare ipotesi troppo fantasiose, viene da chiedersi se per caso, data la pressochè totale identità delle lettere che formano le parole 'starnae' e 'anates' (e 'anatres'), le starne dei nostri manoscritti non siano un errore di lettura che nasconda proprio le anatre, uccelli ben più a loro agio in compagnia di cigni ed oche. Un'imprecisione del genere, data la natura tecnica del nostro testo, può rivelarsi molto interessante dal punto di vista della ricostruzione criticotestuale. Naturalmente, la quantità di dati a mia disposizione è largamente insufficiente per pensare di proporre qualcosa di più di una semplice notazione.

interpretiamo come insicurezza, potrebbe essere inteso anche come un tentativo, portato all'estremo, di mantenere intatto un testo che, modificato, avrebbe potuto perdere di valore e di interesse.

Di certo, stupisce la perfetta aderenza al dettato originale, senza nessun intervento di rilievo che sia riconducibile con certezza al traduttore. In particolare colpisce, almeno per quanto riguarda il decimo libro, la pressoché totale assenza di interventi o aggiunte, forse l'indice più esplicito del grado di partecipazione del traduttore medievale. Traduttore che è soprattutto lettore, prima, e divulgatore, dopo, colui che legge, si spiega e spiega i passaggi più oscuri o più interessanti del testo.

Roma

Marianne ELSAKKERS

THE EARLY MEDIEVAL LATIN AND VERNACULAR VOCABULARY OF ABORTION AND EMBRYOLOGY

0. Introduction*

Some legal, philosophical and theological texts on abortion distinguish between early and late termination of pregnancy[1]. This is done by choosing an important stage of embryological development as a criterion that defines the difference between an abortion that is regarded as murder and one that is regarded as a less serious offence. The criteria used to differentiate between early term and late term abortion are usually based on earlier descriptive and normative sources. In the early medieval period the main sources used are the available classical works on medicine, gynecology, biology and philosophy and the texts and commentaries they inspired[2].

* This article is for my favorite uncle Toon. I would like to thank Einhard and Wilken Engelbrecht, Thea van der Linden, Andy Griffiths, Bertine Bouwman, Louise Cilliers, Liduine Smit, Jan Hastrich, Rob Meens and my 'promotores' for their help. Special thanks are due to the Law Library and the Theology Library of the University of Utrecht for giving me special privileges, and to the Koninklijke Bibliotheek in The Hague for being exceptionally prompt in sending me copies of all kinds of articles.

1. Texts which condemn abortion regardless of the stage of development of the fetus will not be discussed in this study. On abortion, see, for instance: Kapparis K., *Abortion in the Ancient World* (London, 2002); Riddle J.M., *Eve's Herbs; a History of Contraception and Abortion in the West* (Cambridge, Mass., 1997); Schwarz H.W., *Der Schutz des Kindes im Recht des frühen Mittelalters* (Siegburg, 1993); Riddle J.M., *Contraception and Abortion from the Ancient World to the Renaissance* (Cambridge, Mass., 1992); Brundage J.A., *Law, Sex, and Christian Society in Medieval Europe* (Chicago – London, 1987); Gorman M.J., *Abortion and the Early Church: Christian, Jewish and Pagan Attitudes in the Greco-Roman World* (Downers Grove ILL, 1982); Connery J., *Abortion: the Development of the Roman Catholic Perspective* (Chicago, 1977); Noonan Jr. J.T. (ed.), *The Morality of Abortion* (Cambridge, Mass., 1970); Nardi E., *Procurato aborto nel mondo Greco Romano* (Milano, 1971); Noonan Jr. J.T., *Contraception; a History of its Treatment by the Catholic Theologians and Canonists*, enlarged ed. (Cambridge, Mass., 1986; first ed. 1965); Huser R.J., *The Crime of Abortion in Canon Law* (Washington DC, 1942).

2. On medieval embryology, see, for instance: Pahta P., *Medieval Embryology in the Vernacular; the Case of 'De Spermate'*, Mémoires de la Société Néophilologique de Helsinki 53 (Helsinki, 1998); Dunstan G.R. (ed.), *The Human Embryo: Aristotle and the Arabic and European Traditions* (Exeter, 1990); Dunstan G.R. & Seller M.J. (eds.), *The Status of the Human Embryo; Perspectives from Moral Tradition* (London, 1988); Musallam B.F., *Sex and Society in Islam; Birth Control Before the Nineteenth Century*,

In this article I will discuss the terms chosen as abortion criteria in early medieval Latin and vernacular legal texts from the sixth to the eleventh centuries, albeit that there are only a few embryological texts in the vernacular in this period[3]. Both secular and church law will be discussed, but the focus of this study will be on the terminological choices made by the authors of the early medieval penitentials. Many of the terms used were, became, or seem to us to be ambiguous, and, moreover, the meaning of some of these abortion criteria slowly changed as Christianity gained ground.

Whereas descriptive embryological texts distinguish at least three or four important stages of fetal development, normative texts, as a rule, only distinguish two stages, because only the difference between early term and late term abortion is important. This distinction is usually labeled 'formed' – 'unformed'. As soon as a fetus is 'formed', it is considered a 'human being' or a 'person', and this is why late term abortion is usually punished as murder. In descriptive embryology the most important stages of development are formation, movement and birth. In some embryologies vivification is added as a separate stage of fetal development, in others it is implied in either formation or movement, because 'alive' is the way a pregnant woman labels her perception of the first movement of her child. In the writings of the early Church fathers ensoulment is often defined as yet another separate stage of development, but it, too, is often implied in formation, movement or vivification. In normative texts the following labels indicating a stage of fetal growth can be used to denote the dividing line between early term and late term abortion:

abortion criteria	**synonyms**
formation	articulation, gender differentiation, quickening
movement	quickening
sensation	activation of the senses, movement
vivification	life, aliveness, breathing, animation, quickening
ensoulment	animation

Cambridge Studies in Islamic Civilization (Cambridge, 1983); Weisser U., *Zeugung, Vererbung und pränatale Entwicklung in der Medizin des arabisch-islamischen Mittelalters* (Erlangen, 1983); Schöffler H.H., 'Zur mittelalterlichen Embryologie', *Sudhoffs Archiv* 57 (1973), p. 297-314; Lesky E., *Die Zeugungs- und Vererbungslehren der Antike und ihr Nachwirken* (Wiesbaden-Mainz, 1950); Needham J., *A History of Embryology* (New York, 1975; first ed. 1934).

3. There are many more vernacular embryological texts in the post-Salernitan period, cf. Elsakkers M., 'Middeleeuwse embryologie in vogelvlucht; met een paar volkstalige voorbeelden uit de Lage Landen', *W-Mail* 5 (2004), p. 15-26.

The number of different abortion criteria is confusing and the 'synonyms' listed in the right-hand column add to the terminological confusion. These 'synonyms' are not only often used indiscriminately in medieval texts, but also in scholarly discourse[4].

I will first outline some of the descriptive and normative embryological sources available in the pre-Salernitan period, that is, before the School of Salerno started its translation activities in the mid eleventh century, and then move on to the abortion criteria used in the Old Germanic laws and the early medieval penitentials.

1. Classical and late antique heritage

In the Hippocratic Corpus (c. 460-c. 370 BC) we find detailed descriptions of the stages of development of the fetus and many different calculations of the period of gestation, usually in a fixed ratio between

4. The words *anima* and *animatus* are notoriously ambiguous and can denote both 'vivification' or 'ensoulment', or both. Cf. the entries for *anima*, *animatus* and *animus* in the *Oxford Latin Dictionary* (Oxford, 1968), vol. 1, p. 132-135. In discussions of 'animation' the meaning of the word often fluctuates between 'ensoulment', 'vivification' and even 'sensation'; see, for instance, Kapparis, *Abortion in the Ancient World*, p. 48-51. The word 'animation' will be included in single quotes, when referring to the ambiguous meaning 'ensoulment'-'vivification'.

The word 'quickening' is also used in at least two different meanings: 'movement' and 'formation'. It is used in the meaning 'movement' in: Jane M. Oppenheimer, 'When Sense and Life Began: Background for a Remark in Aristotle's Politics (1335b24)', *Arethusa* 8 (1975), p. 331-343 at p. 337, and in Kapparis, *Abortion in the Ancient World*, p. 50. 'Quickening', 'formation' and 'ensoulment' are more or less treated as synonyms in Musallam's discussion of Muslim views on abortion in: *Sex and Society in Islam*, p. 54-59. Lonie uses the word 'quickening' as a synonym of 'articulation' and 'formation', cf. Lonie I.M., *The Hippocratic Treatises 'On Generation', 'On the Nature of the Child', 'Diseases IV'*, Ars Medica; Texte und Untersuchungen zur Quellenkunde der alten Medizin II.7 (Berlin – New York, 1981), p. 190, and Helen King equates 'quickening' and 'movement' on the one hand and 'formation' and 'articulation' on the other, cf. King H., 'Making a Man: Becoming Human in Early Greek Medicine', in: Dunstan (ed.), *The Human Embryo*, p. 10-19, at p. 15-16. See also: McLaren A., 'Policing Pregnancies: Changes in Nineteenth-Century Criminal and Canon Law', in: Dunstan (ed.), *The Human Embryo*, p. 187-207, on the disappearance of the legal terms 'quickening' ('aliveness', the mother's first perception of fetal movement, cf. p. 193) and 'ensoulment' ('animation') in criminal and church law, respectively.

On p. 1 of his *The Crime of Abortion in Canon Law* Huser warns us of the terminological confusion as regards the concept 'formed – unformed': 'Frequently the terms *animated* and *non-animated* are employed as synonymous for *formed* and *non-formed*. Strictly, *animated* and *non-animated* refer to the presence or absence of the human soul. According to the theory of delayed animation, the human soul cannot be present until the fetus has reached the certainty of development mentioned above, and hence the term *animated* becomes synonymous with *formed*. Sometimes the terms *vivified* and *non-vivified* are used'.

formation, movement and birth[5]. The fetus is formed between the thirtieth and fiftieth day after conception, it starts to move between the sixtieth and hundredth day, and is born after 180 to 300 days.

In the discussion of human generation in his *Historia Animalium* Aristotle (384-322 BC) states that the male fetus is formed in forty days and the female fetus in ninety days; formation and the first fetal movement occur at approximately the same time[6]. Aristotle uses embryological terminology in a normative context, when advocating abortion as a method of family planning in his *Politica*. Abortion is allowed before the fetus has developed 'sensation and life'[7].

The Aristotelian criterion for abortion 'formed – unformed' was incorporated into the Greek Septuagint version of the biblical law on abortion (third century BC)[8]:

> Ex. 21.22-23. Si autem litigabunt duo viri, et percusserint mulierem in utero habentem, et exierit infans ejus nondum formatus: detrimentum patietur, quantum indixerit vir mulieris, et dabit cum postulatione. Si autem formatum fuerit, dabit animam pro anima[9].

Vetus Latina translations of the Septuagint version of Exodus 21.22-23 were available from the second century onwards and in use throughout the medieval period[10].

5. Cf. King H., *Hippocrates' Woman; Reading the Female Body in Ancient Greece* (London – New York, 1998), p. 134, and Hippocrates' 'On the Nature of the Child' in: Lonie, *Hippocratic Treatises*, p. 6 ff., p. 9-10, p. 12, p. 190-194. On *De Alimento* and *De Carnibus*, see: Hanson A.E., 'The Eight Months' Child and the Etiquette of Birth: Obsit Omen!', *Bulletin of the History of Medicine* 61 (1987), p. 589-602, at p. 592-593. See also: King, 'Making a Man', p. 16 and Jouanna J., *Hippocrates*, transl. M.B. DeBevoise (Baltimore – London, 1999; orig. ed. 1992).

6. *Historia Animalium*, book 9 (7), 7.3.583b, cf. Aristotle, *History of Animals*, ed. and tr. D.M. Balme, 3 vols., Loeb Classical Library (Cambridge, Mass. – London, 1965-1991), vol. 3, p. 435-437.

7. *Politica*, book 7.14.10.1335b, cf. Aristotle, *Politics*, tr. H. Rackham, Loeb Classical Library (London-Cambridge, Mass., 1967), p. 622-625.

8. On the difference between the Hebrew and Greek versions of Exodus 21.22-23, cf. Isser S., 'Two Traditions: the Law of Exodus 21: 22-23 Revisited', *The Catholic Biblical Quarterly* 52 (1990), p. 30-45; Kapparis, *Abortion in the Ancient World*, p. 46 ff.; Dunstan G.R., 'The Human Embryo in the Western Moral Tradition', in: Dunstan & Seller (eds.), *Status of the Human Embryo*, p. 39-57, and te Lindert J.G., *Over de status van het menselijk embryo in de Joodse en de Christelijke ethiek; een analyse van opvattingen toegelicht aan de hand van verklaringen van Exodus 21: 22vv*, Ph.D. Dissertation, University of Utrecht (s.l., 1998).

9. Sabatier P., *Bibliorum Sacrorum Latinae Versiones Antiquae* (1743; rpr. München, 1976), vol. 1, p. 178. See also: Augustinus Hipponensis, *Quaestionum in Heptateuchum libri VII*, ed. J. Fraipont, Aurelii Augustini Opera 5, CCSL 33.5 (Turnhout, 1958), p. 110-111.

10. Jerome's Vulgate translation of the Bible – based on the Hebrew version of Exodus – slowly became more and more important in the course of the Carolingian period.

The abridged embryology in Pliny's *Naturalis Historia* (1st century AD) illustrates how difficult it was to pinpoint formation and movement. Pliny says that fetal formation starts on the tenth day after conception and that it coincides with the mother's first bout of morning sickness:

> A conceptu decimo die dolores capitis, oculorum vertigines tenebraeque, fastidium in cibis, redundatio stomachi indices sunt hominis inchoati[11].

Pliny does not use verbs like *fingere* or *formare*, 'to shape, form', and he does not tell us when formation is completed, only that movement – *motus in utero* – begins on the fortieth day (for males) or on the ninetieth day (for females)[12].

In classical embryology the most important stages of fetal growth are formation and movement. Formation takes place between thirty and fifty days after conception – usually the fortieth day is named. There is less consensus regarding movement. In some sources it takes twice as long as formation, in others it practically coincides with formation. Classical texts by no means present a coherent description of human embryology; this lack of consistency is reflected in medieval embryological texts.

Classical medical learning flourished in fourth- and fifth-century North Africa. Two important embryologies originated in North Africa: one is in Vindicianus' *Gynaecia* and the other is in Macrobius' *Commentariorum in somnium Scipionis*. A third embryology was written down by the Church father Augustine – also from North Africa.

Vindicianus (late fourth century) discusses human embryology in chapters 19 and 20 of his *Gynaecia*[13]. In chapter 19 he gives us a Hippocratic calculation of the 'due date': if the fetus is formed on day forty, it will start to move on the ninetieth day and be born in the ninth month. Chapter 20 contains a month-by-month overview of the development of the fetus. The table below lists the core elements of Vindicianus' description of the development of the fetus[14].

11. 'As of the tenth day from conception pains in the head, giddiness and blurred sight, distaste for food, and vomiting are symptoms of the beginning of a human being' (slightly emended translation), cf. Gaius Plinius Secundus, *Naturalis Historia – Natural History*, 10 vols., ed. & tr. H. Rackham, Loeb Classical Library (Cambridge, Mass., 1963), vol. 2. Libri III-VII, p. 532-535 (7.5.41 ff.), at p. 532-533.

12. Pliny probably means that the completion of formation triggers movement.

13. See: *Theodori Prisciani Euporiston Libri III cum physicorum fragmento et additamentis pseudo-Theodoreis (accedunt Vindiciani Afri feruntur reliquiae)*, ed. V. Rose (Leipzig, 1894), p. 450-455, p. 465-466, for seven different versions of Chapters 19 and 20 of the *Gynaecia*. The oldest surviving manuscript of Vindicianus' *Gynaecia* is from the eighth century and the youngest from the twelfth century.

14. My list of core elements is based on the editions of nine different manuscript versions of Vindicianus' *Gynaecia*. In the versions used to compile this list, the months and

(0) manifestum est – infantem in utero – sic – formare – quomodo intendit
(1) iactus – seminis – in utero materno – generat – umbillicum – et facit – congregationem sanguinis
(2) expressio – corporis
(3) formatus – invenit – partus – et – tribuit – animam
(4) unguis – accipiat – capillos – ducit – et – motum – facit – infantem – mulier – nausiatur
(5) facit – caracterem – patri aut matri
(6) facit – nervos – confirmationem
(7) medullam – consolidat
(8) ossa – dura facit
(9) maturam – evolvit – infantem – perfecta – naturae – et – de tenebris – ad lucem – adducit.

Vindicianus distinguishes between the beginning of formation in the second month when the body is being formed (*expressio corporis*), and the completion of formation in the third month (*formatus invenit partus*), when the fetus is recognizable as a human being. After formation the fetus (now called *partus*) is 'animated' (*tribuit animam*) – 'animation' should probably be interpreted as vivification in this text. Vivification seems to be a prerequisite for the beginning of fetal movement in the fourth month (*motus facit*). For Vindicianus, formation, 'animation' and movement are separate, well-defined stages of fetal growth. Vindicianus' month-by-month embryology was by no means unknown in the early medieval West. In the post-Salernitan period it was reintroduced as part of the 'Trotula ensemble', incorporated into Vincent of Beauvais' thirteenth-century encyclopedia, and often included as a separate embryological text in medieval medical treatises[15]. It was also translated into various vernaculars; the Old English version will be discussed below.

the corresponding stages vary, especially the stages 'formation', 'movement' and 'vivification'. Cf. Egert F.P., *Gynäkologische Fragmente aus dem frühen Mittelalter; nach einer Petersburger Handschrift aus dem VIII.–IX. Jahrhundert zum ersten Mal gedruckt*, Abhandlungen zur Geschichte der Medizin und der Naturwissenschaften 11 (Berlin, 1936; rpr. Nendeln-Liechtenstein, 1977), Fragment V, p. 49; Schipper J., *Ein neuer Text der Gynaecia des Vindician aus einer Münchener Handschrift des 12. Jahrhunderts – Cod. lat. 4622, Blatt 40-45* (Erlangen, 1921), p. 18; Priscianus, *Euporiston*, p. 452-455, p. 465; Wellmann M. (ed.), *Die Fragmente der Sikelischen Ärzte Akron, Philistion und des Diokles von Karystos*, Fragmentsammlung der Griechischen Ärzte 1 (Berlin, 1901), p. 218-219. Many thanks are due to Louise Cilliers for sending me her unpublished edition and translation of Vindicianus' *Gynaecia*, cf. Louise Cilliers, 'Vindicianus' *Gynaecia*, Text and Translation, Codex Monacensis Clm 4622', *Journal of Medieval Latin* 15 (2005), p. 153-236.

15. See, for instance, Green M.H., *The Trotula; a Medical Compendium of Woman's Medicine* (Philadelphia, 2001), p. 106-107 and Vincentius Bellovacensis, *Speculum Naturale* (Duaci, 1624; rpr. Graz, 1964), p. 2324 ff. (Liber 31.43 ff.).

The early fifth-century embryology included in Macrobius' *Commentary on the Dream of Scipio* was popular during the entire medieval period. It is based on the number seven, which means that each stage of fetal development takes seven days[16]. Using words like *fingere*, 'to shape, form' and *lineamenta*, 'outlines' or 'features', Macrobius defines formation as the stage in which the *humana figura*, 'human shape', is formed:

> quinta vero interdum fingi in ipsa substantia umoris humanam figuram, magnitudine quidem apis sed ut in illa brevitate membra omnia et designata totius corporis liniamenta consistant[17].

The process of formation is complete in either the fifth, sixth, or seventh period of seven days, that is, between the 35th and the 49th day after conception (depending on the sex of the child). The formed fetus – called *conceptus* – moves on the seventieth or ninetieth day. Formation, movement and the length of gestation are linked in a fixed ratio[18].

In his *De diversis quaestionibus* the Church father Augustine of Hippo (354-430 AD) recorded a short embryological text in the *quaestio* that deals with the 'rebuilding of the temple' (John 2.20-21). This embryology was popular during the entire medieval period and its authorship is often ascribed to Augustine.

> LXXXIII.i.lvi. De annis quadraginta sex aedificati templi.
> Sex, novem, duodecim, decem et octo, haec in unum fiunt quadraginta quinque.
>
> Adde ergo ipsum unum, fiunt quadraginta sex: hoc sexies, fiunt ducenta septuaginta sex.
> Dicitur autem conceptio humana sic procedere et perfici, ut
> (1) primis sex diebus quasi lactis habeat similitudinem,
> (2) sequentibus novem diebus convertatur in sanguinem,
> (3) deinde duodecim diebus solidetur,
> (4) reliquis decem et octo diebus formetur usque ad perfecta lineamenta omnium membrorum,
>
> et hinc jam reliquo tempore usque ad tempus partus magnitudine augeatur. (...) Non ergo absurde quadraginta sex annis dicitur fabricatum

16. Cf. Ambrosius Aurelius Theodosius Macrobius, *Commentariorum in somnium Scipionis libri duo*, introd., testo, trad. e note a cura di L. Scarpa (Padova, 1981), p. 106, p. 122-123 (1.6.14-17;1.6.62-66) and Ambrosius Aurelius Theodosius Macrobius, *Commentary on the Dream of Scipio*, tr. W.H. Stahl (New York etc., 1966), p. 102-103, 112-113.

17. Scarpa, *Commentariorum in somnium Scipionis*, p. 122 (1.6.65); 'on the fifth hebdomad a human shape is being molded in the substance of the humor, no larger than a bee in fact, but in such a manner that on that small scale all the limbs and the distinct contour of the whole body stand out', Stahl, *Commentary on the Dream of Scipio*, p. 113.

18. Cf. Scarpa, *Commentariorum in somnium Scipionis*, p. 106 (1.6.17).

> esse templum, quod corpus ejus significabat: ut quot anni fuerunt in fabricatione templi, tot dies fuerint in corporis Dominici perfectione[19].

Augustine's embryology is based on the number six. It distinguishes the stages 'milk' (6 days), 'blood' (9 days), 'flesh' (12 days) and 'formation' (18 days); the fetus is formed on day 46 (45 days plus one)[20]. The process of formation is described using the verb *formare* and the phrase *usque ad perfecta lineamenta omnium membrorum*, 'until there is perfect shape in all the members'.

The embryologies by Vindicianus, Macrobius and Augustine were all ultimately derived from older, classical sources. In his *De die natali*, 'On the day of birth', the third-century Roman grammarian Censorinus, gathered together various embryological theories that were known in his day[21]. Censorinus names Hippocrates and Pythagoras as sources for books IX and XI, the books in which he describes the development of the *partus minor*, 'small fetus' and the *partus maior*, 'large fetus'. The development of the former is based on the number six, as in Augustine's embryology, and that of the latter on the number seven, as in Macrobius' embryology.

Embryological terminology in a normative context is found in discussions of abortion in Christian moral literature[22]. Aristotle's views on embryology and abortion not only influenced the Greek Septuagint version of Exodus 21.22-23, but they also had a profound influence on Christian thought through the writings of theologians and canonists, even though the Church fathers were divided on the subject of abortion. Caesarius of Arles (469–543) and the eastern Church father Basil of Caesarea (c. 330-379) are important representatives of the strict view that we also find in the Church council canons[23]. The Church fathers Augustine and

19. Extracted from the *Cetedoc Library of Christian Latin Texts*, version 4.0 (Turnhout, 1999), Augustinus Hipponensis – *De diuersis quaestionibus octoginta tribus* – Cl. 0289, quaestio: 56, linea: 2 ff. The 'milky' stage is also called the 'seminal stage'.

20. The due date is calculated by multiplying the amount of days for formation (46 days) by six (276 days).

21. Censorinus, *Betrachtungen zum Tag der Geburt; 'De die natali'*, übers. und ed. von K. Sallmann (Weinheim, 1988), p. 32-35, 38-43. At least seventeen manuscripts of *De die natali* (written 238) have come down to us; the oldest is a seventh-century Merovingian manuscript now in Cologne (p. 5) – indicating that theoretical works on embryology were available in the early medieval West.

22. Cf. note 1.

23. The Church father Basil even explicitly rejects the distinction 'formed' – 'unformed' in the Septuagint, cf. Kapparis, *Abortion in the Ancient World*, p. 48.

Four important Church councils contain canons that condemn abortion: Elvira (c. 300-306) c. 63, in: Vives J. (ed.), *Concilios Visigóticos e Hispano-Romanos* (Barcelona – Madrid, 1963), p. 12; Ancyra (314) c. 21, in: Martínez Díez G. & Rodriquez F. (eds.), *La*

Jerome (c. 350-420) are more lenient, punishing only late term abortion as homicide.

In his *Letter to Algasia* Jerome (c. 348-420) uses the verb *formare* to explain that the fetus is not formed until it has acquired its ultimate human shape:

> sicuti enim semina paulatim formantur in uteris et tam diu non reputatur homicidium, donec elementa confusa suas imagines membra que suscipiant[24].

Augustine's often quoted *quaestio LXXX* on abortion echoes Aristotle and the Septuagint version of Exodus 21.22-23, when he describes the difference between early term and late term abortion.

> Si ergo illud informe puerperium iam quidem fuerat, sed adhuc quodam modo informiter animatum (...) ideo lex noluit ad homicidium pertinere, quia nondum dici potest anima uiua in eo corpore quod sensu caret, si talis est in carne nondum formata, et ideo nondum sensibus praedita[25].

Augustine uses terms such as *informe*, 'unformed', *informiter animatus*, 'unformed being', *in carne nondum formata*, 'not yet formed in the flesh', *nondum anima uiua*, 'not yet a living being' and *sensu caret*, 'lacks sense', to denote a fetus which cannot yet be characterized as a living human being. In terms of descriptive embryology, Augustine equates formation, sensation, vivification and 'animation', thus lumping different stages of fetal growth together. From a Christian point of view, however, what Augustine is probably saying is that formation and 'animation' – here to be interpreted as ensoulment – practically coincide, because the (rational or human) soul can only enter a completely formed body[26].

colección canonica Hispana, vol. 3, *Concilios Griegos y Africanos*, Monumenta Hispaniae Sacra, serie Canónica 3 (Madrid, 1982), p. 101; Lerida (524) c. 2, in: Vives, *Concilios Visigóticos*, p. 55-56, and Braga II (572) c. 77, in: Vives, *Concilios Visigóticos*, p. 104.

24. Extracted from the *Cetedoc Library of Christian Latin Texts*, version 4.0 (Turnhout, 1999), Hieronymus – Epistula [ad Algasiam] – Cl. 0620, epist.: 121, vol.: 56, par.: 4, pag.: 16, linea: 22 [*]. Noonan, *Contraception*, translates: '... seeds are gradually formed in the uterus, and it [abortion] is not reputed homicide, until the scattered elements receive their appearance and members [are formed]', p. 90.

25. Extracted from the *Cetedoc Library of Christian Latin Texts*, version 4.0 (Turnhout, 1999), Augustinus Hipponensis – Quaestionum in heptateuchum libri septem – Cl. 0270, lib.: 2, Quaest. Exodi, quaestio: 80, linea: 1439; 'If what is brought forth is unformed, but at this stage some sort of living, shapeless thing, (...) then the law of homicide would not apply, for it could not be said that there was a living soul in that body, for it lacks all sense, if it be such as is not yet formed and therefore not yet endowed with its senses,' Dunstan, 'The Human Embryo', p. 45.

26. 'Animation' is interpreted as the infusion of the rational or sentient soul, as opposed to the earlier stages of fetal development called vegetative soul and animal soul by Aristotle, cf. Kapparis, *Abortion in the Ancient World*, p. 51.

We will see below that the ambiguous word *anima* ('soul' and/or 'life') recurs in the medieval penitentials.

Whereas Pliny's encyclopedia presents us with rudimentary information on the development of the fetus, the late antique embryologies described above deal with the whole period of gestation. Formation – still the most important stage of fetal development – takes approximately forty days and movement begins between the fortieth and the ninetieth day. The terms and phrases used to define 'formation' differ, but basically a formed fetus is described as a complete miniature human being. The Church fathers introduced the concept ensoulment as a 'synonym' of formation. Despite the fact that Augustine's embryology considers formation to be complete on the forty-sixth day after conception, early Christian authors usually stick to the forty-day period for formation found in most embryological texts. In his *De Opificio Dei*, 'On the workmanship of God', Lactantius (c. 300), an early Christian author also from North Africa, explains that it takes forty days for the fetus to become formed:

> et primum quidem cor hominis effingi, quod in eo sit et uita omnis et sapientia, denique totum opus quadragesimo die consummari[27].

However, as we saw above, some classical sources associate day forty with movement. In his *De Anima*, Cassiodorus (c. 480-575) says that ensoulment occurs directly after the onset of movement; he does not associate it with formation, as Augustine does in his *quaestio LXXX*:

> Medendi autem artifices quadragesimo die humanum ac mortale pecus animam dicunt accipere, cum se in utero matris coeperit commouere[28].

Thus, some Christian authors link movement and ensoulment and others, the majority, link formation and ensoulment. It is clear, however, that 'forty days' is the dividing line between a human being and a being that cannot yet be considered human. It is also clear that 'forty days' is becoming the dividing line between early term and late term abortion.

27. *Patrologia Latina Database* based on J.-P. Migne *Patrologia Latina* (Cambridge, 1993), PL 007: 0055A. 'First, then, the heart of a man is fashioned, because in it resides life and all wisdom, and then the whole is consummated by the fortieth day', cf. Flavius Magnus Aurelius Cassiodorus, *Institutions of Divine and Secular Learning and On the Soul*, tr. J.W. Halporn with an introd. by M. Vessey (Liverpool, 2004), p. 260, note 53. For more Christian texts that link formation and ensoulment, cf. Kapparis, *Abortion in the Ancient World*, p. 48-49.

28. Extracted from the *Cetedoc Library of Christian Latin Texts*, version 4.0 (Turnhout, 1999), Cassiodorus – De anima – Cl. 0897, cap.: 9, linea: 11; 'Physicians, however, say that the human and mortal animal receives a soul on the fortieth day when it has begun to move in its mother's womb', Cassiodorus, *Institutions of Divine and Secular Learning and on the Soul*, p. 260.

2. Early medieval embryology

The seventh-century encyclopedist, Isidore of Seville (c. 570-636), wrote only a few paragraphs on the development of the fetus in book X of his *Etymologiae*. The process of formation is finished by the fortieth day, or as Isidore says: *deinde quadragesimo die totum opus expleri*[29]. Movement is discussed in connection with the lungs and the *pneuma* or 'vital spirit', but no point in time is mentioned for the first movement[30]. A few centuries later Isidore is quoted – nearly verbatim – in Hrabanus Maurus' (780-856) encyclopedia *De Universo*[31]. Late antique and early medieval encyclopedic texts, such as Pliny's *Naturalis Historia* and Isidore's *Etymologiae* – which were popular throughout the medieval period, contain only bits and pieces of classical embryology; they do, however, show us that scholars were acquainted with classical learning on the subject.

Embryological terminology in a normative context is found in secular and church law on abortion. Throughout the medieval period we find two views on abortion: one categorically condemns abortion as murder, whatever the stage of development of the aborted fetus, and the other usually punishes late term abortion as murder, and stipulates a much milder punishment for early term abortion. We shall see below that, when used in a normative context, the terms used to denote the difference between early and late abortion do not always refer to the same stage of embryological development as in the descriptive texts.

2.1. Early medieval secular law

Four early medieval secular laws contain articles on abortion that distinguish between early term and late term abortion. The oldest of these

29. The full quote of Isidore's description of formation: 'Primum autem aiunt cor hominis fingi, quod in eo sit et uita omnis et sapientia; deinde quadragesimo die totum opus expleri, quod ex abortionibus, ut ferunt, collecta sunt' is based on Lactantius (see p. 376), cf. Isidorus Hispalensis, *Etymologiarum sive originum libri XX*, ed. W.M. Lindsay, 2 vols., Scriptorum classicorum bibliotheca Oxoniensis (Oxford, 1966; first ed. 1911), vol. 2, Liber 11.1.143. See also: Sharpe W.D., 'Isidore of Seville: the Medical Writings; an English translation with an introduction and commentary', *Transactions of the American Philosophical Society N.S.* 54 (1964), p. 1-75, at p. 48. It is interesting to note that both Lactantius and Isidore tell us that 'this has been learned, they say, from abortions'.

30. Cf. Lindsay, *Etymologiarum*, vol. 2, 11.1.124.

31. See: *Patrologia Latina Database,* PL 111: 0174C (Liber 6.1).

is an early or mid sixth-century Visigothic law on abortion that was clearly influenced by the Septuagint translation of Exodus 21.22-23:

> Leges Visigothorum 6.3.2. Antiqua. Si ingenuus ingenuam abortare fecerit. Si quis mulierem gravidam percusserit quocumque hictu aut per aliquam occasionem mulierem ingenuam abortare fecerit, et exinde mortua fuerit, pro homicidio puniatur. Si autem tantumodo partus excutiatur, et mulier in nullo debilitata fuerit, et ingenuus ingenue hoc intulisse cognoscitur, si formatum infantem extincxit, CL solidos reddat; si vero informem, C solidos pro facto restituat[32].

Like the Septuagint it employs the Aristotelian criterion 'formed – unformed', and regards late term abortion as a more serious crime[33].

The late sixth-century laws of the Salian Franks contain two provisions on violent abortion that implicitly differentiate between early term and late term abortion[34]. These laws demand the 'wergeld', that is, monetary compensation for murder, if the sex of the fetus is recognizable – which is another way of saying that the fetus is 'formed'.

> PLS 65e.1. Si quis mulierem grauidam occiderit, XXIIIIM den. qui faciunt solid. DC culpabilis iudicetur. Et si probatum fuerit, quod partus ille puer fuerit, simili conditione pro ipso puero DC solid. culpabilis iudicetur[35].

32. Cf. *Leges Visigothorum*, ed. K. Zeumer, MGH, Leges Nationum Germanicarum 1 (Hannover – Leipzig, 1902), p. 261; '6.3.2. Old law. If a free man causes a free woman to abort. If anyone strikes a pregnant woman by any blow whatever or through any circumstance causes a free woman to abort, and from this she dies, let him be punished for homicide. If, however, only the fetus is destroyed, and the woman is in no way debilitated, and a free man is recognized as having inflicted this to a free woman, if he has destroyed a developed [formed] fetus, let him pay 150 solidi; if it is actually an undeveloped [unformed] fetus, let him pay 100 solidi in restitution for the deed', Amundsen D.W., 'Visigothic Medical Legislation', *Bulletin of the History of Medicine* 45 (1971), p. 553-569, at p. 567.

33. Note that this Old Germanic law does not include the phrase *animam pro anima* (cf. Ex. 21.22-23). The correspondence between this Visigothic law on abortion, the Gothic Bible and the Septuagint version of Exodus 21: 22-23 is discussed in Elsakkers M., 'Gothic Bible, Vetus Latina and Visigothic Law: Evidence for a Septuagint-based Gothic Version of Exodus', *Sacris Erudiri* 44 (2005), p. 37-76.

34. See: Elsakkers M., 'Abortion, Poisoning, Magic and Contraception in Eckhardt's *Pactus Legis Salicae*', in: W. Pijnenburg, A. Quak & T. Schoonheim (eds.), *Quod vulgo dicitur; Studien zum Altniederländischen – Amsterdamer Beiträge zur Älteren Germanistik* 57 (2003), p. 233-267 at p. 245-247.

35. Cf. Eckhardt K.A. (ed.), *Pactus Legis Salicae*, MGH, Leges Nationum Germanicarum, 4.1 (Hannover, 1962), p. 235; '65e.1. He who kills a pregnant woman shall be liable to pay twenty-four thousand denarii (i.e., six hundred solidi). And if it is proved that the fetus was a boy, he shall also be liable to pay six hundred solidi for the child', Drew K.F. (tr./ed.), *The Laws of the Salian Franks* (Philadelphia, 1991), p. 127. We find approximately the same articles on abortion in the laws of the Ripuarian Franks.

The *Pactus Legis Salicae* (PLS 65.e.1) punishes killing a male fetus with 600 solidi, i.e., the wergeld for a young boy (PLS 24.1)[36]. The second provision, a late sixth-century Merovingian capitulary, which probably supplements this article, demands 2400 solidi for killing a female fetus (PLS III.104.8)[37].

Early seventh-century Alamannic law also uses the criterion of formation, and defines the difference between early term and late abortion in terms of gender differentiation:

> Lex Alamannorum 88.1. (A) Si quis aliquis mulierem prignantem aborsum fecerit, ita ut iam cognoscere possis, utrum vir an femina fuisset: si vir debuit esse, cum 12 solidis conponat; si autem femina cum 24.
>
> Lex Alamannorum 88.2. (A) Si nec utrum cognoscere potest, et iam non fuit formatus in liniamenta corporis, 12 solidos conponat. Si amplius requiret, cum sacramentalis se edoniet[38].

Lex Alamannorum 88.2.A adds an explanation of its abortion criterion and explicitly states that not being able to recognize the gender of the fetus means that the fetus is not yet 'formed': *iam non fuit formatus in liniamenta corporis*, '[the fetus] is not yet formed in [its] bodily features', that is, not yet recognizable as a human being. The wording of this abortion criterion is reminiscent of Macrobius' description of a formed fetus – and perhaps also of Augustine's embryology.

Eighth-century Bavarian law on abortion distinguishes between a fetus that is *vivus*, 'alive' at the time of the abortion, and one that was not yet 'alive', *vivus non fuit*.

> Lex Baiwariorum 8.19. Vario modo dicit de avorso. Si quis mulieri ictu quolibet avorsum fecerit, si mulier mortua fuerit, tamquam homicida teneatur. Si autem tantum partus extinguitur, si adhuc partus vivus

36. Eckhardt, *Pactus Legis Salicae*, p. 89.

37. Cf. articles III.104.4 – III.104.8 in the Third Merovingian capitulary, Eckhardt, *Pactus Legis Salicae*, p. 260.

38. Cf. Lehmann K. & Eckhardt K.A. (eds.), *Leges Alamannorum*, MGH, Leges Nationum Germanicarum 5.1 (Hannover, 1966), p. 150-151; '88.1. If anyone causes an abortion in a pregnant woman so that you can immediately recognize whether [it] would have been a boy or a girl: if it was to be a boy, let him compensate with twelve solidi; however, if a girl, [let him compensate] with twenty-four; 88.2. If whether [the fetus is male or female] cannot be immediately recognized, and [the fetus] was not formed in [its] bodily features let him compensate with twelve solidi. If he seeks more, let him clear himself with oathtakers', Rivers T.J. (tr./ed.), *Laws of the Alamans and Bavarians* (Philadelphia, 1977), p. 98 (slightly emended). The Salian and Alamannic laws both impose a higher fine for the abortion of a female fetus.

> non fuit, XX sold conponat. Si autem iam vivens fuit, weregeldum persolvat L et III sold et tremisse[39].

Abortion of a 'living' fetus is homicide and punished with the wergeld.

In these four early medieval Germanic laws, embryological terminology is used to differentiate between early and late abortion; abortion is condemned in terms that remind us of the biblical law on abortion[40]. Three laws define the difference between early term and late term abortion using formation as an abortion criterion; two of these, the Salian and Alamannic laws, explain formation in terms of gender distinction. The relatively young Bavarian law opts for the criterion of vivification.

2.2. Early medieval penitentials

Only a small group of *literati* would have been acquainted with patristic and canonical writings on Christian morals. Yet, other clerics and laymen were probably not completely ignorant of the Church's views on abortion. They heard them expounded in sermons and when they went to confession. The medieval penitentials confessors used were rooted in everyday life; they dealt with the sins clerics and ordinary laymen committed. The penances usually require daily fasting – probably from sun dawn to sunset – during a certain period of time. Most medieval penitentials follow the hard-line view on abortion, often quoting the Church council canons verbatim[41]. But there are also a number of popular and influential penitentials with articles that distinguish between early term and late term abortion[42]. These articles practically all ultimately derive

39. Schwind E. von (ed.), *Lex Baiwariorum*, MGH, Leges Nationum Germanicarum 5.2 (Hannover, 1926), p. 362-363; '8.19. Various cases of abortion. If anyone causes an abortion in a woman through any blow, if the woman dies, let it be considered the same as a homicide. However, if the child alone is killed, let him compensate twenty solidi if the child does not come forth alive. If, however, it was living [at the time of the abortion], let him pay the wergeld', Rivers, *Laws of the Alamans and Bavarians*, p. 141.

40. The other early medieval Germanic laws on abortion do not distinguish between early and late abortion.

41. Cf. note 23.

42. On the penitentials, see, for instance: Kottje R., 'Erfassung und Untersuchung der frühmittelalterlichen kontinentalen Bussbücher; Probleme, Ergebnisse, Aufgaben eines Forschungsprojektes an der Universität Bonn', *Studi Medievali* 26 (1985), p. 941-950; Kottje R., 'Busspraxis und Bussritus', *Settimane di studio del Centro italiano di studi sull'alto Medioevo* 33.1 (1987), p. 369-403; McNeill J.T. & Gamer H.M. (eds./tr.), *Medieval Handbooks of Penance; a Translation of the Principal Libri Poenitentiales and Selections from Related Documents* (New York, 1938; rpr. 1990); Meens R., *Het tripartite boeteboek; overlevering en betekenis van vroegmiddeleeuwse biechtvoorschriften met editie en*

from the late seventh-century so-called Anglo-Saxon penitential attributed to Theodore of Canterbury († 690), a penitential that was well known on the continent[43].

Theodore's penitential contains two different articles on abortion. In four of the five recensions (Co, D, G, U) we find a short article in which the dividing line between early term and late term abortion is the forty-day period (*XL dies*) that usually denotes the completion of formation in descriptive embryology:

> Iudicia Theodori U.1.14.27. Mulier quae concepit et occidit infantem suum in utero ante XL dies I annum peniteat. Si vero post XL dies ut homicida peniteat[44].

The longer article in recensions B, Co and U defines early term abortion as *antequam animam habeat*, 'before it is alive' or, alternatively, 'before it is ensouled' – using the ambiguous word *anima*. Late term abortion is defined as *post id est post XL dies accepti seminis*, 'after, that is, forty days after conception (accepting the seed)'. This means that the fetus is considered to be 'animated' on day forty:

> Iudicia Theodori U.1.14.24. Mulieres quae abortivum faciunt antequam animam habeat I annum vel III XL mas vel XL diebus iuxta qualitatem culpae peniteant. et post id est post XL dies accepti seminis ut homicidae peniteant, id est III annos, in IIII et VI feria et III bus XL mis. Hoc secundum canones decennium iudicatur[45].

vertaling van vier tripartita, Dissertatie Nijmegen (Hilversum, 1994); Schmitz H.J. (ed.), *Die Bussbücher und die Bussdisciplin der Kirche; die Bussbücher und das kanonische Bussverfahren*, 2 Bde. (Mainz – Düsseldorf, 1883-1898; rpr. Graz, 1958); Wasserschleben F.W.H. (ed.), *Die Bussordnungen der abendländischen Kirche* (Halle, 1851; rpr. Graz, 1958). Indispensable is: Kéry L., *Canonical Collections of the Early Middle Ages (ca. 400-1140); a Bibliographical Guide to the Manuscripts and Literature* (Washington DC, 1999).

43. There are five recensions of the *Paenitentiale Theodori*: Canones Basilienses (B), Canones Cottoniani (Co), Capitula Dacheriana (D), Canones Gregorii (G) and Discipulus Umbrensius (U); they are collectively called *Iudicia Theodori*. Cf. Wasserschleben, *Bussordnungen*, p. 145-219; Finsterwalder P.W., *Die Canones Theodori Cantuariensis und ihre Überlieferungsformen*, Untersuchungen zu den Bussbüchern des 7., 8. und 9. Jahrhunderts 1 (Weimar, 1929) (D, G, Co, U); Asbach F.B., *Das poenitentiale Remense und der sogen. Excarpsus Cummeani; Überlieferung, Quellen und Entwicklung zweier kontinentaler Bussbücher aus der 1. Hälfte des 8. Jahrhunderts*, Inaugural-Dissertation (Regensburg, 1975), p. 78-89 (B). A translation of recension U is in: McNeill & Gamer, *Medieval Handbooks of Penance*, p. 179-217.

44. Finsterwalder, *Canones Theodori*, p. 248 (D), p. 263 (G), p. 280 (Co), p. 310 (U). 'U.1.14.27. A woman who conceives and slays her child in the womb within forty days shall do penance for one year; but if later than forty days, she shall do penance as a murderess', McNeill & Gamer, *Medieval Handbooks of Penance*, p. 197.

45. Finsterwalder, *Canones Theodori*, p. 280 (Co), p. 309 (U); Asbach, *Das poenitentiale Remense*, Anhang, p. 85 (B). 'U.1.14.24. Women who commit abortion before [the

If *XL dies* was indeed associated with fetal formation in U.1.14.27, then the fact that recensions Co and U contain versions of both articles seems to imply that formation and 'animation' were thought to occur at the same time. Use of formation and 'animation' as synonymous terms tallies with Augustine's *quaestio* on abortion, but it is not consistent with descriptive embryology, where 'animation' – if mentioned at all – is situated after formation.

Theodore's penitential contains one more article that is implicitly concerned with abortion.

> Iudicia Theodori U.2.12.3. Mulier tres menses debet se abstinere a viro suo quando concepit ante partum (...)[46].

Article U.2.12.3 prescribes sexual abstinence for women during the last three months of pregnancy and is concerned with the risk of spontaneous abortion during the latter stages of pregnancy. I will return to this later.

Theodore's penitential was one of the three main sources for the tripartite penitentials that start to appear in the eighth century[47]. Characteristic for the tripartite penitentials is the fact that they usually contain a number of different articles on the same subject, so the confessor can decide for himself which article is applicable. The *Paenitentiale Remense* (c. 700-725), one of the earliest tripartite penitentials – possibly the earliest – contains both of the Theodorian articles on abortion[48]. The other eighth-century tripartites choose the shorter article with the *ante – post XL dies* criterion (U.1.14.27)[49].

fetus] has life, shall do penance for one year or for the three forty-day periods or for forty days, according to the nature of the offense; and if later, that is, more than forty days after conception, they shall do penance as murderesses, that is for three years on Wednesdays and Fridays and in the three forty-day periods. This according to the canons is judged [punishable by] ten years', McNeill & Gamer, *Medieval Handbooks of Penance*, p. 197.

46. Finsterwalder, *Canones Theodori*, p. 326. 'When she has conceived a woman ought to abstain from her husband for three months before the birth (...)', McNeill & Gamer, *Medieval Handbooks of Penance*, p. 208. This article was probably ultimately based on article 46 of the Irish *P. Vinniani* (Finnian's penitential), cf. Bieler L. (ed.), *The Irish Penitentials*, with an appendix by D.A. Binchy, Scriptores Latini Hiberniae 5 (Dublin, 1963), p. 90-93.

47. On the tripartite penitentials, see: Meens, *Het tripartite boeteboek*, passim. Most of the tripartite penitentials contain three different articles on abortion. One of these is derived from Theodore's penitential and therefore differentiates between early term and late term abortion. The other two condemn abortion whatever the stage of development of the fetus.

48. Cf. articles 8.26 and 8.46 in: Asbach, *Das poenitentiale Remense*, Anhang, p. 52-53.

49. The following eighth-century tripartitite penitentials contain the shorter Theodorian article on abortion with the *XL dies* criterion: *P. Remense,* art. 8.46 (ante XL dies – post XL dies), Asbach, *Das poenitentiale Remense*, Anhang, p. 53; *Excarpsus Cummeani*,

Early ninth-century Church councils condemned the 'old', anonymous penitentials with their inconsistencies and varying penances for the same sin[50]. Only a few ninth-century penitentials retain an abortion criterion, besides the group of penitentials attributed to Bede and Egbert and their derivatives[51]. The *Paenitentiale Ps.-Bedae* and the *Paenitentiale Ps.-Egberti* (both late eighth-century) were gradually combined into a new penitential during the ninth century, as Haggenmüller has demonstrated[52]. The last two stages of this process, the *Paenitentiale Additivum Ps.-Bedae-Egberti* and the *Paenitentiale Mixtum Ps.-Bedae-Egberti*, contain an innovative

art. 6.11 (ante XL dies – post XL dies post conceptionem), Wasserschleben, *Bussordnungen*, p. 479; *P. Merseburgense A*, art. 1.64 (ante XL dies – post XL dies conceptionis), Wasserschleben, *Bussordnungen*, p. 407; *P. Vindobonense B*, art. 33.9 (ante XL dies – post XL dies conceptus), Meens, *Het tripartite boeteboek*, p. 408-409; *P. Sangallense*, art. 2.5 (ante XL dies – post XL dies), Meens, *Het tripartite boeteboek*, p. 336-337; *P. Capitula Iudiciorum*, art. 3.2a (ante XL dies – post XL dies), Meens, *Het tripartite boeteboek*, p. 440-441.

50. Most of the penitentials written after the ninth-century Carolingian reform reflect a more conservative attitude towards abortion, cf. Meens, *Het tripartite boeteboek*, p. 60-69. They quote the early Church council canons on abortion extensively, and do not distinguish between early term and late term abortion. See, for instance, the two penitentials written by Hrabanus Maurus, *P. ad Otgarium* (caput 841-842), in: *Patrologia Latina Database*, PL 112: 1410C-1411A (caput. 11) and *P. ad Heribaldum* (c. 853-856), in: *Patrologia Latina Database*, PL 110: 0474B-0474C (c. 8-9). See also note 23.

51. Other ninth-century penitentials containing one of the Theodorian abortion criteria – besides those related to the *Ps.-Bedae-Egberti* penitentials – are: (a) *P. Ps. Gregorii*, art. 17 (*ante XL dies – post XL dies*), Kerff F., 'Das Paenitentiale Pseudo-Gregorii; eine kritische Edition', in: H. Mordek (ed.), *Aus Archiven und Bibliotheken; Festschrift für Raymund Kottje zum 65. Geburtstag*, Freiburger Beiträge zur mittelalterlicher Geschichte; Studien und Texte 3 (Frankfurt am Main, 1992), p. 161-188 at p. 177-178; (b) *P. Vallicellianum II*, art. 29 (*ante XL dies – post XL dies*), Schmitz, *Bussbücher*, Bd. 1, p. 361; (c) *P. Vigilianum*, arts. 10.57-58 (*antequam animam habeat – post*), Körntgen L. & Bezler F. (eds.), *Paenitentialia Hispaniae*, Paenitentialia Franciae, Italiae et Hispaniae saeculi VIII-XI.2, CCSL 156A (Turnhout, 1998), p. 8.

52. Cf. Haggenmüller R., *Die Überlieferung der Beda und Egbert zugeschriebenen Bussbücher*, Europäische Hochschulschriften 3.3.461 (Frankfurt am Main, 1991) and Haggenmüller R., 'Zur Rezeption der Beda und Egbert zugeschriebene Bussbücher', in: Mordek (ed.), *Aus Archiven und Bibliotheken*, p. 149-159. Haggenmüller distinguishes fives stages: (1) *P. Ps.-Bedae* (late 8th century), (2) *P. Ps.-Egberti* (late 8th century), (3) 'Vorstufe' of the *P. Additivum Ps.-Bedae-Egberti* (early 9th century), (4) *P. Additivum Ps.-Bedae-Egberti* (800-825), and (5) *P. Mixtum Ps.-Bedae-Egberti* (870-880).

The *P. Ps.-Bedae* contains an article on abortion (4.12) that is derived from Theodore's penitential (U.I.14.27): '*P. Ps.-Bedae 4.12*. Mulier qui occidit filium suum in utero ante dies XL, I annum peniteat. Si vero post dies XL, III annos. Sed distat multum, utrum paupercula pro difficultate nutriendi an fornicaria causa sui sceleris celandi faciat', Wasserschleben, *Bussordnungen*, p. 225. The *P. Ps.-Egberti* does not have an article on abortion, only an article that labels infanticide as homicide and assigns a ten-years' penance: '*P. Ps.-Egberti 7.9*. Mulier si occiderit filium suum per homicidium, X annos peniteat', Wasserschleben, *Bussordnungen*, p. 239.

article on abortion that combines the two Theodorian articles on abortion into a three-tier article[53].

> P. Mixtum 14.1. Si qua mulier partum suum ante XL dies in utero sponte perdiderit, I annum peniteat. Si vero post XL dies eum occiderit, III annos peniteat; si vero postquam animatus fuerit eum perdiderit, quasi homicida peniteat, sed distat multum utrum paupercula pro difficultate nutriendi aut fornicaria causa sit aut pro suis sceleribus celandi faciat[54].

Article 14.1 postulates three stages of fetal development with corresponding penances: (a) within the first forty days (one year), (b) after forty days (three years) and (c) when the fetus is *animatus*, 'alive' or 'ensouled' (homicide)[55]. This article regards *XL dies* and 'animation' as different stages of embryonic development – although it would be difficult to assign a penance for homicide, because the time of 'animation' is not specified. In this article *XL dies* probably refers to formation and *animatus* should probably be interpreted as 'ensouled'. Whereas formation and 'animation' seem to be lumped together in Theodore's penitential and Augustine's *quaestio LXXX*, they are clearly differentiated in the *P. Mixtum*[56]. Three or more stages of fetal growth in medieval normative texts on abortion is rare; to my knowledge only the slightly older Old Irish penitential (c. 800), which will be discussed below, and the much younger Old Frisian laws mention more than two stages[57].

53. The three-tier articles on abortion in the *P. Additivum* and the *P. Mixtum* seem to have gone largely unobserved; to my knowledge only Connery noticed the three stages in these articles, cf. Connery, *Abortion: the Development of the Roman Catholic Perspective*, p. 80. The version in the *P. Additivum* is slightly shorter: '*P. Additivum 2.11*. Mulier partum suum ante dies XL sponte perdens, unum annum poeniteat. Si vero post XL dies, III annos poeniteat. Si vero postquam natus [animatus] fuit, quasi homicida. Sed distat multum utrum paupercula per difficultatem nutriendi, an fornicaria, causa sceleris celandi faciat', cf. *Patrologia Latina Database*, PL 94: 0571A, and Albers B., 'Wann sind die Beda-Egbert'schen Bussbücher verfasst worden, und wer ist ihr Verfasser?', *Archiv für Katholisches Kirchenrecht*, 81 (1901), p. 393-420 at p. 404.

54. Wasserschleben, *Bussordnungen*, p. 265-266. Note that both the *P. Additivum* and the *P. Mixtum* have *postquam animatus fuerit* instead of *postquam animam habeat*. Article 14.1 in the *P. Mixtum* is followed by an article taken from the *P. Ps.-Egberti* that assigns a ten years' penance for infanticide: '*14.2*. Mulier si occiderit filium suum per homicidium, X annos peniteat' (cf. note 52).

55. The penalty for homicide would probably be equivalent to the penalty for infanticide, that is, a penance of ten years, cf. notes 52 and 54.

56. One wonders if the author of this article was acquainted with descriptive embryology or theological discussions that considered formation a prerequisite for ensoulment and thus a stage that precedes ensoulment.

57. Two other, three-tier articles on abortion are based on or related to the three-tier articles in the *P. Additivum* and the *P. Mixtum* article: (a) art. 6.4 in the ninth-century

The *P. Mixtum* opens with an *Ordo ad dandam penitentiam* and a list of questions to ask the penitents[58]. Question 31 deals with infanticide and abortion and employs the criterion *ante – post conceptum*, to distinguish between early term and late term abortion.

> P. Mixtum 31. Necasti partus tuos? X annos, et si filium aut filiam occidisti, XII annos poeniteas, et si in utero ante conceptum, annum I poeniteas, si post conceptum, III annos[59].

The penances (one or three years) suggest that question 31 was based on the *antequam animam habeat – postquam XL dies* article on abortion in the *Iudicia Theodori* (U.1.14.24). The meaning of the criterion *ante – post conceptum* is not immediately clear. *Post conceptum* may refer to 'conception'; it would then be short for *post XL dies a conceptione seminis, post XL dies post conceptionem* or *post XL dies conceptus* – which are all variant forms of the phrase *post XL dies accepti seminis*, we found in article U.1.14.24 of Theodore's penitential[60]. However, *post conceptum* may also be short for *post conceptum spiritum*, 'after ensoulment', a phrase found in Burchard's two-stage article on abortion which will be discussed below. *Ante – post conceptum* can probably best be interpreted as 'before – after ensoulment'[61]. Other ninth-century *ordines* contain

P. Ps. Theodori (Wasserschleben, *Bussordnungen*, p. 586-587) and (b) art. 43 in the late ninth-century *P. Martenianum* (Hörmann W., 'Bussbücherstudien VI. Das sog. poenitentiale Martenianum. Anhang I', *Zeitschrift der Savigny-Stiftung für Rechtsgeschichte* 35. *Kanonistische Abteilung* 4 (1914), p. 358-483 at p. 379). The *P. Martenianum* will be discussed below.

On Old Frisian abortion, cf. my '*Her anda neylar*: An Intriguing Criterion for Abortion in Old Frisian Law', *Scientiarum Historia* 30 (2004), p. 107-154.

58. The *P. Mixtum* seems to have evolved into a tripartite penitential, because it contains four different articles on abortion. Besides the question in article 31 and the article on abortion in the penitential proper (14.1), there are also two articles on abortion in the section *De diversis causis*. Article 39.1 contains one of the Church council canons on abortion and article 39.2 quotes an article on voluntary abortion derived from the Frankish penitentials, cf. Wasserschleben, *Bussordnungen*, p. 274.

59. Wasserschleben, *Bussordnungen*, p. 255; '*P. Mixtum* 31. Did you kill your newborn child? Ten years, and if you killed your son or daughter, you must do penance for twelve years, and if [it is] in the uterus *ante conceptum*, you must do penance for one year, if [it is] *post conceptum*, three years'.

60. See also: *P. Remense* 8.26, in: Asbach, *Das poenitentiale Remense*, Anhang, p. 52, and note 49.

61. A third possibility is that the word *conceptus* was used in the meaning 'formed fetus' as in Macrobius' embryology. This possibility seems far-fetched, because early medieval normative embryology hardly seems to have any links with descriptive embryology, but on the other hand, if, as in some normative texts, formation and ensoulment are considered coincidental, then *conceptus* might indicate a 'formed and ensouled fetus' in this text.

See also: note 65 on Hartmann's translation of the same article in Regino's penitential.

the same question on abortion with the same *ante – post conceptum* criterion[62].

The *P. Mixtum* also has two articles on continence during pregnancy[63]. A husband is required to be sexually continent as soon as it is clear that his wife is pregnant: *a conceptione manifestata*, 'after pregnancy is obvious' (art. 5.1).

> P. Mixtum 5.1. Uxoratus contineat se (...) a conceptione manifestata usque post natum sobolem. Vir autem contineat se ab uxore III menses. (...)
>
> P. Mixtum 5.2. Mulier abstineat se a viro III menses, quando concepta est, antequam pariat et post partum XL dies (...)[64].

This rather vague criterion seems to refer to the outward signs of pregnancy. However, the *P. Mixtum* proceeds to define the period of continence for both men (5.1) and women (5.2), dividing the pregnancy into two parts: the first six months and the last three months during which continence is mandatory.

Regino of Prüm's *Libri Duo De Synodalibus Causis et Disciplinis Ecclesiasticis* (c. 906), a handbook for Church discipline, repeats the articles on abortion in the *P. Mixtum* (31; 14.1), including the three-tier article[65]. Regino also includes an article on sexual abstinence taken from

62. The *ordines* are a new form of penitential literature, which first appear in the ninth century. Cf. Schmitz, *Bussbücher*, Bd. 1, p. 749.

63. Article 5.1 was derived from the *P. Ps.-Bedae* and article 5.2 is from the *P. Ps.-Egbert*. Cf. *P. Ps.-Bedae* art. 3.37. 'Uxoratus contineat se (...) a conceptione manifestata usque post natam sobolem (...)', Wasserschleben, *Bussordnungen*, p. 224, and *P. Ps.-Egberti* articles 6.1. 'Mulier se abstineat se a viro tres menses, quando concepta est antequam pareat, et post partum XL dies' and 6.2. 'Qui autem nupserit his diebus, XL diebus peniteat vel XXX vel XX', Wasserschleben, *Bussordnungen*, p. 238. *P. Mixtum* 5.2 was based on *Iudicia Theodori* U.2.12.3.

64. Wasserschleben, *Bussordnungen*, p. 261-262.

65. Regino's penitential in two books is also called *Libellus de Ecclesiasticis Disciplinis et Religione Christiana*. There are three important editions of Regino's penitential. The oldest editions are by Migne and Wasserschleben; Wasserschleben's nineteenth-century edition has recently been reedited and translated by Wilfried Hartmann. See: Wasserschleben F.W.H. (ed.), *Die Irische Kanonensammlung*, 2. Aufl. (Leipzig, 1885; rpr. Aalen, 1966); *Patrologia Latina Database*, PL 132: 0187-0400B; Reginonis Prumiensis, *Das Sendhandbuch des Regino von Prüm*, ed. und übers. von W. Hartmann unter Benutzung der Edition von F.W.H. Wasserschleben, Ausgewählte Quellen zur deutsche Geschichte des Mittelalters 42 (Darmstadt, 2004).

The article numbers in the various editions do not correspond. The articles quoted here are from Hartmann's edition; Migne's article numbers have been added between brackets.

For Regino's version of *P. Mixtum* 31, see: Hartmann, *Das Sendbuch*, p. 164-165, article 1.304 [1.300]. Hartmann translates *in utero ante conceptum* as 'im Mutterleib vor

the *Collectio Canonum Hibernensis*, a late seventh- or early eighth-century Irish collection of Church council canons that was well known on the continent. This article was not incorporated into any of the earlier medieval penitentials.

> Regino of Prüm 1.338 [328]. Synodus Helibernensium dicit: (...) conjugales continere se debent. Item nec in illis diebus copulari, quandiu gravata fuerit uxor, id est, a quo die filius in utero motum fecerit usque ad partum. (...)[66].

Article 1.328 forbids intercourse *a quo die filius in utero motum fecerit usque ad partum*, 'from the day the child starts to move in the womb until birth', thus distinguishing between fetal development before and after movement. Regino is the first penitential author to mention movement, and thus Regino seems to be reintroducing one of the other important stages of fetal development we found in descriptive embryology. Regino's other two articles on continence were derived from the *Iudicia Theodori* (U.2.12.3) and the *P. Mixtum* (5.1-5.2).

> Regino of Prüm 1.339 [329]. Ut uxoratus contineat se (...) a conceptione manifestata usque post natam sobolem [*Vir autem contineat se ab uxore III menses. Uxor post natam sobolem*] abstineat se ab ecclesia, (...). Mulier se abstineat a viro tres menses ante partum (...) viginti dies poeniteat[67].

Article 1.339 forbids intercourse *a conceptione manifestata*, which is defined as the last three months of pregnancy[68]. We find practically the same article, condemning intercourse *post conceptionem manifestatam*, in Regino's article 1.304.

den Empfängnis', which does not seem to make sense. As noted above, I think *ante conceptum* should be interpreted as *ante conceptum spiritum*.

For Regino's version of the three-tier article in *P. Mixtum* 14.1, see: Hartmann, *Das Sendbuch*, p. 282-283, article 2.65 [2.66]. Article 2.65 is preceded by three articles on abortion that quote the most important Church council canons on abortion (Ancyra, Lerida and Elvira). See also: note 23.

66. Hartmann, *Das Sendbuch*, p. 180-181. Cf. *Collectio Canonum Hibernensis*, article 46.11.a. '(...) conjuges continere se debent. 46.11.b. Item. In omnibus solemnitatibus et in illis diebus, quibus uxor praegnans, hoc est, a die, quo filius in utero ejus motum fecerit, usque ad partus sui diem', Wasserschleben, *Die Irische Kanonensammlung*, p. 187-188. Regino wrongly refers to the council of Elvira (Synodus Helibernensium).

67. Hartmann, *Das Sendbuch*, p. 180-181.

68. This article is 'verderbt', as Hartmann rightly remarks in note 289 (*Das Sendbuch*, p. 181). The passage linking *a conceptione manifestata usque post natum sobolem* with continence is missing, obviously due to an eye skip on *post natum sobolem*. The missing text – taken from the *P. Mixtum* 5.1 – has been included between brackets.

Regino of Prüm 1.304 [300]. Coinguinatus es cum uxore tua in quadragesima? (...) et post conceptionem manifestatam vir contineat se ab uxore[69].

Because article 1.339 directly follows 1.338, it seems as if Regino is equating movement and *conceptio manifestata*. This would in itself not be surprising, because to an expectant mother fetal movement is proof that her child is alive and that she is indeed pregnant. However, the period of three months' abstinence before birth mentioned in article 1.329 (i.e. from about 170-180 days after conception until birth) does not correspond to any point in time mentioned in the early descriptive embryologies for the onset of fetal movement. It does, however, correspond to the period of pregnancy during which fetal movement is apparent, and intercourse might induce premature labor and cause a spontaneous abortion. The articles on sexual continence in the early medieval penitentials assign a mild penance for incontinence, and usage of the words *uxoratus* and *uxor* indicates that these articles should probably be interpreted as a guideline for young couples to help them prevent a miscarriage.

Burchard of Worms (965-1025) wrote his *Decretum* in the early eleventh century (1008-1012), just before the Gregorian reform movement. Book 19, also called *Corrector et Medicus*, is the penitential proper[70]. Burchard cites the major Church council decisions on abortion in book 17 and includes lists of questions for penitents in books 1 and 19[71]. Burchard's two articles on continence were taken from Regino's penitential, but turned into questions and each provided with a different penance – albeit that Burchard does not mention the three months' period before birth during which husband and wife should abstain from sexual relations.

Burchard of Worms 19.5.54. Concubuisti cum uxore tua postquam infans motum in utero fecerat? vel saltem XL dies ante partum? Si fecisti, XX dies in pane et aqua poeniteas.

69. Hartmann, *Das Sendbuch*, p. 162-163.

70. Book 19 of Burchard's *Decretum* was edited by Migne, Schmitz and Wasserschleben, cf. *Patrologia Latina Database*, PL 140: 0537A-1058C; Schmitz, *Bussbücher*, Bd. 2, p. 403-467, and Wasserschleben, *Bussordnungen*, p. 624-682. For a translation of parts of Book 19, cf. McNeill & Gamer, *Medieval Handbooks of Penance*, p. 321-345. The texts quoted in this study are from Migne's *Patrologia Latina*; the article numbers are from Migne and Schmitz. See also: Hartmann W. (ed.), *Bischof Burchard von Worms 1000-1025* (Mainz, 2000).

71. Cf. Book 17, arts. 51-54 on the Church council canons on abortion, *Patrologia Latina Database*, PL 140: 0931A-0931D. The *interrogatio* on abortion in Book 1 is worth quoting in full, because it seems to distinguish between a *partus* and a *conceptus*: '1.94. Interrogatio 5. Est aliquis vel aliqua qui alterius partum excusserit, vel si ipsa femina propria voluntate suum partum vel conceptum excusserit, et abortivum fecerit?', cf. *Patrologia Latina Database*, PL 140: 0573B.

> Burchard of Worms 19.5.55. Concubuisti cum uxore tua post manifestatam conceptionem? decem dies in pane et aqua poeniteas[72].

The variation in the penances introduced by Burchard indicates that he intentionally emended Regino's articles[73]. By clearly differentiating between *conceptio manifestata* and movement Burchard is telling us that in his opinion fetal movement starts some time after it is clear that the woman is pregnant.

Burchard's article 19.147 is a long, orthodox condemnation of fornication, contraception and abortion that echoes the hard-line views found in the sermons of Caesarius of Arles and the Church council canons[74]. Three articles further on, however, Burchard includes an article that differentiates between early term and late term abortion. Although Burchard borrowed a lot from Regino, he apparently decided against the three-tier article on abortion Regino had copied from the *P. Mixtum*, and chose to write a two-stage abortion article of his own – perhaps inspired by the question on abortion in Regino's penitential (1.304)[75].

> Burchard of Worms 19.5.150. Excussisti conceptum tuum antequam vivificaretur? Si fecisti, unum annum per legitimas ferias poenitere debes: fecisti post conceptum spiritum, tres annos per legitimas ferias poenitere debes[76].

This article was ultimately based on Theodore's article on abortion with the *antequam animam habeat* criterion (U.1.14.24). However, Burchard uses different criteria; he defines early term abortion as *antequam vivificaretur*, 'before the *conceptus* (that which was conceived) became

72. *Patrologia Latina Database*, PL 140: 0959D; Schmitz, *Bussbücher*, Bd. 2, p. 421; see also: Wasserschleben, *Bussordnungen*, p. 642. Migne's edition contains a second version of article 19.5.54, which retains the wording of Regino's article 1.338, cf. article 19.155, *Patrologia Latina Database*, PL 140: 1013C.

73. The penances in the various manuscripts and editions vary, but intercourse after the fetus has started to move is always punished with a slightly higher penance in Burchard's penitential; see, for example: Schmitz, *Bussbücher*, Bd. 2, p. 421 and Wasserschleben, *Bussordnungen*, p. 642.

74. *Patrologia Latina Database*, PL 140: 0972D; Wasserschleben, *Bussordnungen*, p. 659. See also: note 23.

75. I have not been able to find a source for this article. Hoffmann and Pokorny state that the source for article 5 in *caput* 19 is 'unidentifiziert, erstmals fassbar bei Regino (stark erweitert und umgestaltet) ... RP I 304', cf. Hoffmann H. & Pokorny R., *Das Dekret des Bischofs Burchard von Worms; Textstufen – Frühe Verbreitung – Vorlagen*, MGH Hilfsmittel 12 (München, 1991), p. 233. However, even though the penances are the same, Regino's two-stage article – discussed above – uses different abortion criteria (1.304 [300]).

76. *Patrologia Latina Database*, PL 140: 0972D; see also: Wasserschleben, *Bussordnungen*, p. 659.

alive', and late term abortion as *post conceptum spiritum*, 'after the spirit was conceived', that is, 'after ensoulment'. Burchard avoids using the word *anima(tus)*, although he does – purposely? – render both aspects of the word's meaning, 'alive' and 'ensouled'. It is evident that he treats vivification and ensoulment as synonyms, that is, stages of fetal development that are identical or occur simultaneously.

Fulbert of Chartres (c. 960-1028), a contemporary of Burchard, composed the *Paenitentiale Fulberti*, an ultra-short penitential that has a two-stage article on abortion with the criterion *vivus – non vivus*[77]:

> P. Fulberti 9. Si qua partum necat aut excutit iam uiuum sponte, XII annos peniteat. Si non uiuum, VI annos, et sempre sexta feria ieiunet[78].

Fulbert does not even try to elucidate the word *anima(tus)*. Instead he chooses one of the two meanings of *anima*, the unambiguous word *vivus*, to explain the difference between early term and late term abortion. Whereas Burchard retains the original penances mentioned in Theodore's penitential (one year – three years), Fulbert now doles out the extremely harsh penances of six years for early term abortion, and twelve years for late term abortion. Fulbert does not do this without reason; he has added the word *sponte*, 'deliberately', to his article on abortion, so that, in fact, he is explicitly punishing voluntary abortion[79].

In the course of the five centuries during which penitential practice flourished in the early medieval West, the terms used and their interpretation changed every so often. Theodore's late seventh-century penitential

77. Kerff F., 'Das sogenannte Paenitentiale Fulberti; Überlieferung, Verfasserfrage, Edition', *Zeitschrift der Savigny-Stiftung für Rechtsgeschichte 104. Kanonistische Abteilung* 72 (1987), p. 1-40. See also: Meens, *Het tripartite boeteboek*, p. 200-205.

The *Paenitentiale Parisiense Compositum* (c. 1025-1075) contains the same article on abortion (art. 57), adding only the subclause *humanius VII uel III* after *si uiuum XII*. As Meens has demonstrated, this article was taken from the *P. Fulberti*, cf. Meens, *Het tripartite boeteboek*, p. 177-219 at p. 185 and p. 494-495. The following article on abortion in the *P. Parisiense Compositum* (art. 58) was taken from the *Iudicia Theodori* (U.1.14.27) and assigns the usual penances for abortion of one and three years.

78. Kerff, 'Paenitentiale Fulberti', p. 30. In the older editions of the *P. Fulberti* this is article 4, cf. Wasserschleben, *Bussordnungen*, p. 624 and Schmitz, *Bussbücher*, Bd. 1, p. 773. In his notes Kerff shows us that some scholars still associate 'aliveness' with both movement and formation without differentiating between these stages of fetal development: 'Eine deutlich längere Busszeit für die werdende Mutter, die nach dem Verspüren der ersten Kindesbewegungen die Leibesfrucht abgetrieben hat, sehen eine Reihe von Bussbüchern vor, legen jedoch den Beginn des Lebens auf den 40. Tag nach der Empfängenis fest ...' (p. 36).

79. The word *sponte*, 'deliberately', is first found in the so-called 'Vorstufe' of the *P. Additivum Ps.-Bedae-Egberti* (stage (3); cf. note 52). It also occurs in the *P. Additivum*, the *P. Mixtum* and in Regino's penitential (cf. note 65).

is the oldest penitential to differentiate between early term and late term abortion. It has two articles on abortion, one with the abortion criterion *XL dies* (U.1.14.27) and another with the criterion *antequam – postquam animam habeat* (U.1.14.24). Theodore seems to equate 'animation' and *XL dies*, thus linking the 'forty days' usually associated with the completion of formation in descriptive embryology and 'animation'. Two different two-tier articles on abortion in one and the same penitential was apparently considered confusing, because almost all of the eighth-century tripartite penitentials choose the Theodorian article with the straightforward and more practical *XL dies* criterion (U.1.14.27). The fuzziness of the abortion criteria *XL dies* and 'animation' was resolved in the three-tier article on abortion we find in the ninth-century *P. Mixtum* (14.1). This article combines both of the Theodorian articles on abortion and explicitly distinguishes between *XL dies* (formation) and 'animation'. Also new in the ninth century is the criterion *ante – post conceptum* in the *P. Mixtum* (31), where the word *conceptus* probably denotes *conceptus spiritus*, that is, ensoulment.

Regino of Prüm borrowed extensively from the *P. Mixtum* and quotes both the two-stage and the three-stage article on abortion in his early tenth-century *Libri Duo De Synodalibus Causis* (1.304; 2.65). Writing a century later, Burchard of Worms used Regino as one of the main sources for his *Decretum*, but he did not include the three-tier article on abortion. Instead, he devised a two-stage article of his own in which he defined early term abortion as *antequam vivificaretur* and late term abortion as *post conceptum spiritum*. Burchard's contemporary Fulbert of Chartres – undoubtedly a more practical-minded pastor – opts for the simple criterion *vivus – non vivus*, a criterion that would be intelligible to all, but not burdened with the ambiguity of the word *anima*.

Regino also copied the articles that forbid intercourse during pregnancy from the *P. Mixtum*. Abstinence is required when pregnancy is manifest (*conceptio manifestata*), that is, during the last three months of pregnancy (after the sixth month). To these articles Regino added an older Irish canon that demands continence after the onset of fetal movement – which according to descriptive embryology would mean that abstinence was mandatory after about the second or third month of pregnancy instead of after the sixth month. The provisions on continence were undoubtedly added in order to prevent spontaneous abortion. Burchard recycled Regino's articles on sexual abstinence, and he suggests that movement is a separate stage of fetal growth that starts some time after it is clear that the woman is pregnant. However, he gives us no clue as to when fetal movement starts or when a woman is considered to be pregnant. Obviously the term *conceptio manifestata* – which is not found in descriptive embryology – is open to many different interpretations. Although the articles on continence do not betray any knowledge of descriptive embryology, they do indicate that movement was apparently considered an important stage of fetal development.

The articles on abortion in the penitentials use a variety of terms to denote 'animation', sometimes focussing on ensoulment and sometimes on vivification, sometimes on both. The term *anima* was not taken from descriptive embryology – even though it occurs in Vindicianus' embryology. Movement is mentioned, but there is no apparent relation with descriptive embryology. The only term borrowed from descriptive embryology by the penitential authors is *XL dies*, and, although this term refers to formation, it acquired the additional meaning 'ensoulment' in the penitentials. The association with formation was, however, not lost, as we saw in the three-tier article in the *P. Mixtum*.

2.3. Early medieval embryology in the vernacular

The Old English 'formation of the fetus' is the only early medieval descriptive embryological text in the vernacular that I know of[80]. It was probably translated into Old English in tenth-century Anglo-Saxon England and it is found in only one mid eleventh-century manuscript.

> Her onginð secgan ymbe mannes gecynde hu he on his modor innoþe to men gewyrðeð.
>
> (1) ærest þæs mannes brægen bið geworden on his moder innoþe þonne biþ þæt brægen utan mid reaman bewefen on þære syxtan wucan.

80. The Old English Embryology in MS London, British Library, Cotton Tiberius A.iii, f. 40v-41r was first published and translated in: Cockayne O., *Leechdoms, Wortcunning and Starcraft of Early England*, 3 vols., Rerum Brittanicarum Medii Aevi Scriptores (London, 1864-1866; rpr. London, 1961), vol. 3, p. 146-147. The most recent edition is: Chardonnens L.S., 'A New Edition of the Old English "Formation of the Foetus"', *Notes and Queries* 47 (2000), p. 10-11. See also: Cameron M.L., *Anglo-Saxon Medicine* (Cambridge, 1993), p. 184, and Deegan M., 'Pregnancy and Childbirth in the Anglo-Saxon Medical Texts: a Preliminary Survey', in: M. Deegan & D.G. Scragg (eds.), *Medicine in Early Medieval England* (Manchester, 1987; rev. ed. 1989), p. 17-26 at p. 22-24.

The Old English embryology is part of a group of prognostic texts, but only the very last sentence, which is not in the main source text, can in fact be labelled prognostic. The prognostic texts in MS London, BL, Cotton Tiberius A.iii are listed in: Liuzza R.M., 'Anglo-Saxon Prognostics in Context: a Survey and Handlist of Manuscripts', *Anglo-Saxon England* 30 (2001), p. 181-230 at p. 216-218.

(2) On oðrum monðe þa ædran beoð geworden. on lxv. 7 þreo hundræd scy[r]tran 7 lengran hi beoð todælede 7 þæt blod þonne floweð on þa fet 7 uppan þa handa. 7 be þonne byþ on limum todæled. 7 tosomne gea[r]wað.
(3) On þam þriddan monþe he biþ man butan sawle.
(4) On þam feorþan monþe he biþ on limum staþolfæst.
(5) On þam fiftan monþe he bið cwicu. 7 weaxeð, 7 seo modur lið witleas. 7 þonne þa ribb beoð geworden. þonne gelimpð þæræ manigfeald sar þonne þæs byrþ[n]es lic on hire innoþe styrigende bið.
(6) On þam syxtan monþe he byþ gehyd. 7 ban beoð weaxende.
(7) On þam seofoþan monþe þa tan. 7 þa fingras. beoð weaxende.
(8) On þam eahtoþan monþe him beoð þa breostþing wexende. 7 heorte. 7 blod. 7 he bið eall staþolfæstlice. geseted.
(9) On þam nigoþan monþe witodlice wifum bið cuð hwæðer hi cennan magon.
(10) On þam teoþan monþe þæt wif hit ne gedigð hyre feore. gif þæt bearn accenned ne biþ. forþam þe hit in þam magan wyrð hire to feorhadle oftost on tiwesniht[81].

This strange vernacular embryology looks like an Old English translation of Vindicianus' month-by-month embryology, but only a few stages more or less correspond to Vindicianus' text. This means that the author must have used a number of additional sources, when compiling this embryology. The text is much more detailed than Vindicianus' text, some stages are missing, and new stages and extra information have been added.

However, if we look a little closer at both texts and focus on the most important stages of development in Vindicianus' embryology, i.e., formation, 'animation' and movement, we see that underneath a thick layer of textual embellishment, more of Vindicianus' embryology is visible than we thought at first. The Old English Embryology does not mention

81. Chardonnens, 'A new edition', p. 10-11. Chardonnens' edition contains a number of important corrections, so that past translations are no longer accurate. Cameron, *Anglo-Saxon Medicine*, p. 184 gives the following summary of the Old English text: '... the brain is formed first and by the sixth week its membranes are developing; in the second month blood vessels, three hundred and sixty-five of them, large and small, and blood flows through them and the limbs appear; even by the third month the foetus does not have a soul; by the fourth month the limbs are well developed; in the fifth the foetus begins to move and to grow more rapidly and the mother is 'witless', the ribs form and the mother has many afflictions; in the sixth skin and bones are formed; in the seventh the fingers and toes develop; in the eighth the foetus is fully developed with heart and blood; in the ninth the woman knows for certain whether she can bring forth; in the tenth the child must be born or it dies in the womb and the woman does not escape with her life; this happens most often on the eve of Tuesday'. Cameron's version of the fifth month is now incorrect; the same goes for Deegan's translation in: Deegan, 'Pregnancy and Childbirth', p. 23.

formation explicitly, but in the third month it says: *he biþ man butan sawle*, 'he [the fetus] is a person without a soul'. This is in fact a creative, but roundabout way of saying that the fetus is formed, because it now looks like a *man*, i.e., a human being. Although formation and 'animation' (*tribuit animam*) are usually both – but not always – situated in the third month in Vindicianus' embryology, the Old English Embryology explicitly states that the formed fetus is *butan sawle*, 'without a soul', i.e., 'not yet animated', in month three. Use of the word *sawol* indicates that the Old English translator must be translating the notoriously ambiguous word *anima*, as in the phrase *tribuit animam*, and that he opted for the meaning 'ensouled'.

Tribuit animam, 'animation', and movement are both situated in the fifth month in the Old English Embryology. 'Animation' is described as *he bið cwicu*, 'he is alive', and movement as *þæs byrþ[n]es lic on hire innoþe styrigende bið*, 'the body of the child inside her belly is moving'. Again, the Old English translator seems to be translating the phrase *tribuit animam*, but here he decided to translate *anima* as *cwic*, 'alive'. This means that Vindicianus' *tribuit animam* was translated in two different ways in this text: 'ensouled' and 'alive' or 'vivified'.[82] Apparently, the Old English translator made a conscious decision to render both aspects of the word *anima*. This, in turn, means that he considered a *man buton sawle* to be not yet *cwicu*, and that *anima ~ sawol* and *anima ~ cwic* must be interpreted as synonyms, albeit ambiguous synonyms, because *sawol* and *cwic* each translate one of the meanings of *anima*. The Old English translator was well aware of the ambiguity of the word *anima*, and it almost seems as if he took the easy way out by rendering both meanings, instead of chosing between 'ensouled' and 'alive'.

The Old English Embryology is in a class of its own. Not all of the mysteries of this text can be unravelled here, because an in-depth analysis is beyond the scope of this article. However, one more observation can be made. When rereading this text for the umpteenth time it struck me that the author must have been acquainted with many other scientific works on embryology and gynecology, because the text is littered with tidbits of information[83]. The

82. Unless, of course, *cwicu* is a translation of the word *vivus*, 'alive, vivified', a word that does not occur in Vindicianus' embryology, but which could have been introduced into the Latin source text through an intermediate source.

83. On science and learning in the tenth century, see, for instance: Lendinara P., 'The World of Anglo-Saxon Learning', in: M. Godden & M. Lapidge (eds.), *The Cambridge Companion to Old English Literature* (Cambridge, 1991), p. 264-281 and Lapidge M., 'Schools, Learning and Literature in Tenth-Century England', in: M. Lapidge, *Anglo-*

roundabout way the various stages of development of the fetus are explained and the added details suddenly sounded just like a teacher explaining to his class what embryology is all about, embellishing the text at hand as he went along. For instance, when the formed fetus is described as 'a person, not yet ensouled', it is as if you hear the teacher saying, 'be patient, I will discuss ensoulment presently'. This leads me to the conclusion that this text might originally have been in a teacher's or student's notebook.

The one thing that is really missing in the Old English Embryology is a calculation of formation, movement and the due date in days, like the one in chapter 19 of Vindicianus' *Gynaecia*, so that formation would be linked to the 'forty-day' period mentioned in other descriptive embryologies. This unique, early medieval embryology in the vernacular is atypical in its descriptions of fetal development, but the important stages, that is formation, 'animation' and movement, are all present.

2.4. Early medieval penitentials in the vernacular

Four early medieval penitential texts that were translated into the vernacular have come down to us. The earliest penitential written in the vernacular is the Old Irish Penitential, usually dated c. 800[84]. This penitential is interesting because it contains an article on abortion that describes four stages of fetal growth.

Old Irish Penitential

5.6. Banscal togluasti lind comperta o rufothaichedar i mbru teora bliadna .i. colleth.	5.6. A woman who aborts seminal liquid after it has become established in the womb, three years and a half [of penance].
Ma rofeolaigther isecht mbliadnai.	If the flesh has formed, it is seven years.
Math do thuidicht anim ind .xiiii. bliadna pendite.	If it is after the soul has entered it, fourteen years' penance[85].

Saxon Literature 900-1066 (London, 1993), p. 1-48 (first published in *Settimane di studio del Centro italiano di studi sull'alto Medioevo* 37 (1991), p. 951-998).

84. The Old Irish Penitential – also called *Paenitentiale Hibernicum* – was edited and translated by E.J. Gwynn, cf. Gwynn E.J., 'An Irish Penitential', *Eriu* 7 (1913), p. 121-195. See also: Gwynn E.J., 'Notes on the Irish Penitential', *Eriu* 12 (1938), p. 245-249. Gwynn's translation was reprinted (and revised) in McNeill & Gamer, *Medieval Handbooks of Penance*, p. 155-168 ff., and in Bieler, *The Irish Penitentials*, p. 258-277.

85. Gwynn, *Irish Penitential*, p. 166-167; the translation has been slightly emended in order to reflect the Irish text more precisely. Article 5.6 ends with a provision that mentions

The Old Irish Penitential was based on seventh- or early eighth-century Hiberno-Latin penitential sources that distinguish two stages of fetal development in their articles on abortion: 'liquid material' and 'body and soul'[86]. The Latin text in the Bigotian Penitential (c. 700-725) is as follows:

Paenitentiale quod dicitur bigotianum

4.2.2. Penitentia perditionis liquoris materiae infantis in utero mulieris .iii. anni in pane et aqua.

4.2.3. Penitentia perditionis carnis et animae in utero: .xiiii. peniteat in pane et aqua.

4.2.4. Praetium animae mulieris morientis de perditione carnis cum anima: .xiiii. ancelle[87].

The article on abortion in the Old Irish Penitential seems to be just as atypical as the Old English Embryology. It distinguishes four stages of fetal development, three of which are important in determining the penance for abortion. The stages mentioned are not new, but no other text mentions these stages together in this way. The first stage of fetal growth 'seminal liquid', that is, when the seeds mix, is not mentioned explicitly; it is implied in the text, because it necessarily precedes the stage in which 'seminal liquid' becomes 'established in the womb'. This stage is called 'milk' in Augustine's embryology[88]. The next stage, the stage in which the 'liquid material' is 'established in the womb', is probably a reference to implantation, and an indication that the woman was

the compensation due if the pregnant women dies as a result of the abortion: 'Ma atbela a mbanscal din togluasacht .i. bas coirp 7 anma cethora cumala .x. il-lúag a hanma du dia nó cethora bliadna deac pendi'; 'If the woman should die due to the abortion, that is, death of body and soul, fourteen cumals [must be given] to God as compensation of her souls, or fourteen years' penance'. A 'cumal' is 'a unit of value usually equivalent to a milk cow'. See also: McNeill & Gamer, *Medieval Handbooks of Penance*, p. 166, and Bieler, *The Irish Penitentials*, p. 272.

I would like to thank Andy Griffiths for providing me with a word-for-word translation of the Old Irish text.

86. The mid seventh-century *Canones Hibernensis* and the late seventh- or early eighth-century Bigotian Penitential or *Paenitentiale quod dicitur Bigotianum* are both closely related to the Old Irish Penitential. For the article on abortion in *Canones Hibernensis I*, see: Bieler, *The Irish Penitentials*, p. 160-161 (articles 6-10).

87. Bieler, *Irish Penitentials*, p. 228; '4.2.2. 'The penance for the destruction of the embryo [liquid material] of a child in the mother's womb, three years on bread and water. 4.2.3. The penance for the destruction of flesh and spirit in the womb: to do penance for fourteen years on bread and water. 4.2.4. The life price for a woman who dies of the destruction of flesh and spirit: fourteen female slaves', Bieler, *Irish Penitentials*, p. 229. See also: Wasserschleben, *Bussordnungen*, p. 441-460, Schmitz, *Bussbücher*, Bd. 1, p. 707-711, and McNeill & Gamer, *Medieval Handbooks of Penance*, p. 148-155.

88. See above and note 19.

now considered to be pregnant[89]. The third stage, 'formation of the flesh', describes the completion of formation: the fleshy parts of the body are no longer amorphous and the fetus now looks like a child – a description that seems to combine the stages 'flesh' and 'formation' in Augustine's embryology. The fourth and last stage is when the soul has entered the body; the Irish text uses the Old Irish word *ainim*, 'soul, life', a cognate of *anima* that is just as ambiguous – although it probably means 'soul' in this context. Here, the Old Irish text departs significantly from its Hiberno-Latin sources, where formation and ensoulment – *caro et anima* – are defined as one stage, and not as two different stages (cf. Bigotian Penitential 4.2.3). This Old Irish text is reminiscent of the embryology attributed to Augustine, which probably influenced its Hiberno-Latin sources. There are, however, important differences between Augustine's embryology and the Old Irish Penitential, the most striking differences being that the Old Irish text omits the stage 'blood' and defines 'animation' as a separate stage. The Old Irish Penitential describes three different abortion criteria and four stages of fetal development. It does, however, not say how long each stage takes, and this omission would not facilitate the use of this text in actual penitential practice.

Three Old English penitential texts have come down to us, but only one of them has an article on abortion that distinguishes between early term and late term abortion[90]. Articles 16.19.i – 16.19.k in the *Confessionale Pseudo Egberti* or Old English Confessional (c. 950-1000) seem to be a garbled remnant of a three-stage article on abortion like the one in the *P. Mixtum*.

> Confessionale Pseudo-Egberti
> 16.19.i. Wif seo ðe to æwyrpe gedo hire geeacnunga in hire hrife *** [and] cwelle ymbe XL nihta þæs ðe heo þam sæde onfo, *ær-ðon hit gesawlad wære,*
> swa se myrðra fæste III winter, [and] æghwylcere wucan þa twegan dagas to æfenes, [and] ðreo æfestenu.

89. I know of no other embryological text that mentions implantation, except the much younger Old Frisian embryology, cf. Elsakkers '*Her anda neylar*', p. 126.

90. The *Penitentiale Pseudo-Ecgberti* or Old English Penitential contains two articles that condemn abortion without making a distinction between early term and late term abortion, cf. Raith J. (ed.), *Die altenglische Version des Halitgar'schen Bussbuches; sog. Poenitentiale pseudo-Ecgberti*, Bibliothek der angelsächsischen Prosa 13 (Hamburg, 1933; rpr. Darmstadt, 1964), p. 16, p. 55. One of the articles on abortion in the Old English Penitential is also found in the Old English Handbook for a Confessor, cf. Fowler R., 'A Late Old English Handbook for the Use of a Confessor', *Anglia* 83 (1965), p. 1-34 at p. 21. On Old English penitential literature in general, see: Frantzen A.J., 'The tradition of penitentials in Anglo-Saxon England', *Anglo-Saxon England* 11 (1983), p. 23-56.

16.19.k. Gif heo beorðer forleose [*ante XL dies*], I gear oððe III æfestenu
[*si vero post XL dies ut homicida peniteat*][91].

However, a comparison with the *Paenitentiale Martenianum* (c. 850-900) shows us that the Old English text is related to the *P. Martenianum* and not to the *P. Mixtum*, and that it is not a three-tier article on abortion, but a two-tier article[92].

P. Martenianum.
43. Mulieres quae abortivum faciunt antequam animam habeat [*I annum (...) peniteant*]
et [*si*] postea id est post XL dies accepti seminis
ut homicidae peniteant id est III annos in IIII feria et VI et in tribus XL
mis [*hoc secundum canones decennium iudicatur*]
si ante XL dies I annum peniteant
[*si vero post XL dies ut homicida peniteat*][93].

The article in the *P. Martenianum* seems to be a clumsy attempt at combining both Theodorian articles on abortion (U.1.14.24; U.1.14.27)[94]. If we read this article *with* the missing text (added between brackets), it becomes much more intelligible. The *P. Martenianum* and the *Confessionale Pseudo Egberti* both contain the same textual corruptions. It is therefore likely that the articles on abortion in the Old English Confessional

91. Spindler R. (ed.), *Das altenglische Bussbuch (sog. Confessionale Pseudo-Egberti). Ein Beitrag zu den kirchlichen Gesetzen der Angelsachsen* (Leipzig, 1934), p. 184. The unemended Old English text can be translated as follows: 'Old English Confessional. 19.i. A woman who causes an abortion of the fetus (conception) in her womb, and kills [it] after forty days after she received the seed, before it was ensouled, shall fast as a murderess for three years each week on two days till evening [and] in three [forty-day] periods. 19.k. If she loses the child (fetus) [she shall fast] for one year or during three fasting periods [i.e. periods of forty days].'

The text between brackets was added by the editor. The italicized Latin text was added between brackets by the author of this paper from the Theodorian articles on abortion. The text between asterisks in line two belongs in line one at the three asterisks.

92. The *P. Martenianum* was edited by Hörmann, 'Bussbücherstudien VI', p. 358-483. It has recently been redated to the second half of the ninth century, cf. Meens, *Het tripartite boeteboek*, p. 54-55; see also: Hörmann W., 'Bussbücherstudien VIII. Das sog. poenitentiale Martenianum. Schluss', *Zeitschrift der Savigny-Stiftung für Rechtsgeschichte* 34. *Kanonistische Abteilung* 3 (1913), p. 413-492. Hörmann (p. 461 ff.) notes that this penitential was written in north(-eastern) France and influenced by the Anglo-Saxon and Irish penitential tradition (p. 464). Because this penitential was subsequently used as a source for the Old English Confessional, there must have also been a copy in England.

93. Hörmann, 'Bussbücherstudien VI', p. 379. The missing parts of the Theodorian articles on abortion have been printed in italics and put between brackets.

94. Spindler has also pointed out the correspondence between the two Theodorian articles on abortion, the *P. Mixtum* and articles 16.19.i-k in the Old English Confessional, cf. Spindler, *Das altenglische Bussbuch*, p. 51-52.

were translated from the *P. Martenianum*[95]. The Old English text became further corrupted, when a scribe copied *ær-ðon hit gesawlad wære*, 'before it was ensouled' (line 2) in the wrong place and left out the translation of the crucial phrase *ante XL dies* in 16.19.k[96]. Notwithstanding its textual corruptions, the Old English Confessional does help us understand the meanings of the word *anima* in penitential literature, because in this text the Old English scribe translated this ambiguous word quite unambiguously as *gesawlad*, 'ensouled'. Although the sources of both penitentials were well known on the continent, it is in England and Ireland that the first penitentials were translated into the vernacular. By a strange coincidence one of the manuscripts containing part of the Old English Confessional is the same manuscript that contains the sole copy of the Old English Embryology. The two Old English texts themselves do not seem to be related, but both Old English translations are probably products of the so-called Benedictine reform (tenth century) and possibly also originated in the same area.

3. Conclusion

The criteria employed to distinguish between early term and late term abortion in the early medieval legal texts discussed above are for the most part based on the earlier descriptive and normative sources discussed in the first section of this contribution. In the early medieval West, embryological terminology is mainly found in a normative context. The scope of the embryological terminology in normative texts is limited, because these texts only describe the stage of fetal development that is named as an abortion criterion. In the biblical law on abortion (Exodus 21.22-23) and in the teachings of the Church Fathers, the Aristotelian concept 'formed – unformed' is used as an abortion criterion. In his *quaestio LXXX* Augustine presents us with a number of synonyms for formation, when describing

95. Both texts open with the Theodorian article containing the *antequam animam habeat* criterion (U.1.14.24), and both leave out the penance for early term abortion (*I annum … peniteant*) – probably because the scribe planned to put it in later. The phrase on the Church council canons on abortion (*hoc secundum …*), which follows the penance for late term abortion in U.1.14.24 was also omitted. The first Theodorian article is followed by the first part of the second Theodorian article (U.1.14.27). The second part of U.1.14.27 with the criterion *post XL dies* to denote late term abortion is missing in both texts. The omissions both texts have in common prove their interdependence.

96. *Ær-ðon hit gesawlad wære* (line two) should follow *geeacnunga in hire hrife* in line one (see: asterisks), which makes more sense than assigning the standard three years' penance for murder, if the fetus is *not* yet ensouled.

the unformed fetus; for instance, *nondum anima viva*, *in carne nondum formata* and *informiter animatus*. It is probably largely due to Augustine's influential *quaestio* that the ambiguous term *anima* slowly developed into a Christian criterion for abortion with the meaning 'ensoulment'. Originally the word *anima* means 'life' or 'breath of life', as in the biblical law on abortion which demands *animam pro anima*, 'life for life', the penalty for murder. Later *anima* (*animatus*) also takes on the meaning of the word *animus*, 'soul', thus becoming an ambiguous word meaning 'life' ('alive'), 'soul' ('ensouled') or both. The word *anima* is not often used in descriptive embryology, and, when it is used, it probably retains its old meaning and denotes vivification, as in Vindicianus' embryology (*animam tribuit*).

In early medieval secular law, formation is the main criterion for abortion. The Visigothic law on abortion is an almost literal quotation of the Septuagint version of Exodus, with its 'formed – unformed' abortion criterion. Salian and Alamannic law both explain formation as the stage in which the gender of the fetus can be distinguished, and Alamannic law adds an extra explanation: *formatus in liniamenta corporis*, showing us that other sources, such as Macrobius and Augustine, were also known to the Old Germanic lawmakers. Gender distinction as an explanation of formation and as an abortion criterion is unique; we do not find it in other medieval texts[97]. The slightly younger Bavarian law on abortion (eighth century) is the only secular law that uses the abortion criterion *vivus*.

A sizeable portion of this study is devoted to the embryological terminology used in penitential literature. The penitentials were an important part of practical Christianity in the early medieval period, because one of the primary concerns of the Christian Church was teaching people how to be good Christians. The penitentials with their catalogs of sins and matching penances were one of the instruments of practical Christianity. We find a variety of embryological terminology in the articles on abortion that differentiate between early term and late term abortion, and, to a lesser extent, in the articles on sexual abstinence during pregnancy.

Theodore of Canterbury's late seventh-century penitential contains two abortion criteria: *XL dies* and *animam habuat*. Between the seventh and eleventh centuries, penitential authors regularly change the term used to

97. In classical embryology gender differentiation is mainly discussed in relation to the rate of development of a male and a female fetus – females are considered to develop more slowly.

denote the criterion for abortion, although ultimately, all the abortion criteria mentioned go back to the criteria formulated in Theodore's penitential. We find: *ante – post XL dies*, *antequam – post animam habuat*, *ante – post conceptum*, *antequam vivificatur – postquam conceptum spiritum*, and finally *non vivus – vivus*. The most frequently used criterion is *XL dies*.

The *XL dies* criterion in Theodore's penitential was still associated with formation in the late seventh century. *Animam habuat* probably meant 'has life' or 'is alive', as in the biblical law, but it is very likely that it was already in some way associated with ensoulment. Both Theodorian criteria are concerned with the difference between an amorphous being on the one hand and its transformation into a living human being on the other, so that both terms could more or less function as synonyms. This would explain the use of *XL dies* in both Theodorian articles. As Christianity took firmer root, however, the word *anima* acquired a more ambiguous meaning, and *animam habuat* and *XL dies* both came to definitely be associated with ensoulment. *XL dies* had also acquired a special meaning in Christian liturgy, because fasting periods in the liturgical calendar consisted of periods of forty days (*quadragesima*), the same forty-day periods during which the penances assigned in the penitentials had to be carried out. This link with Church liturgy is probably the reason that the number 'forty' is fixed in the articles on abortion in the penitentials; there is no variation whatsoever in the number of days, as there is in classical and late antique descriptive embryology.

The three-tier article on abortion in the ninth-century *P. Mixtum* shows us that not everyone associated *XL dies* with ensoulment. This article must have been devised by a cleric who was familiar with the rudimentary principles of descriptive embryology and who therefore knew that 'forty days' was associated with formation. The author of the *P. Mixtum* sought to disentangle the abortion criteria *XL dies* and *animam habuat* by distinguishing three abortion criteria: *ante XL dies*, *post XL dies* and *animatus*. Although the article in the *P. Mixtum* is found in the *P. Ps. Theodori* and in Regino of Prüm's *Libri Duo De Synodalibus Causis et Disciplinis Ecclesiasticis*, it was probably felt to be an impractical abortion article, because distinguishing three different stages of fetal development would involve too much guesswork. Later the three-tier article on abortion in the *P. Mixtum* was subtly rejected by Burchard, who wrote a two-stage article on abortion of his own. Burchard incorporated both meanings of the word *anima* (*vivificatio*, 'vivification' and *conceptus spiritus*, 'ensoulment') into this article, without using the ambiguous word *anima*. His contemporary Fulbert's choice of the simple word *vivus* may have been based on more practical considerations and suggests that there may have been an ongoing debate on the meaning of the word *anima*.

The three vernacular embryological texts compiled in the pre-Salernitan period were written in insular scriptoria; one is descriptive and two are translations of penitential texts. The key word *anima* is rendered differently in all three texts, thus clearly confirming its ambiguity. Like Burchard, the Old English Embryology gives us a double rendering of the word *anima*, translating it as 'ensouled' in the third month and 'alive' in the fifth month. The Old Irish Penitential with its four-tier article on abortion does not resolve the ambiguity of the word *anima*, and translates it with the word *ainim*, 'soul, life', which is just as ambiguous. The Old English Confessional has a two-tier article on abortion and, although the text is garbled and corrupt, it opts for clarity regarding the abortion criterion by choosing the Old English word *gesawlad*, 'ensouled' to render *anima*[98].

Although we would expect authors devising abortion criteria to consult descriptive texts on embryology, or at least be acquainted with the information in Pliny's or Isidore's encyclopedia, only a few early medieval normative texts betray any knowledge of descriptive embryology. Augustine's and Macrobius' description of formation using the word *lineamenta* recurs in the early seventh-century Alamannic laws. The Septuagint was the source used in Visigothic law, the only normative text, besides Alamannic law, with the word *formatus*. The penitentials were hardly influenced by descriptive embryology; only *XL dies* was originally taken from descriptive embryology. The sources for the abortion criterion *animam habuat* are undoubtedly the Septuagint version of Exodus 21.22-23 and patristic writings on the soul and on abortion, such as Augustine's *quaestio*, but the author of the *Iudicia Theodori* may also have been influenced by other classical sources available in seventh-century Anglo-Saxon England. Under influence of Christian moral literature *XL dies* and *animam habuat* developed into the new Christian abortion criterion of ensoulment. All the other terms chosen to denote the difference between early term and late term abortion in the penitentials are in some way related to the criterion *animam habuat*. The variation in terminology

98. Centuries later, the word *anima* was still felt to have a double meaning. In the fourteenth-century Old Frisian translation of Vindicianus' embryology that was added to Old Frisian abortion law in order to help the authorities determine the fine for abortion, the Old Frisian translator resolved the ambiguity of the word *anima* by translating both aspects of the word: *om,* 'breath (of life)' and *sele*, 'soul', cf. Elsakkers '*Her anda neylar*', passim.

is probably due to the ambiguity of the word *anima*, and this is probably the reason the word *vivus* was chosen by Burchard and Fulbert[99].

Movement, also an important stage of fetal development in descriptive embryology, is not used as an abortion criterion, even though it is occasionally associated with abortion by some Christian authors. The only reference to movement that shows that it was acknowledged as an important stage of fetal development is in the penitential articles on sexual continence during pregnancy. But, as we saw above, there is no direct link with descriptive embryology.

Medieval embryological texts are fraught with inconsistencies. Christianity added to the confusion by introducing ensoulment as an abortion criterion. In this article I hope to have shown that it is difficult to pinpoint the sources used in the normative texts, and that these texts only occasionally seem to betray a knowledge of descriptive embryology. Christian moral writings were the main sources for the normative texts. Although the abortion criteria used in the various early medieval legal texts discussed above are not exactly the same, the criteria used undoubtedly all refer to approximately the same stage of fetal growth, the stage in which the fetus looks like or feels like a living human being, that is, the stage in which committing abortion would imply killing a human being. Whether these abortion criteria could actually be implemented is open to debate. The only seemingly practicable criteria are *XL dies* and *vivus*, because 'forty days' sounds like a rational criterion, and because 'aliveness' would be one of the reasons for a woman to definitely consider herself pregnant. This brings us to the conclusion that these texts on abortion – devised no doubt by men – could not be used in actual judicial practice without relying on women's testimonies to decide whether the abortion was early term or late term. Birth and babies were women's business and only women would be able to determine the approximate age of the aborted fetus. It does not seem likely that women would implicate themselves or their 'sisters' as murderesses[100].

Utrecht

99. Besides its connection with *anima*, there may also be a link with Bavarian law – which also has the criterion *vivus*, because Burchard's penitential and Bavarian law both originated in German-speaking areas.

100. In late medieval Old Frisian law, women were appointed as witnesses in abortion cases; they determined the stage of fetal growth. The fact that younger versions of these Old Frisian abortion laws stipulate that the female witnesses were to be accompanied by a priest seems to speak volumes, cf. my '*Her anda neylar*', p. 135.

Erwin HUIZENGA

UNINTENDED SIGNATURES: MIDDLE DUTCH TRANSLATORS OF SURGICAL WORKS

1. The importance of written surgical knowledge: from Latin to the vernacular[1]

Around 1300 there lived in the small town of Ypres, in Flanders, a surgeon who set down his professional expertise in an original Middle Dutch work. His name was Jan Yperman, and his work was simply called *Cyrurgie* ('Surgery')[2]. Modern medical-historical research in the Netherlands considers him to be the first surgeon to write his work in his vernacular. But this does not mean that in doing so, Yperman became detached from the Latin tradition, which up until then had dominated surgery. The question is, what it meant to write such a text in the vernacular. The beginning of Yperman's text reads:

> Hier beghint die *Cirurgie* des meesters Ioannis Ypermans, den welke dat hi uten Latine, ende ute syn selves verstandenisse ende leringe ende uut sine goeder geloefder werken dit troc hi, ende maecte in dyetscher talen de welke dat hi bestont te maken om zyns selves zone binnen zyns selves live.
>
> 'Here begins the Surgery of master Jan Yperman, which he compiled both from Latin sources, from his own knowledge and experience, and from authoritative works, and which he wrote in Dutch, and he made it for the benefit of his son[3]'.

So, Ypermam wrote in Middle Dutch, but, as he mentions, he used Latin sources as well as his own experience. The following complicates

1. I want to thank Prof. dr. Michael McVaugh for his kind and useful remarks and for checking the English.

2. Edited by Van Leersum E.C., *De «cyrurgie» van Meester Jan Yperman. Naar de Handschriften van Brussel, Cambridge, Gent en London* (Leiden, 1912). An Italian translation can be found in Tabanelli M., *Jehan Yperman. Padre della Chirurgia Fiamminga*, Bibliotheca della "Rivista di storia delle scienze mediche e naturali" 16 (Florence, 1969). More information about Yperman's life and work in Huizenga E., *Tussen autoriteit en empirie. De Middelnederlandse chirurgieën in de veertiende en vijftiende eeuw eeuw en hun maatschappelijke context*, Artesliteratuur in de Nederlanden 2 (Hilversum, 2003), p. 133-143, with many references.

3. Van Leersum, *De «cyrurgie»*..., p. 5.

this statement a bit further. The text of the *Surgery* is transmitted in five manuscripts[4]. In two of these, this Middle Dutch introduction is preceded by the corresponding Latin words[5]. The Brussels manuscript, for instance, reads: 'Hic incipit cyrurgia magistri Johannis domini Ypermans quam ipse compilavit et in teutonico redigit filio suo'. And in the Cambridge codex we see:

> Hic est practica et doctrina composita a magistro iohanne ypermanni quam ipse tractavit in flamingo ad utilitatem filij sui in tempore vite sue sane et voluit quod ipse haberet aliquid de opere suo et doctrina sua a multis magistris scilicet de lanfranco et a quatuor magistris de salerno et a galieno et a rolando et a rogero et a brutto et a raso et a magistro hugone de luckes et a magistro albucaso.
>
> 'This is the practice and learning put together by master Jan Yperman which he himself had treated in Flemish, on behalf of his son, at a time in his life when he was still sane. And he (i.e. Yperman senior) wanted that he (Yperman junior) should have something of his work and learning, (brought together) from many (learned) masters like Lanfranc, the Four Masters of Salerno, Galen, Roland of Parma, and from Roger Frugardi, Bruno Longoburgensis, Rhazes, master Hugo of Luca, and master Albucasis[6].'

We see in these manuscripts a double incipit, one in Latin, followed by (roughly) the same in Middle Dutch. This has led some scholars in the past to think that in spite of everything perhaps Yperman originally díd write in Latin[7]. This, however, is not the case[8]. But it strikingly illustrates how complex the relations were that existed between Latin and vernacular in late-medieval culture. Certainly, they were different linguistic systems, but they were not completely separated from each other.

4. They are: Brussels, Royal Library Albert I, 15.624-41, f. 107r-147r; Cambridge, St. John's College, A 19, I, f. 1r-81v; Ghent, University Library, ms. 1273, f. 2ra-125ra; Lincoln, Cathedral Chapter Library, ms. 127, f. 46r-62v (*partim*, Cap. 1-11); London, British Library, Harley 1684, f. 143r-173r. Descriptions of all these manuscripts (and many more) can be found in Jansen-Sieben R., *Repertorium van de Middelnederlandse artesliteratuur* (Utrecht, 1989), resp. p. 258: n°. B 750, 2.18.; p. 281: n°. C 20, 2.; p. 305: n°. G 330, 2.1.; p. 384: n°. L 680, 2.9.; and p. 392: n°. L 890, 2.6, together with numerous references to other catalogues.

5. I.e., the Brussels and the Cambridge manuscripts.

6. Van Leersum, *De «cyrurgie»...*, p. 5. The references in the Cambridge manuscript to the authors that Yperman had read and used are absent in the other codices.

7. See Häser H., *Lehrbuch der Geschichte der Medicin und der epidemischen Krankheiten* (Jena, 1875). Vol. I: *Geschichte der Medicin im Alterthum und Mittelalter*, p. 769; Pagel J.L., *Handbuch der Geschichte der Medizin* (Jena, 1905). Vol. I: *Altertum und Mittelalter*, p. 738; and Neuburger M., *Geschichte der Medizin* (Stuttgart, 1911). Vol. II, p. 519.

8. The continuation of the incipit mentions explicitly that Yperman wrote in Dutch for his son who did not understand Latin – see the quoted passage hereafter.

Medieval surgeons showed a far greater interest in books in general, and in written medical-surgical knowledge in particular, than one might expect, considering that the command of manual skills was the most important part of their trade. This interest in books is undoubtedly related to the surgeons' continual efforts to get their profession acknowledged as a *science*, and to affiliate themselves with the learned classical literary tradition of the physicians, the *doctores medicinae*[9].

This article presents a study of the way in which highly specialized information from the Latin surgical tradition was transmitted by translators in Middle Dutch surgical works. I will mainly use examples in Middle Dutch, but sometimes I will make a brief excursion to other vernaculars as well. Focus will be on the motives of the intermediaries (in most cases translators, but sometimes also compilers, and, of course, sometimes a combination of both): why did they translate, select or compile from the large reservoir of surgical knowledge as they did? What can be learned about their personality, level of skill or intended audience? When one takes a close look at the Middle Dutch surgical manuscripts of the fourteenth and fifteenth centuries, the changes in these surgical works that can be seen during the processes of translation and vernacularisation may be considered as some sort of mark, a signature of the maker that was not intended as such.

2. Knowledge transmission in the Netherlands: a chronological overview

Scientific-utilitarian knowledge was transmitted in the late Middle Ages through a whole range of texts, in the Netherlands often designated as *artes*-literature. The oldest extant of such *artes*-writings in Middle Dutch date from the first half of the thirteenth century. They are marginal notes or page-filler in Latin manuscripts. One manuscript for example contains a list of Latin plant names, followed by their Dutch synonyms[10]. Another

9. See extensively Huizenga, *Tussen autoriteit en empirie...* For a briefer introduction to this subject, see Huizenga E., 'Middelnederlandse chirurgieën en hun maatschappelijke context. Een introductie op het belang van oud-Nederlandse, medische teksten', *Literatuur* 16 (1999), p. 273-283. See also Huizenga E., *Een nuttelike practijke van cirurgien. Geneeskunde en astrologie in het Middelnederlandse handschrift Wenen, Österreichische Nationalbibliothek, 2818*, Middeleeuwse Studies en Bronnen 54 (Hilversum, 1997), p. 302-305; and Huizenga E., 'De diepe kloof. Over het onderscheid tussen ideaal en realiteit van de middeleeuwse chirurg, in het bijzonder in de Nederlanden', *Millennium* 14 (2000), p. 83-84.

10. Edition: 'Plantenglossarium uit Vlaams Artesië of Frans-Vlaanderen', in: M. Gysseling, *Corpus van Middelnederlandse teksten (tot en met het jaar 1300). M.m.v. en van*

early example is the so-called *Nederbergse recipes*, Dutch medical recipes copied on the empty pages of a manuscript containing Latin medical treatises[11]. Finally, a monthly health regimen, written in Middle Dutch, was added in the margins and open spaces of a Latin calendar that was written in 1253[12]. In these manuscripts, Middle Dutch still played a marginal role when compared to the Latin treatises. But their very existence suggests that at that point Middle Dutch had a written status that was comparable to that of Latin. The texts show the beginning of a new development that gradually led to the convergence of Latin and Middle Dutch. They make it clear that a vernacular such as Middle Dutch was becoming accepted as a new medium of written communication.

After 1250, a different type of Middle Dutch *artes*-literature arose. The purpose of these new texts was to inform or educate a general public that had little or no formal schooling in Latin. There is, for example, the first vernacular translation of Thomas of Cantimpré's Latin encyclopaedia on natural history[13], in Middle Dutch titled *Der naturen bloeme* ('The flowers of nature'), by the Flemish poet Jacob van Maerlant[14]. Another thirteenth-century treatise is a rhymed Middle Dutch cosmological text, called the *Natuurkunde van het geheelal* ('The nature of the cosmos')[15]. These tracts, and others, were all written using verse.

Thus, in this period we see two developments in the way in which knowledge was transmitted in the vernacular. On the one hand, there were scholars familiar with Latin who started to use Middle Dutch as an alternative medium for writing. On the other hand there were educational texts for laymen. In these tracts, the learned Latin-educated poets like Maerlant mediated between the Latin and the vernacular tradition. Of course, Middle Dutch

woordindices voorzien door W. Pijnenburg (The Hague, 1980), Series II: *Literaire handschriften*. Vol. 1: *Fragmenten*, p. 333-335.

11. 'Nederbergse geneeskundige recepten', in: Gysseling, *Corpus van Middelnederlandse teksten...*, p. 343-344.

12. 'Noordlimburgse gezondheidsregels', in: Gysseling, *Corpus van Middelnederlandse teksten...*, p. 345-347.

13. Thomas Cantimpratensis, *Liber de Natura Rerum. Editio princeps secundum codices manuscriptos*, ed. H. Boese, Vol. 1: *Text* (Berlin [etc.], 1973).

14. This text was edited in Verwijs E., *Jacob van Maerlant's Naturen Bloeme*, Bibliotheek van Middelnederlandsche Letterkunde (Arnhem, 1980); *'Der Naturen Bloeme'*, in: Gysseling, *Corpus van Middelnederlandse teksten...*, Vol. 2, p. 16-435. The literature about Jacob van Maerlant is massive. For the most complete work on his life and works, see Van Oostrom F., *Maerlants wereld* (Amsterdam, 1996), with numerous references.

15. Edited in Jansen-Sieben R., *De natuurkunde van het geheelal. Een 13de-eeuws Middelnederlands leerdicht* (Diss. Brussels, 1968), vol. II, p. 274-433.

is not the only vernacular where such processes can be observed. They took place everywhere in Europe.

In the fourteenth century, the importance of the vernacular as the medium of written *artes*-literature increased. One reason was that, because of a different valuation of the status of knowledge, *artes*-literature became more and more popular among different strata of the population[16]. A new and large lay public felt a growing need for information or instruction on secular knowledge. This led to an increase in literacy and to a demand for more specialised vernacular texts. But there were also professionals and other literate intellectuals who wished to improve their knowledge.

The fourteenth century also marks the beginning of a tradition of surgical texts in the vernacular. From the twelfth century onwards, there had been a strong Latin tradition of books on medicine and surgery. A few words on this Latin tradition may be appropriate here.

An important centre for the study of surgery was Bologna, where many great surgeons studied or practised, like Bruno Longoburgensis[17], Theodoric Borgognoni[18], and William of Saliceto[19]. At the end of the thirteenth century, the centre of both academic teaching and the practice of surgery moved from Northern Italy to Germany, the Netherlands, and in particular to Paris, under the influence of the great surgeon Lanfranc of Milan[20]. Lanfranc's work greatly stimulated the professionalization of

16. Haage B.D., 'Anwendungsmöglichkeiten und bisherige Anwendung von philologisch-historischen Methoden bei der Erforschung der Fachsprachen der Artes', in: L. Hoffmann, H. Kalverkämper & H.E. Wiegand (eds.), together with C. Galinski & W. Hüllen, *Fachsprachen. Ein internationales Handbuch zur Fachsprachenforschung und Terminologiewissenschaft. Languages for Special Purposes. An International Handbook of Special-Language and Terminology Research* (Berlin, 1998), vol. 1, p. 271.

17. About Bruno's life and work, see Keil G., 'Bruno von Longoburgo', in: K. Ruh e.a. (eds.), *Die deutsche Literatur des Mittelalters. Verfasserlexikon* (Berlin/New York[2], 1978-), vol. 1 (1978), col. 1070-1071; Tabanelli M., *Un chirurgo Italiano del 1200: Bruno da Longoburgo* (Florence, 1970). About the Middle Dutch translation of Bruno's work, see Huizenga, *Een nuttelike practijke...*, p. 159-176; and Huizenga, *Tussen autoriteit en empirie...*, p. 113-116.

18. Gurlt E., *Geschichte der Chirurgie und ihrer Ausübung* (Berlin, 1898), vol. I, p. 740-742; Wallace W.A., 'Borgognoni of Lucca, Theodoric', in: C.C. Gillispie e.a. (eds.), *Dictionary of Scientific Biography* (New York, 1970-1980), vol. 2 (1970), col. 314-315. For an English edition of his Latin surgery, see Campbell E. & Colton J. (eds.), *The Surgery of Theodoric. Ca. A.D. 1267* (New York, 1955 and 1960), 2 vols.

19. On William, see Gurlt, *Geschichte der Chirurgie...*, vol. 1, p. 754-765; Agrimi J. & Crisciani C., 'The science and practice of medicine in the thirteenth century according to Guglielmo da Saliceto, Italian surgeon', in: L. García-Ballester e.a. (eds.), *Practical Medicine from Salerno to the Black Death* (Cambridge, 1994), p. 60-87.

20. See for his life and works Gurlt, *Geschichte der Chirurgie...*, vol. 1, p. 765-791; Keil G., 'Lanfrank von Mailand', in: Ruh e.a. (eds.), *Die deutsche Literatur...*, vol. 5 (1985), col. 560-572. On the Middle Dutch translation of Lanfranc's *Chirurgia magna*, see:

surgery. The French surgeons Henri de Mondeville[21] and Guy de Chauliac, whose *Chirurgia magna* dates from 1363[22], also wrote important textbooks. Many of the works mentioned above were translated, and these translations were, in turn, adapted and summarized.

The Flemish surgeon Jan Yperman played an important role in this development with regard to the Middle Dutch tradition. As we saw at the beginning, his handbook on surgery provided valuable information, based on a thorough understanding of the Latin surgical tradition, as well as his own practical experience. What makes the opening of his *Surgery* all the more interesting is its continuation, where Yperman explains his decision to write in Middle Dutch:

> (…) ende hi maecte dit werck in dietsche om die minne van zyn zoen die soe ionc was dat hi hem niet wel verstant in grammarien al zoe die boeken leerden die zyns vaders waren ende dair ute dat hi wrochte ende die hi gehoirt hadde lesen ende dair hi ute zyn werck voldede. Ende die bouken en hadden sinen zoen niet te goede geworden. Ende dair om maecte hi dit in Vlaemscher talen om dat hi begeerde dat syn soen profiteerde dair mede ende hem bleve van synre leringe die men zeer prisende is.
>
> '(…) and he wrote this work in Dutch for the love of his son, who was so young that he did not understand the grammar of the Latin books that belonged to his father and from which he (i.e. Yperman senior) worked, and which he (Yperman junior) had heard reading from, and which he (Yperman senior) used to operate. And these books could not fully be used by his son. For this reason he wrote in the Flemish language, because he wished his son to profit from it and take to heart the valuable lessons that it provided[23]'.

Reynaert J., 'Over medische kennis in de late Middeleeuwen. De Middelnederlandse vertaling van Lanfrancs *Chirurgia magna*', *Millennium* 13 (1999), p. 21-30; Huizenga E. & Reynaert J., 'De Middelnederlandse vertalingen van de *Chirurgia magna* van Lanfranc van Milaan. Een vergelijkende editie van de preliminaire hoofdstukken', *Verslagen en Mededelingen van de Koninklijke Akademie voor Nederlandse Taal- en Letterkunde* n.r. (2002), p. 229-369; Huizenga, *Tussen autoriteit en empirie…*, p. 120-126.

21. About Henri's life and work, see Gurlt, *Geschichte der Chirurgie…*, vol. 2, p. 34-77; Bullough V.L., 'Henry of Mondeville', in: Gillispie e.a. (eds.), *Dictionary of Scientific Biography…*, vol. 6 (1972), col. 276-277. Edition of the Latin text of the *Cirurgia*: Pagel J.L., *Die Chirurgie des Heinrich von Mondeville (Hermondaville). Nach Berliner, Erfurter und Pariser Codices zum ersten Male herausgegeben* (Berlin, 1892). About the Middle Dutch translation: Huizenga, *Tussen autoriteit en empirie…*, p. 126-128.

22. About Guy's life and works, see Gurlt, *Geschichte der Chirurgie…*, vol. 2, p. 77-107; Keil G., 'Guy de Chauliac (Guigo de Chaulhaco)', in: Ruh e.a. (eds.), *Die deutsche Literatur…*, vol. 3 (1981), col. 347-353. For an edition of the Latin text of the *Chirurgia magna*, see: McVaugh M.R., *Guigonis de Caulhiaco (Guy de Chauliac). Inventarium sive Chirurgia Magna*, vol. 1: *Text*; vol. 2 (with M.S. Ogden, ed.): *Commentary* (Leiden [etc.], 1997). The Middle Dutch translations: Huizenga, *Tussen autoriteit en empirie…*, p. 128-133.

23. Van Leersum, *De «cyrurgie»…*, p. 5, n. 1.

Yperman wanted to share his knowledge and experience with a new generation of medical professionals. His work shows that there were two categories of professionals. One category (to which Yperman belonged) was academically trained in Latin and had access to a vast reservoir of highly specialized knowledge. The other category was literate, but only in the vernacular. They had at best enjoyed partial academic schooling, and in most cases had received only practical training from a tutor. The production of professional handbooks in the vernacular, by writers like Yperman, made Latin knowledge available for these medical workers. At the same time their professional knowledge and skills were improved. Thus Latin as the medium of *artes*-literature was gradually replaced by Middle Dutch. At this point, the vast majority of the surgical texts was written in the vernacular.

In the fifteenth and sixteenth centuries, when manuscripts came to coexist with printed books, there was a growing demand for vernacular *artes*-literature. Popularised encyclopaedic literature for a general public was produced along with the more specialized, professional handbooks. If we compare the Middle Dutch *artes*-literature in this period in general, and medical-surgical literature in particular, to the marginal vernacular scientific knowledge of the thirteenth century, it is clear that there was a sharp increase in literacy among the lay population, and that the vernacular as the medium for scientific writings had achieved a really strong position. But in spite of all this, for a long time Latin did not disappear completely in these texts. It was restricted, but it continued to play an important role based on the long academic tradition in which it rooted.

3. Some principles and characteristics of translations, adaptations, and compilations of medical-surgical texts

In no other genre do the relations between Latin and Middle Dutch seem to have been as complex and intense as in that of science. The growing urban public of the late Middle Ages wanted to know how the world and nature around them functioned, and how people fitted in this mirror of God's work. Latin, however, was not suited to describe these natural phenomena properly for them. Being the language of the church, it had for centuries been the common language of European clergymen, scholars and other intellectuals. As this new public had not been educated in Latin, the need for translations and adaptations arose. Such processes of translation had to cope with their own problems: how to translate Latin words, terms and conceptions into a vernacular with a still inadequate vocabulary? New words had to be developed, and the function of existing ones had to be adapted.

3.1. Translation

In the Middle Ages, people were very aware of the great differences that could exist between translations, adaptations and compilations. For medical-surgical works such interventions in the sources were the rule rather than the exception. Already in the thirteenth century, Bonaventura wrote about the four ways to 'make a book' (*faciendi librum*), making the following distinction between scribes, compilers, commentators and authors:

> The way to make a book is fourfold. [1] Somebody copies something from another person, without adding or changing anything; and such a person is merely called *scribe*. [2] Somebody copies something from another person and adds something to this, but not from himself; and such a person is called *compiler*. [3] Somebody copies something from another person and adds something from himself, whereby the other person's words are most important, and his own words are an addition to prove something; and such a person is called a *commentator who is no author*. [4] Somebody writes himself [something new] and adds something from another person, whereby his own words are most important, and the words of the other person are an addition to confirm something; such a person must be called *author*[24].

A single translator could unite all of these functions except perhaps those of the *author*, with the difference being that an extra round of adaptation preceded, that is, the conversion of his source into another language. In most cases, the translator wanted to be as loyal as possible to his source, and in doing so paid little attention to his style[25]. Depending on what he

24. Bonaventura, *In primum librum sententiarum*, proëm. quaest. iv; ed. Bonaventura, *Opera omnia iussu et auctoritate Bernardini a Portu Romatini. Edita studio et cura PP. Colegii a S. Bonaventura*. Vol. I: *Commentaria in quattuor libros Sententiarum* (Florence, 1882), p. 14, col. 2: '(...) quadruplex est modum faciendi librum. Aliquis enim scribit aliena, nihil addendo vel mutando; et iste mere dicitur scriptor. Aliquis scribit aliena addendo, sed non de suo; et iste compilator dicitur. Aliquis scribit et aliena et sua, sed aliena tamquam principalia, et sua tamquam annexa ad evidentiam; et iste dicitur commentator non auctor. Aliquis scribit et sua et aliena, sed sua tamquam principalia, aliena tamquam annexa ad confirmationem et debet dici auctor'. About this passage see also Parkes M.B., 'The Influence of the Concepts of *Ordinatio* and *Compilatio* on the Development of the book', in: J.J.G. Alexander & M.T. Gibson (eds.), *Medieval Learning and Literature. Essays Presented to Richard William Hunt* (Oxford, 1976), p. 127-128.

25. Grant E., *The Foundations of Modern Science in the Middle Ages. Their Religious, Institutional, and Intellectual Contexts*, Cambridge History of Science (Cambridge, 1996), p. 25-26.

wanted to do with it, the original could be treated in several ways. First of all, the translator could make a literal word-by-word translation; a second possibility was to make a translation which was in its content true to the original text; and finally he could focus completely on the content and in doing so give up on a literal similarity[26]. Often there was a conflict between the desire to be as transparent as possible, and striving for loyalty towards the words; in most cases the need for transparency prevailed[27].

With their translations, the translators normally wanted to stay as close to their own language as possible. In trying to accomplish this, they sometimes named this language and in doing so signified their origin – part of an unintended signature. A scribe of the text of the *Jonghe Lanfranc*[28], for instance, clarifies a Latin expression in the following way:

> (...) ende comt dese aposteme ofte canckere int tansichte, soe heetent die mesters *noli me tangere* (dats te segghen in Wlamsche: ne gheraect mi niet).
>
> '(...) and when such a swelling or cancer appears in the face, then the masters call it *noli me tangere* (that is in Flemish: don't touch me)[29].'

That, even between the distinct regional variants of the Middle Dutch in the borderlands, a Babylonian confusion of tongues lurked, is shown by the following remark in ms. Heverlee, Parkabbey, Cod. 4, in a medical-surgical vademecum:

> Eyn gewis stuck dat mannichmael probiert is vur den roeden bueck, dat men noempt in Brabant *dat roet menisoen*.
>
> 'A certain matter which has been proven many times against the red belly, which in Brabant is called the red diarrhoea[30].'

26. See for the English situation Ellis R., 'The Choices of the Translator in the Late Middle English Period', in: M. Glasscoe (ed.), *The Medieval Mystical Tradition in England. Papers Read at Dartington Hall, July 1982* (Exeter, 1982), *passim*; and Renevey D., 'The Choices of the Compiler. Vernacular Hermeneutics in *A Talkyng of þe Loue of God*', in: R. Ellis, R. Tixier & B. Weitemeier (eds.), *The Medieval Translator. Traduire au Moyen Age*. Vol. 6: *Proceedings of the International Conference of Göttingen (22-25 july 1996). Actes du Colloque international de Göttingen (22-25 juillet 1996)* (Turnhout, 1998), p. 244 ff.

27. Burnley J.D., 'Late Medieval English Translation: Types and Reflections', in: R. Ellis (ed.), *The Medieval Translator. The Theory and Practice of Translation in the Middle Ages. Papers Read at a Conference held 20-23 August 1987 at the University of Wales Conference Centre, Gregynog Hall* (Cambridge, 1989), p. 52.

28. More about this text below in this paragraph.

29. *Jonghe Lanfranc*; ms. Cambridge, St. John's College, A 19, II, f. 12va; ed. Müller R., *Der 'Jonghe Lanfranc'*, Altdeutsche Lanfranc Übersetzungen 1 (diss. Bonn, 1968), p. 95. This explanation is peculiar for this scribe; in the other textual witnesses of the *Jonghe Lanfranc*, it does not appear.

30. Ms. Heverlee, Parkabbey, Cod. 4, f. 32r-v; ed. Braekman W.L., 'Een Middelnederlands medisch vademecum uit het handschrift van de Norbertijnerabdij te Heverlee',

There were also gradual changes between the different types of written transmission of knowledge. Especially in the case of the Middle Dutch surgical tracts, it is often hard to determine whether we are dealing with a translation, adaptation or compilation. In the last two cases changes are obviously intentional. Such alterations could range from small modifications to more drastic changes (by means of omissions, or by means of additions)[31]. Particularly when the translator/compiler wanted to make major changes in his text (for instance by removing large parts of his source), the differences between compilation and an original, authorised new text could be very subtle[32].

It is very rare that two scientific manuscripts can be identified where one is the Latin source and the other is demonstrably its translation in a copy written by the translator himself. In such a case, it would be possible to determine the precise interventions and motives of the translator, without the intervention of one or more scribes. There are no Middle Dutch examples of this known, but there are two manuscripts in the British Library that can be mentioned here[33].

Regionally, or even locally, the reception of Latin surgeries in the vernaculars sometimes showed show great differences. To explain this, William Crossgrove has proposed that the translators of surgeries wanted to translate texts that were as recent as possible[34]. Once this was done, such a relatively new text was copied again and again, while older texts received less attention and thus had less chance of being translated[35].

Verslagen en Mededelingen van de Koninklijke Academie voor Nederlandse Taal- en Letterkunde (1971), p. 300.

31. Renevey, 'The Choices...', p. 235-243.

32. Renevey, 'The Choices...', p. 233.

33. Mss. London, British Library, Harley 3371 (Latin source), and Sloane 767 (English translation). These manuscripts contain, among others, texts by John of Arderne, an English surgeon who wrote his works in Latin in the second half of the fourteenth century. By comparing both codices meticulously, Peter Murray Jones arrives at the seemingly irrefutable conclusion that we are dealing with translations which were written by the translator himself; these translations were at the same time adaptations, through which the translator met the needs of his public, young students of surgery. Jones P.M., 'British Library MS Sloane 767: A Translator's Holograph', in: L.L. Brownrigg (ed.), *Medieval Book Production. Assessing the Evidence. Proceedings of the Second Conference of The Seminar in the History of the Book to 1500. Oxford, July 1988* (Oxford, 1990), p. 27-29.

34. I refer in relation to this to Jan Yperman. When he wrote his *Cyrurgie*, he already included information from the *Lilium medicinae*, a famous work by the physician Bernard de Gordon, working in Montpellier. This *Lilium medicinae* was written in 1305 or shortly afterwards, and must therefore almost immediately have gotten into the hands of Jan: cf. Van Leersum, *De «cyrurgie»...*, p. 182b, VII(58). See Huizenga, *Tussen autoriteit en empirie...*, p. 190.

35. Crossgrove W., 'The Vernacularization of Science, Medicine, and Technology in Late Medieval Europe: Broadening our Perspectives', *Early Science and Medicine* 5 (2000), p. 55.

3.2. Adaptation and compilation

One of the most important consequences of the way that scientific and utilitarian knowledge was transmitted from the twelfth century onwards was that people began systematically to try to collect all new knowledge available, and to classify it. The twelfth and thirteenth centuries were the period of the great encyclopaedias and synopses.

However, the fourteenth and fifteenth centuries increasingly witnessed a preference for shorter texts that gave as concisely as possible the essence of what was important for the user or reader. It is this aspect to which the fifteenth century Dutch surgeon Jan Bertand refers in his *Rogerglosse*, when he explains his reasons for abbreviating his source:

> Dat ic dusgedane groet werck ende grote scone leringhe (...) hebbe getoecht met dus corten woerden, ende die bescreven hebbe, dat en was niet om dat my der pinen verdroet, maer het was omme die ghene te voerderen die nu sijn ende die hierna wesen sullen van deser sciencien. Want beter es te verstane een corte redene ende openbaer, dan een lange die becommert es.
>
> 'The reason why (...) I wrote such a great work and very useful lesson in so few words was not because I didn't like the efforts, but to help those who'll have to use this science, now and in the future. For it is better to understand a short and clear text, than a long one which is hard to grasp[36].'

Naturally, once the Middle Dutch as a medium for *artes*-literature was fully accepted (see paragraph 2 above), many new texts and new combinations of texts were produced. Every individual user started to gather important knowledge in his (or her) field of expertise. Everyone had his own reasons for creating a compilation, and some of them made those reasons public[37].

36. Jan Bertand, *Rogerglosse*, III.introd.; ms. The Hague, Royal Library, 78 C 27, f. 120vb. About the life and works of Bertand see Huizenga, *Tussen autoriteit en empirie...*, p. 163-166.

37. In the first half of the fourteenth century, for instance, the Italian surgeon Niccolò Bertucci explained in the foreword of his *Collectorium* why he had compiled his work in the way he did. First of all, he wrote, it was to satisfy the intellectual needs of those who wanted to practise medicine. Secondly, it was to prevent the loss of memory that every aging person suffered. Moreover, Niccolò wanted to meet the many requests of his friends. Being succinct in writing was also the fashion of his time, he states, and, as he gives his somewhat curious but maybe very practical reason, he also did it 'because travellers are bending heavily under the weight of piles of books'... See Demaitre L., 'Scholasticism in Compendia of Practical Medicine, 1250-1450', *Manuscripta* 20 (1976), p. 83-84 [Special issue: *Science, Medicine and the University, 1200-1550. Essays in Honor of Pearl Kibre.*]

Another surgical text, the *Chirurgia parva* by Guy de Chauliac, actually consists of two originally separated texts. They were brought together by Guy because he thought that the first, *Formularium auxiliorum apostematum et pustularum* ('Guide as help to (the treatment of) swellings and vesicles'), did not receive enough attention[38]. In the Middle Dutch translation of the *Chirurgia parva* this is verbalized as follows:

> Ende ic willet continueren (dat is versamenen) metten formulier van der hulpen der apostemen ende puijsten, het welck niet lange te Parijs gelesen ende ter gelaten is, om sonderlinge saken willen.
>
> 'And I want to unite it with the text about the treatment of swellings and vesicles, which was read and composed in Paris not long ago, for the sake of some things which need attention[39].'

An important part in the process of compilation, in particular when considering the medical field, is the fusion of many shorter tracts (often consisting of recipes or collections of recipes) into one single text. The reception of the important medical-surgical text *Jonghe Lanfranc* shows this very clearly[40]. This text is a Middle Dutch compilation of four different Latin sources that treat varied aspects of late medieval surgery. These sources dealt with such matters as how to stop bleeding, the treatment of open wounds, wounds to the head and tumours, and the preparation of drugs and potions. The reception of the text is extremely complex, since this Middle Dutch text has a very rich textual tradition: there are at least ten manuscripts known bearing the tract. Among these manuscripts there are great, sometimes even very great, differences in the way they hand down the *Jonghe Lanfranc*.

The following may be said with regard to the intentions of the scribes and compilers of the text. First of all, there is a large diversity in the way that the different parts of the tract were changed, adapted and supplemented. In general, it can be said that the translators of surgical works in the thirteenth and fourteenth centuries were at first simply satisfied with translating (see above, paragraph 2), and with thereby making specialised surgical knowledge accessible for those sections of the population for

38. See Huizenga, *Tussen autoriteit en empirie…*, p. 83. More about the *Chirurgia parva* in paragraph 7.

39. Ms. Middelburg, Zeeuwse Bibliotheek, 6353, p. 41.

40. For the following, see Huizenga E., 'Het wonderbaarlijke leven van meester Jan Framons. Handschriftelijke context, structuur, en intentie van de Middelnederlandse overlevering van de Jonghe Lanfranc', in: O.S.H. Lie & J. Reynaert (eds.), *Artes in context. Opstellen over het handschriftelijke milieu van Middelnederlandse artesteksten*, Artesliteratuur in de Nederlanden 3 (Hilversum, 2004), p. 99-148.

whom it was not available before[41]. But as time went by, and the new compilation was copied more often, the scribes of the *Jonghe Lanfranc* turned more critical, and became real editors. They selected, deleted what they didn't like, criticized their source, and added things they thought were lacking. The different Middle Dutch parts of the *Jonghe Lanfranc*, translated from the Latin, were in turn used for new selections, or were made part of larger text conglomerates and gathered texts, by a group of surgeons and practitioners of lesser prestige, for whom the selections and choices they made were always aimed at their practice: how could one best treat a certain disease or illness[42]?

During this process, theory fell more and more into the background, and practice gained more importance. The *Jonghe Lanfranc* illustrates this process strikingly. Very quickly, the more theoretical parts of the sources were left out, in favour of the more practical parts. As the *Jonghe Lanfranc* became more and more accessible to different layers of the population, everyone pursued his or her own interests. In doing so, every scribe of the text gave his copy of the *Jonghe Lanfranc* his own identity and provided his own, unintended signature. The four building blocks of the compilation were taken up from their context and led an independent life elsewhere.

Thus, what we see here is that the way in which surgical literature was written, used and received during the late Middle Ages could vary enormously. On the one hand, there were surgical authors whose influence was limited to their autograph alone, used only by themselves. Quite different, on the other hand, were the famous academic surgeons who had influence far beyond national boundaries. A famous authority like Lanfranc of Milan, for instance, could influence many others, be it in his writings or in his teachings, and in different ways, on different levels and in different degrees of immediacy[43].

41. See about this, for the Middle Dutch surgical tracts for instance, Huizenga & Reynaert, 'De Middelnederlandse vertalingen…', in which the different translations of Lanfranc's *Chirurgia magna* and their peculiarities are described comprehensively.

42. See about this, for the Netherlands, also Van Leeuwen C.G., 'Opvattingen omtrent ziektes en zieken in de Nederlanden gedurende de late middeleeuwen. Een eerste aanzet', in: N. Lettinck & J.J. van Moolenbroek (eds.), *In de schaduw van de eeuwigheid. Tien studies over religie en samenleving in laatmiddeleeuws Nederland aangeboden aan prof. dr. A.H. Bredero* (Utrecht, 1986), p. 134-166, 282-287.

43. See Keil G. & Müller R., 'Deutsche Lanfrank-Übersetzungen des 14. und 15. Jahrhunderts. Zur Wertung der Lanfrank-Zitate in Brunschwigs «Chirurgie»', in: H.H. Eulner e.a. (eds.), *Medizingeschichte in unserer Zeit. Festgabe für Edith Heischkel-Artelt und Walter Artelt zum 65. Geburtstag* (Stuttgart, 1971), p. 100-102.

4. Bi- and multilingualism in Middle Dutch surgeries

Almost all of the famous surgical textbooks of the Italian and French schools were translated into Middle Dutch, most of them relatively early, and often in different adaptations as well as in several copies[44]. Of one of the most important medieval surgical textbooks, the *Chirurgia magna* of Lanfranc of Milan, there are no fewer than three independent Middle Dutch translations, something unique of its kind with regard to Middle Dutch surgeries[45]. The earliest translation dates from the fourteenth century and was made in Flanders. The second translation is from the first half of the fifteenth century and has been traced to Guelders, while the third dates from shortly after the middle of the fifteenth century and was made in the north-eastern part of the Netherlands.

However, Middle Dutch surgical literature does not consist solely of integral translations. There were also a number of Dutch authors who showed remarkable originality and independence, like Jan Yperman, who wrote his *Cyrurgie* around 1310, and Thomas Scellinck of Tienen, whose *Boec van surgien* dates from 1343[46]. Their work shows how, in the Netherlands, many Latin works were translated or at least assimilated very early.

Quite rapidly, Middle Dutch became the most important language in which Dutch surgeons wrote down their knowledge, replacing Latin. This is illustrated by a Utrecht manuscript, written in the first half of the fifteenth century, containing the Middle Dutch translation of the *Chirurgia magna* of Guy de Chauliac[47]. This huge work had been translated relatively early, around 1400. In the Utrecht codex, almost all the chapter titles, as well as the main and intermediate headings, are bilingual, that is to say that they are given first in Latin and then in Middle Dutch. The Latin versions are written in a different script and are highlighted by means of a stroke in rubric. The routine manner in which the scribe deals with Latin and with its usual abbreviations suggests that using Latin must

44. See exhaustively Huizenga, *Tussen autoriteit en empirie…*, p. 103-191.

45. See in far greater detail about these translations, Huizenga & Reynaert, 'De Middelnederlandse vertalingen…', and cf. also par. 6.

46. See Van Leersum, *De «cyrurgie»…*, and Van Leersum E.C., *Het 'Boeck van Surgien' van Meester Thomas Scellinck van Thienen. Naar de handschriften van de Koninklijke Bibliotheek te 's-Gravenhage en het British Museum te Londen uitgegeven*, Opuscula selecta Neerlandicorum de arte medica (…) (Amsterdam, 1928).

47. Ms. Utrecht, University Library, 1356; description in Jansen-Sieben, *Repertorium…*, p. 467, nr. U 300; see also Van der Horst K., *Illuminated and Decorated Medieval Manuscripts in the University Library, Utrecht. An Illustrated Catalogue* (The Hague, 1989), p. 38-39, nr. 139, ills. nr. 602-603.

have been quite normal for him. The beginning of book four, about ulcers, for instance, reads:

> [Latin] Hic incipit tractatus quartus et est de ulceribus, cuius sunt due doctrine. Doctrina prima est de ulceribus prout insunt membris cimplicibus. Doctrina 2ª est prout insunt membris compositis. Prima doctrina quinque habet capitula. Capittulum primum est sermo universalis de vulneribus.
>
> [Middle Dutch] Hier beghint die vierde tractaet ende es van den ulceren (zweren), des welcx siin 2 leeren. De eerste leere es van den ulceren by also dat sy siin in den simpelen leden. De ander leere es by also dat sij siin in den compoesten (vergaderden) leden.[48] De eerste leere heeft v capittelen. Deirste capittel es een ghemeen sermoen van zweren et caetera.
>
> 'Here begins the fourth tract, and it is about ulcers (that is, sore), and there are two parts. The first part treats ulcers in simple limbs; the second treats them in compound (that is, composite) limbs. The first part has five chapters. The first chapter is a general sermon about ulcers, etc.[49].'

It is not clear why the scribe decided to give the same information twice, in Latin as well as in Dutch, but his choice reveals him as a person balancing between two languages. One reason for his bilingualism might be the increasing importance of Middle Dutch, instead of Latin, as a medium for scientific texts. The Utrecht codex would then represent a transitional phase in this process of Dutch taking over from Latin. Note in relation to this that the Middle Dutch translator added two synonyms where the meaning of the Latin loanwords might not have been immediately clear to his readers: *ulceren → (zweren)*; *compoesten→ (vergaderden)*. Particularly in the vernacular medical-surgical literature, writers and compilers often struggled with problems that were connected to the process of translation. This was caused by their Latin sources, products of the pre-eminently scientific and academic environment of the *doctores medicinae*. They were physicians with a university degree in medicine, who were specifically engaged on theoretical backgrounds[50].

48. In his *Chirurgia magna*, Guy makes a distinction between simple and compound limbs. Simple limbs cannot be counted as being part of another limb, he says, because each one of those parts is similar to the whole – they are homogeneous in composition. Such limbs are skin, muscles, nerves, ligament, connective tissue, veins, bones, nails and hairs. Compound limbs are heterogeneous in composition – they are accordingly the head, neck, back, arms, shoulders, breasts, belly, hips, legs and feet. The idea refers to an old Galenic doctrine – see McVaugh, *Guigonis de Caulhiaco…*, vol. I, p. 25-26.

49. Ms. Utrecht, University Library, 1356, f. 134va.

50. Cf. Jones P.M., 'Four Middle English Translations of John of Arderne', in: A.J. Minnis (ed.), *Latin and Vernacular. Studies in Late-Medieval Texts and Manuscripts* (Cambridge, 1989), p. 64.

Bi- or multilingualism is the rule rather than the exception for the medieval manuscripts in which the medical and surgical texts have come down to us. In fact, it occurs far more frequently than is often realised[51]. We rarely meet a codex with only vernacular or only Latin texts. In these surgical texts, Latin and the vernacular met, influenced and enriched each other. These complex interactions between the two can be observed on different levels: volume proportions (for instance, much Latin and little Middle Dutch, or the reverse) show a great deal of diversity, and this diversity can also be observed in the way in which the languages functioned with respect to each other. Bilingual medieval manuscripts can be divided into two groups: manuscripts with a combination of languages, drafted in that way by the scribe (or his commissioner) – cf. the Utrecht Chauliac-manuscript; and manuscripts where the bi- or multilingualism is caused by later additions or changes, and is coincidental[52].

The translator of a medieval surgery had to deal with a very specific technical language, in which, moreover, it was necessary to describe the manual operations that his text discussed as comprehensibly as possible, so as to allow each reader to put them into practice. In places where a translator did not succeed, we find vernacular medical works where the Latin technical terminology is maintained. An important phenomenon is that, in his wish to be as clear as possible, the translator might use both the Latin and the vernacular (in this case, Middle Dutch) technical term – often immediately after each other[53].

This phenomenon is generally called the *.i.-periphrasis*, where the *.i.* stands for the Latin *id est*. Such a linguistic feature in medical-surgical tracts is important, because it shows us the transitional phase from Latin towards the vernacular. It also appears frequently in the Middle Dutch surgical literature. The introduction of synonyms almost always points to

51. Voigts L.E., 'What's the Word? Bilingualism in Late-Medieval England', *Speculum* 71 (1966), p. 819.

52. See Huizenga E., '"Hoe mense noemt in Latijnen metten Duytsche". Tweetaligheid in een Middelnederlands artes-handschrift', in: P. Wackers e.a., *Verraders en bruggenbouwers. Verkenningen naar de relatie tussen Latinitas en Middelnederlandse letterkunde*, Nederlandse Literatuur en Cultuur in de Middeleeuwen 15 (Amsterdam, 1996), p. 151.

53. We can see this process in other vernaculars, and even in some Latin tracts. In the Latin text of the *Practica* of the English surgeon John Arderne for instance, written around 1370, the Latin technical terms are regularly glossed with their English or Anglo-Norman counterparts. These glosses, most likely by Arderne himself, show clearly that the author did not have just a Latin-oriented public in mind. See Jones, 'Four Middle English Translations…', p. 68-89.

the fact that the translator or compiler meant to communicate the specific technical language of his text as clearly as he could[54].

We encounter such bilingualism in different places in medieval vernacular medical texts. It can very often be witnessed, for instance, in medical prescriptions. It is very common to find a vernacular prescription followed by a Latin final line, such as *sanabitur*, *videbis mirabilia*, or *probatum est* (meaning 'to be healed', 'you will see miracles' or 'it has proven its value'). Research by Claire Jones on English prescriptions has shown that such so-called 'efficacy phrases', meant to attest to the value of a given remedy, were often written down because of the authoritative effect that Latin still had, which made these small texts seem more believable[55]. Apart from that, the Latin phrases also made a clear distinction between a text and a part of it. One must realise that in medieval manuscripts making this distinction is not always as simple as it looks, because parchment and paper were expensive, and had to be used as economically as possible; for this reason, texts were often written extremely close to one another, and this most certainly accounts for the utilitarian codices that were made up by the *artes*-texts by their nature[56]. Latin often appears at the fringe areas of vernacular *artes*-texts anyway, that is, at the beginning and at the end of tracts, in incipits and explicits, in chapter headings, and in owners notes. We have already seen this at the beginning of Yperman's surgery in two of the surviving text witnesses[57].

But even more interesting, because it is more substantial, are those places where Latin and Middle Dutch mix within a text. In the Middle Dutch translation of the *Chirurgia parva* of Lanfranc of Milan, for example, we find a fragment of the original Latin source immediately followed by the Middle Dutch equivalent: 'Hec enim cura, et cetera' (Which refers to the Latin source-text 'Hec enim cura causa certa est'), followed by a Middle Dutch sentence which can be translated as: 'this cure is certain, because (…)'[58]. What makes this remark interesting is that we find it only

54. Kool M., *Die conste vanden getale. Een studie over Nederlandstalige rekenboeken uit de vijftiende en zestiende eeuw, met een glossarium van rekenkundige termen*, Middeleeuwse Studies en Bronnen 64, Diss. Utrecht (Hilversum, 1999), p. 315.

55. Jones C., 'Formula and Formulation. "Efficacy Phrases" in Medieval English Medical Manuscripts', *Neuphilologische Mitteilungen* 99 (1998), p. 202-203.

56. Jones, 'Formula and Formulation…', p. 200-201.

57. Van Leersum, *De «cyrurgie»…*, p. 5.

58. Ms. London, British Library, Harley 1684, f. 117rb. Latin text: London, British Library, IB 22989 (Venice, 1498), f. 164ra. Ed. of the Dutch text: Scholle S., *Lanfranks «Chirurgia Parva» in mittelniederfränkischer Übertragung*, Altdeutsche Lanfrank-Übersetzungen II, 1 (Diss. Würzburg, 1978), p. 72.

in one of the three Middle Dutch text witnesses[59]. Obviously the scribe of this manuscript had a Latin text at his disposal besides the Middle Dutch layer, perhaps for verification purposes – an indication of his high level of education, and again a signature not intended as such.

In Middle Dutch surgeries there are numerous examples of, or variations on, the *.i.-periphrasis*. Thus, in Yperman's text, the heading of one of the chapters reads as follows: 'Van pusten in dogen ende wit ungula heet in latine' ('About pimples in the white parts of the eyes, that in Latin is called *ungula*')[60]; or, in a chapter about the brain and nerves: '(...) ende dat march heet in latine nucha' ('(...) and the spinal marrow is called in Latin *nucha*')[61].

We also see, in the Middle Dutch surgeries, that a number of Latin loan expressions are accompanied by an explanatory remark. This is the case in the already mentioned Middle Dutch text of Lanfranc's *Chirurgia parva*. This text reads:

> Hier selstu merken, also dicke als een lit wilste morwen, so fomenteert eerst (dat is bayen ende roet maken) tlit mit wermen water, daer in ghesoden is (...).
>
> 'Here you shall take notice, as often as you want to soften a limb, so first foment this (that is to say, bath it until it gets red) limb with hot water, in which is boiled (...)[62].'

Here, the Dutch gives a periphrasis of the loanword of Latin origin *fomenteren, fomentatio* ('to warm'), before continuing the description of the treatment.

In spite of the decreasing importance of Latin for the *artes*-literature, its use continued for a long time. One reason for this is that in the fifteenth and even in the sixteenth century, Latin was still considered to be the language of the *literati*. An excursion to the German linguistic region can illustrate this[63]. There, in this period, many surgeons and practitioners tried to earn a living by travelling around from village to village. Because they could only stay for a short while in each place, and had to recruit as

59. The other two text witnesses are Paris, Bibliothèque de l'Arsenal, 8216, f. 59r-73v; and an early print: London, British Library, I A.49307 (Leuven, 24th november 1481), f. A1r-E1r.

60. Van Leersum, *De «cyrurgie»...*, p. 77.

61. Van Leersum, *De «cyrurgie»...*, p. 6.

62. Ms. London, British Library, Harley 1684, f. 125va; ed. Scholle, *Lanfranks «Chirurgia Parva»*, p. 92.

63. The next case is described in Keil G., 'Die Niederlassungsankündigung eines Wundarztes aus dem 15. Jahrhundert. Untersuchungen zum ärztlichen Werbe-Formular', *Beiträge zur Geschichte der deutschen Sprache und Literatur* 89 (1967), p. 302-318.

many clients as they could, it was their practice to advertise themselves with posters that they attached to walls or trees with wax. The interesting fact is that such posters were not only written in the vernacular, as one would expect, but also contained the Latin version of the text. As one of them says, they were written *tam litteratis quam ydeotis*, or: 'both for the literate as for the illiterate'[64].

The linguistic concept of 'domains' – meaning here, in the words of Linda Voigts, 'the distinction in a speech community between 'high' and 'low' forms of the language, where 'high' is a learned or acquired form of a language that fulfils a different function from the 'low' vernacular used for ordinary interaction'[65] – remained an important factor for a long time in the interaction of Latin and the vernacular in Middle Dutch surgical literature.

5. Bilingualism in ms. Vienna, ÖNB, 2818

Up to this point, I have discussed translated surgical texts and matters like bilingualism on different levels and in different categories of texts and manuscripts. I now want to look at one Middle Dutch *artes*-codex as a whole, to see precisely how and where this bilingualism manifests itself.

The manuscript in question is codex 2818 of the Österreichische Nationalbibliothek in Vienna, which is considered to be one of the most important Middle Dutch *artes*-manuscripts[66]. In spite of the fact that it dates roughly from the 1490's, we find here, too, many aspects of the relations between Latin and the vernacular. It contains translations of widespread and well-known standard medical works such as the alchemically-orientated text *De consideratione quintae essentiae* ('Considerations about the fifth essence') of John of Rupescissa[67], the *Antidotarium*

64. Keil, 'Die Niederlassungsankündigung...', p. 303.

65. Voigts, 'What's the Word?...', p. 818.

66. A complete and integral edition of the whole manuscript, together with a description of its contents, its scribe and user, its function, plus a codicological analysis can be found in Huizenga E., *Het Weense arteshandschrift. Hs. Wenen, Österreichische Nationalbibliothek, 2818. Diplomatische editie*, Middeleeuwse Verzamelhandschriften uit de Nederlanden 10 (Hilversum, 2004). See also Huizenga, *Een nuttelike practijke...*, and Jansen-Sieben, *Repertorium...*, p. 472-477, nr. W 80.

67. About this tract, see Huizenga E., 'De vijfde essentie. Monniken, armoede, gezondheid en goudkoorts in een Middelnederlands alchemistisch traktaat', *Literatuur* 11 (1994-2), p. 86-94, and Huizenga, *Een nuttelike practijke...*, p. 127-146; for German manuscripts and incunabula with this text (including ms. 2818), see Benzenhöfer U., *Johannes' de Rupescissa Liber de consideratione quintae essentiae omnium rerum deutsch. Studien zur*

Nicolai[68], the *Chirurgia magna* of Bruno Longoburgensis[69], and the *Circa instans*[70]. Along with those texts, the codex contains several glossaries and a large number of short astrological and astronomical tracts and tables[71]. All of these texts are translations from more or less well-known scientific works, and in numerous places in this predominantly Middle Dutch manuscript we encounter Latin, and influences from the Latinitas. I would like to point out several particular features.

First of all, the astrological tables in this codex are completely or partially written in Latin. Technical terms are not translated here, and the names of planets and months are written in their Latinised form[72]. Moreover, there are numerous titles, headings and explicits throughout the whole manuscript that are not translated. As I have said, such relics of Latin are not unusual in the Middle Dutch *artes*-manuscripts[73]. But we

Alchemia medica des 15. bis 17. Jahrhunderts mit kritischer Edition des Textes, Heidelberger Studien zur Naturkunde der frühen Neuzeit 1 (Stuttgart, 1989); ms. 2818 on p. 41-42 and 52.

68. Literature concerning this tract is extensive. I mention only: Goltz D., *Mittelalterliche Pharmazie und Medizin. Dargestellt an Geschichte und Inhalt des Antidotarium Nicolai. Mit einem Nachdruck der Druckfassung von 1471*, Veröffentlichungen der Internationalen Gesellschaft für Geschichte der Pharmazie N.F. 44 (Stuttgart, 1976); Mellbourn G. & Keil G., 'Das "Antidotarium Nicolai in einer sächsichen Fassung des 15. Jahrhunderts', in: G. Keil e.a. (eds.), *Fachprosa-studien. Beiträge zur Mittelalterlichen Wissenschafts- und Geistesgeschichte* (Berlin, 1982), p. 346-362. For the Dutch Antidotarium Nicolai-texts in general, see Van den Berg W.S., *Eene middelnederlandsche vertaling van het Antidotarium Nicolai. (Ms. 15624-15641, Kon. Bibl. Brussel)* (Diss. Leiden, 1917); and Braekman W.L. & Keil G., 'Fünf mittelniederlandische Übersetzungen des "Antidotarium Nicolai". Untersuchungen zum pharmazeutischen Fachschrifttum der mittelalterlichen Niederlände', *Sudhoffs Archiv* 55 (1971), p. 257-320. For the text in this manuscript in particular, see Huizenga, *Een nuttelike practijke…*, p.146-159.

69. See Huizenga, *Een nuttelike practijke…*, p. 159-176.

70. See Blome J., *Transkription, Übersetzung und systematisch-botanische bearbeitung der in der Basler Universitätsbibliothek aufbewahrter Kräuterbuch-Handschrift "Circa Instans" aus dem letzten Viertel des 14. Jahrhunderts* (Diss. Basel, 1978). About the Middle Dutch translations: Vandewiele L.J., *Een Middelnederlandse versie van de Circa Instans van Platearius* (Oudenaarde, 1970). For the text in ms. 2818, see Huizenga, *Een nuttelike practijke…*, p. 71-78.

71. They are edited in Huizenga, *Het Weense arteshandschrift…*, p. 109-441. For further information about some of these astronomical and astrological tracts, see Huizenga E., 'Die 100 capittelen van astronomijen. Tekst en traditie van een Middelnederlands astrologisch traktaat', *Scientiarum Historia* 16 (1990), p. 29-56; Huizenga E., 'Een Middelnederlands centiloquium', *Scientiarum Historia* 18 (1992), p. 49-59 [Special issue: *Astronomie en toegepaste astronomie*]; and Huizenga E., '"Die conste vanden almenack". Astronomie in een Middelnederlands verzamelhandschrift', *Queeste* 1 (1994), p. 12-33.

72. For instance, on f. 1r (a circular zodiacal table), 1vb (division of the astrological mansions), 2rb (a circular figure of the aspects), 5va (astrological mansions), and 12va (a table with the days on which the sun enters a sign: *Sol intrat anno domini (…)*. Huizenga (ed.), *Het Weense arteshandschrift…*, p. 160, lines 74-75.

73. See above, par. 4.

find something different in a tract on the conception of children. At the end of this text, where the four quarters of the moon are discussed, Middle Dutch and Latin alternate in a remarkable way:

> Item luna heeft 4 quartieren. Dierste is sangwijnea, heet ende vocht, *et mulieres sunt pacientes tunc menstrorum sunt sanguinie*. Dat ander quartier is colerica, heet ende droge, *et omnes mulieres sunt colerice complexiones pacientes menstrum*. Die dorde is flematica, cout ende vocht van complexie. Ende die vierde melancolica, cout ende droge, nochtans nyet wederstaende gemeynlick *mulieres paciuntur menstrorum jn fine quia quasi omnes sunt frigide nature et flematice complexiones etcetera*.
>
> 'Item the moon has four quarters. The first is sanguine, hot and humid, *and the women who then menstruate are sanguine*. The other quarter is choleric, hot and dry, *and all women menstruating then have a choleric complexion*. The third is phlegmatic, cold and humid by complexion. And the fourth is melancholic, cold and dry, but yet this does not prevent in general that *women at the end of their menstruation are of a frigid nature and of a phlegmatic complexion etc.*'[74].

The Galenic idea that women are cold by nature appears frequently in Middle Dutch medical literature[75]. What we see here is a switch to Latin as soon as the text mentions the menstrual cycle of women. It seems that, in this case, the Latin has a concealing function within a vernacular context: it is used to withhold certain information from undesirable, incompetent or indiscrete readers.

The Middle Dutch translation of the alchemical-medical *De consideratione quintae essentiae* was characterised by Benzenhöfer as being 'von durchschnittlicher Qualität', whereas 'auffällig häufig (...) die alchemistischen Fachtermini in ihrer lateinischen Form belassen [sind]'[76]. Many of the specific technical alchemical terms are not translated here. To give a few examples: 'daer na graeft dat al heel *in ventre equi*' ('after that, bury it completely *under horse-dung*') (f. 102ra); 'settet dat gedestilleerde water *in distillatorio circulationis*' ('put that distilled water *in a distillation equipment where it can circulate*') (f. 103rb); (to divide pure water

74. Italics here and henceforth indicate Latin phrases and their translation. Ms. Vienna, ÖNB, 2818, f. 92rb; ed. Huizenga, *Het Weense arteshandschrift...*, p. 440-441, lines 48-59. Quotations from this manuscript here and henceforth are slightly adapted in the sense that punctuation is added, and abbreviations are spelled out.

75. Jansen-Sieben R., 'De vrouw in de medische literatuur', in: R.E.V. Stuip & C. Vellekoop (eds.), *Middeleeuwers over vrouwen*, Utrechtse bijdragen tot de mediëvistiek 4 (Utrecht, 1985), vol. 2, p. 160-161.

76. Benzenhöfer, *Johannes' de Rupescissa...*, p. 52.

from impure:) 'want daer sijn twe elementen, te wetene dat element des waters ende des vuers, ende settet in een *distillarorium ad ignem balnei*' ('because there are two elements, that is the element of water and of fire, and put it in a *distillation device near the fire of the bath* (meaning: *au bain-Marie*)') (f. 104ra). There are numerous others. This conservation of the alchemical terminology in its original Latin form can have several causes, again, for example, secrecy[77]. Alchemical texts often used a peculiar symbolic language to hide certain subjects from outsiders. It is possible that here the Latin replaces this symbolic language, and has the same function in relation to Middle Dutch.

The same work shows another striking phenomenon: in many places, we find Middle Dutch synonyms for the Latin names of vegetable medicines written above their Latin counterparts. There are more examples of this in the Middle Dutch *artes*-literature, and again they almost always point to a transitional stage, in which the text exists on the borderland of the two languages. The synonyms functioned as a glossary, as an expedient in dealing with knowledge that was translated into the vernacular, a vernacular that still partly lacked the means to reproduce thís specific knowledge[78].

In the *Antidotarium Nicolai*, a list of prescriptions ordered alphabetically, we see again the same strong mixture of Latin and Middle Dutch. Apart from the many closing formulas like 'mel quod sufficit' ('use enough honey here'), there are also number of passages where Middle Dutch and Latin mix together, for instance, in a sentence like: 'pulueriseret te gader getempert met honich *et fiant pilule in massa*' ('make powder of it, mix it with honey, *and there will be a lot of pills*'). And we find many longer passages that were not translated as well. This important linguistic process of what is called 'code mixing', the juxtaposition of passages belonging to two different grammatical systems or subsystems, obviously plays a large role in this manuscript[79]. The bilingualism in the *Antidotarium Nicolai* provides us with clues that characterize the translator of the text as a person who must have belonged to a group of pharmaceutical specialists, apothecaries and the like: in every place in this tract where highly specialised knowledge is treated (in formulas, names

77. See Huizenga, 'De vijfde essentie...', p. 92-93.

78. About the long tradition of glossaries and their origins, see MacKinney L.C., 'Medieval Medical Dictionaries and Glossaries', in: J.L. Cate & E.N. Anderson (ed.), *Medieval and Historiographical Essays in Honor of James Westfall Thompson* (Chicago, 1938), p. 240-268.

79. Voigts, 'What's the Word?...', p. 818.

of plants, weights and composition of the prescriptions), we see a transition from Middle Dutch to Latin[80].

For the compiler it was obviously of great importance to get as much help as possible when conveying the originally Latin technical language in Middle Dutch, especially with regard to surgery and pharmaceutics. Though, for him, the vernacular was at this point the new 'high domain', he still had to rely on the old one, the Latin in which his sources were written. The existing medical knowledge had to be made understandable by the anonymous surgeon who, in all likelihood, was also the compiler and user of this impressive codex[81].

6. The Middle Dutch translations of the *Chirurgia magna* of Lanfranc of Milan

The most important surgical author of the late Middle Ages was without doubt Lanfranc of Milan (ca. 1245 – before 1306). He was a student of William of Saliceto, and was educated in Bologna. In Milan, he quite soon became very famous and attracted many wealthy patients, nobles and clerics. However he got himself involved in political quarrels between the pope and the emperor and chose the side of the losing imperial party, and was therefore forced to emigrate[82]. From 1290 onwards, he practised for a while in the Provence, notably in Lyon. There he wrote his *Chirurgia parva*, and in the meanwhile, worked on his *Chirurgia magna*, which he would complete in 1296.

The *Chirurgia magna* has been transmitted in Middle Dutch in no less than three translations, independent from each other[83]. In general, the

80. Cf. Keil, 'Zur Frage der kurativ-konsiliarischen Tätigkeit des mittelalterlichen deutschen Apothekers', in: P. Dilg e.a. (eds.), *Perspektiven der Pharmaziegeschichte. Festschrift für Rudolf Schmitz zum 65. Geburtstag* (Graz, 1983), p. 192, n. 29.

81. About this, see in particular Huizenga, *Een nuttelike practijke...*, p. 275-330, and Huizenga (ed.), *Het Weense arteshandschrift...*, p. 65-72.

82. Lanfranc, *Chirurgia magna*, V.7 (epilogue). See also Keil, 'Lanfrank von Mailand...', col. 560-561.

83. The first translation is known from two manuscripts. One of them is rendered only partially, that is in the fourteenth-century fragments Darmstadt, Hessische Landes- und Hochschulbibliothek, zu 2201; Cologne, Historisches Archiv, Fragm. A 57; and Cologne, Universitäts- und Stadtbibliothek, zu Inc. GB V 167. The other manuscript of this first Middle Dutch translation, Ghent, University Library, 1272, does have a complete translation. Each of the other translations is known from one manuscript, both dating from the fifteenth century: codices Uppsala, University Library, Kodex Waller 132 (the second Middle Dutch translation), and Amsterdam, University Library, II F 39 (the third Middle Dutch translation). Furthermore, there are excerpts from Lanfranc's text in manuscripts

three translators followed their Latin text scrupulously, but there are of course differences in the way they approached their source[84].

Some of these differences are striking. For instance, compared to the translation in manuscript Ghent, University Library, 1272 (henceforth *G*), the translators of Amsterdam, University Library II F 39 (*A*) and Uppsala, University Library, Kodex Waller 132 (*U*) made their text much more accessible and user-friendly by adding a table of contents before the actual text. The reader could thus find information about certain subjects easily and quickly. The scribe of *U* did much more than just translate, as compared to the scribes of the other two manuscripts; he was also an editor of his text, as is shown by his continuous simplifications, explanations and additions of didactic interventions[85].

In this manuscript (*U*), we see that the editor consciously left out of the Middle Dutch translation those passages in the Latin original which were thought to be too theoretical, and this also included the supplemental quotations from medical authorities. He did this on behalf of a changing reading public, which, during the fourteenth and fifteenth centuries, was more and more in need of literature that was aimed at the practice of surgery.

This is in sharp contrast to the texts in Ghent and Amsterdam: they are clearly 'translations' in the most literal sense of the word. Both translators apparently wanted to render their example completely and faithfully[86]. The translation in *G* is sometimes a bit blunt, as if the translator was trying to provide a text as short as possible while he was interpreting. The translator of *A*, on the other hand, makes a translation as literal as he can, and moreover at a very high level. This translator was well aware of the essence not just of surgery or medicine, but also, in a slightly broader context, of the *artes liberales* and the sciences of nature in general.

Kassel, Landesbibliothek und Murhardsche Bibliothek, 2° Ms. med. 1, and in Hattem, Municipal Archives, 958. About this Hattem manuscript see Lie O.S.H. & Versélewel de Witt Hamer E.M., 'Beproefde middeleeuwse recepten in het Hattemse Gemeentearchief', *W-mail* 2 (2001), nr. 2, p. 8-15; and, as a reaction, Versélewel de Witt Hamer E.M., 'Geheimschrift in het Hattemse handschrift', *W-mail* 2 (2001), nr. 3, p. 14-15.

84. The following rests upon Huizenga & Reynaert, 'De Middelnederlandse vertalingen...', p. 256-281.

85. One example may elucidate this. On f. 1ra-b, about the concept of 'theory' (that is, of surgery): 'Dat seg ic om theuorijc, die men alleen begrijpt mit sinen eynde ende mit speculacien' [I say this because of the theory, which can only be understood properly from its goal and by contemplation]. There are also numerous references to what authorities like Galen or Avicenna had said that are left out. See about this Huizenga & Reynaert, 'De Middelnederlandse vertalingen...', p. 261.

86. See for the following, Huizenga & Reynaert, 'De Middelnederlandse vertalingen...', p. 280 ff.

The translators of *G* and *A* followed their Latin text closely, more so than the translator of *U*, even in those places where the Latin text deals with difficult matters like the mutual relationship between surgical practice and theory on the one hand, and their relationship to the other scientific disciplines on the other. And yet in the second chapter of the *Chirurgia magna* we see a striking difference between the translations in *G* and *A*. The Amsterdam text, in a literal sense, is closer to the Latin than the Ghent text, which is more interpretative ('how can I provide the essence of the text as best as possible?'). This may be observed in the way in which the terminology of the scientific disciplines is rendered[87]. Even though the translator of *A* does his best to stay as close to the Latin source as possible, sometimes he goes astray. In those places, he does not seem to fully understand the exact meaning or the intention of the Latin[88].

A piquant deviation from the source-text, and one that allows us to get an indication of the profile of the translator of the Amsterdam manuscript, can be found in the second chapter[89]:

Translation of the Latin:	**Ms. *A*, f. 11r:**
What the *surgeon* has to know about complexions in bodies, limbs and medicines, has to be felt through perception, which is proven by experiments[90].	Daer den *mediciin* noetsaeclic is to weten de complexien van den menschen, ende van den leden, ende van der medicinen, dat wert gheprovet by der cirurgien experimenten[91].

So, the Latin text tells us here that it is also necessary for a surgeon to know the significance of the complexions in the human body and in medicines. Within the context of Lanfranc's emphatic pleading for the more thorough theoretical knowledge that characterizes his *Chirurgia magna*, this is a logical assertion. *G* and *U* therefore adopt the word 'surgeon'.

87. About this, see the Latin text of the *Chirurgia magna*, I.i.2. Latin print: Venice, 1498, ex.: London, British Library, IB 22989, f. 167v, and the translation of this chapter in *A* on f. 8v. Edited in Huizenga & Reynaert, 'De Middelnederlandse vertalingen...', p. 299, A.5.

88. See, for instance, the *Chirurgia magna*, I.i.2; ed. Huizenga & Reynaert, 'De Middelnederlandse vertalingen...', p. 303 + 305, C.2-C.9

89. Lanfranc of Milan, *Chirurgia magna*, I.i.2; ed. Huizenga & Reynaert, 'De Middelnederlandse vertalingen...', p. 311, L.2.

90. Latin text on f. 168ra: 'Quod enim cyrurgico scire de complexionibus corporum et membrorum et medicinarum sit necessarium sensualiter sic per experimentum probatur'.

91. Italics added by me, EH.

Manuscript *A*, however, states clearly that such knowledge is necessary for a physician: 'Daer den *mediciin* noetsaeclic is to weten [...]'. This might seem to be a completely superfluous communication, for physicians were always boasting that they were in possession of the indispensable theoretical foundations. The internal medicine to which this sentence refers was their area of expertise. The reasons why the translator of *A* uses *medicijn* [physician] here are not very clear; on other occasions he uses 'ordinarily' *cirurgijn* [surgeon], just like both other Middle Dutch translators. Perhaps he was not a surgeon in the strict sense of the word, but was someone who possessed more general medical skills, in the realm of internal medicine. It is possible that the scribe/translator left out those passages about which he thought himself, as a physician, sufficiently informed[92], for his general scientific knowledge was of a high level, and he must have had a good basic understanding of surgery; otherwise it seems hard to explain why he should have undertaken a task of such gigantic proportions: translating one of the largest medieval surgeries.

The purposes and intended public were no doubt partly responsible for the different ways in which Lanfranc's text was translated in Middle Dutch in the three translations. This appears most clearly in *U*, which is more of an adaptation than a strict translation. Here the text is simplified and elucidated, and it shows didactic features. The other two translations, on the other hand, are real translations in the full sense of the word: both *G* and *A* clearly want to reproduce their source text completely and loyally, and in this way they stand in contrast to *U*.

In its turn, however, the text in *G* stands in opposition to both other translations. Both the prolix theologically tinted prologue, which opens Lanfranc's surgery, as well as several autobiographical remarks of Lanfranc, were resolutely left out by the translators of *A* and *U*. Apparently their attitude towards the text was different from that of the translator of the Ghent text. Those aspects still had a certain amount of relevance for the fourteenth century translator of *G*; but for the translators (or scribes?) of *A* and *U*, a century later, this was no longer the case. The text in the Ghent manuscript was created relatively shortly after the Latin original, at a moment when everybody, and certainly somebody close to or involved with the profession of a surgeon, could remember his person

92. The manuscript was made in two steps: firstly there were 76 chapters copied on the first 162 folia (of which some only partially) – the Latin original counts 91 chapters. The second step was when, on f. 162v-188v, fourteen of the chapters which were initially left out were copied, and a table of contents was added for the part until f. 162, in which (wrongly) 80 chapters were distinguished.

still very well. Both translators of the others texts lived a hundred years or more later, and for them the remembrance of Lanfranc had faded considerably. Omitting the theological beginning is also in keeping with the approach of the translators of *A* and *U*, which was aimed much more at direct medical practice.

The three translations of Lanfranc's *Chirurgia magna* show certain things very clearly. First of all, the text itself became one of the great authoritative surgical texts during the two centuries that followed its genesis – a text that was worthwhile translating or copying entirely. This is remarkable in a period in which surgical literature became more and more short and practical, as numerous surviving collections of prescriptions prove. It appears that within the vernacular surgical tradition there was a target group to whom Lanfranc's pleading for bringing together theory and empiricism strongly appealed. But besides this the translations show that as years passed, the too theoretical, philosophical or theological passages were appreciated less and less by the translators, and were left out or not understood anymore. More and more the text was adapted to individual needs, as the authority of the *Chirurgia magna* slowly came to an end.

7. The Middle Dutch translations of the *Chirurgia parva* of Lanfranc of Milan: the anthrax-segment

Lanfranc wrote his *Chirurgia parva* before the *Chirurgia magna*, around the year 1290. He dedicated it to his friend, a certain Bernard[93]. His intention with this small text was to provide in a nutshell all of the information he thought necessary for daily surgical practice. In Middle Dutch, the tract is rendered in two codices. These manuscripts are: London, British Library, Harley 1684, f. 105r-126v (henceforth *L*); and Paris, Bibliothèque de l'Arsenal, 8216, 59r-73v (*P*). There is also an early printed edition of the text[94].

The translator of Lanfranc's *Chirurgia parva* in the London manuscript handled his Latin source freely. He inserted new passages, conveying a didactic point of view: he wanted to provide his reader above all with practical knowledge[95]. So he put a lot of emphasis on surgical practice,

93. See Huizenga, *Tussen autoriteit en empirie...*, p. 74-75.
94. Print: Londen, British Library, I A.49307, f. Ai^{r}-Ei^{r} (printed by Conrad Braem in Louvain in 1481).
95. Scholle S., *Lanfranks «Chirurgia Parva» in mittelniederfränkischer Übertragung*, Altdeutsche Lanfrank-Übersetzungen II, 1 (Diss. Würzburg, 1978), p. 24-28.

but paid relatively little attention to theoretical humoural-pathological problems. The editor of this Middle Dutch surgical text, Scholle, thinks that it is likely that the translator himself was a surgeon, who wanted to protect his profession[96]. The first version of the Middle Dutch translation originated in Flanders, most likely very early, possibly in the first half of the fourteenth century[97].

When we take a closer look at the changes that were made in the text by the translator, we see that in his choices for shortening the text or leaving out passages, as Scholle has already noticed, the translator/ editor shows a remarkable repugnance towards pieces he thought to be too theoretical or circuitous[98]. To this category belongs an anecdote told by Lanfranc about the miraculous cure of a boy in Milan, who suffered from anthrax on his head[99]. He was cured by *theriac*, a panacea that was thought throughout the Middle Ages to possess miraculous healing powers[100].

With this passage, left out by the translator, something curious is going on. For although it does not appear in the Middle Dutch manuscripts of Lanfranc's *Chirurgia parva*, we certainly do know the story in Middle Dutch: it appears in a strongly abbreviated and therefore often hardly recognisable adaptation of the *Boec van surgien* by Thomas Scellinck of Tienen in manuscript Mainz, Stadtbibliothek, ms. I 514, f. 67r-116v (henceforth *M*). In this manuscript, it has to be a personal addition by the compiler of this tract, because it is missing in the other manuscripts containing the text of the *Boec van surgien*[101]. I quote here first the translation of the Latin text of Lanfranc's *Chirurgia parva*, and next to it the text of the passage in *M*:

96. Scholle, *Lanfranks «Chirurgia Parva»*, p. 27-28.

97. Scholle, *Lanfranks «Chirurgia Parva»*, p. 28-35.

98. Scholle, *Lanfranks «Chirurgia Parva»*, p. 26, n. 89.

99. Nowadays anthrax is known to be a contagious disease, caused by the anthrax-bacterium. In the Middle Ages anthrax was often confused with carbuncle, a large ulcer which arises when several smaller boilings are flowing together. As a result, the skin dies and turns a blackish blue colour, the colour of coals (*carbo*). Someone like Thomas Scellinck however does distinguish between anthrax and carbuncle in his *Boec van surgien*, II.4. Cf. Van Leersum, *Het 'Boeck van Surgien'*..., p. 100b-101a.

100. About theriac, see Holste T., *Der Theriakkrämer. Ein Beitrag zur Frühgeschichte der Arzneimittelwerbung*, Würzburger Medizinhistorische Forschungen 5 (Pattensen [etc.], 1976); and Bierman A.I., 'Theriak. Het mirakel triakel', *Geschiedenis der Geneeskunde* 1 (1993), p. 11-15.

101. About Thomas Scellinck of Tienen and his *Boec van surgien*, see Huizenga, *Tussen autoriteit en empirie*..., p. 143-150.

Lanfranc of Milan, *Chirurgia parva*, 28. Translation of Latin text[102]

And I say that in the city of Milan a young man came to me with anthrax in parts of his face, and the evil had spread so far that his whole head and face and neck were swollen more than you would believe. And a certain apothecary, a friend of the patient, brought him therefore to me. And the patient already showed signs of death, because he had no pulse, and was deeply unconscious. And while he was dying I told the apothecary to take him to his house. And the apothecary asked: 'Is there no better cure?' And I answered, without thinking: 'There is one for sure, if you only had the right theriac. And you have to use it in large quantities.' Maybe by accident he stayed alive until now, but I don't believe that. [It would have been so] if the apothecary, full of doubt, had not brought him home in time, and if he not had given him the three drams of the theriac of which it is proven that it works. And he put him on a bed and covered him. Shortly afterwards he began to sweat all over his body, and his

Thomas Scellinck of Tienen, *Boec van surgien*, II.4. Ms. *M*, ch. 14, fol. 75r-76v

Ende ic segghe dat in de stede van Avinioen quam ein jonc man mi toe mit einen antrax in siin aensicht, ende sinen als sere gheswollen, de welke ein apoteker, siin guede vruent, tot mi brochte. Ende in den zieken waren cuemen teikene der doot, want in hum en waren ghene pulsen, ende he beswijmde. Ende ic tot dem apoteker sachte: 'Leiten huus, want he sel stappans sterven'. Doe vraechde mi die apoteker ofter gheinen troest toe en werde, bedachtic mi van der triakel, ende sachte: 'Addestu van der bester driakel, en duse hum gaefs in groter quantiteit, hi mochte ter aventuren becomen, maer het es seer ghevreist'. Doe gaf hi hem van der bester driakel II drachmen, ende decten wel up ein bedde. Ende corts daer naer begonde te zweiten. Eerst begant omtrent dat oeft daer dat antrax was, ende daer naer over al siin lijf, ende doe quam die puls weder in hum. Ende dat herte was wat ghestarct. Ende doe hi huter bezwijmde quam, gaf men hum noch der selver driakel I dragme. Ende

102. Latin text: print in London, British Library, IB 22989, f. 163va-b: 'Et ego dico quod ad me venit in civitate Mediolanen[sis] iuvenis cum antrace in partibus facie, cuius malicia adeo augmentata erat, quod totum caput et facies et collum erat ultra credibile inflatum. Quem adduxerat ad me quidam apothecarius amicus egri. Et eger erat iam in signis mortis, quia non erat in eo pulsus, et sincopizabat iam. Ita quod dixi apothecario quod duceret eum ad domum suam quia moritur. Et apothecarius dixit: 'Non est in amplius remedium?'. Et ego parum cogitans dixi: 'Certe si tu haberes de optima tyriaca. Et tu dares ei in magna quantitate.' Forte posset adhuc vivere, sed non credo. Nisi valde dubitanter apothecarius duxit eum in domum sum vix euntem, et dedit ei de tiriaca approbata circa drachma II. Et posuit eum in lecto et cooperuit eum. Et incontinenti arripuit eum sudor, incipiens circa caput et locum egritudinis. Et post paululum [l. paulatim] fuit sudor universalis, et redit in eo pulsus et confortatus est. Et post exiccationem sudoris apothecarius dedit ei proprio motu iterum drachma I eiusdem tiriace in eodem die. Et sanus factus est, nisi quod remansit in eo quoddam ulcus vesicosum, quod postea leviter est curatum. Et ego nunquam vidi alium, qui postquam fuit in sincopi et tremore cordis et specialiter sine pulsu, qui evaderet. Sed de his qui loquentes, et quasi nullum malum sentientes iudicati sunt ad mortem per me, et qui mortui sunt admirantibus multum amicis de pronosticatione, vidi ego infinitos. Hec dicta sunt tibi de tyriaca, quia forte aliquam posses adhuc per ista verba liberare a morte (...)'.

pulse came back and strengthened his heart. After the sweat had dried up, the same day the apothecary gave him in the same way once again a dram of the same theriac. And he recovered, apart from a certain swollen tumour that was cured easily later on. And I have never seen someone else who escaped like that, after he got unconscious, with heart beatings, and one who in particular had no pulse. Others however, who talked about it as if they had experienced no harm, were condemned to their death by me, and those who died after this prediction surprised many friends, as I have witnessed often. I have said this about the theriac, in order that you may be able to free more from death with the help of these words (…).

binnen den selven daghe wort hi ghenesen, sonder dat daer bleef ein groet ulcus in die stede, dat corts daer naer ghenas. Ende noit en sagh ic gheinen andren die van dem antrax ghenas, die also tierde sonder puls ende mit bevinghe der herten, sonder desen alleine. Maer van den genen die spraken of hun niet ghedoert en adde, hebbic einich ter doot ghewijst, die aesteliken storven, so dat dem luden seere verwonderde. ¶ Dese dinghen siin hier ghesacht van der triakel tot einen exempel, dat noch hiement mochte gheolpen worden na die confortacie der herten, ende na dat die materi van binnen niet gheidelt en is.

The fact that this passage in Middle Dutch can only be found in this manuscript shows us once again some remarkable aspects of the Middle Dutch surgical literature. First of all there is the fact that this kind of case, hardly more than shreds of surgical anecdotes, apparently functioned as an independent element in late medieval surgical literature. It was fitted into a compilation at a point where the composer thought it to be appropriate, for instance based on the topic at hand. From which source this new element came evidently didn't matter much: the editor of *M* added the text seamlessly between the other pieces of Scellinck's text, without revealing his source (or he was possibly ignorant of it). Had not the origin of the fragment been so apparent from Lanfranc's *Chirurgia parva*, one might have thought that here the editor – or perhaps Scellinck himself – was speaking in the first person.

What is also remarkable is the modification of the place name in the Mainz manuscript: whereas the Latin logically names the city of Milan (for it was there that the meeting between Lanfranc and his patient had taken place), the Middle Dutch text names *Avinioen* [Avignon]. It is possible that the compiler of *M* (or the source from which he got this case) thought that the extract derived from the *Chirurgia magna* by Guy de Chauliac, who had served three popes in Avignon as a physician. In any case, the fact that such a fragment was inserted into the text of Thomas Scellinck of Tienen tells us that the written tradition and the use of authoritative surgical sources manifested itself strongly for a long time, in spite of the emerging changes in late medieval science. It also shows that even a vernacular surgical compiler with relatively modest training could easily maintain himself in the ebbing waves of reception of the great fourteenth-century surgeries by using more than one written source.

Finally, omitting this extensive and illustrative passage tells us something about the translator, who focused above all on the practice of surgery: for Lanfranc's story only mentions that it is possible to cure this severe illness and does not say how – for example, no prescription for the preparation of theriac is given. This is possibly the reason why the translator left it out.

But at the same time this translator does not hesitate to add new elements to the *Chirurgia parva* whenever he thinks necessary. This can already be seen in the second chapter, where one of the most exceptional interpolations of this text occurs. While he is dealing with the ways in which to treat ulcers, the translator bursts out in a lengthy addition to the original, in very vehement language that shows him to be very indignant. He vents his anger about the unskilled intervention of *sommighe sotten* [some fools] who make patients sicker rather than better. Because of such ignorant practicioners, eager for money, many of these patients die or become lame.

This interpolation comes in a context where the subject, the treatment of wounds, often divided medieval surgeons into two camps[103]. The conservatives (among whom included Lanfranc) pleaded mostly for a large production of pus. This so-called *pus laudabile* or 'praiseworthy pus' would help to drain the bad humours by which the body was afflicted. The progressives, on the other hand, like Henri de Mondeville, thought that it was much better to dry the wound by means of alcoholic substances. Clearly, the translator belonged to the progressive party: he does not like the use of greasy medicines or warm water, for they provided only more *wule materien* [polluted matters]. The interpolation thus characterizes the translator as a skilled surgeon, well acquainted with the debates of his time, not afraid to insert his own practical knowledge when he thought it necessary, but respectful of the learned written tradition (after all, he translated from the Latin). Most likely he belonged to the upper class of the surgical profession, that is, the surgeons who worried about their reputation and warned against the spreading inexpertness and swindling in their trade[104].

103. See, in particular, Huizenga, *Tussen autoriteit en empirie*..., p. 283-285.

104. About this subject, see Huizenga, *Tussen autoriteit en empirie*..., p. 238-247; and Huizenga, 'De diepe kloof...', *passim*.

The translator's actions thus reveal something of his personality and skills – an unintended signature.

Not only the translator, but also the scribes of the Middle Dutch texts of Lanfranc's *Chirurgia parva*, made some notable changes in their source text. We see for instance that in *L* a useful hint to determine whether the skull bone is broken or not is lacking, whereas it does appear in *P* and in the Latin text[105]. Elsewhere, towards the end of the *Chirurgia parva*, the scribe of *L* has added some prescriptions to the series of purgative and corrosive medicines that the Latin source already had. These prescriptions are also lacking in *P*, so it is almost certain that they derive from the scribe of *L*[106]. This scribe thus reveals himself to be more independent and possibly also more proficient or experienced than the scribe of *P*, who seems to have copied the text of the translation relatively faithfully and without further changes.

All in all, the changes in the Middle Dutch tradition of the *Chirurgia parva* of Lanfranc of Milan show very clearly some important features of vernacular surgical texts in the late Middle Ages, even though there are only two surviving textual witnesses: source texts were adapted, supplemented or stripped according to the individual needs and skills of the translator, editor or scribe. The great Latin surgeries were treasures of learning, conveying a range of reading and experience, which could be used to one's heart's content or even plundered. And this was precisely what happened: there was plenty of space in the vernacular reception of Latin surgical works to adapt the level and density of information to the intended public or to one's own skills – no work was sacred in this respect. More than any other medieval textual genre, *artes*-literature in general and surgical specialised literature in particular was characterized by flexibility and adaptation.

105. 'Item slaet opt hoeft mit enen lichten droghen wilghen of pijnbomen stocke ende legt vore ant hoeft van den zieken. Ende is dat been gans, zo seldi horen stomme gheluut nader comparacie van eenre heelre clocken; ende is tbeen ghescoert of ghebroken, zo ludet na eenre ghescoerder clocken; ende is tbeen gans zo seldi horen gansen gheluut' [Hit the head with a light dry stick made of willow or pine tree on the front of the patients' head. For if the bone is not broken, you will hear a quiet sound, similar to the sound of a large clock; and when the bone is ripped or broken, it will sound like a cracked clock; and if it is intact, you will hear an intact sound]. Lanfranc of Milan, *Chirurgia parva*, 7; ms. *P*, f. 63ra; Latin: f. 161vb. Cf. ms. *L*, f. 109rb (ed. Scholle, *Lanfranks «Chirurgia Parva»*, p. 53).

106. Lanfranc of Milan, *Chirurgia parva*, 46 and 49; ms. *L*, f. 124rb and 126ra (ed. Scholle, *Lanfranks «Chirurgia Parva»*, p. 89 and 94).

8. Conclusion

The relations between Latin and the vernacular in the Middle Dutch surgical literature were complex and manifold. In many different ways the codices show how the search for a new vocabulary for a language which was continuously developing could coexist fruitfully with the old, Latin one. The text of the *Cyrurgie* by Jan Yperman is a classic example of this.

'Quam ipse compilavit et in teutonico redigit filio suo', was the Latin incipit of this Middle Dutch surgical text, discussed at the beginning of this article. We saw that Yperman did not write in Latin himself. It seems that the Latin incipit resulted from the status of the Latin as the 'high domain', with a different function than the vernacular. It is likely that originally the Latin incipit did not belong to Yperman's text, but was added later on by a scribe of the tract – making the Latin a translation of the first Middle Dutch words of the original work. It was done to give the *Cyrurgie* a status that it lacked in the eyes of the scribe. This would explain why the same introduction appears twice in some of the codices containing Yperman's text, in Latin as well as in Middle Dutch. Moreover, there are some other textual arguments for Middle Dutch as the original medium for the *Cyrurgie*. One passage in one of the text witnesses says:

> Hier willic eynde maken vanden anderen boeke dye es vanden halze, wairbij ick bydde alle denghenen diere an zullen zien ofte leeren dat sij zullen bydden over den ghenen, die desen boec troc uten latyne int Vlaemsche, want ick deedt bij mijnre van den genen die achter mij soude bliven.
>
> 'Here I will end the other book which is about the throat, whereby I pray all those who will see it or be taught by it to pray for him, who turned this book from Latin into Flemish, for I did it for love of those who would remain behind (that is: for his son)[107]'.

So 'book' refers here to one of the Latin sources which Yperman himself translated, and not to a possible Dutch translation of a Latin version

107. Ms. Ghent, University Library, 1273, f. 105ra; see Van Leersum, *De «cyrurgie»...*, p. 137, n. 1; Van Leersum E.C., 'Notes concerning the life of Jan Yperman', *Janus* 18 (1913), p. 15. Some documents, like wills, show that late-medieval surgeons, who often taught their own sons, hoped that their pupils might profit from this learned written tradition. Both Yperman and Scellinck wrote in the first half of the fourteenth century surgical works for their sons, who did not know Latin, by translating Latin sources into the vernacular and inserting them into their own compilations. In the Middle Dutch surgical tradition it was common to dedicate one's book on surgery to one's son. See Van Herwaarden J., 'Medici in de Nederlandse samenleving in de late Middeleeuwen (veertiende-zestiende eeuw)', *Tijdschrift voor Geschiedenis* 96 (1983), p. 377. See also Huizenga, *Tussen autoriteit en empirie...*, p. 158-159 and 256.

of Yperman's work by some anonymous scribe. Furthermore, there is a passage in the same manuscript where it is explained that according to Galen, humours run to a sore place ('loopen ter onsochter steden'), followed by this addition: 'and so he [that is, Galen] puts it in Latin: "ad locum dolorosum confluent humores"', a sentence which would have no sense at all in a Latin work[108].

Jan Yperman wrote in his vernacular, but he lived in a world where his language was still in the process of developing. As a matter of fact, he belonged to two worlds: the world of the Latinitas, of the scientist, and the world of the vernacular, of the lay practitioner. As such he combined the best of both.

We know Yperman's name, his work and much about his life, for he was well known in his time, even famous, and scribes copied his work for a while. The same can be said of Thomas Scellinck of Tienen. Among many other accomplishment, Jan and Thomas translated Latin surgical works into their vernacular. Yet there were many other translators whose names are now lost. We cannot be as familiar with them as we can with Yperman, but we can observe them from a distance, through their unintended signatures.

Huygens Instituut, Den Haag

108. Ghent, University Library, ms. 1273, f. 3vb; ed. Van Leersum, *De «cyrurgie»*..., p. 145.

Orlanda S.H. LIE

WOMEN'S MEDICINE IN MIDDLE DUTCH

0. Introduction

In the thirteenth century, an anonymous Latin scholar wrote a treatise on the secrets of the female body that has survived in more than 80 manuscripts and more than 120 printed editions. The ideas propagated in this immensely popular text left an indelible mark on the medieval notion of sexuality and the tradition of women's medicine. The female body is described as a flaw of nature, a deformed male[1]:

> Thus, according to Aristotle in the 16th book On Animals, woman is a failed male, that is, the matter that forms a human being will not result in a girl except when nature is impeded in her actions because of the disposition of the matter and of natural heat, for a particular nature always intends to produce a male and not a female. If a female results, this is because of certain factors hindering the disposition of the matter, and thus it has been said that woman is not human, but a monster in nature.

Women menstruate because they are by nature cold and humid, whereas men are hot and dry. 'Because the heat in women is weaker than that in men, and all their foods cannot be converted into flesh, nature [...] leaves the excess in the place where the menses are kept'[2]. Due to this natural defect, women are less suitable for learning than men:

> Women are purged by means of their menstrual periods because the excess humidity is removed, and if it remained coarse humors would be formed from it. Therefore they are not suited for learning because of the coarseness of their spirits which stop up and deaden the brain[3].

In addition to being regarded as a waste product, menstrual fluid is also presented as a highly dangerous substance:

> From this it follows that the menses of women is extremely poisonous and infects the body, so that Albert tells us that if menses touch the twig

1. I am citing the Latin text in the English translation of Lemay H.R., *Women's Secrets. A Translation of Pseudo-Albertus Magnus's De Secretis Mulierum with Commentaries* (Albany N.Y., 1992), here p. 106.
2. Lemay, *Women's Secrets*..., p. 74.
3. Lemay, *Women's Secrets*..., p. 70.

> of a green tree it immediately dries up, therefore nature makes a great effort to expel this matter[4].

Another consequence of their innate imperfection is that, unlike men, women will benefit greatly from sexual intercourse:

> And note that coitus is beneficial for women because through it they lose their superfluous cold and receive heat, and this tempers their frigid natures; however the opposite is true in men. Thus women who have much sexual intercourse do not have their lives shortened as men do, and this is proved by the nature of the swallow who dies because he has sex too often[5].

As a medievalist, I have often wondered how medieval women reacted to this type of information. Did they laugh? Did they get angry? Did they dismiss the whole matter as a bag of lies, as suggested by Christine de Pizan, the first professional women writer of the Middle Ages[6]? These questions prompted me to look more closely at the way Latin knowledge of the female body was handed down in vernacular writings. In this article I will focus on the Middle Dutch gynecological tradition: what does the extant body of Middle Dutch texts on women's medicine tell us about the intricate process of knowledge transmission?

1. The Latin Tradition

Before we take a look at the transmission of women's medicine in the vernacular, let us look more closely at the Latin tradition that shaped the vernacular tradition. I have already introduced one influential text on this subject: the anonymous Latin treatise on the mysterious workings of the female body, the so-called *Secreta mulierum*, falsely attributed to the thirteenth-century Dominican scholar, Albertus Magnus. This pseudo-Albertan text originated in a monastic milieu. According to its author, the book had a manifold purpose: it could help women and provide a remedy for their afflictions; but it was also helpful in providing guidelines for the confessor in whom women will confide their ailments; moreover, it could also serve as a prophylactic: to warn its male readership for the shortcomings of the female body and their impurities[7]:

4. Lemay, *Women's Secrets...*, p. 75.
5. Lemay, *Women's Secrets...*, p. 70.
6. See Christine de Pizan, *The Book of the City of Ladies*, translated by E.J. Richards (New York, 1982), p. 21-24, especially p. 22-23.
7. See the preface of the 1580 Lyons edition, cited in Lemay, *Women's Secrets...*, p. 59-60 (citation p. 60).

> The reason for this is that women are so full of venom in the time of their menstruation that they poison animals by their glance; they infect children in the cradle; they spot the cleanest mirror; and whenever men have sexual intercourse with them they are made leprous and sometimes cancerous. And because an evil cannot be avoided unless it is known, those who wish to avoid it must abstain from this unclean coitus, and from many other things which are taught in this book.

It is safe to conclude, therefore, that the immense popularity of the *Secreta mulierum*, attested by a substantial amount of extant copies and translations into various vernacular languages, exerted considerable influence on the dissemination and transmission of gynecological writings in the vernacular.

In studying the transmission process of this body of Latin knowledge into the vernacular, it should be kept in mind that in the twelfth century, when the vernacular developed into a written language, there were different approaches to women's physiology. Generally speaking, the pseudo-Albertan *Secreta mulierum* provides a more philosophical and theoretical knowledge of the female body and female sexuality within the framework of natural science. A second approach is from a more medical and therapeutic perspective. A prominent representative of this tradition is the so-called *Trotula*, a compendium on women's medicine, named after Trota, a famous eleventh-century woman physician, who taught and worked in the Italian city of Salerno. The leading expert on the *Trotula*-tradition is the American historian Monica Green. Thanks to her in-depth studies and especially her recent edition of the Latin text, accompanied by an English translation, the Latin *Trotula* has become quite accessible to scholars of the vernacular tradition[8].

Compared to the *Secreta mulierum*, the original *Trotula*-tekst was more woman-friendly; it emphasized the anatomy, physiology, and pathologies of the female body and aimed at the maintenance of health and the treatment of diseases. While the *Trotula*-compendium also proceeded from the premise that woman are by nature weaker than man, it did not dwell on menstrual fluid as a corruptible influence and impure matter[9]. The dissemination of the Latin *Trotula* was extensive. There are 122 extant Latin manuscripts and twenty-five different vernacular translations, including translations and adaptations into Middle Dutch. Green's analysis of the manuscript tradition shows that 'the Latin texts existed in fifteen

8. See, e.g., Green M.H., *Women's Healthcare in the Medieval West. Texts and Contexts* (Aldershot & Burlington, 2000); Green M.H. (ed. & tr.), *The Trotula. A Medieval Compendium of Women's Medicine* (Philadelphia, 2001).

9. See Green (ed. & tr.), *The Trotula...*, p. 70-71.

distinct versions, with the various vernacular translations branching off from this 'tree' at differing stages'[10]. Later on, in the fourteenth century, the misogynistic philosophic tradition of the *Secreta mulierum* and the woman-friendly *Trotula* tradition moved toward each other and became indistinguishably mixed[11].

After this quick and superficial survey of the Latin tradition, I will now focus on the transmission process of gynecological knowledge by looking at the body of Middle Dutch texts on women's medicine.

2. The Middle Dutch Tradition

The Middle Dutch gynecological tradition shows an interesting range of manuscripts and printed texts. According to the inventory of Middle Dutch *artes* literature published in 1989 by Ria Jansen-Sieben, the extant Middle Dutch gynecological tradition is represented by more than twenty manuscripts and thirty printed texts[12]. Comparison with the Latin gynecological tradition reveals that the transmission of women's medicine in Middle Dutch is manifold and polymorphous. Both the *Secreta mulierum* and the *Trotula* tradition have placed their stamps on the Middle Dutch writings.

2.1. Women's medicine in the manuscript tradition

It is possible to classify the extant manuscript tradition into four groups[13]:

The first group (group A) presents an abridged translation or adaptation of the first book of the *Trotula*, dealing only with a few basic chapters on

10. Green M.H., 'The development of the *Trotula*', *Revue d'Histoire des Textes* 26 (1996), p. 119-203, especially p. 123. This essay was also published in Green, *Women's Healthcare...*, V, p. 119-203.

11. Cf. Green M.H., ''Traittié tout de mençonges'; The *Secrés des dames,* 'Trotula' and Attitudes Towards Women's Medicine in Fourteenth- and Early Fifteenth-Century France', in: M. Desmond (ed.), *Christine de Pizan and the Categories of Difference* (Minneapolis, 1998), p. 146-178, also published in Green, *Women's Healthcare...*, VI, p. 146-178.

12. Jansen-Sieben R., *Repertorium van de Middelnederlandse Artes-literatuur* (Utrecht, 1989), p. 63-67.

13. My classification is greatly indebted to Monica Green's analysis of the Latin and vernacular manuscript tradition: see Green M.H., 'The Development...'. Green M.H., 'A Handlist of Latin and Vernacular Manuscripts of the So-called *Trotula* Texts', *Scriptorium* 50 (1996), p. 137-175 provides an inventory of the Latin manuscripts; the vernacular manuscripts are analyzed in the next volume: *Scriptorium* 51 (1997), p. 80-103, especially p. 81-84.

the female physiology, menstruation and related problems, recipes and various treatments of gynecological disorders. About twelve manuscripts belong to this group, but so far I have only analyzed four of them more closely: Utrecht, UB, 1328; Vienna, ÖNB, 2818; Hattem, GA, 958; Amsterdam, UB, II E 42[14]. The following observations are based on this first sampling.

The Middle Dutch tracts in these four manuscripts are written in prose; they have survived in miscellany manuscripts from the fifteenth and sixteenth century. Judging from the predominant medical and pharmaceutical contents of the surrounding texts in these codices, I would suggest that the gynecological information was probably consulted by pharmacists and medical practitioners. Looking at the provenance of these four manuscripts, it is interesting to note that two of these manuscripts, both dating from the sixteenth century, show traces of a female owner or user of the text. Manuscript Hattem, GA, 958 features the signature of Anna Sebastiaens, and a reference to the countess Juliana of Nassau as the source of a remedy against epileptic seizure. A seventeenth-century owner of manuscript Amsterdam, UB, II E 42 mentions that he received this book from Maria, countess of Berge and Culemborch in 1608.

The second group (group B) is a hybrid text. Its main portion consists of a translation of the first two books of the *Trotula*-compendium, but it also shows the influence of other gynecological writings, such as the pseudo-Albertan *Secreta mulierum*, pseudo-Aristotle's *Secretum secretorum* and the pseudo-Galenic *De spermate*[15]. These tracts deal with both the male and female sexual parts and related ailments. The Middle Dutch representative of group B is written in verse and has survived in two miscellany manuscripts: Brussels, KB, 16.421-42 and Vienna, ÖNB, 2818[16]. Recent studies by Dutch medievalists show that both manuscripts were in the hands

14. For the Utrecht-*Trotula* see Daems W.F. (ed.), *Boec van medicinen in dietsche. Een Middelnederlandse compilatie van medisch-farmaceutische literatuur* (Leiden, 1967), p. 179- 181; the Viennese codex is available in a diplomatic edition: Huizenga E. (ed.), *Het Weense arteshandschrift. Hs. Wenen, Österreichische Nationalbibliothek, 2818. Diplomatische editie* (Hilversum, 2004), 2 vols., the *Trotula*-tract is in vol. 2, p. 1019-1026. See the database *Medieval Manuscripts in Dutch Collections* (www.mmdc.nl) for a codicological description of the Hattem and Amsterdam codex. For the Hattem manuscript, see also Lie O.S.H. & Versélewel de Witt Hamer N., 'Beproefde middeleeuwse recepten in het Hattemse gemeentearchief', *Heemkunde Hattem. Tijdschrift van de Vereniging Heemkunde Hattem* 81 (1999), p. 199-204, especially p. 198.

15. According to Green, the Middle Dutch *Trotula*-text of group B goes back to the version of the so-called *meretrices*-group, see Green, 'A Handlist of Vernacular Manuscripts...', p. 82.

16. For an edition of the Middle Dutch text of group B, see 'Der mannen ende der vrouwen heimelijkheid', in: N. De Pauw (ed.), *Middelnederlandsche gedichten en fragmenten* (Gent, 1893), vol. 1, p. 121-190.

of male professionals. The Brussels manuscript dates from 1351; it was compiled and used by a fourteenth-century surgeon, Jan van Aalter[17]. The fifteenth-century Viennese codex was also consulted by a surgeon[18].

Group C represents a free translation and adaptation of the gynecological and obstetrical selections from the *Trotula*-standardized ensemble, supplemented by material from other gynecological sources. Three Middle Dutch versions are extant: Brugge, SB, 593; Copenhagen, GKS, 1657, 4°; Hamburg, SB, cod. med. 798. All three versions are written in prose and have survived in fifteenth-century manuscripts. In all three texts the author is addressing a female audience or female readership:

> Ic bidde alle vrauwen die dit sullen sien ofte horen lesen hem biddic dat si my niet ne vloucken want ic ne hebbe hem dit niet gescreven te scande maer te horen profite
>
> 'I pray all women who will see or listen to this that they not curse me, for I did not write this to their shame but to their benefit[19]'.

Furthermore, all three texts are accompanied by illustrations of pessaries and instruments for the treatment of menstrual and vaginal disorders. The presence of these pictorial instructions points at their practical aims and purposes: these texts were probably intended as a health guide to benefit women, either to inform them about their own ailments and/or to instruct medical practitioners. While the Bruges and Copenhagen manuscripts contain explicit references to Trotula as an authoritative source, the Middle Dutch version in the Hamburg manuscript has omitted all references to Trotula. Instead the text is presented as the work of Albertus Magnus (see the opening page: 'Hic incipit alberti liber de secretis mulierum') and is combined with a translation of the pseudo-Albertan *Secreta mulierum*. In this part of the text, the translator also propagates the pseudo-Albertan view that menstrual fluid will harm dogs, kill trees, engender a serpent, cast spells on men. The remaining contents of this miscellany manuscript clearly suggest that the codex was intended for

17. Biemans J.A.A.M., 'Het chirurgijns boek van Jan van Aalter. Over schaalvergroting en nieuwe toepassingen bij de productie en vormgeving van het handgeschreven boek in de veertiende eeuw', *Jaarboek voor Nederlandse Boekgeschiedenis* 6 (1999), p. 67-86.

18. Huizenga E., *Een nuttelike practijke van cirurgien. Geneeskunde en astrologie in het Middelnederlandse handschrift Wenen, Österreichische Nationalbibliothek, 2818* (Hilversum, 1998), p. 301-305.

19. The example is from the Bruges manuscript (Bruges, SB, 593, f. 1r.). For an edition see Delva A., *Vrouwengeneeskunde in Vlaanderen tijdens de late middeleeuwen*, Vlaamse Historische Studies 2 (Brugge, 1983). The Copenhagen *Trotula* has been edited by Kusche B., *Frauenaufklärung im Spätmittelalter. Eine philologische-medizinhistorische Untersuchung und Edition des gynäkologisch-obstetrischen GKS. 1657, Kopenhagen* (Stockholm, 1990).

professional usage, for example by surgeons and other medical practitioners[20].

The fourth variant (Group D) is witness to yet another branch of the medieval gynecological tree. This Middle Dutch version presents a rhymed translation of the pseudo-Albertan *Secreta mulierum*, but it also shows influence of the work of another clerical author, Thomas of Cantimpre's *Liber de natura rerum*[21]. The Middle Dutch translator of the text preserved in the Ghent University Library, claimed to have used a Latin source for his translation: 'Alsoe als't in't lattijn es bescreven/ Sal ic 't in Dietsche voert vertrecken nu' ('I will translate it into Dutch, in accordance with the way it has been written in Latin')[22]. An intriguing characteristic of this rhymed tract is the interweaving of lyrical interludes, in which the 'I' proclaims his love for a lady. He presents himself as a steadfast lover, who suffers innumerable hardships in the name of love and is rewarded at the end, by obtaining his lady's love.[23] Scholars have made the suggestion, based on an acrostics, that the original translation was intended for a hitherto unidentified lady: Margareta Godevarts[24]. Further analysis of the text has led to the conclusion that this Middle Dutch tract was interesting to both professionals (midwives, medical practitioners) and non-specialists (women and men)[25]. In the

20. The codex contains Latin medical tracts and surgical manuals, including Middle Dutch excerpts from the surgical writings of Albucasis. No edition exists of the Hamburg *Trotula*.

21. For an edition of the Middle Dutch gynecology that is representative of group D (ms. Ghent, UB, 444), see Blommaert Ph., *Der vrouwen heimelijkheid. Dichtwerk der XIVe eeuw* (Gent, 1846). An excerpt with obstetrical information, based on Thomas of Cantimpré's *Liber de natura rerum* has survived in manuscript Brussels, KB, 19308 and is edited by Elaut L., *Van smeinscen lede. Een middelnederlands geneeskundig geschrift. Zijn betekenis in het raam van de medische literatuur der dertiende eeuw* (Sint-Niklaas, 1956); ed. Elaut 1956, v. 1580-1681 corresponds with ed. Blommaert 1846, v. 823-925.

22. Blommaert, *Der vrouwen heimelijkheid...*, v. 37-38:

23. See Lie O.S.H., 'De verliefde leraar. Liefdesverzen in Middelnederlandse artesteksten', in: R. Sleiderink, V. Uyttersprot & B. Besamusca (eds.), *Maar er is meer. Avontuurlijk lezen in de epiek van de Lage Landen. Studies voor Jozef D. Janssens* (Leuven, 2005), p. 239-256, here p. 249-251. Lie suggests that the dual function of the first person narrator, i.e. as a lover and as a teacher, is a rhetorical device, inspired by the courtly literature tradition.

24. Van Doorn M. & Kuiper W., 'Der vrouwen heimlicheid', *Spektator* 6 (1976-1977), p. 539-551; cf. also Claassens G.M.H., 'Een dichter lijdt. De verborgen liefdesgeschiedenis in "Der vrouwen heimelijcheit" (Gent UB., 444)', in: K. Porteman, W. Verbeke & F. Willaert (eds.), *Tegendraads genot. Opstellen over de kwaliteit van middeleeuwse teksten* (Leuven, 1996), p. 93-102 and Claassens G.H.M., 'From *Ars* to *Amor*. Ms. Gent, U.B., 444 as a labour of love', in: G. Claassens & W. Verbeke (eds.), *Manuscripts in transition. Tradition and creative recycling* (Leuven, 2006), p. 97-111.

25. Poel D.E. van der, 'Mijn lieve joncfrouwe heft mi gebeden iet te dichtene', *Nederlandse letterkunde* 1 (1996), p. 249-260.

seventeenth century the codex was owned by a member of a religious order for men[26].

One of the distinguishing features of this Middle Dutch rhymed version is its adherence to the menstruation myth, conspicuously absent in the Middle Dutch *Trotula*-versions of the other three groups. In this fourth group, menstruation is presented as impure matter: a menstruating woman will contaminate her own child simply by looking at it; mirrors will stain when a menstruating woman looks in them; a dog who eats from the menstrual fluid will go berserk; a leafy tree will dry up when menstrual fluid is poured over its roots, etc. All these misogynistic elements can be recognized as part of the pseudo-Albertan *Secreta mulierum* tradition.

Interestingly, this Middle Dutch tract has had an impact on other Middle Dutch writings, both in the medical and non-medical field. A recent study by the Belgian medievalist Joris Reynaert shows that excerpts of this Middle Dutch rhymed gynecology were later incorporated into a surgical tract, a Middle Dutch translation of the *Chirurgia magna*, written by the fourteenth-century scholar and surgeon Lanfranc of Milan. This surgical tract was translated into Middle Dutch on three different occasions. One of these translations (preserved in Ms. Ghent, UB, 1272) has expanded the chapter on anatomy with excerpts from *Der vrouwen heimelijcheit*[27]. Apparently, the Middle Dutch translator of the *Chirurgia magna* considered the gynecological information from this rhymed tract useful and worthy of incorporation into his prose surgical manual.

A second Middle Dutch author who was familiar with the tradition of *Der vrouwen heimelijcheit* for gynecological and obstetrical information, is the anonymous fourteenth-century translator of the *Historiebijbel*, based on the *Historia Scolastica* of Petrus Comestor (+ 1179), commissioned by

26. For a detailed codicological description of this codex see Reynaert J., *Catalogus van de Middelnederlandse handschriften in de Bibliotheek van de Rijksuniversiteit te Gent. 1. De handschriften verworven vóór 1852* (Gent, 1984), p. 111-113. Aside from the Middle Dutch text, there is also a Low German variant (Berlin Germ. 8°, 187) that belongs to this tradition. Comparison between these two texts shows that both are rhymed and have a prologue that is addressed to a beloved lady. Interestingly, however, the rest of the German text shows no sign of the lyrical interruptions that are so characteristic of the Dutch *Der vrouwen heimelijcheit*. Cf. Van Doorn & Kuiper, 'Der vrouwen heimlicheid' and Lie O.S.H., *Vrouwengeheimen. Geneeskunst en beeldvorming in de Middelnederlandse artesliteratuur*, Achtste Bert van Selm-lezing (Amsterdam,1999), p. 26-33.

27. Reynaert J., 'Over medische kennis in de late Middeleeuwen', *Millennium* 13 (1999), p. 21-30; Reynaert J., 'Der vrouwen heimelijcheit als secundaire bron in de Zuid-Nederlandse bewerking van de 'Chirurgia Magna' van Lanfranc van Milaan', *Koninklijke Academie voor Nederlandse Taal- en Letterkunde* 2001/1, p. 165-188.

Jan Taye, a patrician from Brussels[28]. His commentary of *Genesis* 18:11 ('Both Abraham and Sarah had grown very old, and Sarah was past the age of childbearing.') features a little digression on menstruation and menopause which can be traced back to *Der vrouwen heimelijcheit*[29]. In a similar vein, the account of Tamar who gave birth to twins (*Genesis* 38:27-30) is accompanied by a discussion on the seven-chambered uterus, multiple births and the process of the development of the fetus. Some of this information is paralleled in *Der vrouwen heimelijcheit*[30].

2.2. Women's medicine in the age of printing

With the invention of the printing press, a new group of recipients came into view. These potential readers must be looked for in an urban setting, especially among the literate members of the bourgeoisie (male and female), the (lower) clergy and the aristocracy.[31] At the same time, a growing group of professionals, such as medical practitioners, also increased the demand for information manuals in the vernacular.

In the era of printed books, gynecological and obstetrical information continued to be propagated. To give a first impression of this printed tradition, I will briefly introduce five late medieval texts that deal with women's medicine:

1. *Der vrouwen natuere ende complexie*
2. *Tprofijt der vrouwen*
3. *Den Roseghaert vanden bevruchten vrouwen*
4. *T'Boeck vande Vroet-Wijfs*
5. John of Ketham, *Fasciculus medicinae*

2.2.1. Der vrouwen natuere ende complexie ('Of womens' nature and complexion')

The *Repertorium* lists nine sixteenth-century editions of this text, printed in Antwerp, Utrecht, and Amsterdam[32]. In 1980 the Belgian scholar W. Braekman published a facsimile edition of the 1538 Utrecht

28. Sherwood-Smith M.C., 'God and Gynaecology: Women's Secrets in the Dutch *Historiebijbel van 1360*', *German Life and Letters* 50 (1997), p. 390-402. For recent information on the *Historiebijbel* see Kors M., 'De genese van de *Historiebijbel* van 1361. De bewerkingstechniek in Genesis 1-3', *Nederlandse letterkunde* 8 (2003), p. 217-255.
29. Sherwood-Smith, 'God and Gynaecology...', p. 396-398.
30. Sherwood-Smith, 'God and Gynaecology...', p. 398-401.
31. Cf. Franssen P.J.A., *Tussen tekst en publiek* (Amsterdam, 1990), p. 165ff.
32. Cf. Jansen-Sieben, *Repertorium...*, p. 64-67.

edition by Jan Berntsz[33]. The author/printer opens his book with the promise that he will disclose information on a new and remarkable subject: the complexion and nature of women[34].

> Indien die menschen begeeren nieuwe ende vremde dinghen te hooren ende te weten. So sal ick huer wat scrijven van dye complexie ende natuere der vrouwen/ ende wat vrouwen dat heet van natueren zijn/ ende begeerlijc/ ende wat vrouwen dat cout van natueren zijn ende niet begeerlic. Om dat te kennen bi sekere teykenen van buten te sien. Hier bi mogen die mannen weten hoe si met haer vrouwen leven souden/ ende die vriendelicke te tracteren na haer begeerte ende natuere/ want si schamen hem te bidden. Doer dit mach dye man leeren ende kennen die vrouwe/ ende doen haer gerief/ want het maect groote vrienscap tusschen man ende wijf/ ende tusschen twee ghelievekens ende amoreuskens. Men plach oock te seggen inden tijt voorleden/ Een wel gemint wijf is een huijs vol vreden.

> 'If people desire to hear and learn more about new and extraordinary things I shall write for them about the complexion and nature of women. About what type of women has a hot temperament and is suitable for love and what type is cold and unsuitable for love. How to recognize these types of women from the way they look. How men will know how to live with their wives and how to please them sexually, for the women are ashamed to ask [for this favor]. With this information men can learn about the nature of women and fulfill their needs, since this matter generates great love between man and wife and between lovers. A well-known saying from the past is: a wife well-loved makes a peaceful home.'

In other words: the printer is advertising his work as a guide book for men. It provides information that will help them in their amorous dealings with members of the opposite sex. The subject matter is sexuality and the female body. Women are discussed as sex objects; a woman is sexually insatiable, menstruation is considered a highly dangerous and corruptible substance. Men who consult this manual will gain a better insight into woman's nature and know better how to please her. This type of knowledge will engender great love between husband and wife, or between lovers, and make an important contribution to domestic peace.

33. Braekman W.L., *Der vrouwen natuere ende complexie. Een volksboek, naar de Utrechtse druk van Jan Berntsz, van omstreeks 1538* (Sint-Niklaas, 1980). The oldest extant edition, however, dates from 1531 and was issued by the Antwerp printer Jan van Doesborch, see Vriesema P., 'Recensie van W.L. Braekman. *Der vrouwen natuere ende complexie*', *Dokumentaal* 10 (1981), p. 128-132.

34. Braekman, *Der vrouwen natuere…*, Die Prologhe.

The epilogue corroborates the suggestion that the book is intended for a male readership[35]:

> Dits ghedaen om dat elc man sou weten/ Hem te regeren na wisen raet/ Op dat hi niet en worde gesmeten/ vanden vrouwen met moede quaet/ Dus neemt it goede die hem groots vermeten/Laet staen de woorden/ volbringt die daet/ Dit doch gade slaet.
>
> 'This was done [written] in order for each man to know how he should behave himself wisely, so that he will not be beaten up by evil-minded women. Take the good things to heart, you, who have great ambitions, forget about words, perform the act; take note of this.'

Interestingly, subsequent editions have left out this epilogue, together with the preceding paragraph on the male private parts ('Die tekenen & complexion der secreter ballen'; 'The characteristics and complexion of the testicles'). Scholars have suggested that these adjustments were made to transgress its original exclusive male readership in order to accommodate a growing group of female readers as well[36].

The source of this intriguing book is the *Liber Physionomiae* of Michael Scotus, who wrote the text in 1220 for the German emperor Frederic the Second. It was also known by different names, such as *De secretis naturae*, or *De secretis mulierum*. The original Latin text consists of three books. The first and second book deal with the nature and the complexion of women; the third book is devoted to the physiognomy of both sexes. The Middle Dutch text is a translation of the first two books. Its opinion on women, their body, their complexion and their sexuality, concurs with the Pseudo-Albertan tradition of the *Secreta mulierum*. The emphasis is more on the curious characteristics of the female body and female sexuality, and less on the diagnosis and treatment of women ailments. The menstruation myth is propagated once again:

> Van dat menstruum scrijftmen dat tet een hont laepte/ hi wort rasende/ spreyt ment op groen cruyden/ si sullen verdorren gheeft ment eenen persoon teeten/ hy wordt sodt/ ende het maket hem lazarus. Dat menstruum ys bequaem om die vrouwe te doen ontfanghen/ want en is daer geen menstruum in die cellekens der moedere/ soe en dooch die generacie niet/ ende isser te veel van dat menstruum/ so crijcht de vrucht een teyken int quaet Ende geschiet die generacie inden tijt des vloets van dat menstruum/ soe wort die creatuer geteykent int quaet/ als een bult met lazarie/ blint/ crepel/ lam/ in armen/ in handen/ in voeten of in vingeren[37].

35. This information is absent in Braekman's edition. I am citing the text from the oldest extant edition of Jan van Doesborch (1531), see Vriesema, 'Recensie...', p. 131.
36. Cf. Vriesema, 'Recensie...'; Franssen, *Tussen tekst en publiek...*, p. 33-34.
37. Braekman, *Der vrouwen natuere...*

'It is written about menstruation that if a dog eats from it he will go berserk; if it is poured on green herbs, they will wither; if administered to a person, he will go mad and become leprous. Menstruation is indispensable for women to allow them to conceive, because without menstruation in the womb, conception is not possible; if there is an excess of menstrual fluid, the fetus will get a deformity. And if conception takes place during menstruation, the fetus will get a deformity, such as a leprous growth, blindness, crippleness, lameness in arms, hands, feet or fingers.'

2.2.2. *Tprofijt der vrouwen ('The profit of women')*

Of this text, five printed editions from the sixteenth century are extant[38]. It was written to benefit both women and men: '*The profit of women*, in which is taught the remedy for all ailments of women, widows, girls and all other persons. It also teaches how to live chastely. And for the man whose nature has dried up, it will inform him how to strengthen and restore his nature'[39].

The prologue features a dialogue between the printer and a young girl on the issue of misogynistic literature. The girl reprimands the printer for producing works that are degrading for women. She even refers to a specific work by name that is extant today: *Tbedroch der vrouwen* ('The deceit of women')[40]. But the printer disagrees by arguing that this type of books should in fact aid women in coping with husbands who are drunk, adulterous, addicted to gambling, and quarrelsome. Realizing that the conversation is going nowhere, the girl changes the subject and inquires about the name of his new book: 'Het profijt der vrouwen/ daermen klaerlijcken zal mogen aenschouen, raet en remedie voor alle ghebreken / die den vrouwen onder de leden steken.' ('The profit of women / where one can find good advice and remedy for all ailments that afflict women.')

The printer/author concludes the prologue with his plan for the book. In order to accomplish his goal and to make the contents of the book more readily available, he will give several recommendations and remedies for each affliction, so that each person can chose the treatment that is available. In the closing statement of the prologue he reiterates once again that he is writing this book for the benefit and physical well-being

38. Jansen-Sieben, *Repertorium...*, p. 66-67

39. 'Tprofijt der vrouwen in het welcke geleert wort de remedie teghen alle ghebreken der vrouwen, weduwen, meyskens ende allen anderen persoonen. Ende om kuysschelijc te leven. Den verdrooghden mans, om haer natuere te versterckene ende te restaureren.' No edition of the text is available. I am citing from the edition printed by Jacob Bosselaer of Antwerp in 1595, preserved at the University Library of Amsterdam University, ex. 652 J 27.

40. For a facsimile edition see Braekman W.L., *Dat bedroch der vrouwen. Naar het unieke, volledige exemplaar van de Utrechtse druk van Jan Bernts. Van circa 1532* (Brugge, 1983).

of women and he humbly asks his female readership for understanding and forgiveness, since he is going to divulge their intimate secrets in this work:

> Maer ic submittere my der goedertierenheyt ende zaechtmoedicheyt der vrouwen en ic bidde hun ootmoedelijcken dat sy hen op my niet en stooren noch vergrammen, omdat ik hier van haren secreten ende gebreckelijkcheden vermane, het welcke ic niet en doe tot haren spijte, noch verwijte, maer tot haren profijte, om dat sy zouden hebben welvaren en ghesontheyt.
>
> 'But I submit myself to the goodness and gentleness of women and I humbly pray that they do not get angry or annoyed with me for disclosing their secrets and defects. I do not do this to harm or blame them, but to benefit them so that they may enjoy wellbeing and good health.'

By integrating this traditional feature of gynecological writings (in which the author apologizes for his disclosure of intimate female matters) into his dialogue with the girl, the printer acknowledges his intended female readership, emphasizing once again that his book is beneficial to all women and that it offers a wealth of useful information to them.

Tprofijt der vrouwen provides treatment and recipes for more than thirty ailments in the field of women's medicine. A few examples may suffice: 'How to keep the breasts of virgins and ladies small, how to achieve firm breasts', 'How to provoke the menses'; 'How to contain excessive menstrual flow', 'How to treat afflictions of the womb', 'How to make women fertile', 'How men and women can live a chaste life'; 'What is good and bad for pregnant women', 'How to extract a dead fetus', 'How to heal a ruptured vagina after delivery'; 'How nursing mothers can increase their milk'; various recipes for beauty treatments (such as facials and body scrubs). But there is also attention for male concerns, such as: 'How old or sexually worn-out men can regain their sexual vigor'.

The contents of this treatise suggest that the author intended to impart knowledge to a general readership of both laywomen and laymen. Unlike *Der vrouwen natuere ende complexie*, the primary focus of *Tprofijt der vrouwen* is on the practical side of women's medicine. No mention is made of the poisonous quality of menstrual fluid; instead, ample space is reserved for the treatment of ailments that effect the female body, with some attention being paid to male issues, such as impotence due to old age, or remedies to increase sexual performance. Around 1540 *Tprofijt der vrouwen* featured as an appendix in the Middle Dutch translation of

Eucharius Rösslin's *Den Roseghaert vanden bevruchten vrouwen*, a manual for midwives[41]. To this work I will now turn.

2.2.3. Den Roseghaert vanden bevruchten vrouwen ('The rosegarden of pregnant women')

In 1516, three years after the original text written by the German physician Eucharius Rösslin was published, the Brussels printer Thomas van der Noot produced the Dutch translation. The substantial number of extant printed editions, sixteen in the sixteenth century, and eight from the seventeenth century, underlines its popularity[42].

According to the prologue the manual was written for pregnant women and midwives. The author makes it very clear that the experts on women's medicine are male philosophers and physicians. Even though it is evident that gynecological and obstetrical information from the *Trotula*-tradition has been incorporated along the way, no tribute is paid to any female authorities. Instead, the text issues a clear warning against ignorant midwives who endanger the lives of pregnant women[43]:

> Beminde leser dit teghenwoerdige boeck is nu opt nieuwe herprent / ende neerstelijc met groten arbeyt ende coste gecorrigeert ende vermeerdert uut die boecken van die alder excellenste scrivers die van naturlicke ende secrete dingen der mannen ende vrouwen ghesecreven hebben/ ende van dat ontfangen / baren / ende hantierengen der vrouwen/ de welcke hier vore opten titel meest genoemt staen / tot troost ende onderstant der eersamiger vrouwen die met kinde bevrucht sijn / opdat si daeruut mogen weten hoe men met hem lieden omgaen sal/ wantter vele in perikel haers levens comen doer die ongeleerde vroevrouwen / Item tot onderwisinge ende leeringe der vroevrouwen die haer daer mede becommeren / opdat sy den eerliken vroukens als hem den last nakende is / mogen weten te subvenieren / want leeringen van experte meesters sijn [...]. Ende [wi] bidden alle personen dient in handen comen sal / dat si dat so eerlijck ende secretelijc houden (gelijck si dat int uuterste oerdeel voer den oegen gods verantwoerden willen) op dat niet en come in handen der kinderen oft vileynen / diet liever lesen souden den vrouwen te verwijte oft tot onaerdicheyt Daerom en salmen dese boeken niet alle personen vercopen/ noch thoonen noch te coope vinden tot enighen boeck vercopers/ dan alleene in die coopstat van Antwerpen bi Symon Cock ende Jacob van Liesvelt diese gheprint hebben int iaer 1528.

41. Franssen, *Tekst en publiek...*, p. 34.
42. Jansen-Sieben, *Repertorium...*, p. 64-67; Franssen, *Tekst en publiek...*, p. 34.
43. No edition is available; I cite the text from the 1528 edition printed in Antwerp by Simon Cock and Jacob van Liesvelt, signature Amsterdam, UB, Ned. Inc. 446.

'Dear Reader: this book is a reprint, with great care and cost, corrected and expanded with material from books written by authoritative writers, named previously on the title page, who are experts in the field of men and women's secrets, of gynecology and obstetrics[44]. It is printed to comfort and support honorable, pregnant women, so that they can learn from it how they should be treated when they are with child, because their live is at risk due to the (actions of) illiterate (ignorant) midwives. This book is also to instruct midwives so that they may learn from the teachings of learned men in order to know how to treat the honorable ladies when their time has come to deliver a child. [...] And [we] therefore pray all persons who receive a copy of this book that they take care not to let it fall into the hands of children or villains, who would read it in order to blame or disparage women. For this reason this book should not be sold to all persons, nor should it be shown or sold by any bookseller, except in the merchant city of Antwerp, by: Symon Cock and Jacob van Liesvelt, who printed it in 1528.'

2.2.4. T'boeck vande Vroet-Wijfs ('The Book of the Midwives')

Another authoritative sixteenth-century manual for midwives, based on personal experience, is *T'boeck vande Vroet-Wijfs* of the Swiss physician, Jakob Rüff. The text was translated into Middle Dutch in 1591 by Martyn Everaert. The *Repertorium* lists three extant editions from the sixteenth century. A salient feature of this manual is its paternalistic and moralistic tone. By incorporating the reference to *Genesis* 3 on the title page, the author clearly suggests that labor pains and a wife's obedience to her husband are part of the divine plan[45].

T'boeck vande Vroet-Wijfs. Int welcke men mach leeren alle heymelicheden vande vrouwen/ ende in wat ghestalte de Mensche in zijn moeders lichaem ontfanghen/ groeyet ende gheboren wort. Voorts hoe alderleye siecten die den Craemvrouwen lichtelick overcomen/ met costelicke Medicijnen mogen voorcomen ende ghenesen worden. Alle t'samen uut eygen ervarentheyt vanden seer vermaerden Jacob Ruffen Stadt Medicijn tot Zurich/ eertijts in druck uutgegheven/ ende nu ter tijt op een nieu verbetert/ ende met schoone figueren verciert. Daer by ghevoecht zijnde een profytelicke leeringhe, van het voesteren ende handelen vande nieugheboren kinderen. Overgeset uutten Hoochduytsche in ons Nederlantsche sprake / Deur Martyn Everaert B.

Genesis III. Godt die Heere sprack totten Wyve, ick sal u smerten vermenichvuldighen, te weten als ghy bevrucht zijt, Ghy sult met smerte

44. These are Albertus Magnus, Aristotle, Plinius, Avicenna, Marcus Varro, Jason a Pratis, and others.

45. I am citing the text from the 1591 edition, printed by Cornelis Claesz. in Amsterdam; signature Amsterdam, UB, 976 G 42.

u kinderen baren, ende uwen wille sal uwen Man onderworpen zijn, ende hy sal u Heere zijn.

'The book of the midwives. Herein one may learn all the secrets of women and (everything about) the various shapes in which the fetus in its mother's body is growing and is delivered. In addition (information) about how to prevent and cure with precious medication all sorts of ailments that could afflict the mother of the newly born. All these matters are based on the personal experience of the famous Jacob Ruffen, city physician of Zurich; published in print before, but this time printed in a new, improved version and decorated with beautiful illustrations. Added is also a profitable treatise about breast feeding and the nursing of newly born children, converted from the high German into our Dutch tongue by Martyn Everaert B.

Genesis III: God, the Lord, said to the woman: "I will increase your labour and your groaning, and in labour you shall bear children. You shall be eager for your husband, and he shall be your master"'.

The treatise also contains a chapter in which the author fulminates against those frivolous and sinful ladies, young girls and widows, who having indulged in lascivious behavior, have become pregnant. To cover their shame they go at great lengths to harm their unborn child. Prompted by the devil they seek help from an old, evil woman who is an expert in this sinful business[46].

2.2.5. John of Ketham, Fasciculus medicinae ('Medical Treatise')

The last printed text to be discussed is the Middle Dutch translation of a medical treatise written by the fifteenth-century German Physician, John of Ketham. His *Fasciculus medicinae* was the first printed book with anatomical illustrations. In 1512 a Middle Dutch translation by Petrus Antonianus came from the press in Antwerp. The *Repertorium* lists five extant copies of the Middle Dutch version[47]. A facsimile edition of the Middle Dutch translation in the Antwerp edition of 1512 was published in 2003[48].

46. 'Daernae gaen sy deur ingeven des duyvels ende uut een ongeregelt boos gemoet ende herte tot een eerloos boos snoode ende in desen dingen gheleert out ende wel ervaren wijf die veel meestert ende alle haer leven lanck in sulken saken haer ghebruyct ende geoeffent heeft die oock daer voor over al langen tijt ende veel iaren heeft openbaerlick befaemt ende bekent geweest.'

47. Jansen-Sieben, *Repertorium...*, p. 100.

48. *Joannes de Ketham. Fasciculus medicine houdende in hem dese nauolghende tractaten...vertaald door Petrus Antonianus. Fotografische herdruk van de uitgave door Claes de Grave te Antwerpen uit 1512 (Koninklijke Bibliotheek van België, II 12.786 A (RP),* (Brussel, 2003).

Ketham's treatise 'which is recommended to surgeons and other persons' ('die allen cyrurginen ende andere menschen te wetene seere profitelijc ende nootsakelijc zijn'), presents treatments and cures for a wide range of afflictions, including two chapters on women's medicine: 'Een tractate teghen die ghebreken der vrouwen' ('a treatise on the afflictions of women') and 'Sommighe probleumata ende natuerlike vraghen van die leden der generatien' ('some issues and questions on the generative parts').

A comparison between the contents of Ketham's tract and the contemporary gynecological tradition makes it clear that Ketham is following the natural philosophical pseudo-Albertan *Secreta mulierum* tradition. Here, too, attention is paid to the dangers of menstruating women, and women who have entered menopause:

> Ende daerom vergaderen si in haer een quade materie. Want si onsuiver worden also dat si met haren ademe die kinderkens dicwils ontsteken. Ende daeromme hebben si dicwils vuyle catarren ende hoesten op die borst ende ander vele coude siecten. Ende hierom raden alle philozophen dat hem iegelijck van oude vrouwen wachten sal (f. 27v).
>
> 'And for that reason they retain the bad substance (i.e.: menstrual fluid) in their body. They then become so contaminated that they often infect children with their breath. And this is why they often have terrible cataracts and coughing from the chest, and many other ailments caused by their cold nature. And for this reason, all philosophers advise that everyone should be careful of old women.'

3. Conclusion

The transmission and propagation of gynecological and obstetrical knowledge in Middle Dutch took place in different ways and in different forms. It becomes apparent that the extant corpus of Middle Dutch texts bears the stamp of both the clerical Latin tradition of the pseudo-Albertan *Secreta mulierum* and the more practical and therapeutic tradition of the *Trotula*.

The Middle Dutch manuscript tradition of women's medicine attests to the fact that written knowledge about the female body and female sexuality was copied, transmitted and propagated in various forms and combinations. In most texts, general gynecology is combined with therapeutic and/or obstetrical information. In other words, the Middle Dutch texts bear witness to the intermingling of both the pseudo-Albertan and the *Trotula* tradition. At this stage of transmission the distinction between a misogynistic and a woman friendly tradition was no longer perceptible in the written vernacular texts.

Middle Dutch tracts on women's medicine were written both in verse and in prose. Judging from the contents and codicological milieu of gynecological tracts, this type of knowledge is propagated both within a professional context (among the group of medical practitioners, such as surgeons, pharmacists, midwives, etc.) and in a non-professional setting (laymen and -women, (semi)-religious groups). The general assumption for the miscellany manuscripts is that they were owned or used by men. So far, only two sixteenth-century manuscripts show some involvement of (aristocratic) women in the transmission of gynecological writings[49].

In the age of printing, the transmission and propagation of gynecological and obstetrical knowledge were continued both at a faster pace and at a larger scale. While some texts were clearly written for usage in the professional medical field, others were directed at a non-specialist readership, a lay public consisting of the urban bourgeoisie, the lower clergy and the aristocracy.

The printed works on women's medicine show many different approaches, ranging from guidebooks for men, medical encyclopedias for women, to manuals for midwives. By studying the texts in their codicological and historical context, some insight has been gained into the intricate process of knowledge transmission into the vernacular. This survey clearly suggests that the field of women's medicine, originally the private domain of women, was gradually appropriated by men, who slowly changed it into a men-controlled activity. This type of written medical knowledge was especially favorable to literate, academically trained male practitioners, but placed women at a disadvantage. At the same time, we must take to heart the warning sounded by Monica Green that any serious study of medical literacy among women must also take non-written dissemination of medical knowledge into account[50]. But that is the subject for another article.

Utrecht University

49. For a different approach, see Demaitre L., 'Domesticity in Middle Dutch 'Secrets of Men and Women', *Social History of Medicine* 1 (2001), p. 1-25. His thesis is that the Middle Dutch compilers and translators of these 'Secrets' texts reveal a 'special appreciation for the 'domestic' values of sexual harmony, enjoyment, and privacy' (p. 1).

50. See her inspiring article: 'The Possibilities of Literacy and the Limits of Reading: Women and the Gendering of Medical Literacy', in: Green M.H., *Women's Healthcare in the Medieval West* (Aldershot/Great Britain & Burlington/USA, 2000), VII, p. 1-76.

INDEX MANU SCRIPTORUM

INDEX AUCTORUM OPERUMQUE ANONYMORUM

PRINTED ON PERMANENT PAPER • IMPRIME SUR PAPIER PERMANENT • GEDRUKT OP DUURZAAM PAPIER - ISO 9706
N.V. PEETERS S.A., WAROTSTRAAT 50, B-3020 HERENT